Sertoli Cell Biology

Second Edition

Edited by

Michael D. Griswold
Center for Reproductive Biology
School of Molecular Biosciences
College of Veterinary Medicine
Washington State University
Pullman, WA, USA

ELSEVIER

AMSTERDAM • BOSTON • HEIDELBERG • LONDON • NEW YORK • OXFORD
PARIS • SAN DIEGO • SAN FRANCISCO • SYDNEY • TOKYO

Elsevier
Radarweg 29, PO Box 211, 1000 AE Amsterdam, Netherlands
The Boulevard, Langford Lane, Kidlington, Oxford OX5 1GB, UK
225 Wyman Street, Waltham, MA 02451, USA

First edition 2005
Second edition 2015

ISBN: 978-0-12-417047-6

British Library Cataloguing-in-Publication Data
A catalogue record for this book is available from the British Library

Library of Congress Cataloging-in-Publication Data
A catalog record for this book is available from the Library of Congress

For Information on all Elsevier publications
visit our website at http://store.elsevier.com/

Contents

List of contributors

Valentine A. Agbor Department of Molecular and Integrative Physiology, University of Kansas Medical Center, Kansas City, KS

Blanche Capel Department of Cell Biology, Duke University Medical Center, Durham, NC

C. Yan Cheng The Mary M. Wohlford Laboratory for Male Contraceptive Research, Center for Biomedical Research, Population Council, New York, NY

Luiz H. De Castro Assis Department Biology, Science Faculty, Utrecht University, Utrecht, The Netherlands

Dirk G. de Rooij Center for Reproductive Medicine, Academic Medical Center, University of Amsterdam, Amsterdam, The Netherlands; Department of Endocrinology and Metabolism, Faculty of Science, Utrecht University, Utrecht, The Netherlands

Luiz R. França Department of Morphology, Institute of Biological Sciences, Federal University of Minas Gerais, Belo Horizonte-MG, Brazil

Heather Franco Reproductive Developmental Biology Group, Laboratory of Reproductive and Developmental Toxicology, NIEHS/NIH, Research Triangle Park, NC

Michael D. Griswold School of Molecular Biosciences, Center for Reproductive Biology, Washington State University, Pullman, WA

Jenna T. Haverfield Monash Institute of Medical Research-Prince Henry's Institute, Clayton, Victoria, Australia; Department of Anatomy and Developmental Biology, Monash University, Victoria, Australia

Leslie L. Heckert Department of Molecular and Integrative Physiology, University of Kansas Medical Center, Kansas City, KS

Mark P. Hedger MIMR-PHI Institute of Medical Research, Clayton, Victoria, Australia

Rex A. Hess Reproductive Biology and Toxicology, Department of Comparative Biosciences, University of Illinois, Urbana, IL

Cathryn Hogarth School of Molecular Biosciences and the Center for Reproductive Biology, Washington State University, Pullman, WA

Kate L. Loveland Departments of Biochemistry and Molecular Biology & Anatomy and Developmental Biology, Monash University, Clayton, Victoria, Australia

Sarah J. Meachem Monash Institute of Medical Research-Prince Henry's Institute, Clayton, Victoria, Australia; Department of Anatomy and Developmental Biology, Monash University, Victoria, Australia

Roberto D.V.S. Morais Department Biology, Science Faculty, Utrecht University, Utrecht, The Netherlands

Dolores D. Mruk The Mary M. Wohlford Laboratory for Male Contraceptive Research, Center for Biomedical Research, Population Council, New York, NY

Rafael H. Nóbrega Department Biology, Science Faculty, Utrecht University, Utrecht, The Netherlands

Liza O'Donnell Monash Institute of Medical Research-Prince Henry's Institute, Clayton, Victoria, Australia

Jon M. Oatley Center for Reproductive Biology, School of Molecular Biosciences, College of Veterinary Medicine, Washington State University, Pullman, WA

Madhuvanthi Ramaiah Department of Reproductive Medicine, University of California San Diego, La Jolla, CA

Rüdiger W. Schulz Department Biology, Science Faculty, Utrecht University, Utrecht, The Netherlands

Lee B. Smith MRC Centre for Reproductive Health, University of Edinburgh, The Queen's Medical Research Institute, Edinburgh, UK

Peter G. Stanton Monash Institute of Medical Research-Prince Henry's Institute, Clayton, Victoria, Australia; Department of Biochemistry, Monash University, Victoria, Australia

Erica Ungewitter Reproductive Developmental Biology Group, Laboratory of Reproductive and Developmental Toxicology, NIEHS/NIH, Research Triangle Park, NC

A. Wayne Vogl Department of Cellular and Physiological Sciences, Faculty of Medicine, The University of British Columbia, Vancouver, BC

William H. Walker Department of Obstetrics, Gynecology and Reproductive Sciences, Magee Womens Research Institute, University of Pittsburgh, School of Medicine, Pittsburgh, PA

Miles F. Wilkinson Department of Reproductive Medicine, University of California San Diego, La Jolla, CA

William W. Wright Division of Reproductive Biology, Department of Biochemistry and Molecular Biology, Johns Hopkins Bloomberg School of Public Health, Baltimore, MD

Wei Yan Department of Physiology and Cell Biology, University of Nevada School of Medicine, Reno, NV

Qi-En Yang Center for Reproductive Biology, School of Molecular Biosciences, College of Veterinary Medicine, Washington State University, Pullman, WA

Humphrey Hung-Chang Yao Reproductive Developmental Biology Group, Laboratory of Reproductive and Developmental Toxicology, NIEHS/NIH, Research Triangle Park, NC

Preface

This book, *Sertoli Cell Biology vol. 2* is the third book in a series dedicated to understanding the Sertoli cells. Lonnie Russell and I edited the first book, entitled *The Sertoli Cell* and published by Cache River Press in 1993. This first book generated a lot of interest in our favorite cell, and we received very positive feedback on the utility of the information provided by many authors. After Lonnie's untimely death in 2001, the second book, entitled *Sertoli Cell Biology* and published in 2005 by Elsevier, was conceived and delivered. I co-edited this book along with Michael Skinner, and we dedicated it to Lonnie Russell. In *The Sertoli Cell,* Lonnie noted a linear increase in publications relating to Sertoli cells from the mid-1960s to the early 1990s. Hess and Franca, in *Sertoli Cell Biology,* continued this analysis and showed that publications relating to Sertoli cells rose from about 250 per year in 1990 to about 350 per year in 2003. A similar analysis shows that since 2003, the number of publications concerning Sertoli cells has risen to between 400 and 450 per year and has remained relatively constant through 2013. Thus, we can conclude that the scientific interest in these fascinating cells remains very strong.

In the preface to *The Sertoli Cell,* Lonnie and I mused over "dogmatic religious Sertologists" and the "germ cell worshiping cult." These two categorizations were an attempt to underscore the lack of information about how the Sertoli cells and the germ cells interacted and which one, if either, was most influential in directing the process. Working with Lonnie was a joy and a challenge. The joy came from his fun-loving yet warm embrace of science and life in general. The challenge for me was to keep up with his visionary approach to the biology of the testis.

In my opinion, the major advances in the field between 1993 and 2005 involved the ability to transplant spermatogonia in murine testes and establish long-term cultures of spermatogonia. These experiments revealed that much of germ cell development was cell autonomous and tipped the scale toward the germ cell worshipping cult. In addition, contributing to the disappointment of the dogmatic religious Sertologists was a list of important genes that had been genetically knocked out in Sertoli cells but still allowed germ cells to develop. Altogether, this information supported a notion that had been proposed in a review by Richard Sharpe in 1994 that the function of Sertoli cells was to allow the efficient but autonomous development of germ cells [1]. This idea was termed the "modified permissive view."

In an introductory chapter in *Sertoli Cell Biology,* I made two predictions about what would happen in the next 10 years: (i) *We would have nearly complete knowledge about genes expressed in Sertoli cells throughout development.* This first prediction was a safe one since the advent of array technologies was being applied to the testis. We do have a lot more information about gene expression in the testis and in Sertoli cells than we had 10 years ago and even newer technologies such as RNAseq and RiboTag mice will complete this story. (ii) *We would have the capability to maintain most germ cells in culture and then transition them into haploid cells in vitro.* This second prediction seemed to be a safe one because of reports that some level of success had already been obtained. However, nearly 10 years later, the *in vitro* culture of stem cells and spermatogonia is well documented, but the progression through meiosis and the formation of spermatids has proven to be difficult. Thus far, the ultimate differentiation of cultured stem cells into functional gametes has not been demonstrated with any efficiency and reproducibility [2]. It appears that one of the barriers to spermatogenesis in culture is the requirement for Sertoli cells to provide both biochemical and structural components for this process. Score one for the dogmatic religious Sertologists.

Outside of any predictions, there were important advances in our understanding of Sertoli cells since 2005. Sertoli cell-specific deletion of the androgen receptor (SCARKO) by several groups showed that the initiation of meiosis in germ cells was independent of androgen action, but successful completion of meiosis and formation of round spermatids was dependent on functional androgen receptors in Sertoli cells (for review see Ref. [3]). It was also shown that a major defect in SCARKO mice involved components of the tight junctions. The role of microRNAs and noncoding RNAs in the testis and specifically in Sertoli cells became an area of interest for several research groups (for review see Refs. [4,5]). The role of the Sertoli cells in the initiation of meiosis via control of retinoic acid metabolism has become a research focus for other groups [6,7]. Prior to publication of the first book, *The Sertoli Cell*, Lonnie Russell published a model of a rat Sertoli cell that was assembled as a result of serial electron micrographs. This model at the time allowed the visualization of the complex structure of the Sertoli cell and was idolized by the dogmatic religious Sertologists [8–10]. In 2012, Smith and Braun, using more advanced technologies, described in detail how the tight junctions of the Sertoli cells open and close and allow the movement of the preleptotene spermatocytes from outside the blood–testis barrier to the interior [11]. This study underscored the complex structural interplay between the Sertoli cells and the germ cells.

I will not make any predictions for the next 10 years. It does appear that the differing views of the dogmatic religious Sertologists and the germ cell worshiping cult have been reconciled by what Lonnie Russell termed the "Presbyterian view." This was a middle-of-the-road view that recognized the importance of the Sertoli cells in semiautonomous germ cell development. Much germ cell development does appear autonomous, but the completion of meiosis clearly requires something from Sertoli cells: Is it structural or a signaling event? As we approach 150 years since Enrico Sertoli described a new cell type and speculated that this cell was "linked to the production of spermatozoa," we are still attempting to define this link.

Michael D. Griswold
Regents Professor, Washington State University, Pullman, WA

References

[1] Sharpe RM. Regulation of spermatogenesis. In: Knobil E, Neill JD, editors. The Physiology of Reproduction. 2nd ed. New York, NY: Raven Press Ltd; 1994. p. 1363–433.

[2] Griswold MD. Making male gametes in culture. Proc Natl Acad Sci USA 2012;109:16762–3.

[3] De Gendt K, Verhoeven G. Tissue- and cell-specific functions of the androgen receptor revealed through conditional knockout models in mice. Mol Cell Endocrinol 2012;352:13–25.

[4] Hawkins SM, Buchold GM, Matzuk MM. Minireview: the roles of small RNA pathways in reproductive medicine. Mol Endocrinol 2011;25:1257–79.

[5] McIver SC, Roman SD, Nixon B, McLaughlin EA. miRNA and mammalian male germ cells. Hum Reprod Update 2012;18:44–59.

[6] Hogarth CA, Amory JK, Griswold MD. Inhibiting vitamin A metabolism as an approach to male contraception. Trends Endocrinol Metab 2011;22:136–44.

[7] Hogarth CA, Griswold MD. The key role of vitamin A in spermatogenesis. J Clin Invest 2010;120:956–62.

[8] Russell LD, Tallon DM, Weber JE, Wong V, Peterson RN. Three-dimensional reconstruction of a rat stage V Sertoli cell: III. A study of specific cellular relationships. Am J Anat 1983;167:181–92.

[9] Weber JE, Russell LD, Wong V, Peterson RN. Three-dimensional reconstruction of a rat stage V Sertoli cell: II. Morphometry of Sertoli–Sertoli and Sertoli–germ-cell relationships. Am J Anat 1983;167:163–79.

[10] Wong V, Russell LD. Three-dimensional reconstruction of a rat stage V Sertoli cell: I. Methods, basic configuration, and dimensions. Am J Anat 1983;167:143–61.

[11] Smith BE, Braun RE. Germ cell migration across Sertoli cell tight junctions. Science 2012;338:798–802.

Sertoli cell anatomy and cytoskeleton

1

Rex A. Hess[a] *and A. Wayne Vogl*[b]
[a]Reproductive Biology and Toxicology, Department of Comparative Biosciences, University of Illinois, Urbana, IL, [b]Department of Cellular and Physiological Sciences, Faculty of Medicine, The University of British Columbia, Vancouver, BC

I. Introduction

Numerous and extensive reviews have been written about basic morphology of the mammalian Sertoli cell [1−9]. The purpose of this chapter is not to repeat all that has been covered in the past, but rather to ask how do we deal with the plethora of new data being generated using morphological techniques previously unavailable in the study of this cell [10]. The first book, titled *The Sertoli Cell*, was filled with photomicrographs illustrating Sertoli cell morphology [11], which was an appropriate tribute to Enrico Sertoli, the first scientist to publish drawings of the cell, later to be given his family name [12−14]. It took nearly an additional 100 years before electron microscopy revealed the intricate complexities of the Sertoli cell within the seminiferous epithelium [15]. The second book, *Sertoli Cell Biology*, included a review of the morphological variations in cellular organelles [9]; however, much of the book was devoted to Sertoli cell physiology and molecular biology [16]. So, with regard to Sertoli cell anatomy, what has changed during the past 10 years?

Basic Sertoli cell anatomy began with crude drawings published in 1865 [9,13], showing cellular extensions, described as "...branched out that touch two cells..." and holding germ cells in "...the canaliculi, or free, and still shut away in the mother cells." Thus, the concept of "cellule madri" or "mother cell" was born and subsequent publications have shown the finer details, with descriptions of the Sertoli cell as "...not unlike trees..." with their cytoplasmic arms surrounding germ cells like long branches [17].

These earlier studies attempted to leave the reader with a three-dimensional view of the Sertoli cell (Figure 1.1), sending its thin cytoplasmic processes to envelope germ cells as they moved up and down through the seminiferous epithelium, from basement membrane to the luminal surface. Approximately 40% of the Sertoli cell membrane contacts the surface of the elongated spermatids [19], which results in the extension of thin strands of cytoplasm, sometimes reaching a minimum width of less than 50 nm. The cell's unique morphology made it difficult to observe intimate relationships between cells with routine histology. Ultrastructural studies later helped to

Sertoli Cell Biology. DOI: http://dx.doi.org/10.1016/B978-0-12-417047-6.00001-6

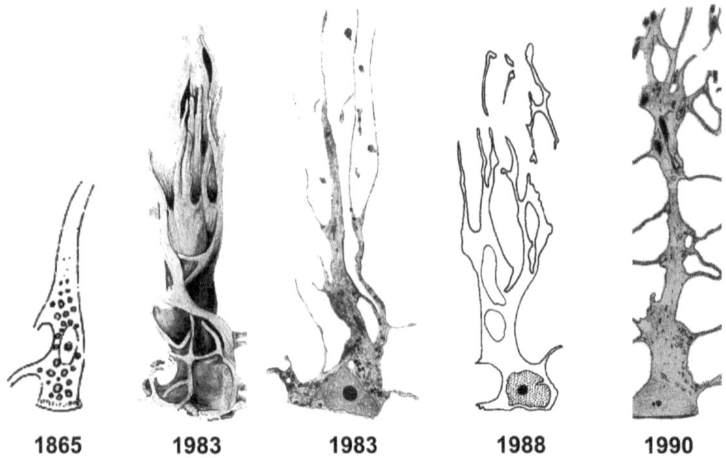

Figure 1.1 Sertoli cell illustrations of three-dimensional-like projections of its cytoplasm. Each illustration was adapted from an original figure, and used with permission of the publisher. 1865: Sertoli; [13] 1993: Russell; [5] 1988: Kerr; [18] 1990: Ueno [6].

fill the gaps in our understanding of junctional complexes, the blood–testis barrier, spermiation, and Sertoli cell's phagocytosis of the residual body [10].

Long ago, Lonnie Russell recognized the importance of improving morphological techniques for observing Sertoli and germ cell interactions. He was one of the first to use thick, plastic-embedded tissue sections of testis for light microscopy, in addition to using thin sections for electron microscopy [20]. During the past decade, scientists have uncovered a wealth of information on genes and proteins expressed in the testis. These advances in basic knowledge were made possible in part because DNA sequencing of the mouse genome was completed. This sequence of data permitted the identification of potentially important gene products for the production of antibodies, which then could be used to localize the proteins in the testis. Thus, since 2005, two techniques have led the way in the study of reproductive morphology. First, the use of immunohistochemistry became the method of choice for identifying and localizing proteins in the cell. Use of this powerful technique has grown exponentially, as evidenced by a recent publication specifically focused on this technique for the study of spermatogenesis [21]. Second, the development of laser-scanning confocal microscopy provided the ability to three-dimensionally image Sertoli–germ cell interactions with relative ease using immunofluorescence.

Our review examines the more general morphological features of Sertoli cells using immunohistochemical and fluorescent markers (Figure 1.2), with a special focus on the cytoskeleton. Immunolocalizations of proteins in the nucleus are fairly simple to interpret if the protein of interest is restricted to the Sertoli cell within the seminiferous epithelium. However, careful interpretation is required for the staining of membrane-associated structures, in which proteins are positioned at the Sertoli–Sertoli junction, the ectoplasmic specialization or the disengagement complex during spermiation. These structural zones of the cytoplasm and membrane

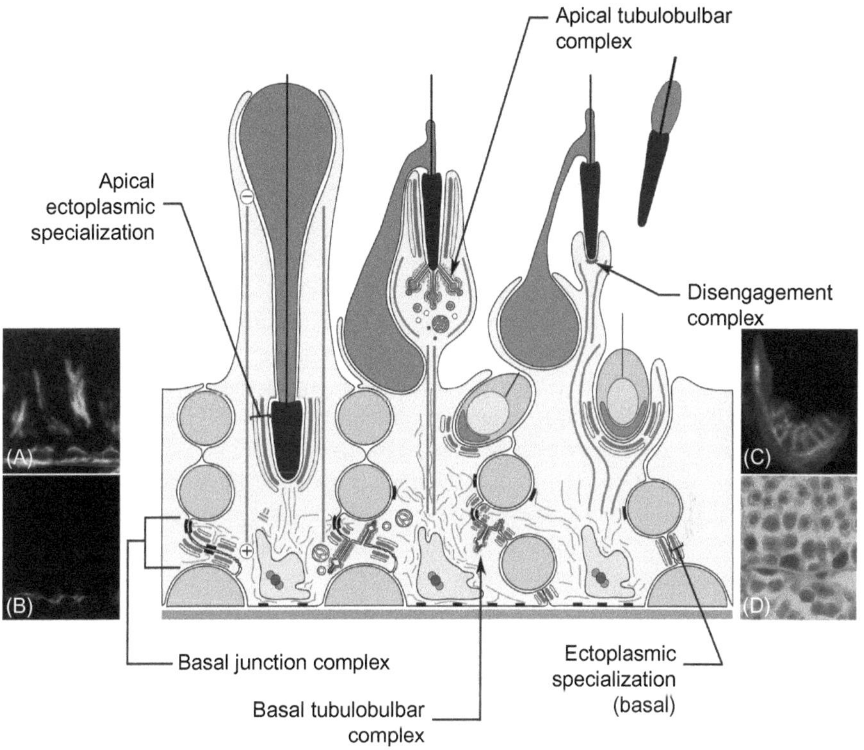

Figure 1.2 Schematic illustration of cytoskeletal distribution in Sertoli cells at different stages during spermatogenesis. Sertoli cells are illustrated in yellow, and spermatogenic cells are in gray. Actin filaments are in red, microtubules are in green, intermediate filaments are in blue, and endoplasmic reticulum is in yellow. This illustrates Sertoli cell's relationship with germ cell movement within the seminiferous epithelium. Photographic examples are presented to demonstrate how immunohistochemistry and immunofluorescence are helping to expand our understanding of the cell's anatomy and biochemistry and their contribution to the physiology of spermatogenesis. (A) Actin filaments (green) are seen along the basal junctions but also lining the heads of elongated spermatids; (B) Claudin-11 (red) stains only the basal junctional complex; (C) Actin (green), Rab5 (red) and DAPI (blue for nucleus) show the intricate relationship of these proteins to the tubulobulbar complex; (D) Androgen receptor (brown) stains only the Sertoli cell nucleus in the hamster seminiferous epithelium. (For interpretation of the references to color in this figure legend, the reader is referred to the web version of this book.)

show dynamic changes not only during development, but also in a cyclical manner during spermatogenesis [22]. Thus, an accurate interpretation of immunolocalization often requires information from additional methodologies, which can include dual staining of overlapping proteins [23–27], immunoelectron microscopy for precise organelle or membrane identification [23,27–29], *in situ* hybridization to determine cell-specific mRNA production [30,31], and isolation and culture of Sertoli and germ cells [32–36].

II. Sertoli cell morphology

A. The nucleus

In light microscopy, the Sertoli cell nucleus is a "trademark" structure, easily recognized in the adult testis (Figure 1.3) but less distinguishable from spermatogonia in the perinatal period [9]. This is true across all species studied to date. The nucleus is large in size and takes on numerous shapes but is positioned either parallel or perpendicular to the basement membrane [37]. Often, textbooks describe the nucleus as being triangular in shape [5]. Early observations in rodents suggested large variations in nuclear shape by stage of the seminiferous epithelial cycle [22,37]. However, it is best not to use this feature in any effort to recognize specific stages because all shapes have been observed in all stages in the mouse (Figure 1.3) and the shape does not appear to change significantly in the primate testis [2].

The nucleus is also described as residing near the basement membrane [38]—and, most often, that is correct. However, in some species, the nucleus is more often located some distance from the base, closer to the lumen, as is common in stages surrounding spermiation in rodents. When staining the seminiferous epithelium for a Sertoli cell nuclear protein, such as androgen receptor (AR), the more apical nuclei are easily recognized. However, when a protein is found in nuclei of both Sertoli and spermatogonial cells, such as the E2f family of transcription factors [39], careful evaluation is required because stages immediately following spermiation have fewer spermatogonia, and nuclear shape or the presence of a large nucleolus may be required for discrimination.

Electron microscopic imaging reveals two unique nuclear structures (Figure 1.4): (i) deep indentations or clefts [40] and (ii) distinctly large and tripartite nucleoli [2,5,9]. The nuclear clefts can be exceptionally deep but are difficult to observe in light microscopy without good fixation and plastic embedment [9] and are rarely observed with immunohistochemical staining. The Sertoli cell nucleolus is large enough for cellular identification with light microscopy [9], although the plane of section will determine the number of visible components. Although most nucleoli

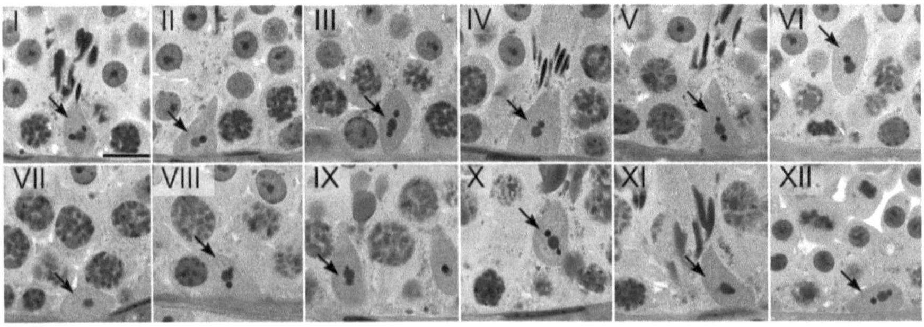

Figure 1.3 Illustration of adult mouse Sertoli cell nuclei (arrows) across all 12 stages. Nuclei perpendicular to the basement membrane are seen in stages I–VI, VIII–XI, while nuclei parallel to the basement membrane are seen in stages VII and XII. Bar = 10 μm.

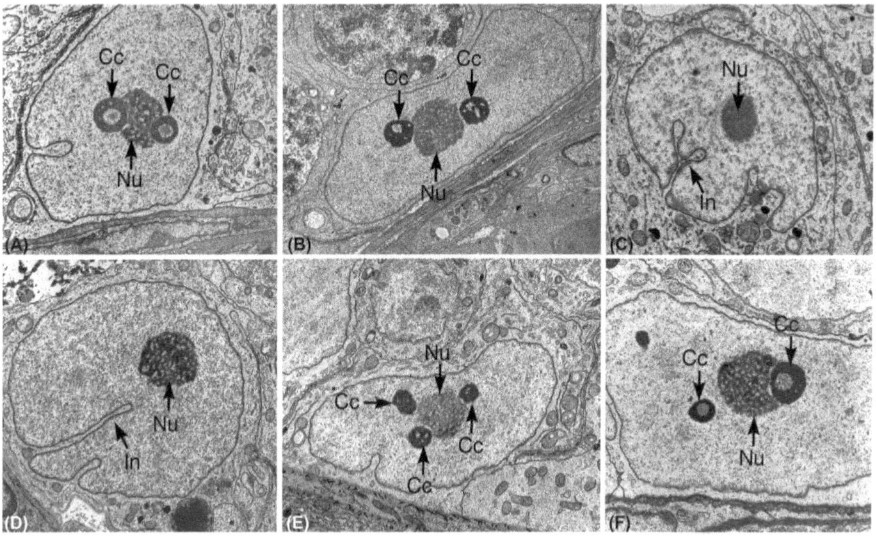

Figure 1.4 Electron microscopy of Sertoli cell nuclei from the mouse (A–C) and rat (D–F) testes. Indentations (In) or clefts in the nuclear envelope are often observed. The nucleolus (Nu) is large in size and has a granular appearance. Satellite chromocenters (Cc) are typically paired and perinucleolar but can be absent due to sectioning (C and D) or in triplicate on rare occasions (E).

have two satellite chromocenters, at least in rodent species three chromocenters are found occasionally in a small percentage of the mouse [41,42] and rat Sertoli cells (Figure 1.4). In some species, the satellite chromocenters form donut shapes, but these structures often are out of the plane of section.

Immunolocalization is now a common method for observing the Sertoli cell nucleus (Table 1.1) because a few proteins have been identified that are specific to the Sertoli cell nucleus (Figure 1.5). These specific proteins are commonly used with dual staining to either colocalize with other proteins or to ensure that germ cell staining is accurately interpreted. Age-specific expression is also important to consider. WT1 is present in the Sertoli cell nucleus throughout all developmental ages, while AR shows increasing expression only after the onset of puberty [43]. GATA4 also is a good marker for Sertoli cells because it is expressed throughout development [72] and does not vary with the cycle of the seminiferous epithelium in the adult. In addition, it is not inhibited by the presence of germ cells, which is a problem with GATA1 expression [157].

B. Cytoplasm and membrane interactions

Sertoli cell anatomy embodies one of the most complex, three-dimensional structures in biology. This tall, columnar cell is polarized [158–161] and extends the entire height of the seminiferous epithelium [162]. It rests on the basement

Table 1.1 **Immunolocalization of proteins in the Sertoli cell nucleus**[a]

Protein annotation	Description	Reference
Major immunohistochemical markers of the Sertoli cell nucleus		
AR	Transcription factor	[43–61]
WT1	Wilm's tumor 1 transcription factor	[43,53,59,62–71]
GATA1,4	GATA-binding protein 1, 4; transcription factor	[53,71–81]
SOX9	SRY-box containing gene 9 transcription factor	[52,53,68,82–86]
CDKN1b	p27(Kip1), Cyclin-dependent kinase inhibitor	[43,53,59,67,69,87–89]
Other proteins found in the Sertoli cell nucleus		
AKT1	Thymoma viral proto-oncogene 1; AKT1 kinase	[90]
ATRX	ATP-dependent helicase; transcriptional regulator ATRX	[91]
ANKRD7	Ankyrin repeat domain 7	[92]
CCND2	Cyclin D2	[93,94]
CDKN1C	P57(Kip2); cyclin-dependent kinase inhibitors	[95]
CDKN2D	Cyclin-dependent kinase inhibitor 2D	[96]
CUX1	Cut-like homeobox 1	[97]
DIAP3	Diaphanous-related formin-3	[98]
DICER	Dcr-1; endoribonuclease	[99]
DMRT1	Doublesex- and mab-3-related transcription factor 1	[66,68,100,101]
E2F-2, 3, 4	E2F transcription factors	[39]
EGR4	Early growth response protein 4	[31]
ESR1	Estrogen receptor 1 (alpha)	[46,102–107]
ESR2	Estrogen receptor 2 (beta)	[103,105,108–113]
EPAS1	HIF-1, hypoxia inducible transcription factor 2alpha; endothelial PAS domain-containing protein 1	[79]
FBL	Fibrillarin, rRNA 2′-O-methyltransferase	[114]
FOXL2	Forkhead box L2	[82]
GATA2	GATA-binding protein 2	[47]
IL1	Interleukin 1 complex	[80,115,116]
HDAC1	Histone deacetylase 1	[117]
HES1	Hairy and enhancer of split 1 transcription factor	[71]
HSPD1	HSP60, heat shock protein 1 (chaperonin)	[118]
INHBB	Inhibin beta-B (Activin beta-B chain)	[119]
KPNA4	Karyopherin (importin) alpha 4	[120,121]
MKI67	Nucleolar protein interacting with the FHA domain of pKI-67	[54,70,80]
MTA1	Metastasis-associated protein 1	[122]

(Continued)

Table 1.1 (Continued)

Protein annotation	Description	Reference
NCOA1,2,3	Nuclear receptor coactivator proteins 1 (SRC1), 2 (SRC2; GRIP-3),3 (SRC3; TRAM1)	[44,49,123]
NONO	Non-POU-domain-containing, octamer binding protein	[124]
NOTCH1	Notch gene homolog 1	[71,85,125,126]
NR3C1	GR, Glucocorticoid receptor, nuclear receptor subfamily 3, group C, member 1	[127]
NR0B1	DAX-1, nuclear receptor subfamily 0, group B, member 1	[128,129]
NR5A1	SF1, steroidogenic factor 1	[53,68,128]
NXF3	Nuclear RNA export factor 3	[130]
ODF2	Outer dense fiber of sperm tails 2	[131]
PALLD	Palladin, cytoskeletal associated protein	[132]
PARP1	Poly (ADP-ribose) polymerase family, member 1	[88]
PCNA	Proliferating cell nuclear antigen	[73,75,76,80,133,134]
PTEN	Phosphatase and tensin homolog	[90]
RAD54L2	RAD54-like protein 2, androgen receptor-interacting protein 4	[57]
RARA,B	Retinoic acid receptor alpha, beta	[135–137]
RXRG	Retinoid X receptor, gamma	[137]
RB1	Retinoblastoma 1	[138]
S100A6	Calcyclin, S100 calcium binding protein A6	[84,139,140]
SLC11A2	DMT1, solute carrier family 11, member 2	[141]
SKIL	SnoN2, Ski-related oncogene	[142]
SMAD1/2/3/5/8	MAD homologs, BMP signaling	[142–144]
SOX8	SRY box containing gene 8	[145]
STAT3	Signal transducer and activator of transcription 3	[146]
STRAP	Serine-threonine kinase receptor-associated protein	[130]
SENP2	SUMO/sentrin specific peptidase 2	[114]
SIN3A	Transcriptional corepressor	[147]
THRA	Thyroid hormone receptor alpha	[134,148–150]
TSPYL2	DENTT; TSPY-like 2	[151]
UBE2B	Ubiquitin-conjugating enzyme E2B	[152]
VCP	Valosin-containing protein	[153]
VDR	Vitamin D receptor	[45,154]
XRCC1	X-ray repair complementing defective repair in Chinese hamster cells 1	[88,155]
ZFPM1	FOG-1, Friend of GATA protein 1, Zinc finger protein multitype 1	[156]

[a]References are from 2004–2013, except for a few proteins requiring earlier reports.

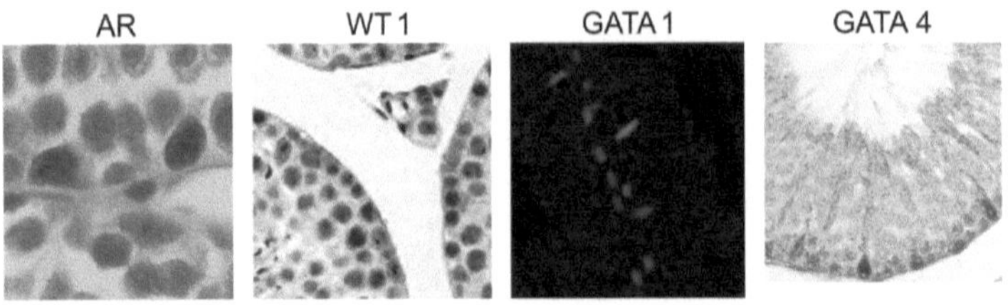

Figure 1.5 Major immunomarkers for the Sertoli cell nucleus. Androgen receptor (AR) in adult hamster; Wilm's tumor protein 1 (WT1) in adult mouse; GATA-binding factors: GATA1 in adult hamster; GATA4 in adult rat.
WT1, modified from Wang et al. [64]; GATA1, modified from Tarulli et al. [73]; GATA4, modified from McClusky et al. [77]. Used with permission of the publishers.

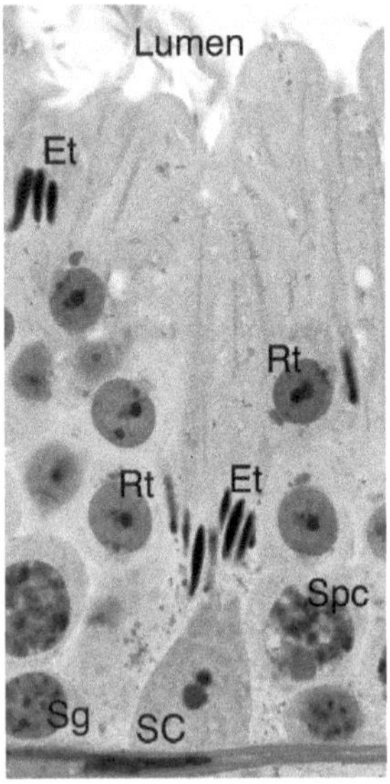

Figure 1.6 Stage IV in the cycle of the seminiferous epithelium of the mouse showing the large Sertoli cell nucleus (SC) and four generations of germ cells: spermatogonia (Sg), spermatocytes (Spc), round spermatids (Rt) and elongated spermatids (Et) buried deep within cupped recesses of the Sertoli cell's membrane.

membrane and projects to the lumen. Arms of a single Sertoli cell extend around four different generations of developing germ cells (Figure 1.6), forming intricate membrane associations that differ between basal and apical locations (see also Section III). This anatomical complexity, which has been described in detail in previous editions of this book, is now the focus of numerous studies to help us understand how different functions are organized both within the cytoplasm and along the plasma membrane. Total surface area of the Sertoli cell plasma membrane varies between 12,000 and 16,000 μm^2, depending on the stage of the cycle of spermatogenesis [19]. Morphological changes in Sertoli cells over the course of the cycle are dramatic and involve the translocation of numerous organelles, expression of a large number of proteins for specific functions, and changes in the location of these proteins throughout the cycle [5,7,22,36,163–165].

Immunohistochemistry has revolutionized morphological studies of the Sertoli cell (Table 1.2). For many years, a few proteins have been used as key markers of the cell's cytoplasm—that is, anti-Mullerian hormone (AMH), vimentin, α,β-tubulin, and zona occludens-1 (ZO-1; TJP1). Three of these are illustrated in Figure 1.7. AMH is an early Sertoli cell cytoplasmic protein that is expressed *in utero* [43]. It has become a standard marker for recognizing immature Sertoli cells because it is down-regulated at the onset of meiosis and Sertoli cell maturation [305]. Cytokeratin 18 (KRT18), a Type I intermediate filament, is also a marker of immature Sertoli cells [73] but does not remain in the adult testis. However, KRT18 has been used in biopsies of adult human testes because it has proven to be a marker for Sertoli cells only and spermatogenic arrest at the spermatogonium, indicating an immaturity of the Sertoli cell [258]. The Type III intermediate filament, vimentin, although also expressed during development [169], has become a standard for Sertoli cell identification in the adult because its staining is intense and surrounds the nucleus [7,139,269]. Tubulins (α,β) are also considered cytoplasmic markers because they assemble into microtubules that are abundant in Sertoli cells [5,7]. Early in the study of Sertoli cell anatomy, microtubules were recognized for their important role in maintaining Sertoli cell shape, transportation, and positioning of organelles and in the translocation of elongated spermatids into deep crypts of the cytoplasm and then back to the lumen for release during spermiation [158,306–313]. Another classical structure now recognized by immunolocalization is the specialized Sertoli–Sertoli tight junction, first thoroughly described by Flickinger in 1967 [314]. It took nearly 20 more years before the tight junction protein ZO-1 was localized at the junctional complex [315,316]. Now, ZO-1 is commonly used for co-localization of more recently discovered proteins (Figure 1.7), such as the Claudins (Table 1.2).

In the adult testis, immunolocalization reveals several basic staining patterns for the seminiferous tubules (Figure 1.8). Protein distribution in each pattern is consistent with unique morphological features of Sertoli cells in stage-specific expressions (as discussed elsewhere in this volume). The polarized pattern is consistent with the appearance of a protein that is distributed from basement membrane toward the lumen, as seen with microtubule tracts that extend to ectoplasmic specialization attachments on the heads of elongated spermatids lining the tubule lumen [7,295]. This pattern can also appear with proteins having a homogenous, but punctate distribution in specific stages, as was found for prosaposin (SGP1) [281] and LAMP2 [29]. Such a distribution would be expected for proteins associated with Sertoli cell lysosomes and involved in the elimination of residual bodies in a stage-specific manner [22]. Vimentin staining also shows polarity consistent with its attachment to elongated spermatids, but the most useful appearance is its thick perinuclear presence [7], which has been used with dual staining to confirm that a nuclear

Table 1.2 Immunolocalization of proteins in Sertoli cell cytoplasm and membrane[a]

Protein annotation	Description	Reference
AAGAB	Alpha- and gamma-adaptin binding protein	[36,166]
ABCD1	ALDP; adrenoleukodystrophy protein	[167]
ABCG2	BCRP1 breast cancer resistance protein	[168]
Actin	Filamentous actin, labeled either by phalloidin solution or anti-actin antibody	[7,8,27,36,73,74, 169–180]
ACTN1	Actinin, alpha 1	[181]
AP2S1	Adaptor-related protein complex 2, sigma 1 subunit	[179]
AMH	MIS; anti-Muellerian hormone	[43,50–53,55,60,68, 83,86,91,139,169, 182–185]
AMHR2	Anti-Mullerian hormone type 2 receptor	[50,51]
AMPH	AMPH1, amphiphysin	[172,186]
ANXA11	Annexin A11	[24,84]
AQP0,8	Aquaporins 0 (MIP),8	[187–189]
ARHGAP42	Rho GTPase activating protein 42	[179]
CYP19A1	Aromatase; cytochrome P450, family 19, subfamily a, polypeptide 1	[190–192]
ARP2/3	Actin-related protein 2/3 complex, subunit 3	[28,193,194]
ATRN	Attractin	[195]
BAX	BCL2-associated X protein	[196]
BCAR1	Breast cancer anti-estrogen resistance 1; MRP1 Multidrug resistance protein	[197–199]
BMP4	Bone morphogenetic protein 4	[116]
CAV1	Caveolin-1	[200]
CCND2	Cyclin D2	[93,94]
CD1D1	Antigen complex	[201]
CDH2	N-cadherin	[202–205]
		[26,194,200,206–213]
CEACAM6	Carcinoembryonic antigen-related cell adhesion molecule 6	[214]
CFTR	Cystic fibrosis transmembrane conductance regulator	[215,216]
CLTC	Clathrin, heavy polypeptide (Hc)	[28,164,200]
CLDN3,5,11	Claudins-3,5,11	[33,36,73,75,80, 145,171,200,205, 211,217]
CLIP-170	p150, EB1	[218]
Caspase-3	Cleaved	[138]
CLU	Clusterin, SGP-2	[43,102]
CFL1	Cofilin 1	[28,219]
CTTN	Cortactin	[179,206,220,221]

(Continued)

Table 1.2 **(Continued)**

Protein annotation	Description	Reference
CREB	cAMP-responsive element binding protein	[215]
CST12	CYSTATIN	[222]
CRTC2	Transducer of CREB protein 2	[176,223]
CTNNA1,B1	Catenins, cadherin associated protein, alpha 1, beta 1	[26,178,194,208,209]
CTSL	Cathepsin L; CP-2 cyclic protein 2, proenzyme of cathepsin L	[224,225]
CXADR	CAR, coxsackie and adenovirus receptor	[161,203,226]
DBNL	Drebrin-like, actin-binding protein 1	[32]
DES	Desmin	[182]
DLL4	Delta-like protein 4	[126]
DOCK1	Dedicator of cytokinesis protein 1	[227]
DNM2, 3	Dynamin 2, 3	[172,228]; [28,179,193]
DSC2	Desmocollin-2	[203]
DSG2	Desmoglein-2	[203]
DYNC1H1	Cytoplasmic dynein 1 heavy chain 1	[229,230]
EEA1	Early endosome antigen 1	[8,36,200]
EEF1A1	Eukaryotic translation elongation factor 1 alpha 1	[231]
EGR4	Early growth response 4	[31]
EHD1	EH-domain-containing 1	[30]
ELMO1	Engulfment and cell motility protein 1	[227,232]
ERBB2	HER2, v-erb-b2 avian erythroblastic leukemia viral oncogene homolog 2	[233]
ESP8	Epidermal growth factor receptor pathway substrate 8	[168,177,234]
ESPN	Espin, ectoplasmic specialization protein	[67,73,138,174,209, 219,235–237]
ESR1	Estrogen receptor 1; ERα	[102,238] [239–241] [242]
ESR2	Estrogen receptor 2; ERβ	[102,113]
EPAS1	HIF2A; endothelial PAS domain protein 1	[243]
F11R	JAM1, JAM-A, Junctional adhesion molecule-1	[75,161,165,207]
FSHR	Follicle stimulating hormone receptor	[244]
FYN	Proto-oncogene, tyrosine-protein kinase Fyn	[245]
GDNF	Glial cell line derived neurotrophic factor	[116]
GGPS1	GGPPS; geranylgeranyl diphosphate synthase 1	[246]
GJA1, 6	Connexin-43, CX43, gap junction protein, alpha 1; Connexin-33	[36,50,51,69,70,75, 205,211,212,221, 247–250]

(*Continued*)

Table 1.2 **(Continued)**

Protein annotation	Description	Reference
GPER1	G protein-coupled estrogen receptor 1; GPR30	[105,239–241]
GUCY1B3	Guanylate cyclase soluble subunit beta-3	[204]
HSD11B2	11beta-hydroxysteroid-dehydrogenase 2	[231]
HSD17B1	7-beta-hydroxysteroid dehydrogenase type 1	[251]
HSP90AB1	Heat shock protein 90 kD alpha (cytosolic), class B member 1	[231]
IGF1	Insulin-like growth factor I, somatomedin	[252]
IGF1R	Insulin-like growth factor 1 receptor	[78]
INSR	Insulin receptor	[78]
ICAM-2	Intercellular adhesion molecule-2	[8,24,36,170,219]
INHA,BA	Inhibin α; β-A	[182,253,254]
IL6	Interleukin 6, interferon, beta 2	[255]
ITGA6, B1	Integrin, α6, β1	[26,170,174,198,209, 235,256]
IQGAP1	IQ motif containing GTPase activating protein 1	[257]
JUP	Junction plakoglobin	[203]
KATNAL1	Katnin p60 subunit A-like 1	[166]
KPNA3, B1	Karyopherin (importin) alpha 3; beta 1	[101,120,121]
KRT8,10,18	Cytokeratin 8, 10, 18	[73,129,182,185, 258,259]
KITL	Kit ligand, stem cell factor, SCF, Steel factor	[116,183]
LAMP1,2	Lysosomal-associated membrane protein 1,2	[29,219]
MIP	AQP0, aquaporin 0	[188]
MRO	Maestro	[260]
PTGES	mPGES-1, prostaglandin E synthase	[261]
MTOR	Mechanistic target of rapamycin, RAPT1, Serine/threonine-protein kinase mTOR	[223,262]
MTMR2	Myotubularin related protein 2	[202,263]
NAMPT	Nicotinamide phosphoribosyltransferase	[264]
NKAP	NF kappaB activating protein	[255,265]
NPHS2	Nephrosis 2, podocin	[155]
NXF3	Nuclear RNA export factor 3	[130]
OCLN	Occludin	[26,33,145,160,165, 175,176,194,200, 203,207,211–213, 266]
P2X2	Purinergic receptor P2X	[267]
PALLD	Palladin, cytoskeletal associated protein	[132,180,268]
PARD3	Par-3 family cell polarity regulator	[213]

(Continued)

Table 1.2 **(Continued)**

Protein annotation	Description	Reference
PAR5	Par5/14-3-3, partitioning defective protein 5	[178]
PLEC	Plectin, plectin 1	[27,256,269,270]
PPP1R11	Protein phosphatase 1, regulatory (inhibitor) subunit 11	[271]
PRKCA	Protein kinase C, alpha, PKC-A	[219]
PTPN11	SHP2 protein tyrosine phosphatase, non-receptor type 1	[272]
PTEM	Phosphatase and tensin homolog	[90]
PTK2	Protein tyrosine kinase 2, focal adhesion kinase 1 (FAK)	[235,236,273,274]
PVRL2	Nectin-2 poliovirus receptor-related 2	[8,24,36,170,219]
PXN	Paxillin	[275]
RAB5A,11A,13	RAS oncogene family	[8,36,170,200,237]
RDX	Radixin, ESP10	[236]
RLN1	Relaxin 1	[276]
RICTOR	Rapamycin-insensitive companion of mTOR	[176,223]
RPS6	Ribosomal protein 6	[33,277]
RHOA,B	Ras homolog family member A,B	[198,201]
RPTOR	Raptor, regulatory associated protein of MTOR, complex 1	[223]
S100A6	Calcyclin; calcium binding protein A6	[84,278,279]
SCRIB	Scribbled planar cell polarity protein	[280]
PSAP	Prosaposin, SGP1, PSAP	[219,281–283]
SHH	Sonic Hedge hog	[243]
SLC11A2	DMT1, Divalent cation transporter 1 (cytopl during develop)	[141]
SLC22A8	Organic anion transporter	[26]
SMAD1/2/3/4/5/8	MAD homologs	[142–144,284–286]
SNON2	Transcriptional repressor SnoN (Ski-like)	[142]
SPNB2	Spectrin beta 2	[27]
SRC	Proto-oncogene c-Src	[202,203,274]
STAT3	Signal transducer and activator of transcription 3 (acute-phase response factor)	[146]
STRAP	Serine/threonine kinase receptor-associated protein	[130]
STK11	LKB1, Serine/threonine-protein kinase 11	[262]
SVIP	Small VCP/p97-interacting protein	[153]
SYNE3	Nesprin-3, spectrin repeat containing, nuclear envelope family member 3	[256]

(Continued)

Table 1.2 (Continued)

Protein annotation	Description	Reference
TBFBR2	Transforming growth factor, beta receptor II	[287]
TIMP2	TIMP metallopeptidase inhibitor 2	[288]
TJP1	Tight junction protein 1, ZO-1	[26,33,69,73–75,78, 80,160,165,176, 178,194,203,211, 212,289–291]
TLR3	Toll-like receptor 3	[255]
TMEM184A	Transmembrane protein 184A, SDMG1	[292]
TF	Transferrin	[102,293,294]
TSPYL2	DENTT; TSPY-like 2	[151]
TUBB3, TUBA1A	Tubulin, beta-3 class III, alpha 1a	[4,67,235,295–297]
VCL	Vinculin	[172,174,198,209,298]
VIM	Vimentin, intermediate filament type III	[7,54,74,116,118, 139,169,182,185, 210,214,270,290, 299–301]
WASL	Wiskott–Aldrich syndrome-like, N-WASP	[28,179,302]
YES1	v-yes-1, Yamaguchi sarcoma viral oncogene homolog 1	[175,303]
ZYX	Zyxin	[35,304]

[a]References are from 2004–2013, except for a few proteins requiring earlier reports.

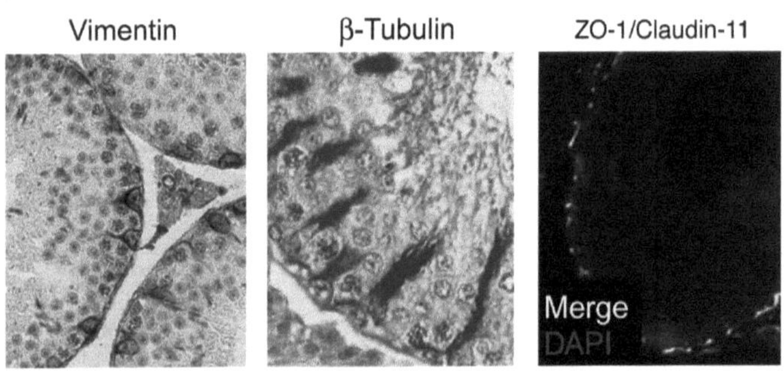

Figure 1.7 Immunolocalization of three major proteins used as markers of Sertoli cell cytoplasm. Vimentin, a Type III intermediate filament, shows an intense perinuclear staining in the adult Sertoli cell cytoplasm. β-Tubulin is a marker for Sertoli cell microtubules that are essential for maintenance of mature Sertoli cell cytoplasm morphology. ZO-1/Claudin-11, the Sertoli–Sertoli tight junction, participates in the formation of a blood–testis barrier. ZO-1 (TJP1; tight junction protein 1) and Claudin-11 (CLDN11) are significant, co-incident markers for this important structure.
Vimentin, modified from Weider et al. [139]; β-Tubulin, modified from De Gendt et al. [296]; ZO-1/Claudin-11, modified from Xu et al. [291]. Used with permission of the publishers.

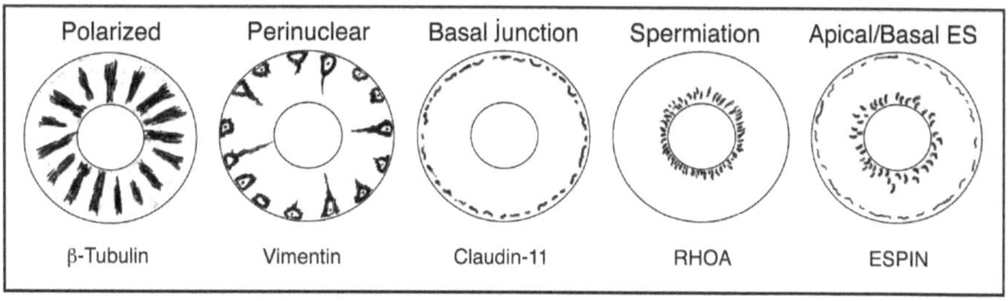

Figure 1.8 Drawings to illustrate examples of staining patterns observed with immunohistochemistry (Table 1.2). A representative protein is indicated for each pattern.

protein was indeed Sertoli cell specific [44]. Some proteins can be expressed only briefly in a perinuclear pattern, such as EGR4, which was present only on day-10 post partum [31]. Proteins associated only with the basal junctional complex appear as thin bands covering the layer of germ cells resting on the basement membrane. Some representative proteins include the claudins, occludins, zona occludens-1, junctional adhesion molecule-1, and scribbled. Other proteins, such as RHOA, are associated with only the apical disengagement complex found in the small cytoplasmic extensions of Sertoli cells that interact with small finger-like tubulobulbar complexes in the heads of elongated spermatids [35,170,206]. Several proteins, such as F-Actin, ESPIN, connexin-43 and vinculin, have been localized at both the apical and basal ectoplasmic specializations, usually giving this appearance in a stage-specific manner.

The interpretation of Sertoli cell morphology based on immunolocalization requires the consideration of numerous factors, including good fixation, well-tested antibodies, proper controls, and other technical aspects of this technique [21]. However, careful evaluation of the data cannot be overemphasized and, at times, may become the major challenge. Several examples are presented.

Vitamin D (1α, 25-dihydroxyvitamin D3) has been long been considered to be one of the factors participating in the regulation of testicular function, specifically acting through the Sertoli cell [317]. The first evidence of vitamin D having a direct effect came from ^{3}H-vitamin D binding, using either autoradiography in tissue sections [318,319] or analysis of isolated Sertoli cell fractions [320]. Based on binding data, and the fact that the receptor, VDR, is a nuclear steroid receptor, we would expect that immunohistochemistry would be consistent with these earlier studies. However, there have been few studies to report VDR staining in testis [45,154]. In two studies, VDR was found in the Sertoli cell nucleus and cytoplasm, as well as nuclei of spermatogonia and spermatocytes. Leydig cells were negative. In contrast, ^{3}H-vitamin D binding was intense over the nucleus of the Sertoli cell; there was no binding in germ cells or the Leydig cell [318]. Thus, an interpretation becomes complicated when some data agree, as with Sertoli cell nuclear presence for VDR, but do not agree with other data, such as the binding studies. Recent studies have also shown non-genomic rapid effects of vitamin D on Sertoli cell membranes [321], which may help to answer some questions that have interfered with our ability to understand the role this vitamin plays in spermatogenesis.

Estrogen receptor-α (ESR1) is another example that has been a challenge to interpret. Similar to vitamin D, it has been known that estrogen binds to proteins in male reproductive tissues, especially the epididymis [322,323]. Immunohistochemical localization of ESR1 in Sertoli cells has been reported but is inconsistent and controversial. However, with careful attention to tissue preservation and well-designed *in vitro* studies, immunolocalization of ESR1 has been demonstrated in the nucleus, cytoplasm, and membrane of the Sertoli cell [102]. For many years, the only reaction considered positive for a nuclear steroid receptor was that observed in the nucleus because cytoplasmic staining was rejected [324]. Nonetheless, changes in the technique, as well as improvements in antibodies and the discovery of rapid effects of estrogen at the membrane have led us to conclude that ESR1 presence in Sertoli cells must be interpreted cautiously [323] and depends on species and age of the animal [46,103,104].

Other problems are also arising with immunolocalization data. The observed expression of a protein may not support its presumed function as reported in other organs or predicted by DNA sequencing. For example, SLC11A2 (DMT1) is a divalent metal transporter protein that was expected to act on the transferrin uptake system in Sertoli cells, as in the intestinal brush border. However, in the adult Sertoli cell, DMT1 was localized to the nucleus and not cytoplasm or the membrane [141]. Another example is the expected presence of the sodium-dependent organic anion transporter (SOAT) in Sertoli cell membranes because sulfated steroids must be transported from Sertoli to germ cells. However, SOAT was recently shown to only localize in germ cells [325]. In other cases, *in situ* hybridization data have contradicted observations of immunohistochemistry, as noted for cystic fibrosis transmembrane conductance regulator (CFTR), which was localized in Sertoli cell cytoplasm [215,216,326]. In contrast, *in situ* hybridization gave a positive signal only in pachytene spermatocytes and spermatids [327].

Finally, proteins associated with Sertoli cell membranes are easily confused with germ cell proteins, particularly in the apical tubulobulbar complex and ectoplasmic specialization areas. The best example of this difficulty is the discovery of a heterotypic interaction between the transmembrane adhesion molecules, nectins-2 and -3 [23,328]. F-actin is an important structural marker of the ectoplasmic specialization that lines the head of elongating spermatids, thus giving a complete fluorescent covering of the spermatid head [28]. Nectins-2 and -3 were found to colocalize with actin at the apical tubulobulbar complex in the concave side of the head. Thus, it was impossible to determine which nectin belonged to Sertoli and which to the germ cell (or to both) with immunolocalization. However, using immunoelectron microscopy, nectin-2 was localized to the Sertoli cell and nectin-3 to the spermatid membrane [23]. The comprehensiveness of Ozaki-Kuroda's seminal paper laid the foundation for accurate localizations of other important proteins associated with the structure and function of the apical Sertoli/spermatid disengagement complex [8,24,25,27,170,206].

III. Sertoli cell cytoskeleton

Elements of the Sertoli cell cytoskeleton in mammals are abundant and have cell-specific patterns of distribution that change in a regular and cyclic fashion during spermatogenesis (Figure 1.2). Moreover, many parts of the cytoskeleton are related to morphogenetic events such as the attachment, release, and translocation of adjacent spermatogenic cells within the epithelium and the transport and positioning of organelles within the Sertoli cell itself. Although, generally, the cytoskeletal system has features in common with similar systems in other cells, some features are specific to and are characteristic of only mature terminally differentiated Sertoli cells. Some of these features include the presence of hexagonally packed actin filaments in bundles adjacent to intercellular adhesion junctions, the presence of actin network cuffs that surround the cores of tubulobulbar complexes, the association of vimentin intermediate filaments with desmosome-like and hemidesmosome-like junctions, and the linking of microtubule-based vesicular transport systems to adhesion junctions.

Because of the fundamental importance of intercellular junctions and the cytoskeleton to male fertility, there currently is great interest in defining the structure, function, regulation, and interrelationship of these two systems. Understanding the basic biology of junctions and their related cytoskeletal elements in Sertoli cells not only will help us better understand the process of spermatogenesis, but may provide insight into the underlying causes of certain forms of male infertility. It could also lead to the identification of molecular targets for the development of male contraceptives [166,231,329].

A. Actin filaments

Although there is morphological evidence suggesting that at least some actin filaments are present generally throughout the cell [1], particularly in association with the cell cortex, the filaments are mainly concentrated within two Sertoli-cell-specific structures—ectoplasmic specializations and tubulobulbar complexes (Figure 1.2). Both structures are associated with intercellular junctions.

1. Ectoplasmic specializations

Ectoplasmic specializations, named by Russell in 1977 [330], are primarily a form of intercellular adhesion junction [298,331]. The structures are characterized by a layer of as many as six or more actin filaments that lies between an attached cistern of endoplasmic reticulum on one side of the layer and regions of the plasma membrane involved with intercellular attachment on the other (Figure 1.9) [1]. Within each structure, the actin filaments occur in bundles arranged parallel to the plasma membrane, and within each bundle, the actin filaments are unipolar in orientation [332,333]. In addition, the filaments are close-packed into hexagonal arrays or paracrystals [1,331], much like the arrangement of actin filaments in the cores of microvilli that form the brush borders of other epithelia.

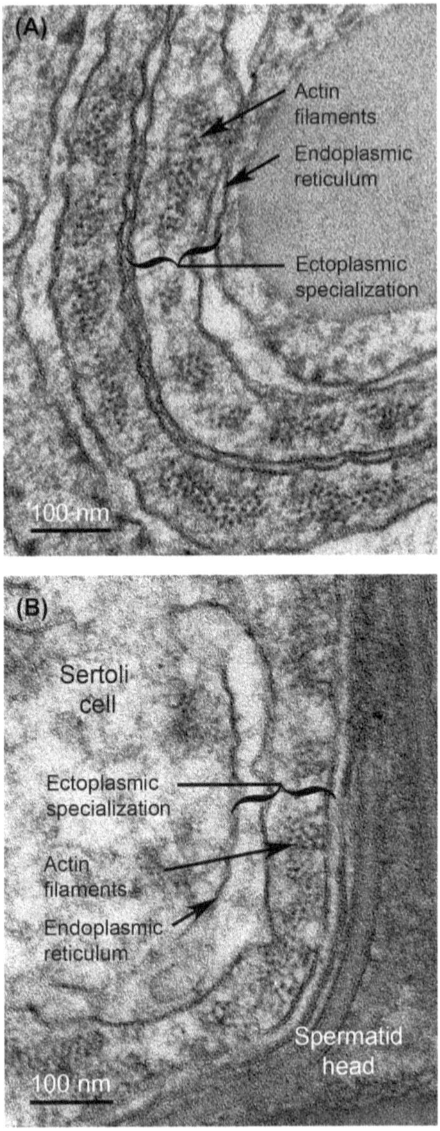

Figure 1.9 Electron micrographs of ectoplasmic specializations at basal sites of attachment between neighboring Sertoli cells (A) and at apical sites of attachment between Sertoli cells and spermatids (B).

Ectoplasmic specializations occur at two major sites in the cell. In basal regions (Figure 1.9A), they occur together with other junction types to form large belt-like junction complexes between adjacent Sertoli cells. Because the adhesion junctions within these complexes are homotypic, a filament mass and cistern of reticulum occur on each side of the junction. Tight and gap junctions overlap with and are integrated into the adhesion zones, and linear tracts of desmosome-like junctions occur within the junction complexes where there are gaps in the ectoplasmic specializations. Tight junctions within the junction complexes contribute to the blood–testis barrier that separates the adluminal compartment from the basal compartment of the seminiferous epithelium [334].

In apical regions of the Sertoli cells, ectoplasmic specializations occur at sites of attachment to spermatids (Figure 1.9B). The structures appear in regions adjacent to where the developing acrosome becomes juxtaposed to the plasma membrane of spermatids as these cells begin to polarize and elongate. Ectoplasmic specializations are the primary mechanism of Sertoli cell attachment to elongated spermatids until spermiation, when the structures are replaced by tubulobulbar complexes [335,336]. The last sites of attachment between Sertoli cells and spermatids during spermiation are small integrin-based disengagement complexes (Figure 1.2) [235] (see fig. 21 in Ref. [335]). Adhesion junctions between Sertoli cells are heterotypic, and actin filaments—along with the related cistern of endoplasmic reticulum—occur only on the Sertoli cell side of the junction.

Actin filaments in ectoplasmic specializations both at apical and at basal sites in the seminiferous epithelium may function to reinforce junction domains in the Sertoli cell plasma membrane. The observation that actin disruption—either by cytochalasin D [337] or by drugs that presumably target actin at the junctions [231,329,338–340]—results in the loss of intercellular adhesion is consistent with the hypothesis that the actin filaments have a role in establishing and reinforcing adhesion domains in the membrane. Disruption of actin filaments at basal sites with cytochalasin D also disrupts the blood–testis barrier [341]. Unlike at basal sites, where the bundles of actin filaments appear to circle the perimeter of the Sertoli cell throughout all stages of spermatogenesis, the bundles of actin filaments in apical ectoplasmic specializations undergo remarkable changes in overall pattern during spermatogenesis [342,343]. These changes may function to adjust the contours of the Sertoli cell cortex to the irregular head shapes that are developing in the adjacent spermatids. In addition to actin filaments themselves, many related proteins have now been reported to occur at ectoplasmic specializations. These include the adaptor proteins vinculin [298,331] and afadin [23] that link integral membrane adhesion molecules to the filament layer and numerous proteins that are related to regulating the arrangement and dynamics of filaments within the actin layer. These latter proteins include α-actinin [344], fimbrin [331], Eps8 [345], cortactin [346], Keap1 [347], espin [348], myosin VIIa [349], paxillin [275], palladin [268], and zyxin [304]. In addition, a number of kinases (FAK [273,274], Src [202,350], ILK [275,351]) and phosphatases (SHP2 [272], MTMR2 [202]) have been localized to the sites as components of signaling cascades that regulate the filaments and junctions [352]. One protein that is conspicuous by its absence from ectoplasmic specializations is myosin II [343]. As a consequence, the actin bundles in ectoplasmic specializations are not contractile [343], unlike the more apically positioned contractile ring associated with the zona adherens junctions of epithelial cells in general.

Integral membrane adhesion molecules at ectoplasmic specializations include nectin-2 [23] and α6β1-integrin [353,354]. The ligand for nectin-2 is another nectin-2 at basal sites of attachment to a neighboring Sertoli cell or nectin-3 at apical sites of attachment to the adjacent spermatid [23]. The ligand for integrins is reported to be γ3-laminin [355]. Although ectoplasmic specializations appear to be mainly nectin- and integrin-based junctions, many other integral membrane adhesion molecules more recently have been reported to occur at these sites. Among these are junctional adhesion molecules (JAMs) [356] and coxsackie and adenovirus receptors (CARs) [226]. In addition, intercellular adhesion molecule-2 (ICAM-2) is reported to occur at apical sites [24], and N-cadherin and its adaptor β-catenin also are reported to occur at basal sites [207].

Ectoplasmic specializations are found only in morphologically mature or terminally differentiated Sertoli cells, they form only at intercellular junctions, and they appear and disappear predominantly at certain times and at specific locations in the cells during spermatogenesis. Turnover of ectoplasmic specializations occurs both at apical and at basal sites in the seminiferous epithelium during spermatogenesis. At basal sites, ectoplasmic specializations, together with other components of the junction complex, are degraded above primary spermatocytes as these cells translocate between basal and adluminal compartments of the epithelium [357]. Simultaneously, new junctions form below these cells as they move upward through the junction complex [171,357]. There now is evidence that junction internalization above spermatocytes begins much earlier than previously thought and occurs well before the actual upward translocation of spermatocytes through the blood–testis barrier [36,164]. At apical sites, ectoplasmic specializations are degraded as part of the mechanism of sperm release, and new ectoplasmic specializations form in association with the next generation of elongated spermatids deeper in the epithelium.

2. Tubulobulbar complexes

Tubulobulbar complexes, like ectoplasmic specializations, form only in morphologically mature Sertoli cells; they form only at intercellular junctions, and they appear and disappear predominantly at certain times and at specific locations in the cells during spermatogenesis. The structures form at the base of the epithelium at junction complexes between neighboring Sertoli cells and in apical regions of the epithelium where Sertoli cells are attached to mature spermatids. Their formation at basal sites precedes spermatocyte translocation [358], and at apical sites, the structures appear prior to sperm release [359].

Tubulobulbar complexes are elongated tubular evaginations of one cell, either a Sertoli cell or a spermatid, that project into corresponding invaginations of the adjacent Sertoli cell [359] (Figure 1.2). In other words, the structures consist of a double membrane core that consists of the plasma membranes of the two attached cells (Figures 1.10A–C and 1.11). Tubulobulbar complexes develop in regions previously occupied by ectoplasmic specializations [219,359]. Each tubulobulbar complex is classically described as consisting of an elongated proximal tubular component, a large bulb or swelling, and a short distal tubular segment that is capped at its end by a "bristle-coated" pit [359,360] (Figure 1.10B). The proximal and distal tubular parts of each complex are cuffed by a three-dimensional network

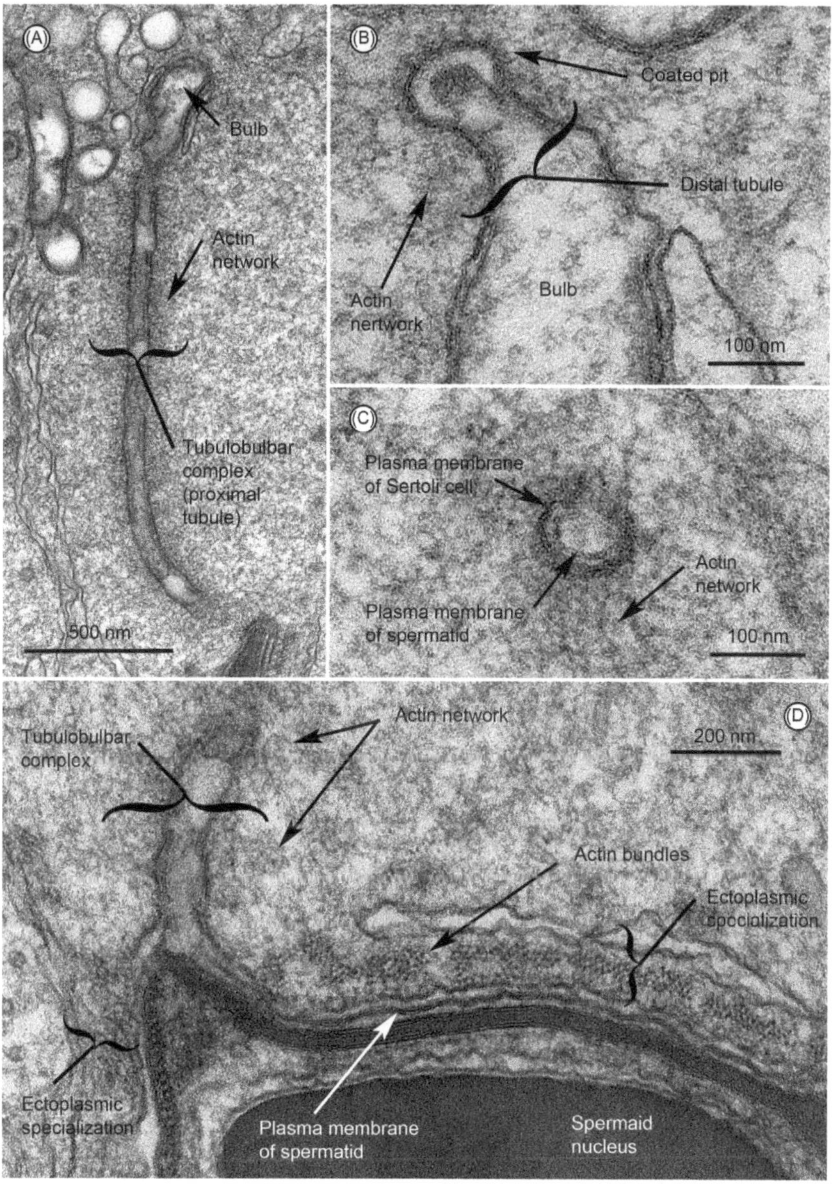

Figure 1.10 Electron micrographs of tubulobulbar complexes that occur at apical sites of attachment between Sertoli cells and spermatids. Shown in (A) is an image of the elongated proximal tubule and bulb region of a tubulobulbar complex. In (B) is a magnified image of the distal tubule and coated pit at the terminal end of a complex. A cross section of the proximal tubule region (C) clearly illustrates the two plasma membranes of the attached cells and the surrounding actin network. Shown in (D) is a tubulobulbar complex with an ectoplasmic specialization on either side of its site of origin. The close-packed filaments in bundles within ectoplasmic specializations and the actin network around the tubulobulbar complex are clearly visible.

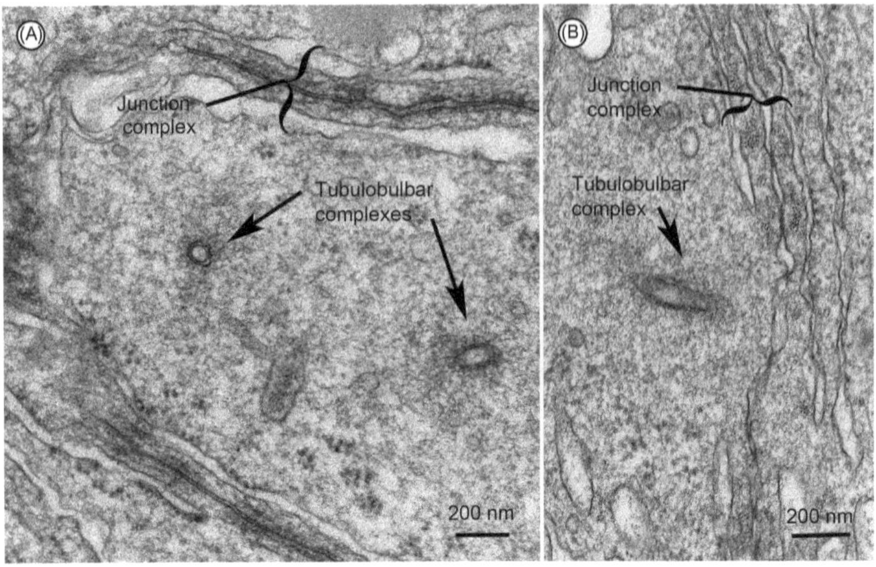

Figure 1.11 Electron micrographs of tubulobulbar complexes that develop at basal junctions between adjacent Sertoli cells. In (A), two tubulobulbar complexes are cut in cross section, whereas in (B), a single complex is cut obliquely through the proximal tubule region. Basal tubulobulbar complexes develop at junction complexes between Sertoli cells.

of actin filaments (Figure 1.10C), unlike the linear bundles of unipolar hexagonally packed actin filaments that characterized ectoplasmic specializations. In fact, there is a distinct zone of transition from a bundled pattern of actin filaments in ectoplasmic specializations to a three-dimensional network where tubulobulbar complexes form at the junctions (Figure 1.10D). The bulb region of tubulobulbar complexes lacks a surrounding cuff of actin filaments. In this region, the Sertoli cell plasma membrane becomes intimately related to a cistern of endoplasmic reticulum (Figure 1.10A). The bulbs eventually separate or "bud" from the complexes and are degraded by the Sertoli cells [360]. Multiple generations of complexes develop at apical sites of attachment to spermatids [360], and similar generations likely develop at basal sites of attachment between Sertoli cells as well.

Numerous molecular components of tubulobulbar complexes have now been identified and localized to the structures. These include elements such as Arp2/3 [193], N-WASP [28], paxillin [275], cofilin [361], cortactin [28], Eps8 [362], and espin [219], which are associated with actin-based structures generally in cells. Also present are amphiphysin [172], dynamin 2 [172], and dynamin 3 [193], elements associated with membrane tubulation and scission. The focal adhesion proteins zyxin and vinculin are present as well [35]. In addition, a network of plectin and spectrin forms a shell around the actin cuffs and may function to support the actin networks or link adjacent tubulobulbar complexes to each other [27,270]. Components that characterize clathrin-coated pits (clathrin [28], AP2 [36], and EPs15 [363]) are present at the terminal end the complexes.

The function of the actin network that cuffs tubulobulbar complexes is likely involved with the formation and support of the dramatically elongated tubular membrane cores of the complexes that resemble the much shorter necks, generally forming clathrin-coated vesicles in cells. The observation that knocking down cortactin—a protein that promotes actin filament nucleation and branching by Arp2/3 in addition to stabilizing the branches [364]—results in shorter tubulobulbar complexes is consistent with this view. The actin networks also may be involved with scission and budding events, but this remains to be verified experimentally.

Numerous functions have been proposed for tubulobulbar complexes since they were first described. Among these are that the structures are simply attachment or anchoring devices that take the place of ectoplasmic specializations as the latter structures disassemble [359]. Another is that they function to eliminate excess cytoplasm from the heads of spermatids [365,366]. Yet others are that the structures eliminate material from the acrosome of spermatid heads [367] or participate in the development of spermatid head shape [368]. All of these hypotheses involve spermatogenic cells and do not account for the presence of the structures at basal junction complexes between neighboring Sertoli cells. A primary function that accounts for the formation of tubulobulbar complexes both at apical and basal sites in the epithelium is the structure's involvement with junction turnover [337].

There is a growing body of evidence to support the hypothesis that tubulobulbar complexes are subcellular machines that internalize intercellular junctions as part of the mechanisms both of sperm release and spermatocyte translocation [8,164]. First, tubulobulbar complexes develop only in regions previously occupied by intercellular junctions [219,359]. Second, tubulobulbar complexes contain representative components of each of the protein domains recognized for clathrin-based endocytosis machinery generally in mammalian cells [369]. In fact, the basic architecture of a tubulobulbar complex resembles a clathrin-coated pit that does not immediately undergo scission and forms a very long neck. However, unlike in typical clathrin-based endocytosis, the cores of tubulobulbar complexes are formed by two attached plasma membranes. In addition, swellings develop along the complexes, and it is these enlarged "bulbs" that separate or "bud" from the structures. Third, junction proteins and endocytosis markers occur together at tubulobulbar complexes and their related vesicles. At apical sites of attachment to spermatids, these structures label with probes for adhesion molecules (nectin-2 [170,206,219], nectin-3 [219], $\alpha6\beta1$-integrin [170,206]) and for endosomal/lysosomal markers (EEA1 [206], Rab5 [170], LAMP [219], SGP1 [219], acid phosphatase [360]). In addition, the structures also label with probes for polyubiquitin [370]. Similarly, basal complexes contain intercellular junction proteins (claudin-11, connexin-43, nectin-2), and endocytic markers (EEA1 and Rab5) are found in similar locations [36]. A final piece of evidence is that intact tight and gap junctions are present in basal complexes when evaluated by electron microscopy [36,358]. All of these observations are consistent with the conclusion that tubulobulbar complexes internalize intercellular junctions and that the internalized material enters endocytic compartments of the Sertoli cell. Moreover, the observation that the two plasma membranes in the core of the complexes remain attached and that tight and gap junctions are morphologically visible within basal complexes indicates that integral membrane junction molecules in adjacent membranes remain "engaged" as the junctions are drawn into the endocytic machinery.

If junction internalization by tubulobulbar complexes is part of the mechanism both of sperm release and of spermatocyte translocation, then interfering with tubulobulbar structure/function should result in delaying or preventing both of these processes. At least in the case of apical tubulobulbar complexes, this appears to be the case. In amphiphysin knockout mice [172] or in rats treated with 17b-estradiol [371], apical tubulobulbar complexes do not form and spermiation failure occurs. In rats treated with siRNA duplexes to knock down cortactin (a component of the actin network of tubulobulbar complexes), apical tubulobulbar complexes are dramatically shorter than in controls, and spermiation fails or is delayed [220]. Effects on basal tubulobulbar complexes and spermatocyte translocation have not yet been evaluated in these models.

B. Microtubules

Microtubules are a conspicuous element of the Sertoli cell cytoskeleton (Figure 1.2), have a pattern of distribution that changes during spermatogenesis, and have multiple functions in the seminiferous epithelium [1,4,7]. In addition to being a determinant of the overall columnar shape of morphologically mature Sertoli cells, they likely are the tracts along which organelles such as mitochondria, elements of the endoplasmic reticulum, and other cargo are moved in the cells. They also are part of the mechanism by which elongated spermatids are positioned and transported within the seminiferous epithelium.

In mature morphologically differentiated Sertoli cells, microtubules are generally oriented parallel to the long axis of the cells [1,372]. The patterns within this overall parallel or radial organization change during spermatogenesis. Generally, these changes reflect the position and developmental step of spermatids with which the Sertoli cell is associated [295] (Figure 1.12). For example, when elongated spermatids are embedded deep in Sertoli cell crypts (stage V in the rat) and the epithelium is labeled for tubulin, the microtubules in each Sertoli cell appear as a thick column in which the crypts are embedded. When late spermatids are moved to the apex of the epithelium (stage VII in the rat), microtubules appear "cable-like" in the body of the cell with a tree-like branching pattern apically. Each branch consists of a bundle of microtubules within a stalk that supports an apical lobe of Sertoli cell cytoplasm

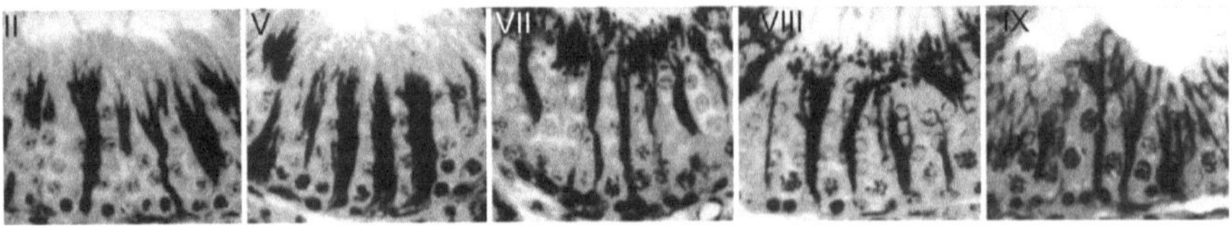

Figure 1.12 Changes in the overall microtubule pattern in Sertoli cells of the rat at different stages of spermatogenesis. The microtubules are labeled immunohistochemically for tubulin. Adapted from Wenz and Hess [295].

surrounding a single spermatid. Within each apical Sertoli cell lobe, the microtubules have a species-specific pattern that conforms to the shape of the lobe and to the associated spermatid head [373,374]. Microtubules are also abundant in Sertoli cell processes that penetrate the residual cytoplasm of spermatids [375].

Unlike in most cultured cells, where microtubules are nucleated at the centrosome with the fast-growing plus ends directed toward the periphery, microtubules in mature morphologically differentiated columnar Sertoli cells are predominantly nucleated apically or peripherally in the cells [376] with their plus ends oriented toward the base of the cell [158], where the nucleus and centrioles are located. This pattern develops before tight junctions form and the blood–testis barrier is established [159]. Although the exact identity and molecular composition of microtubule nucleation sites at the periphery have not clearly been defined in Sertoli cells, gamma-tubulin (a canonical microtubule nucleator) is a likely candidate for one of the proteins involved [377]. The observation that many microtubules in Sertoli cells are tyrosinated [378] indicates the structures are highly dynamic.

Microtubules have multiple functions in Sertoli cells. They support the secretion of seminiferous tubule fluid through intracellular microtubule-dependent membrane trafficking events [307,379]. Motor-driven microtubule-based transport is also likely involved with positioning organelles such as mitochondria and elements of the endoplasmic reticulum in the Sertoli cell because many of these elements are aligned along microtubules and can dramatically change position during spermatogenesis [15,373]. Through a mechanism where a motor-driven transport system is coupled through an adhesion junction to an adjacent spermatogenic cell, elongated spermatids are positioned and moved “up and down” in the epithelium along Sertoli cell microtubule tracts [310,380–383]. Microtubules also play a role in determining overall cell and epithelial architecture because perturbation of Sertoli cell microtubules and their dynamics leads to changes in Sertoli cell shape, epithelial integrity, and a disruption of spermatogenesis. In the rat, treatment with colchicine results in extensive sloughing of apical regions of Sertoli cells together with attached spermatogenic cells [311]. This is possibly because the narrow columnar regions of the cells apical to the nucleus, and the apical stalks that suspend maturing spermatids from the apex of the epithelium, lose their structural support. In the ground squirrel, the same drug results in Sertoli cells becoming more cuboidal in shape, with major concentrations of organelles positioned abnormally at the base of the cells [384]. A mutation in KATNAL1, a microtubule severing protein, leads to a disruption of microtubule patterns in Sertoli cells, resulting in abnormal epithelium with an almost complete lack of elongated spermatids [166]. The overexpression of EB1, a plus end-binding protein, also leads to a disruption in Sertoli cell microtubule patterns as well as to there being fewer spermatogenic cells present in the epithelium relative to controls [218]. It is clear that an intact microtubule cytoskeleton is essential for maintaining spermatogenesis.

C. *Intermediate filaments*

Intermediate filaments, like actin filaments and microtubules, are prominent elements of the Sertoli cell cytoskeleton (Figure 1.2); however, unlike actin filaments and microtubules, their function is somewhat of an enigma because complete absence of the filaments has no obvious effect on spermatogenesis or fertility [385].

Intermediate filaments in mature Sertoli cells are of the vimentin type [386]. Although cytokeratins are expressed together with vimentin in undifferentiated cells, the cytokeratins gradually decrease as the cells mature at puberty [387].

In mature mammalian Sertoli cells, intermediate filaments are concentrated in a network around the basally situated nucleus (Figure 1.13). Here, they form a thin layer or halo around the nucleus that excludes most organelles. From the nucleus, bundles of filaments are distributed peripherally where they associate with desmosome-like attachments to neighboring Sertoli cells and spermatogenic cells—and, with as-yet uncharacterized attachment sites, to elongated spermatids. Plectin, a protein that links different cytoskeletal elements to each other and to membrane-associated proteins, is concentrated at these sites [269]. Intermediate filaments also extend from the nucleus to form a layer juxtaposed to the basal plasma membrane of the Sertoli cell where the filaments associate with small focal hemidesmosome-like plaques.

Intermediate filaments appear to be nonessential cytoskeletal elements for the development and normal function of the mature mammalian testis [385]. In Sertoli cells of vimentin null mice, nuclei are positioned basally in the cell, and other organelles also appear normally distributed [388]. In addition, desmosome-like and hemidesmosome-like plaques are present in the appropriate positions, and gaps occur in ectoplasmic specializations associated with elongated spermatids where intermediate filaments would normally be located [388]. However, cytoplasmic intermediate filaments of any type are completely absent from these cells [388] (Figure 1.14). Recent observations of nesprin-3 knockout mice are consistent with the conclusion that intermediate filaments are not required for positioning the nucleus basally in Sertoli cells [256]. Nesprin-3 is a KASH protein that is part of the SUN/KASH complex that links the nucleoskeleton to the cytoskeleton through the nuclear membrane. This linkage also requires plectin, which links the intermediate filaments to nesprin-3. In the nesprin-3 knockout mice, neither plectin nor intermediate filaments are recruited to the nuclear membrane, yet Sertoli cell nuclear position and testicular morphology are the same as in wild-type animals [256]. Although Sertoli cell intermediate filaments appear nonessential for spermatogenesis and for the development and maintenance of cell architecture under normal circumstances, their overall arrangement and association with cell/cell and cell/substratum attachment sites suggests that they may be involved with reinforcing overall epithelial integrity when tubules are mechanically stressed. Biomechanical and mechanosensory functions of intermediate filament networks in Sertoli cells are still viable hypotheses and have yet to be adequately tested experimentally.

1. Desmosome-like junctions and intermediate filaments

Desmosome-like junctions occur between neighboring Sertoli cells as part of basal junction complexes (Figure 1.15A) and between Sertoli cells and most stages of spermatogenic cells (Figure 1.15B), except for elongated spermatids [389], where the junctions are superseded by ectoplasmic specializations. Desmosome-like

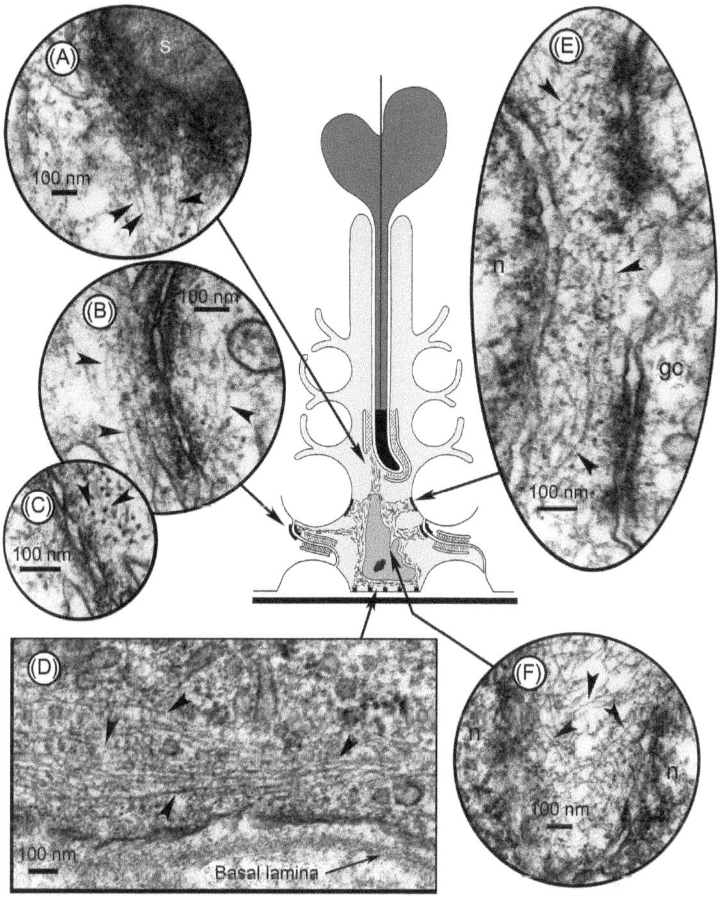

Figure 1.13 Distribution of intermediate filaments in electron micrographs of Sertoli cells from a normal or wild-type mouse. Intermediate filaments (arrowheads) are associated with as regions associated with elongated spermatids (A), with desmosome-like junctions between Sertoli cells (B, C), with hemidesmosome-like densities at the base of the cell (D), with the nuclear surface (E) and with desmosome-like junctions between germ cells and Sertoli cells (F). s, spermatid; n, nucleus; gc, germ cell.
From Vogl et al. [388].

junctions are described as associated with or containing gap junctions [389,390]. In Sertoli cells, a subsurface density that is connected to vimentin intermediate filaments characterizes the structures. In spermatogenic cells, the density is thinner and much less apparent. An intermediate dense line between the plasma membranes is indistinct and often discontinuous [389]. Although desmosome-like junctions have a much less organized structure than typical desmosomes, they are reported to contain similar molecular components [203,391,392]. In addition, plectin is present in the structures [269], and there is some evidence that N-cadherin and actin may

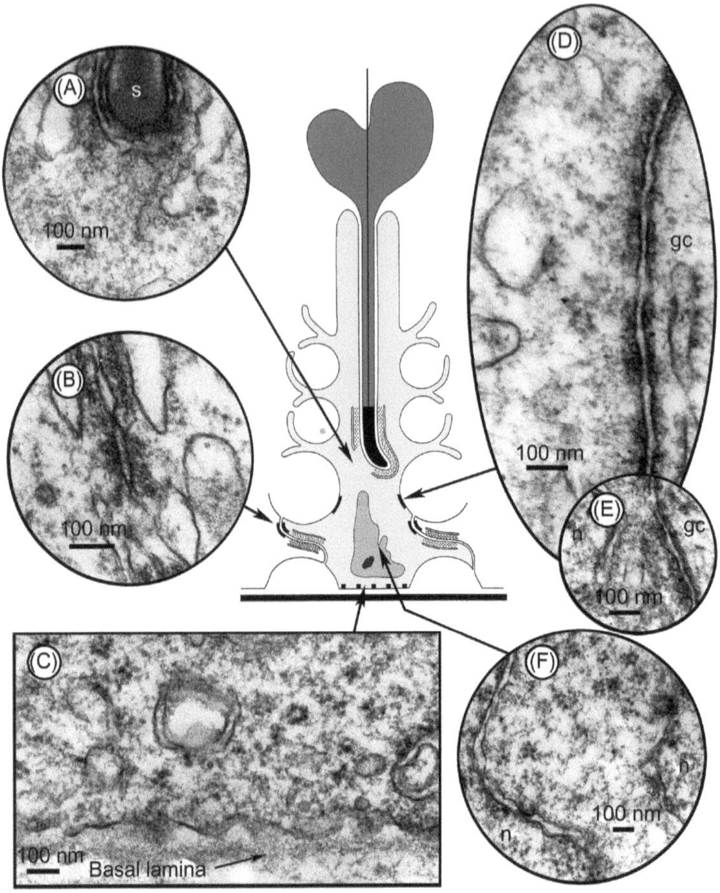

Figure 1.14 Shown here are electron microscopic images similar to those in Figure 1.6, but from a mouse lacking vimentin. Cytoplasmic intermediate filaments are completely absent from regions associated with elongated spermatids (A), desmosome-like junctions between neighboring Sertoli cells (B), basal regions of the cell (C), desmosome-like junctions between Sertoli cells and germ cells (D, E) and regions around the nucleus (F). s, spermatid, n, nucleus, gc, germ cell.
From Vogl et al. [388].

occur at the sites as well [206,275]. These results indicate that the structures may be a hybrid type of junction that shares properties both with canonical adherens junctions and more classically described desmosomes.

2. *Hemidesmosome-like junctions and intermediate filaments*

Hemidesmosome-like structures occur at Sertoli cell sites of attachment to the basal lamina [1,357]. Here, small focal and somewhat indistinct densities occur adjacent

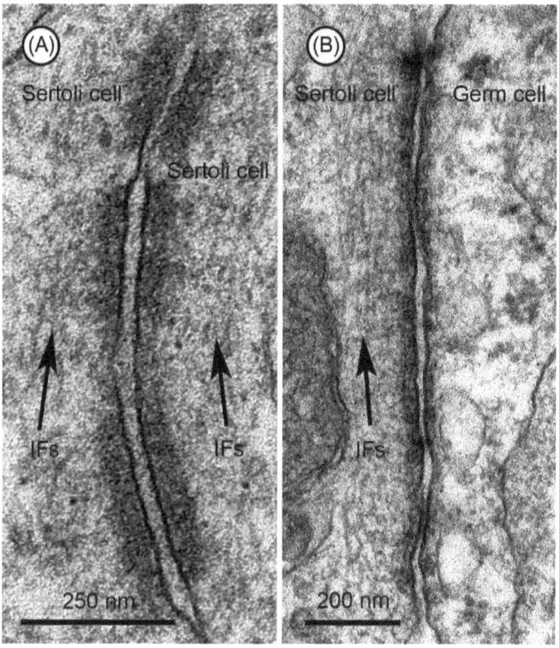

Figure 1.15 Electron micrographs of desmosome-like junctions between neighboring Sertoli cells in a human testis and between a Sertoli cell and a spermatogonia cell in the rat. Intermediate filaments are present in Sertoli cells at the junctions but are absent at the junctions in spermatogenic cells. The image of the human junction is from archived material provided by Dr. Bergmann at the University of Giessen.

to the plasma membrane. These densities are loosely associated with the vimentin intermediate filaments that form a carpet-like layer under the nucleus at the base of the cell. The molecular architecture of these junctions is not known; however, β4 integrin, which is found in typical hemidesmosomes, is likely not present [256]. It is possible that these hemidesmosome-like junctions in Sertoli cells might be a hybrid form of junction that share properties both of actin and of intermediate-filament-related cell/matrix junctions.

D. Regulation of the cytoskeleton

Many of the cytoskeletal features of Sertoli cells are cell-specific, differentiation-specific, site-specific, and time-specific structures; that is, they occur only in morphologically differentiated Sertoli cells, occur only at specific sites in the cell, and change in a cyclical fashion during spermatogenesis. This is particularly true of actin-related ectoplasmic specializations and tubulobulbar complexes and is somewhat less obvious with microtubules and intermediate-filament-related structures. It perhaps is not surprising then that the regulation of cytoskeletal structures and cytoskeletal related events are likely controlled at the global, local, and temporal levels.

Gonadotropins and testosterone are global regulators that likely initiate the programs for cell- and differentiation-specific structures and machinery in Sertoli cells as these cells become morphologically mature during puberty [393,394]. In this regard, the gene (*Tubb3*) that encodes a form of β-tubulin has an androgen response element and is up-regulated as Sertoli cells mature [296]. These hormones probably also regulate some of the changes that occur during spermatogenesis by initiating or modulating signaling cascades [363]. In the case of testosterone, these effects may occur through nonclassical pathways and can occur relatively quickly [393,395]. Many elements of these signaling cascades are present in cytoskeletal-containing features of the Sertoli cell that change during spermatogenesis, most notably the actin-containing ectoplasmic specializations. Key among these elements are the kinases ERK [351], Src [202,350], FAK [396], and ILK [275]. The MAP kinase signaling cascade appears particularly significant in the regulation of ectoplasmic specializations and associated junctions [393,397]. Temporal changes in ARs in the Sertoli cell during spermatogenesis [398] may regulate the timing of certain cytoskeletal events in the epithelium. It likely is not a coincidence that junction turnover occurs, tubulobulbar complexes develop, and ARs become "immunohistochemically detectable" in Sertoli cells [398] during the same time period (stages II–VII in rat seminiferous epithelium).

Factors that regulate site specificity (simultaneous assembly and disassembly at specific locations in apical and basal locations in the epithelium) are still not well understood; however, it is possible that factors secreted by or associated with spermatogenic cells may be involved. In addition, there appears to be extensive "cross-talk" between junction types at the same location and between junctions at different locations [203]. One hypothesis put forward is that biologically active fragments of adhesion-related molecules (γ3-laminin), released or internalized in apical regions in the cell during sperm release, in some way interact with "receptors" at the base of the cell. This then triggers the turnover of basal junction complexes, facilitating spermatocyte translocation [396,399]. The cell biology behind this proposed pathway remains to be elucidated.

IV. Concluding remarks

Our understanding of Sertoli cell "form and function"—or more appropriately stated today as "gene expression and morphology"—has advanced rapidly during the past decade. Much of this achievement was due to the better use of immunohistochemistry and improvements in imaging technology, such as three-dimensional reconstruction of the seminiferous epithelium [171]. Yet, numerous questions remain unanswered and will require even greater skill in adapting current and new methods in the study of Sertoli cell anatomy. Of particular importance will be further development of "live tissue" microscopy and whole organ immunofluorescence, as demonstrated with CLARITY technology [400].

The following are a few unanswered questions for future consideration:

1. Why are gonocytes sequestered in the lumen of the neonate seminiferous tubule [401]? What morphological and signaling mechanisms are involved in the translocation of luminal germ cells to the basement membrane, in both the neonate and when spermatogonia are transplanted into the tubule lumen? It appears that the homing signal mechanism involves Rac1 activation and the adhesion molecule β1-integrin, without which transplanted germ cell migration is inhibited at the Sertoli cell tight junction [402,403]. What is required morphologically for the Sertoli cell to reverse its polarity to accomplish a reverse transport of spermatogonia?
2. Why does the Sertoli cell move elongated spermatid heads deep into the seminiferous epithelium and particularly close to the nucleus in some species (Figure 1.6)? Is there a particular requirement for rapid signaling at that step of spermiogenesis?
3. Why does the Sertoli cell nucleus change its shape so dramatically, showing indentations or clefts that penetrate more than half of its diameter? Is this an advantage of increased surface area or simply a unique design for transporting either proteins or message into and out of the nucleus? Are new proteins yet to be identified associated with the cleft region?
4. How does the Sertoli cell modify its membrane and cytoplasm so rapidly to accommodate thin extensions of its arms to form complex adhesions with four generations of developing germ cells?
5. Why does the Sertoli cell exhibit such strange mitochondrial formations? Are these interesting shapes required for interactions with endoplasmic reticulum, which has been shown to move lipids between the two organelles in other organs [404]?
6. How is the division of labor accomplished within the Sertoli cell? We know a great deal about the genes expressed and which proteins are located at the ectoplasmic specialization and those at the basal junctional complex, but what is the structural basis for compartmentalization? Does the cytoplasm have a unique organization of streets and highways and storage areas that we have not uncovered, due to the limitations of using tissues that are preserved by fixation?
7. How do we interpret Sertoli cell gene expression when the isolation of the cell alters or destroys its defining architecture? Will it be possible to remove one nucleus or cytoplasm from one Sertoli cell at a time for careful analysis?
8. What is the function of intermediate filaments in Sertoli cells? Can we test experimentally the biomechanical and mechanosensory hypotheses?
9. What is the stimulus for initiating junction internalization by tubulobulbar complexes?
10. Does perturbing the tubulobulbar complex structure/function at basal junctions delay or prevent spermatocyte translocation as predicted by the junction internalization hypothesis of tubulobulbar function?
11. What are the subdomains of the endoplasmic reticulum in Sertoli cells? How are they established, what is their function, and how are they interrelated?
12. What is the molecular composition of hemidesmosome-like junctions at the base of Sertoli cells?

References

[1] Vogl AW, Pfeiffer DC, Redenbach DM, Grove BD. Sertoli cell cytoskeleton. In: Russell LD, Griswold MD, editors. The Sertoli cell. Clearwater, FL: Cache River Press; 1993. p. 39–86.

[2] Kerr JB, Loveland KL, O'Bryan MK, de Kretser DM. Cytology of the testis and intrinsic control mechanisms. In: Neill JD, editor. Kobil and Neill's physiology of reproduction, vol. 1. 3rd ed. St. Louis, MO: Elsevier; 2006. p. 827–947. vol. 2.
[3] O'Donnell L, Nicholls PK, O'Bryan MK, McLachlan RI, Stanton PG. Spermiation: the process of sperm release. Spermatogenesis 2011;1:14–35.
[4] O'Donnell L, O'Bryan MK. Microtubules and spermatogenesis. Semin Cell Dev Biol 2014;30:45–54.
[5] Russell LD. Form, dimensions, and cytology of mammalian Sertoli cells. In: Russell LD, Griswold MD, editors. The Sertoli cell. Clearwater, FL: Cache River Press; 1993. p. 1–37.
[6] Ueno H, Mori H. Morphometrical analysis of Sertoli cell ultrastructure during the seminiferous epithelial cycle in rats. Biol Reprod 1990;43:769–76.
[7] Vogl AW, Vaid KS, Guttman JA. The Sertoli cell cytoskeleton. Adv Exp Med Biol 2008;636:186–211.
[8] Vogl AW, Young JS, Du M. New insights into roles of tubulobulbar complexes in sperm release and turnover of blood–testis barrier. Int Rev Cell Mol Biol 2013;303:319–55.
[9] Hess R, França LR. Structure of the Sertoli cell. In: Griswold M, Skinner M, editors. Sertoli cell biology, vol.1. San Diego, CA: Elsevier Academic Press; 2005. p. 19–40.
[10] Clermont Y. Introduction to the Sertoli cell. In: Russell LD, Griswold MD, editors. The Sertoli cell. Clearwater, FL: Cache River Press; 1993. p. xxi–v.
[11] Russell LD, Griswold MD, editors. The Sertoli cell, vol. 1. Clearwater, FL: Cache River Press; 1993.
[12] Hess R, França LR. History of the Sertoli cell discovery. In: Griswold M, Skinner M, editors. Sertoli cell biology, vol. 1. San Diego, CA: Elsevier Academic Press; 2005. p. 3–13.
[13] Sertoli E. Dell'esistenza di particolari cellule ramificate nei canalicoli seminiferi del testicolo umano. Morgagni 1865;7:31–40.
[14] Setchell BP. Some important contributors to our understanding of the male reproductive system: Monesi, Sertoli, Spallanzani and Aubry. Exerpta Medica Int Cong 1986;716:1–12.
[15] Fawcett DW. Ultrastructure and function of the Sertoli cell. In: Hamilton DW, Greep RO, editors. Handbook of physiology, vol. 5. Washington, DC: American Physiology Society; 1975. p. 21–55.
[16] Skinner MK, Griswold MD, editors. Sertoli cell biology. San Diego, CA: Elsevier Academic Press; 2005.
[17] Elftman H. Sertoli cells and testis structure. Am J Anat 1963;113:25–33.
[18] Kerr JB. An ultrastructural and morphometric analysis of the Sertoli cell during the spermatogenic cycle of the rat. Anat Embryol (Berl) 1988;179:191–203.
[19] de Franca LR, Ghosh S, Ye SJ, Russell LD. Surface and surface-to-volume relationships of the Sertoli cell during the cycle of the seminiferous epithelium in the rat. Biol Reprod 1993;49:1215–28.
[20] Russell LD. Sertoli-germ cell interactions: a review. Gamete Res 1980;3:179–202.
[21] Hogarth CA, Griswold MD. Immunohistochemical approaches for the study of spermatogenesis. Methods Mol Biol 2013;927:309–20.
[22] Morales C, Clermont Y. Structural changes of the Sertoli cell during the cycle of the seminiferous epithelium. In: Russell LD, Griswold MD, editors. The Sertoli cell. Clearwater, FL: Cache River Press; 1993. p. 305–29.

[23] Ozaki-Kuroda K, Nakanishi H, Ohta H, Tanaka H, Kurihara H, Mueller S, et al. Nectin couples cell–cell adhesion and the actin scaffold at heterotypic testicular junctions. Curr Biol 2002;12:1145–50.
[24] Xiao X, Cheng CY, Mruk DD. Intercellular adhesion molecule-2 is involved in apical ectoplasmic specialization dynamics during spermatogenesis in the rat. J Endocrinol 2013;216:73–86.
[25] Wan HT, Mruk DD, Li SY, Mok KW, Lee WM, Wong CK, et al. p-FAK-Tyr(397) regulates spermatid adhesion in the rat testis via its effects on F-actin organization at the ectoplasmic specialization. Am J Physiol Endocrinol Metab 2013;305:E687–99.
[26] Su L, Mruk DD, Lee WM, Cheng CY. Drug transporters and blood–testis barrier function. J Endocrinol 2011;209:337–51.
[27] Aristaeus de Asis M, Pires M, Lyon K, Vogl AW. A network of spectrin and plectin surrounds the actin cuffs of apical tubulobulbar complexes in the rat. Spermatogenesis 2013;3:e25733.
[28] Young JS, Guttman JA, Vaid KS, Shahinian H, Vogl AW. Cortactin (CTTN), N-WASP (WASL), and clathrin (CLTC) are present at podosome-like tubulobulbar complexes in the rat testis. Biol Reprod 2009;80:153–61.
[29] Eid N, Ito Y, Otsuki Y. Enhanced mitophagy in Sertoli cells of ethanol-treated rats: morphological evidence and clinical relevance. J Mol Histol 2012;43:71–80.
[30] Rainey MA, George M, Ying G, Akakura R, Burgess DJ, Siefker E, et al. The endocytic recycling regulator EHD1 is essential for spermatogenesis and male fertility in mice. BMC Dev Biol 2010;10:37.
[31] Hogarth CA, Mitchell D, Small C, Griswold M. EGR4 displays both a cell- and intracellular-specific localization pattern in the developing murine testis. Dev Dyn 2010;239:3106–14.
[32] Li MW, Xiao X, Mruk DD, Lam YL, Lee WM, Lui WY, et al. Actin-binding protein drebrin E is involved in junction dynamics during spermatogenesis. Spermatogenesis 2011;1:123–36.
[33] Mok KW, Mruk DD, Silvestrini B, Cheng CY. rpS6 Regulates blood–testis barrier dynamics by affecting F-actin organization and protein recruitment. Endocrinology 2012;153:5036–48.
[34] Wan HT, Mruk DD, Wong CK, Cheng CY. Perfluorooctanesulfonate (PFOS) perturbs male rat Sertoli cell blood–testis barrier function by affecting F-actin organization via p-FAK-Tyr(407): an *in vitro* study. Endocrinology 2014;155:249–62.
[35] Young JS, Vogl AW. Focal adhesion proteins Zyxin and Vinculin are co-distributed at tubulobulbar complexes. Spermatogenesis 2012;2:63–8.
[36] Du M, Young J, De Asis M, Cipollone J, Roskelley C, Takai Y, et al. A novel subcellular machine contributes to basal junction remodeling in the seminiferous epithelium. Biol Reprod 2013;88(3):60.
[37] Leblond CP, Clermont Y. Definition of the stages of the cycle of the seminiferous epithelium in the rat. Ann N Y Acad Sci 1952;55:548–73.
[38] Russell LD, Ettlin RA, Sinha Hikim AP, Clegg ED. Histological and histopathological evaluation of the testis. Clearwater, FL: Cache River Press; 1990.
[39] El-Darwish KS, Parvinen M, Toppari J. Differential expression of members of the E2F family of transcription factors in rodent testes. Reprod Biol Endocrinol 2006;4:63.
[40] Dym M. The fine structure of the monkey (*Macaca*) Sertoli cell and its role in maintaining the blood–testis barrier. Anat Rec 1973;175:639–56.
[41] Guttenbach M, Martinez-Exposito MJ, Engel W, Schmid M. Interphase chromosome arrangement in Sertoli cells of adult mice. Biol Reprod 1996;54:980–6.

[42] Kushida T, Iijima H, Nagato Y, Kushida H. Studies on thick sections of the nucleus of mouse Sertoli cells using an electron microscope operating at 300 kV. Okajimas Folia Anat Jpn 1993;70:41–50.
[43] Sharpe RM, McKinnell C, Kivlin C, Fisher JS. Proliferation and functional maturation of Sertoli cells, and their relevance to disorders of testis function in adulthood. Reproduction 2003;125:769–84.
[44] Vija L, Meduri G, Comperat E, Vasiliu V, Izard V, Ferlicot S, et al. Expression and characterization of androgen receptor coregulators, SRC-2 and HBO1, during human testis ontogenesis and in androgen signaling deficient patients. Mol Cell Endocrinol 2013;375:140–8.
[45] Oliveira AG, Dornas RA, Kalapothakis E, Hess RA, Mahecha GA, Oliveira CA. Vitamin D3 and androgen receptors in testis and epididymal region of roosters (*Gallus domesticus*) as affected by epididymal lithiasis. Anim Reprod Sci 2008;109:343–55.
[46] Ramesh R, Pearl CA, At-Taras E, Roser JF, Berger T. Ontogeny of androgen and estrogen receptor expression in porcine testis: effect of reducing testicular estrogen synthesis. Anim Reprod Sci 2007;102:286–99.
[47] Bhardwaj A, Rao MK, Kaur R, Buttigieg MR, Wilkinson MF. GATA factors and androgen receptor collaborate to transcriptionally activate the *Rhox5* homeobox gene in Sertoli cells. Mol Cell Biol 2008;28:2138–53.
[48] Benbrahim-Tallaa L, Siddeek B, Bozec A, Tronchon V, Florin A, Friry C, et al. Alterations of Sertoli cell activity in the long-term testicular germ cell death process induced by fetal androgen disruption. J Endocrinol 2008;196:21–31.
[49] Igarashi-Migitaka J, Takeshita A, Koibuchi N, Yamada S, Ohtani-Kaneko R, Hirata K. Differential expression of p160 steroid receptor coactivators in the rat testis and epididymis. Eur J Endocrinol 2005;153:595–604.
[50] Almeida J, Conley AJ, Mathewson L, Ball BA. Expression of anti-Mullerian hormone, cyclin-dependent kinase inhibitor (CDKN1B), androgen receptor, and connexin 43 in equine testes during puberty. Theriogenology 2012;77:847–57.
[51] Almeida J, Conley AJ, Ball BA. Expression of anti-Mullerian hormone, CDKN1B, connexin 43, androgen receptor and steroidogenic enzymes in the equine cryptorchid testis. Equine Vet J 2013;45:538–45.
[52] Lan KC, Chen YT, Chang C, Chang YC, Lin HJ, Huang KE, et al. Up-regulation of SOX9 in Sertoli cells from testiculopathic patients accounts for increasing anti-Mullerian hormone expression via impaired androgen receptor signaling. PLoS One 2013;8:e76303.
[53] Kato T, Esaki M, Matsuzawa A, Ikeda Y. NR5A1 is required for functional maturation of Sertoli cells during postnatal development. Reproduction 2012;143: 663–72.
[54] Brehm R, Rey R, Kliesch S, Steger K, Marks A, Bergmann M. Mitotic activity of Sertoli cells in adult human testis: an immunohistochemical study to characterize Sertoli cells in testicular cords from patients showing testicular dysgenesis syndrome. Anat Embryol (Berl) 2006;211:223–36.
[55] Rey RA, Musse M, Venara M, Chemes HE. Ontogeny of the androgen receptor expression in the fetal and postnatal testis: its relevance on Sertoli cell maturation and the onset of adult spermatogenesis. Microsc Res Tech 2009;72:787–95.
[56] Boukari K, Meduri G, Brailly-Tabard S, Guibourdenche J, Ciampi ML, Massin N, et al. Lack of androgen receptor expression in Sertoli cells accounts for the absence

of anti-Mullerian hormone repression during early human testis development. J Clin Endocrinol Metab 2009;94:1818–25.
[57] Domanskyi A, Zhang FP, Nurmio M, Palvimo JJ, Toppari J, Janne OA. Expression and localization of androgen receptor-interacting protein-4 in the testis. Am J Physiol Endocrinol Metab 2007;292:E513–22.
[58] Vernet N, Dennefeld C, Guillou F, Chambon P, Ghyselinck NB, Mark M. Prepubertal testis development relies on retinoic acid but not rexinoid receptors in Sertoli cells. EMBO J 2006;25:5816–25.
[59] Hutchison GR, Scott HM, Walker M, McKinnell C, Ferrara D, Mahood IK, et al. Sertoli cell development and function in an animal model of testicular dysgenesis syndrome. Biol Reprod 2008;78:352–60.
[60] Racine C, Pask AJ, Wijayanti GE, di Clemente N, Picard JY, Shaw G, et al. Early expression of the androgen receptor in the Sertoli cells of a marsupial coincides with downregulation of anti-Mullerian hormone at the time of urogenital virilization. Sex Dev 2009;3:317–25.
[61] Willems A, De Gendt K, Allemeersch J, Smith LB, Welsh M, Swinnen JV, et al. Early effects of Sertoli cell-selective androgen receptor ablation on testicular gene expression. Int J Androl 2010;33:507–17.
[62] Zheng QS, Wang XN, Wen Q, Zhang Y, Chen SR, Zhang J, et al. Wt1 deficiency causes undifferentiated spermatogonia accumulation and meiotic progression disruption in neonatal mice. Reproduction 2014;147:45–52.
[63] Zhang Z, Shao S, Shetty G, Meistrich ML. Donor Sertoli cells transplanted into irradiated rat testes stimulate partial recovery of endogenous spermatogenesis. Reproduction 2009;137:497–508.
[64] Wang XN, Li ZS, Ren Y, Jiang T, Wang YQ, Chen M, et al. The Wilms tumor gene, *Wt1*, is critical for mouse spermatogenesis via regulation of Sertoli cell polarity and is associated with non-obstructive azoospermia in humans. PLoS Genet 2013;9:e1003645.
[65] Gao F, Maiti S, Alam N, Zhang Z, Deng JM, Behringer RR, et al. The Wilms tumor gene, *Wt1*, is required for Sox9 expression and maintenance of tubular architecture in the developing testis. Proc Natl Acad Sci USA 2006;103:11987–92.
[66] Georg I, Barrionuevo F, Wiech T, Scherer G. Sox9 and Sox8 are required for basal lamina integrity of testis cords and for suppression of FOXL2 during embryonic testis development in mice. Biol Reprod 2012.
[67] Myers M, Ebling FJ, Nwagwu M, Boulton R, Wadhwa K, Stewart J, et al. Atypical development of Sertoli cells and impairment of spermatogenesis in the hypogonadal (hpg) mouse. J Anat 2005;207:797–811.
[68] Dadhich RK, Barrionuevo FJ, Lupianez DG, Real FM, Burgos M, Jimenez R. Expression of genes controlling testicular development in adult testis of the seasonally breeding iberian mole. Sex Dev 2011;5:77–88.
[69] Schimenti KJ, Feuer SK, Griffin LB, Graham NR, Bovet CA, Hartford S, et al. AKAP9 is essential for spermatogenesis and Sertoli cell maturation in mice. Genetics 2013;194:447–57.
[70] Sridharan S, Simon L, Meling DD, Cyr DG, Gutstein DE, Fishman GI, et al. Proliferation of adult Sertoli cells following conditional knockout of the Gap junctional protein GJA1 (connexin 43) in mice. Biol Reprod 2007;76:804–12.
[71] Hasegawa K, Okamura Y, Saga Y. Notch signaling in Sertoli cells regulates cyclical gene expression of Hes1 but is dispensable for mouse spermatogenesis. Mol Cell Biol 2012;32:206–15.

[72] Kyronlahti A, Euler R, Bielinska M, Schoeller EL, Moley KH, Toppari J, et al. GATA4 regulates Sertoli cell function and fertility in adult male mice. Mol Cell Endocrinol 2011;333:85–95.

[73] Tarulli GA, Stanton PG, Lerchl A, Meachem SJ. Adult Sertoli cells are not terminally differentiated in the Djungarian hamster: effect of FSH on proliferation and junction protein organization. Biol Reprod 2006;74:798–806.

[74] Willems A, Batlouni SR, Esnal A, Swinnen JV, Saunders PT, Sharpe RM, et al. Selective ablation of the androgen receptor in mouse Sertoli cells affects Sertoli cell maturation, barrier formation and cytoskeletal development. PLoS One 2010;5:e14168.

[75] Tarulli GA, Stanton PG, Loveland KL, Meyts ER, McLachlan RI, Meachem SJ. A survey of Sertoli cell differentiation in men after gonadotropin suppression and in testicular cancer. Spermatogenesis 2013;3:e24014.

[76] Mital P, Kaur G, Bowlin B, Paniagua NJ, Korbutt GS, Dufour JM. Nondividing, postpubertal rat Sertoli cells resumed proliferation after transplantation. Biol Reprod 2014;90:1–10.

[77] McClusky LM, Patrick S, Barnhoorn IE, van Dyk JC, de Jager C, Bornman MS. Immunohistochemical study of nuclear changes associated with male germ cell death and spermiogenesis. J Mol Histol 2009;40:287–99.

[78] Pitetti JL, Calvel P, Zimmermann C, Conne B, Papaioannou MD, Aubry F, et al. An essential role for insulin and IGF1 receptors in regulating Sertoli cell proliferation, testis size, and FSH action in mice. Mol Endocrinol 2013;27:814–27.

[79] Gruber M, Mathew LK, Runge AC, Garcia JA, Simon MC. EPAS1 is required for spermatogenesis in the postnatal mouse testis. Biol Reprod 2010;82:1227–36.

[80] Nicholls PK, Harrison CA, Rainczuk KE, Wayne Vogl A, Stanton PG. Retinoic acid promotes Sertoli cell differentiation and antagonises activin-induced proliferation. Mol Cell Endocrinol 2013;377:33–43.

[81] McCoard SA, Lunstra DD, Wise TH, Ford JJ. Specific staining of Sertoli cell nuclei and evaluation of Sertoli cell number and proliferative activity in Meishan and White Composite boars during the neonatal period. Biol Reprod 2001;64:689–95.

[82] Matson CK, Murphy MW, Sarver AL, Griswold MD, Bardwell VJ, Zarkower D. DMRT1 prevents female reprogramming in the postnatal mammalian testis. Nature 2011;476:101–4.

[83] Ross DG, Bowles J, Hope M, Lehnert S, Koopman P. Profiles of gonadal gene expression in the developing bovine embryo. Sex Dev 2009;3:273–83.

[84] Williams LH, McClive PJ, Van Den Bergen JA, Sinclair AH. Annexin XI co-localises with calcyclin in proliferating cells of the embryonic mouse testis. Dev Dyn 2005;234:432–7.

[85] Garcia TX, Defalco T, Capel B, Hofmann MC. Constitutive activation of NOTCH1 signaling in Sertoli cells causes gonocyte exit from quiescence. Dev Biol 2013.

[86] Sarraj MA, Escalona RM, Umbers A, Chua HK, Small C, Griswold M, et al. Fetal testis dysgenesis and compromised Leydig cell function in Tgfbr3 (beta glycan) knockout mice. Biol Reprod 2010;82:153–62.

[87] Holsberger DR, Jirawatnotai S, Kiyokawa H, Cooke PS. Thyroid hormone regulates the cell cycle inhibitor p27Kip1 in postnatal murine Sertoli cells. Endocrinology 2003;144:3732–8.

[88] Ahmed EA, Barten-van Rijbroek AD, Kal HB, Sadri-Ardekani H, Mizrak SC, van Pelt AM, et al. Proliferative activity *in vitro* and DNA repair indicate that adult mouse and human Sertoli cells are not terminally differentiated, quiescent cells. Biol Reprod 2009;80:1084–91.

[89] Holsberger DR, Kiesewetter SE, Cooke PS. Regulation of neonatal Sertoli cell development by thyroid hormone receptor alpha1. Biol Reprod 2005;73:396–403.

[90] Wang C, Fu W, Quan C, Yan M, Liu C, Qi S, et al. The role of Pten/Akt signaling pathway involved in BPA-induced apoptosis of rat Sertoli cells. Environ Toxicol 2014. Available from: http://dx.doi.org/10.1002/tox.21958 [Epub ahead of print].

[91] Bagheri-Fam S, Argentaro A, Svingen T, Combes AN, Sinclair AH, Koopman P, et al. Defective survival of proliferating Sertoli cells and androgen receptor function in a mouse model of the ATR-X syndrome. Hum Mol Genet 2011;20:2213–24.

[92] Shi YQ, Du LC, Wang QZ, Han CF. *Ankrd7*, a novel gene specifically expressed in Sertoli cells and its potential roles in Sertoli cell maturation. Mol Cells 2009;27:199–203.

[93] Nakayama H, Nishiyama H, Higuchi T, Kaneko Y, Fukumoto N, Fujita J. Change of cyclin D2 mRNA expression during murine testis development detected by fragmented cDNA subtraction method. Dev Growth Differ 1996;38:141–51.

[94] Beumer TL, Roepers-Gajadien HL, Gademan IS, Kal HB, de Rooij DG. Involvement of the D-type cyclins in germ cell proliferation and differentiation in the mouse. Biol Reprod 2000;63:1893–8.

[95] Kim ST, Gye MC. Expression of p57 in mouse and human testes. Dev Growth Differ 2004;46:495–502.

[96] Buchold GM, Magyar PL, Arumugam R, Lee MM, O'Brien DA. p19Ink4d and p18Ink4c cyclin-dependent kinase inhibitors in the male reproductive axis. Mol Reprod Dev 2007;74:997–1007.

[97] Kroll MR, Viss ES, Lamb J, Horstman J, Powell A, Van Wyk A, et al. Asynchronous expression of the homeodomain protein CUX1 in Sertoli cells and spermatids during spermatogenesis in mice. Biol Reprod 2011;84:455–65.

[98] Mironova E, Millette CF. Expression of the diaphanous-related formin proteins mDia1 and mDia2 in the rat testis. Dev Dyn 2008;237:2170–6.

[99] Kim GJ, Georg I, Scherthan H, Merkenschlager M, Guillou F, Scherer G, et al. Dicer is required for Sertoli cell function and survival. Int J Dev Biol 2010;54:867–75.

[100] Agbor VA, Tao S, Lei N, Heckert LL. A *Wt1-Dmrt1* transgene restores DMRT1 to Sertoli cells of *Dmrt1(−/−)* testes: a novel model of DMRT1-deficient germ cells. Biol Reprod 2013;88:51.

[101] Ying M, Chen B, Tian Y, Hou Y, Li Q, Shang X, et al. Nuclear import of human sexual regulator DMRT1 is mediated by importin-beta. Biochim Biophys Acta 2007;1773:804–13.

[102] Lucas TF, Siu ER, Esteves CA, Monteiro HP, Oliveira CA, Porto CS, et al. 17beta-estradiol induces the translocation of the estrogen receptors ESR1 and ESR2 to the cell membrane, MAPK3/1 phosphorylation and proliferation of cultured immature rat Sertoli cells. Biol Reprod 2008;78:101–14.

[103] Pearl CA, Mason H, Roser JF. Immunolocalization of estrogen receptor alpha, estrogen receptor beta and androgen receptor in the pre-, peri- and post-pubertal stallion testis. Anim Reprod Sci 2011;125:103–11.

[104] Baines H, Nwagwu MO, Hastie GR, Wiles RA, Mayhew TM, Ebling FJ. Effects of estradiol and FSH on maturation of the testis in the hypogonadal (hpg) mouse. Reprod Biol Endocrinol 2008;6:4.

[105] Lucas TF, Royer C, Siu ER, Lazari MF, Porto CS. Expression and signaling of G protein-coupled estrogen receptor 1 (GPER) in rat Sertoli cells. Biol Reprod 2010;83:307–17.

[106] Schoen J, Sharbati S, Ritter J, Jewgenow K. Feline gonads exhibit tissue specific alternative splicing of oestrogen receptor alpha (ESR1). Reprod Domest Anim 2012;47 (Suppl. 6):30–4.

[107] Filipiak E, Suliborska D, Laszczynska M, Walczak-Jedrzejowska R, Oszukowska E, Marchlewska K, et al. Estrogen receptor alpha localization in the testes of men with normal spermatogenesis. Folia Histochem Cytobiol 2012;50:340–5.

[108] Gibson DA, Saunders PT. Estrogen dependent signaling in reproductive tissues—a role for estrogen receptors and estrogen related receptors. Mol Cell Endocrinol 2012;348:361–72.

[109] Saunders PT, Fisher JS, Sharpe RM, Millar MR. Expression of oestrogen receptor beta (ER beta) occurs in multiple cell types, including some germ cells, in the rat testis. J Endocrinol 1998;156:R13–7.

[110] Taylor AH, Al-Azzawi F. Immunolocalisation of oestrogen receptor beta in human tissues. J Mol Endocrinol 2000;24:145–55.

[111] Mutembei H, Pesch S, Schuler G, Hoffmann B. Expression of oestrogen receptors alpha and beta and of aromatase in the testis of immature and mature boars. Reprod Domest Anim 2005;40:228–36.

[112] Zhou Q, Nie R, Prins GS, Saunders PT, Katzenellenbogen BS, Hess RA. Localization of androgen and estrogen receptors in adult male mouse reproductive tract. J Androl 2002;23:870–81.

[113] Gunawan A, Cinar MU, Uddin MJ, Kaewmala K, Tesfaye D, Phatsara C, et al. Investigation on association and expression of ESR2 as a candidate gene for boar sperm quality and fertility. Reprod Domest Anim 2012;47:782–90.

[114] Vigodner M, Ishikawa T, Schlegel PN, Morris PL. SUMO-1 human male germ cell development, and the androgen receptor in the testis of men with normal and abnormal spermatogenesis. Am J Physiol Endocrinol Metab 2006;290:E1022–33.

[115] Sultana T, Svechnikov KV, Gustafsson K, Wahlgren A, Tham E, Weber G, et al. Molecular identity, expression and functional analysis of interleukin-1alpha and its isoforms in rat testis. Asian J Androl 2004;6:149–53.

[116] Ma M, Yang S, Zhang Z, Li P, Gong Y, Liu L, et al. Sertoli cells from non-obstructive azoospermia and obstructive azoospermia patients show distinct morphology, Raman spectrum and biochemical phenotype. Hum Reprod 2013;28:1863–73.

[117] Omisanjo OA, Biermann K, Hartmann S, Heukamp LC, Sonnack V, Hild A, et al. *DNMT1* and *HDAC1* gene expression in impaired spermatogenesis and testicular cancer. Histochem Cell Biol 2007;127:175–81.

[118] Tao SX, Guo J, Zhang XS, Li YC, Hu ZY, Han CS, et al. Germ cell apoptosis induced by experimental cryptorchidism is mediated by multiple molecular pathways in Cynomolgus Macaque. Front Biosci 2006;11:1077–89.

[119] Matsuzaki S, Cruzana MB, Budipitojo T, Hondo E, Watanabe G, Taya K, et al. Immunohistochemical localization of inhibin subunits in the testis of the bull. Anat Histol Embryol 2001;30:375–8.

[120] Whiley PA, Miyamoto Y, McLachlan RI, Jans DA, Loveland KL. Changing subcellular localization of nuclear transport factors during human spermatogenesis. Int J Androl 2012;35:158–69.

[121] Hogarth CA, Jans DA, Loveland KL. Subcellular distribution of importins correlates with germ cell maturation. Dev Dyn 2007;236:2311–20.

[122] Sun H, Yang B, Zhu C, Liu R, Wang H, Li W. Presence of metastasis-associated protein 1 in Sertoli cells is required for proper contact between Sertoli cells and adjacent germ cells. Urology 2013;81:66–73.

[123] Tan JA, Hall SH, Petrusz P, French FS. Thyroid receptor activator molecule, TRAM-1, is an androgen receptor coactivator. Endocrinology 2000;141:3440–50.
[124] Kuwahara S, Ikei A, Taguchi Y, Tabuchi Y, Fujimoto N, Obinata M, et al. PSPC1, NONO, and SFPQ are expressed in mouse Sertoli cells and may function as coregulators of androgen receptor-mediated transcription. Biol Reprod 2006;75:352–9.
[125] Garcia TX, Hofmann MC. NOTCH signaling in Sertoli cells regulates gonocyte fate. Cell Cycle 2013;12:2538–45.
[126] Murta D, Batista M, Silva E, Trindade A, Henrique D, Duarte A, et al. Dynamics of Notch pathway expression during mouse testis post-natal development and along the spermatogenic cycle. PLoS One 2013;8:e72767.
[127] Hazra R, Upton D, Jimenez M, Desai R, Handelsman DJ, Allan CM. *In vivo* actions of the Sertoli cell glucocorticoid receptor. Endocrinology 2014;155:1120–30.
[128] Kojima Y, Sasaki S, Hayashi Y, Umemoto Y, Morohashi K, Kohri K. Role of transcription factors Ad4bp/SF-1 and DAX-1 in steroidogenesis and spermatogenesis in human testicular development and idiopathic azoospermia. Int J Urol 2006;13:785–93.
[129] Guo J, Tao SX, Chen M, Shi YQ, Zhang ZQ, Li YC, et al. Heat treatment induces liver receptor homolog-1 expression in monkey and rat Sertoli cells. Endocrinology 2007;148:1255–65.
[130] Yin Y, Wang G, Liang N, Zhang H, Liu Z, Li W, et al. Nuclear export factor 3 is involved in regulating the expression of TGF-beta3 in an mRNA export activity-independent manner in mouse Sertoli cells. Biochem J 2013;452:67–78.
[131] Rivkin E, Tres LL, Kierszenbaum AL. Genomic origin, processing and developmental expression of testicular outer dense fiber 2 (ODF2) transcripts and a novel nucleolar localization of ODF2 protein. Mol Reprod Dev 2008;75:1591–606.
[132] Niedenberger BA, Chappell VK, Kaye EP, Renegar RH, Geyer CB. Nuclear localization of the actin regulatory protein Palladin in Sertoli cells. Mol Reprod Dev 2013;80:403–13.
[133] Angelopoulou R, Balla M, Lavranos G, Chalikias M, Kitsos C, Baka S, et al. Sertoli cell proliferation in the fetal and neonatal rat testis: a continuous phenomenon? Acta Histochem 2008;110:341–7.
[134] Oluwole OA, Bartlewski PM, Hahnel A. Relationships of serum thyroid hormones and follicle-stimulating hormone concentrations to Sertoli cell differentiation during the first wave of spermatogenesis in euthyroid ram lambs. Reprod Biol 2013; 13:150–60.
[135] Vernet N, Dennefeld C, Rochette-Egly C, Oulad-Abdelghani M, Chambon P, Ghyselinck NB, et al. Retinoic acid metabolism and signaling pathways in the adult and developing mouse testis. Endocrinology 2006;147:96–110.
[136] Childs AJ, Cowan G, Kinnell HL, Anderson RA, Saunders PT. Retinoic acid signalling and the control of meiotic entry in the human fetal gonad. PLoS One 2011;6: e20249.
[137] Dufour JM, Kim KH. Cellular and subcellular localization of six retinoid receptors in rat testis during postnatal development: identification of potential heterodimeric receptors. Biol Reprod 1999;61:1300–8.
[138] Nalam RL, Andreu-Vieyra C, Braun RE, Akiyama H, Matzuk MM. Retinoblastoma protein plays multiple essential roles in the terminal differentiation of Sertoli cells. Mol Endocrinol 2009;23:1900–13.
[139] Weider K, Bergmann M, Giese S, Guillou F, Failing K, Brehm R. Altered differentiation and clustering of Sertoli cells in transgenic mice showing a Sertoli cell specific knockout of the *connexin 43* gene. Differentiation 2011;82:38–49.

[140] O'Hara L, York JP, Zhang P, Smith LB. Targeting of GFP-Cre to the mouse Cyp11a1 locus both drives Cre recombinase expression in steroidogenic cells and permits generation of Cyp11a1 knock out mice. PLoS One 2014;9:e84541.

[141] Griffin KP, Ward DT, Liu W, Stewart G, Morris ID, Smith CP. Differential expression of divalent metal transporter DMT1 (Slc11a2) in the spermatogenic epithelium of the developing and adult rat testis. Am J Physiol Cell Physiol 2005;288:C176–84.

[142] Itman C, Whiley PA, Zhou W, Meistrich M, Sahin Z, Loveland KL. Regulated production of SnoN2 is a feature of testicular differentiation. Microsc Res Tech 2009;72:833–44.

[143] Itman C, Loveland KL. SMAD expression in the testis: an insight into BMP regulation of spermatogenesis. Dev Dyn 2008;237:97–111.

[144] Itman C, Small C, Griswold M, Nagaraja AK, Matzuk MM, Brown CW, et al. Developmentally regulated SMAD2 and SMAD3 utilization directs activin signaling outcomes. Dev Dyn 2009;238:1688–700.

[145] Singh AP, Cummings CA, Mishina Y, Archer TK. SOX8 regulates permeability of the blood–testes barrier that affects adult male fertility in the mouse. Biol Reprod 2013.

[146] Rezende N, Lee MY, Monette S, Mark W, Lu A, Gudas LJ. Rex1 (Zfp42) null mice show impaired testicular function, abnormal testis morphology, and aberrant gene expression. Dev Biol 2011.

[147] Payne CJ, Gallagher SJ, Foreman O, Dannenberg JH, Depinho RA, Braun RE. Sin3a is required by Sertoli cells to establish a niche for undifferentiated spermatogonia, germ cell tumors, and spermatid elongation. Stem Cells 2010;28:1424–34.

[148] Jannini EA, Carosa E, Rucci N, Screponi E, D'Armiento M. Ontogeny and regulation of variant thyroid hormone receptor isoforms in developing rat testis. J Endocrinol Invest 1999;22:843–8.

[149] Falcone M, Miyamoto T, Fierro-Renoy F, Macchia E, DeGroot LJ. Antipeptide polyclonal antibodies specifically recognize each human thyroid hormone receptor isoform. Endocrinology 1992;131:2419–29.

[150] Canale D, Agostini M, Giorgilli G, Caglieresi C, Scartabelli G, Nardini V, et al. Thyroid hormone receptors in neonatal, prepubertal, and adult rat testis. J Androl 2001;22:284–8.

[151] Martinez A, Ozbun LL, Angdisen J, Jakowlew SB. Expression of differentially expressed nucleolar transforming growth factor-beta1 target (DENTT) in adult mouse tissues. Dev Dyn 2002;224:186–99.

[152] Mou L, Zhang Q, Wang Y, Zhang Q, Sun L, Li C, et al. Identification of Ube2b as a novel target of androgen receptor in mouse Sertoli cells. Biol Reprod 2013;89:32.

[153] Cayli S, Ocakli S, Erdemir F, Tas U, Aslan H, Yener T, et al. Developmental expression of p97/VCP (Valosin-containing protein) and Jab1/CSN5 in the rat testis and epididymis. Reprod Biol Endocrinol 2011;9:117.

[154] Johnson JA, Grande JP, Roche PC, Kumar R. Immunohistochemical detection and distribution of the 1,25-dihydroxyvitamin D3 receptor in rat reproductive tissues. Histochem Cell Biol 1996;105:7–15.

[155] Relle M, Cash H, Brochhausen C, Strand D, Menke J, Galle PR, et al. New perspectives on the renal slit diaphragm protein podocin. Mod Pathol 2011;24:1101–10.

[156] Ketola I, Anttonen M, Vaskivuo T, Tapanainen JS, Toppari J, Heikinheimo M. Developmental expression and spermatogenic stage specificity of transcription factors GATA-1 and GATA-4 and their cofactors FOG-1 and FOG-2 in the mouse testis. Eur J Endocrinol 2002;147:397–406.

[157] Yomogida K, Ohtani H, Harigae H, Ito E, Nishimune Y, Engel JD, et al. Developmental stage- and spermatogenic cycle-specific expression of transcription factor GATA-1 in mouse Sertoli cells. Development 1994;120:1759–66.
[158] Redenbach DM, Vogl AW. Microtubule polarity in Sertoli cells: a model for microtubule-based spermatid transport. Eur J Cell Biol 1991;54:277–90.
[159] Redenbach DM, Boekelheide K. Microtubules are oriented with their minus-ends directed apically before tight junction formation in rat Sertoli cells. Eur J Cell Biol 1994;65:246–57.
[160] Su WH, Mruk DD, Wong EW, Lui WY, Cheng CY. Polarity protein complex Scribble/Lgl/Dlg and epithelial cell barriers. Adv Exp Med Biol 2012;763:149–70.
[161] Xiao X, Wong EW, Lie PP, Mruk DD, Wong CK, Cheng CY. Cytokines, polarity proteins, and endosomal protein trafficking and signaling-the Sertoli cell blood–testis barrier system *in vitro* as a study model. Methods Enzymol 2014;534:181–94.
[162] Wong V, Russell LD. Three-dimensional reconstruction of a rat stage V Sertoli cell: I. Methods, basic configuration, and dimensions. Am J Anat 1983;167:143–61.
[163] Parvinen M. Cyclic function of Sertoli cells. In: Russell LD, Griswold MD, editors. The Sertoli cell. Clearwater, FL: Cache River Press; 1993. p. 39–86.
[164] Vogl AW, Du M, Wang XY, Young JS. Novel clathrin/actin-based endocytic machinery associated with junction turnover in the seminiferous epithelium. Semin Cell Dev Biol 2014;30:55–64.
[165] Sarkar O, Mathur PP, Cheng CY, Mruk DD. Interleukin 1 alpha (IL1A) is a novel regulator of the blood–testis barrier in the rat. Biol Reprod 2008;78:445–54.
[166] Smith LB, Milne L, Nelson N, Eddie S, Brown P, Atanassova N, et al. KATNAL1 regulation of Sertoli cell microtubule dynamics is essential for spermiogenesis and male fertility. PLoS Genet 2012;8:e1002697.
[167] Hoftberger R, Kunze M, Weinhofer I, Aboul-Enein F, Voigtlander T, Oezen I, et al. Distribution and cellular localization of adrenoleukodystrophy protein in human tissues: implications for X-linked adrenoleukodystrophy. Neurobiol Dis 2007;28: 165–74.
[168] Qian X, Cheng YH, Jenardhanan P, Mruk DD, Mathur PP, Xia W, et al. Adjudin disrupts spermatogenesis by targeting drug transporters: lesson from the breast cancer resistance protein (BCRP). Spermatogenesis 2013;3:e24993.
[169] Albert S, Wistuba J, Eildermann K, Ehmcke J, Schlatt S, Gromoll J, et al. Comparative marker analysis after isolation and culture of testicular cells from the immature marmoset. Cells Tissues Organs 2012;196:543–54.
[170] Young JS, Takai Y, Kojic KL, Vogl AW. Internalization of adhesion junction proteins and their association with recycling endosome marker proteins in rat seminiferous epithelium. Reproduction 2012;143:347–57.
[171] Smith BE, Braun RE. Germ cell migration across Sertoli cell tight junctions. Science 2012;338:798–802.
[172] Kusumi N, Watanabe M, Yamada H, Li SA, Kashiwakura Y, Matsukawa T, et al. Implication of amphiphysin 1 and dynamin 2 in tubulobulbar complex formation and spermatid release. Cell Struct Funct 2007;32:101–13.
[173] Ito C, Suzuki-Toyota F, Maekawa M, Toyama Y, Yao R, Noda T, et al. Failure to assemble the peri-nuclear structures in GOPC deficient spermatids as found in round-headed spermatozoa. Arch Histol Cytol 2004;67:349–60.
[174] Vogl AW, Pfeiffer DC, Mulholland D, Kimel G, Guttman J. Unique and multifunctional adhesion junctions in the testis: ectoplasmic specializations. Arch Histol Cytol 2000;63:1–15.

[175] Xiao X, Mruk DD, Lee WM, Cheng CY. c-Yes regulates cell adhesion at the blood–testis barrier and the apical ectoplasmic specialization in the seminiferous epithelium of rat testes. Int J Biochem Cell Biol 2011;43:651–65.

[176] Mok KW, Mruk DD, Lee WM, Cheng CY. Rictor/mTORC2 regulates blood–testis barrier dynamics via its effects on gap junction communications and actin filament network. FASEB J 2013;27:1137–52.

[177] Cheng CY, Lie PP, Wong EW, Mruk DD, Silvestrini B. Adjudin disrupts spermatogenesis via the action of some unlikely partners: Eps8, Arp2/3 complex, drebrin E, PAR6 and 14-3-3. Spermatogenesis 2011;1:291–7.

[178] Wong EW, Sun S, Li MW, Lee WM, Cheng CY. 14-3-3 Protein regulates cell adhesion in the seminiferous epithelium of rat testes. Endocrinology 2009;150:4713–23.

[179] Upadhyay RD, Kumar AV, Sonawane S, Gaonkar R, Balasinor NH. Estrogen effects on actin cytoskeletal and endocytic proteins associated with tubulobulbar complex disruption in rat testes. Reprod Sci 2013;20:1162–74.

[180] Qian X, Mruk DD, Cheng YH, Cheng CY. Actin cross-linking protein palladin and spermatogenesis. Spermatogenesis 2013;3:e23473.

[181] Yazama F, Sawada H, Hirosawa K, Hayashi Y, Nishida T. Deep-etch visualization of the Sertoli cell (blood–testis) barrier in the boar. Tissue Cell 1991;23:235–46.

[182] Giudice C, Banco B, Veronesi MC, Ferrari A, Di Nardo A, Grieco V. Immunohistochemical expression of markers of immaturity in Sertoli and seminal cells in canine testicular atrophy. J Comp Pathol 2013;150:208–15.

[183] Kocer A, Gallozzi M, Renault L, Tilly G, Pinheiro I, Le Provost F, et al. Goat PRND expression pattern suggests its involvement in early sex differentiation. Dev Dyn 2007;236:836–42.

[184] Banco B, Veronesi MC, Giudice C, Rota A, Grieco V. Immunohistochemical evaluation of the expression of anti-Mullerian hormone in mature, immature and neoplastic canine Sertoli cells. J Comp Pathol 2012;146:18–23.

[185] Franke FE, Pauls K, Rey R, Marks A, Bergmann M, Steger K. Differentiation markers of Sertoli cells and germ cells in fetal and early postnatal human testis. Anat Embryol (Berl) 2004;209:169–77.

[186] Yamada H, Ohashi E, Abe T, Kusumi N, Li SA, Yoshida Y, et al. Amphiphysin 1 is important for actin polymerization during phagocytosis. Mol Biol Cell 2007;18:4669–80.

[187] Badran HH, Hermo LS. Expression and regulation of aquaporins 1, 8, and 9 in the testis, efferent ducts, and epididymis of adult rats and during postnatal development. J Androl 2002;23:358–73.

[188] Hermo L, Krzeczunowicz D, Ruz R. Cell specificity of aquaporins 0, 3, and 10 expressed in the testis, efferent ducts, and epididymis of adult rats. J Androl 2004;25:494–505.

[189] Hermo L, Smith CE. Thirsty business: cell, region, and membrane specificity of aquaporins in the testis, efferent ducts, and epididymis and factors regulating their expression. J Androl 2011;32:565–75.

[190] Gist DH, Bradshaw S, Morrow CM, Congdon JD, Hess RA. Estrogen response system in the reproductive tract of the male turtle: an immunocytochemical study. Gen Comp Endocrinol 2007;151:27–33.

[191] Kopera I, Szczepanowicz M, Gizejewski Z, Sadowska J, Bilinska B. Immunoexpression of aromatase in immature and adult males of the European bison (*Bison bonasus*, Linnaeus 1758). Reprod Domest Anim 2010;45:269–74.

[192] Bouraima-Lelong H, Vanneste M, Delalande C, Zanatta L, Wolczynski S, Carreau S. Aromatase gene expression in immature rat Sertoli cells: age-related changes in the FSH signalling pathway. Reprod Fertil Dev 2010;22:508–15.

[193] Vaid KS, Guttman JA, Babyak N, Deng W, McNiven MA, Mochizuki N, et al. The role of dynamin 3 in the testis. J Cell Physiol 2007;210:644–54.

[194] Lie PP, Cheng CY, Mruk DD. Interleukin-1alpha is a regulator of the blood–testis barrier. Faseb J 2011;25:1244–53.

[195] Shang X, Liu J, Xu Q, Zhang Q, Ma B, Wang Y, et al. Changes in expression and distribution of attractin in the testes of rats at different developmental stages. Int J Mol Med 2013;32:599–606.

[196] Vera Y, Diaz-Romero M, Rodriguez S, Lue Y, Wang C, Swerdloff RS, et al. Mitochondria-dependent pathway is involved in heat-induced male germ cell death: lessons from mutant mice. Biol Reprod 2004;70:1534–40.

[197] Klein DM, Wright SH, Cherrington NJ. Localization of multidrug resistance-associated proteins along the blood–testis barrier in rat, macaque, and human testis. Drug Metab Dispos 2014;42:89–93.

[198] Siu MK, Wong CH, Xia W, Mruk DD, Lee WM, Cheng CY. The beta1-integrin-p-FAK-p130Cas-DOCK180-RhoA-vinculin is a novel regulatory protein complex at the apical ectoplasmic specialization in adult rat testes. Spermatogenesis 2011;1:73–86.

[199] Qian X, Cheng YH, Mruk DD, Cheng CY. Breast cancer resistance protein (Bcrp) and the testis—an unexpected turn of events. Asian J Androl 2013;15:455–60.

[200] Su L, Mruk DD, Lee WM, Cheng CY. Differential effects of testosterone and TGF-beta3 on endocytic vesicle-mediated protein trafficking events at the blood–testis barrier. Exp Cell Res 2010;316:2945–60.

[201] Adly MA, Abdelwahed Hussein MR. Expression of CD1d protein in human testis showing normal and abnormal spermatogenesis. Ultrastruct Pathol 2011;35:124–9.

[202] Zhang J, Wong CH, Xia W, Mruk DD, Lee NP, Lee WM, et al. Regulation of Sertoli-germ cell adherens junction dynamics via changes in protein–protein interactions of the N-cadherin-beta-catenin protein complex which are possibly mediated by c-Src and myotubularin-related protein 2: an *in vivo* study using an androgen suppression model. Endocrinology 2005;146:1268–84.

[203] Lie PP, Cheng CY, Mruk DD. Crosstalk between desmoglein-2/desmocollin-2/Src kinase and coxsackie and adenovirus receptor/ZO-1 protein complexes, regulates blood–testis barrier dynamics. Int J Biochem Cell Biol 2010;42:975–86.

[204] Sarkar O, Xia W, Mruk DD. Adjudin-mediated junction restructuring in the seminiferous epithelium leads to displacement of soluble guanylate cyclase from adherens junctions. J Cell Physiol 2006;208:175–87.

[205] Bi J, Li Y, Sun F, Saalbach A, Klein C, Miller DJ, et al. Basigin null mutant male mice are sterile and exhibit impaired interactions between germ cells and Sertoli cells. Dev Biol 2013;380:145–56.

[206] Young JS, Guttman JA, Vaid KS, Vogl AW. Tubulobulbar complexes are intercellular podosome-like structures that internalize intact intercellular junctions during epithelial remodeling events in the rat testis. Biol Reprod 2009;80:162–74.

[207] Yan HH, Cheng CY. Blood–testis barrier dynamics are regulated by an engagement/disengagement mechanism between tight and adherens junctions via peripheral adaptors. Proc Natl Acad Sci USA 2005;102:11722–7.

[208] Lee NP, Mruk D, Lee WM, Cheng CY. Is the cadherin/catenin complex a functional unit of cell-cell actin-based adherens junctions in the rat testis? Biol Reprod 2003;68:489–508.

[209] Beardsley A, O'Donnell L. Characterization of normal spermiation and spermiation failure induced by hormone suppression in adult rats. Biol Reprod 2003;68:1299–307.

[210] Yan N, Lu Y, Sun H, Qiu W, Tao D, Liu Y, et al. Microarray profiling of microRNAs expressed in testis tissues of developing primates. J Assist Reprod Genet 2009;26:179–86.

[211] Mok KW, Mruk DD, Lee WM, Cheng CY. A study to assess the assembly of a functional blood–testis barrier in developing rat testes. Spermatogenesis 2011;1:270–80.

[212] Li MW, Mruk DD, Lee WM, Cheng CY. Connexin 43 is critical to maintain the homeostasis of the blood–testis barrier via its effects on tight junction reassembly. Proc Natl Acad Sci USA 2010;107:17998–8003.

[213] Wong EW, Mruk DD, Lee WM, Cheng CY. Par3/Par6 polarity complex coordinates apical ectoplasmic specialization and blood–testis barrier restructuring during spermatogenesis. Proc Natl Acad Sci USA 2008;105:9657–62.

[214] Kurio H, Hatsuda H, Murayama E, Kaneko T, Iida H. Identification of CEACAM6 as an intermediate filament-associated protein expressed in Sertoli cells of rat testis. Biol Reprod 2011;85:924–33.

[215] Xu WM, Chen J, Chen H, Diao RY, Fok KL, Dong JD, et al. Defective CFTR-dependent CREB activation results in impaired spermatogenesis and azoospermia. PLoS One 2011;6:e19120.

[216] Teixeira S, Sa R, Grangeia A, Silva J, Oliveira C, Ferraz L, et al. Immunohystochemical analysis of CFTR in normal and disrupted spermatogenesis. Syst Biol Reprod Med 2013;59:53–9.

[217] Morrow CM, Tyagi G, Simon L, Carnes K, Murphy KM, Cooke PS, et al. Claudin 5 expression in mouse seminiferous epithelium is dependent upon the transcription factor ets variant 5 and contributes to blood–testis barrier function. Biol Reprod 2009;81:871–9.

[218] Wang F, Zhang Q, Cao J, Huang Q, Zhu X. The microtubule plus end-binding protein EB1 is involved in Sertoli cell plasticity in testicular seminiferous tubules. Exp Cell Res 2008;314:213–26.

[219] Guttman JA, Takai Y, Vogl AW. Evidence that tubulobulbar complexes in the seminiferous epithelium are involved with internalization of adhesion junctions. Biol Reprod 2004;71:548–59.

[220] Young JS, De Asis M, Guttman J, Vogl AW. Cortactin depletion results in short tubulobulbar complexes and spermiation failure in rat testes. Biol Open 2012;1:1069–77.

[221] Vitale ML, Akpovi CD, Pelletier RM. Cortactin/tyrosine-phosphorylated cortactin interaction with connexin 43 in mouse seminiferous tubules. Microsc Res Tech 2009;72:856–67.

[222] Li Y, Putnam-Lawson CA, Knapp-Hoch H, Friel PJ, Mitchell D, Hively R, et al. Immunolocalization and regulation of cystatin 12 in mouse testis and epididymis. Biol Reprod 2005;73:872–80.

[223] Mok KW, Mruk DD, Cheng CY. Regulation of blood–testis barrier (BTB) dynamics during spermatogenesis via the "Yin" and "Yang" effects of mammalian target of rapamycin complex 1 (mTORC1) and mTORC2. Int Rev Cell Mol Biol 2013;301:291–358.

[224] Gye MC, Kim ST. Expression of cathepsin L in human testis under diverse infertility conditions. Arch Androl 2004;50:187–91.

[225] Kim GH, Wright WW. A comparison of the effects of testicular maturation and aging on the stage-specific expression of CP-2/cathepsin L messenger ribonucleic acid by Sertoli cells of the Brown Norway rat. Biol Reprod 1997;57:1467–77.

[226] Wang CQ, Mruk DD, Lee WM, Cheng CY. Coxsackie and adenovirus receptor (CAR) is a product of Sertoli and germ cells in rat testes which is localized at the Sertoli–Sertoli and Sertoli–germ cell interface. Exp Cell Res 2007;313:1373–92.
[227] Elliott MR, Zheng S, Park D, Woodson RI, Reardon MA, Juncadella IJ, et al. Unexpected requirement for ELMO1 in clearance of apoptotic germ cells *in vivo*. Nature 2010;467:333–7.
[228] Lie PP, Xia W, Wang CQ, Mruk DD, Yan HH, Wong CH, et al. Dynamin II interacts with the cadherin- and occludin-based protein complexes at the blood–testis barrier in adult rat testes. J Endocrinol 2006;191:571–86.
[229] Hall ES, Eveleth J, Jiang C, Redenbach DM, Boekelheide K. Distribution of the microtubule-dependent motors cytoplasmic dynein and kinesin in rat testis. Biol Reprod 1992;46:817–28.
[230] Miller MG, Mulholland DJ, Vogl AW, Rat Testis Motor. Proteins associated with spermatid translocation (dynein) and spermatid flagella (kinesin-II). Biol Reprod 1999;60:1047–56.
[231] Tash JS, Chakrasali R, Jakkaraj SR, Hughes J, Smith SK, Hornbaker K, et al. Gamendazole, an orally active indazole carboxylic acid male contraceptive agent, targets HSP90AB1 (HSP90BETA) and EEF1A1 (eEF1A), and stimulates Il1a transcription in rat Sertoli cells. Biol Reprod 2008;78:1139–52.
[232] Elliott MR, Ravichandran KS. ELMO1 signaling in apoptotic germ cell clearance and spermatogenesis. Ann N Y Acad Sci 2010;1209:30–6.
[233] Shin I, Kim HJ, Nah WH, Park HJ, Gye MC, Park HY. Expression of activated HER2 in human testes. Fertil Steril 2011;95:2725–8.
[234] Wunsch A, Strothmann K, Simoni M, Gromoll J, Nieschlag E, Luetjens CM. Epidermal growth factor receptor pathway substrate 8 (Eps8) expression in maturing testis. Asian J Androl 2004;6:195–203.
[235] Beardsley A, Robertson DM, O'Donnell L. A complex containing alpha6beta1-integrin and phosphorylated focal adhesion kinase between Sertoli cells and elongated spermatids during spermatid release from the seminiferous epithelium. J Endocrinol 2006;190:759–70.
[236] Wakayama T, Nakata H, Kurobo M, Sai Y, Iseki S. Expression, localization, and binding activity of the ezrin/radixin/moesin proteins in the mouse testis. J Histochem Cytochem 2009;57:351–62.
[237] Mruk DD, Lau AS. RAB13 participates in ectoplasmic specialization dynamics in the rat testis. Biol Reprod 2009;80:590–601.
[238] Gunawan A, Kaewmala K, Uddin MJ, Cinar MU, Tesfaye D, Phatsara C, et al. Association study and expression analysis of porcine ESR1 as a candidate gene for boar fertility and sperm quality. Anim Reprod Sci 2011;128:11–21.
[239] Chevalier N, Vega A, Bouskine A, Siddeek B, Michiels JF, Chevallier D, et al. GPR30, the non-classical membrane G protein related estrogen receptor, is overexpressed in human seminoma and promotes seminoma cell proliferation. PLoS ONE 2012;7:e34672.
[240] Rago V, Romeo F, Giordano F, Maggiolini M, Carpino A. Identification of the estrogen receptor GPER in neoplastic and non-neoplastic human testes. Reprod Biol Endocrinol 2011;9:135.
[241] Lucas TF, Pimenta MT, Pisolato R, Lazari MF, Porto CS. 17beta-estradiol signaling and regulation of Sertoli cell function. Spermatogenesis 2011;1:318–24.

[242] Rago V, Maggiolini M, Vivacqua A, Palma A, Carpino A. Differential expression of estrogen receptors (ERalpha/ERbeta) in testis of mature and immature pigs. Anat Rec A Discov Mol Cell Evol Biol 2004;281:1234–9.

[243] Dokucu AI, Ozturk H, Ozturk H, Tuncer MC, Yilmaz F. The effects of molsidomine on hypoxia inducible factor alpha and Sonic hedgehog in testicular ischemia/reperfusion injury in rats. Int Urol Nephrol 2009;41:101–8.

[244] Chui K, Trivedi A, Cheng CY, Cherbavaz DB, Dazin PF, Huynh AL, et al. Characterization and functionality of proliferative human Sertoli cells. Cell Transplant 2011;20:619–35.

[245] Maekawa M, Toyama Y, Yasuda M, Yagi T, Yuasa S. Fyn tyrosine kinase in Sertoli cells is involved in mouse spermatogenesis. Biol Reprod 2002;66:211–21.

[246] Wang XX, Ying P, Diao F, Wang Q, Ye D, Jiang C, et al. Altered protein prenylation in Sertoli cells is associated with adult infertility resulting from childhood mumps infection. J Exp Med 2013;210:1559–74.

[247] Marchlewska K, Kula K, Walczak-Jedrzejowska R, Kula W, Oszukowska E, Filipiak E, et al. Maturational changes in connexin 43 expression in the seminiferous tubules may depend on thyroid hormone action. Arch Med Sci 2013;9:139–45.

[248] Li MW, Mruk DD, Lee WM, Cheng CY. Connexin 43 and plakophilin-2 as a protein complex that regulates blood–testis barrier dynamics. Proc Natl Acad Sci USA 2009;106:10213–8.

[249] Decrouy X, Gasc JM, Pointis G, Segretain D. Functional characterization of Cx43 based gap junctions during spermatogenesis. J Cell Physiol 2004;200: 146–54.

[250] Fiorini C, Decrouy X, Defamie N, Segretain D, Pointis G. Opposite regulation of connexin33 and connexin43 by LPS and IL-1alpha in spermatogenesis. Am J Physiol Cell Physiol 2006;290:C733–40.

[251] Carstensen JF, Tesdorpf JG, Kaufmann M, Markus MM, Husen B, Leenders F, et al. Characterization of 17 beta-hydroxysteroid dehydrogenase IV. J Endocrinol 1996;150 (Suppl):S3–12.

[252] Oner J, Oner H, Colakoglu N, Kukner A, Ozan E. The effects of triiodothyronine on rat testis: a morphometric and immunohistochemical study. J Mol Histol 2006;37:9–14.

[253] Okuma Y, Saito K, O'Connor AE, Phillips DJ, de Kretser DM, Hedger MP. Reciprocal regulation of activin A and inhibin B by interleukin-1 (IL-1) and follicle-stimulating hormone (FSH) in rat Sertoli cells *in vitro*. J Endocrinol 2005;185:99–110.

[254] Grieco V, Banco B, Ferrari A, Rota A, Faustini M, Veronesi MC. Inhibin-alpha immunohistochemical expression in mature and immature canine Sertoli and Leydig cells. Reprod Domest Anim 2011;46:920–3.

[255] Sun B, Qi N, Shang T, Wu H, Deng T, Han D. Sertoli cell-initiated testicular innate immune response through toll-like receptor-3 activation is negatively regulated by Tyro3, Axl, and mer receptors. Endocrinology 2010;151:2886–97.

[256] Ketema M, Kreft M, Secades P, Janssen H, Sonnenberg A. Nesprin-3 connects plectin and vimentin to the nuclear envelope of Sertoli cells but is not required for Sertoli cell function in spermatogenesis. Mol Biol Cell 2013;24:2454–66.

[257] Lui WY, Mruk DD, Cheng CY. Interactions among IQGAP1, Cdc42, and the cadherin/catenin protein complex regulate Sertoli–germ cell adherens junction dynamics in the testis. J Cell Physiol 2005;202:49–66.

[258] Kruse R, Eigelshoven S, Kaiser A, Ruzicka T, Neumann NJ. Cytokeratin 18 expression in immature Sertoli cells: co-localization with interstitial lymphocytic infiltrates. Folia Histochem Cytobiol 2009;47:127–30.

[259] Adly MA, Hussein MR. Expression of cytokeratin 10 protein in the human testis showing normal and abnormal spermatogenesis. Ultrastruct Pathol 2011;35:209–13.

[260] Smith L, Willan J, Warr N, Brook FA, Cheeseman M, Sharpe R, et al. The Maestro (*Mro*) gene is dispensable for normal sexual development and fertility in mice. PLoS One 2008;3:e4091.

[261] Lazarus M, Eguchi N, Matsumoto S, Nagata N, Yano T, Killian GJ, et al. Species-specific expression of microsomal prostaglandin E synthase-1 and cyclooxygenases in male monkey reproductive organs. Prostaglandins Leukot Essent Fatty Acids 2004;71:233–40.

[262] Tanwar PS, Kaneko-Tarui T, Zhang L, Teixeira JM. Altered LKB1/AMPK/TSC1/TSC2/mTOR signaling causes disruption of Sertoli cell polarity and spermatogenesis. Hum Mol Genet 2012;21:4394–405.

[263] Li JC, Samy ET, Grima J, Chung SS, Mruk D, Lee WM, et al. Rat testicular myotubularin, a protein tyrosine phosphatase expressed by Sertoli and germ cells, is a potential marker for studying cell–cell interactions in the rat testis. J Cell Physiol 2000;185:366–85.

[264] Ocon-Grove OM, Krzysik-Walker SM, Maddineni SR, Hendricks 3rd GL, Ramachandran R. NAMPT (visfatin) in the chicken testis: influence of sexual maturation on cellular localization, plasma levels and gene and protein expression. Reproduction 2010;139:217–26.

[265] Lucas TF, Lazari MF, Porto CS. Differential role of the estrogen receptors ESR1 and ESR2 on the regulation of proteins involved with proliferation and differentiation of Sertoli cells from 15-day-old rats. Mol Cell Endocrinol 2014;382:84–96.

[266] Gye MC. Expression of occludin in canine testis and epididymis. Reprod Domest Anim 2004;39:43–7.

[267] Veitinger S, Veitinger T, Cainarca S, Fluegge D, Engelhardt CH, Lohmer S, et al. Purinergic signalling mobilizes mitochondrial Ca(2)(+) in mouse Sertoli cells. J Physiol 2011;589:5033–55.

[268] Qian X, Mruk DD, Wong EW, Lie PP, Cheng CY. Palladin is a regulator of actin filament bundles at the ectoplasmic specialization in adult rat testes. Endocrinology 2013;154:1907–20.

[269] Guttman JA, Mulholland DJ, Vogl AW. Plectin is concentrated at intercellular junctions and at the nuclear surface in morphologically differentiated rat Sertoli cells. Anat Rec 1999;254:418–28.

[270] Upadhyay R, D'Souza R, Sonawane S, Gaonkar R, Pathak S, Jhadav A, et al. Altered phosphorylation and distribution status of vimentin in rat seminiferous epithelium following 17beta-estradiol treatment. Histochem Cell Biol 2011;136:543–55.

[271] Han YB, Feng HL, Cheung CK, Lam PM, Wang CC, Haines CJ. Expression of a novel T-complex testis expressed 5 (Tctex5) in mouse testis, epididymis, and spermatozoa. Mol Reprod Dev 2007;74:1132–40.

[272] Puri P, Walker WH. The tyrosine phosphatase SHP2 regulates Sertoli cell junction complexes. Biol Reprod 2013;88:59.

[273] Lie PP, Mruk DD, Mok KW, Su L, Lee WM, Cheng CY. Focal adhesion kinase-Tyr407 and -Tyr397 exhibit antagonistic effects on blood–testis barrier dynamics in the rat. Proc Natl Acad Sci USA 2012;109:12562–7.

[274] Wong CH, Xia W, Lee NP, Mruk DD, Lee WM, Cheng CY. Regulation of ectoplasmic specialization dynamics in the seminiferous epithelium by focal adhesion-associated proteins in testosterone-suppressed rat testes. Endocrinology 2005;146:1192–204.
[275] Mulholland DJ, Dedhar S, Vogl AW. Rat seminiferous epithelium contains a unique junction (ectoplasmic specialization) with signaling properties both of cell/cell and cell/matrix junctions. Biol Reprod 2001;64:396–407.
[276] Cardoso L, Nascimento A, Royer C, Porto C, Lazari M. Locally produced relaxin may affect testis and vas deferens function in rats. Reproduction 2010;139:185–96.
[277] Abid SN, Richardson TE, Powell HM, Jaichander P, Chaudhary J, Chapman KM, et al. A-single spermatogonia heterogeneity and cell cycles synchronize with rat seminiferous epithelium stages VIII–IX. Biol Reprod 2014;90:32.
[278] Amselgruber W, Sinowatz F, Schams D, Lehmann M. S-100 protein immunoreactivity in bovine testis. Andrologia 1992;24:231–5.
[279] Cruzana BC, Hondo E, Kitamura N, Matsuzaki S, Nakagawa M, Yamada J. Differential localization of immunoreactive alpha- and beta-subunits of S-100 protein in feline testis. Anat Histol Embryol 2000;29:83–6.
[280] Su W, Wong EW, Mruk DD, Cheng CY. The Scribble/Lgl/Dlg polarity protein complex is a regulator of blood–testis barrier dynamics and spermatid polarity during spermatogenesis. Endocrinology 2012;153:6041–53.
[281] Morales CR, El-Alfy M, Zhao Q, Igdoura SA. Expression and tissue distribution of rat sulfated glycoprotein-1 (prosaposin). J Histochem Cytochem 1996;44:327–37.
[282] Igdoura SA, Morales CR. Role of sulfated glycoprotein-1 (SGP-1) in the disposal of residual bodies by Sertoli cells of the rat. Mol Reprod Dev 1995;40:91–102.
[283] Millar MR, Sharpe RM, Maguire SM, Gaughan J, Saunders PT. Colocalization of mRNA and protein using *in situ* hybridization and immunohistochemistry in testicular tissue. Microsc Res Tech 1995;32:498–503.
[284] Cayli S, Erdemir F, Ocakli S, Ungor B, Kesici H, Yener T, et al. Interaction between Smad1 and p97/VCP in rat testis and epididymis during the postnatal development. Reprod Sci 2012;19:190–201.
[285] Zhao L, He J, Guo Q, Wen X, Zhang X, Dong C. Expression of growth differentiation factor 9 (GDF9) and its receptor in adult cat testis. Acta Histochem 2011;113:771–6.
[286] Zhang XJ, Wen XX, Zhao L, He JP. Immunolocalization of Smad4 protein in the testis of domestic fowl (*Gallus domesticus*) during postnatal development. Acta Histochem 2012;114:429–33.
[287] Zhang YQ, He XZ, Zhang JS, Wang RA, Zhou J, Xu RJ. Stage-specific localization of transforming growth factor beta1 and beta3 and their receptors during spermatogenesis in men. Asian J Androl 2004;6:105–9.
[288] Yao PL, Lin YC, Richburg JH. Transcriptional suppression of Sertoli cell Timp2 in rodents following mono-(2-ethylhexyl) phthalate exposure is regulated by CEBPA and MYC. Biol Reprod 2011;85:1203–15.
[289] Tarulli GA, Meachem SJ, Schlatt S, Stanton PG. Regulation of testicular tight junctions by gonadotrophins in the adult Djungarian hamster *in vivo*. Reproduction 2008;135:867–77.
[290] Chung SS, Choi C, Wang X, Hallock L, Wolgemuth DJ. Aberrant distribution of junctional complex components in retinoic acid receptor alpha-deficient mice. Microsc Res Tech 2010;73:583–96.
[291] Xu J, Anuar F, Ali SM, Ng MY, Phua DC, Hunziker W. Zona occludens-2 is critical for blood–testis barrier integrity and male fertility. Mol Biol Cell 2009;20:4268–77.

[292] Best D, Sahlender DA, Walther N, Peden AA, Adams IR. Sdmg1 is a conserved transmembrane protein associated with germ cell sex determination and germline–soma interactions in mice. Development 2008;135:1415–25.
[293] Castilla-Cortazar I, Diez N, Garcia-Fernandez M, Puche JE, Diez-Caballero F, Quiroga J, et al. Hematotesticular barrier is altered from early stages of liver cirrhosis: effect of insulin-like growth factor 1. World J Gastroenterol 2004;10:2529–34.
[294] Brilhante O, Okada FK, Sasso-Cerri E, Stumpp T, Miraglia SM. Late morfofunctional alterations of the Sertoli cell caused by doxorubicin administered to prepubertal rats. Reprod Biol Endocrinol 2012;10:79.
[295] Wenz JR, Hess RA. Characterization of stage-specific tyrosinated alpha-tubulin immunoperoxidase staining patterns in Sertoli cells of rat seminiferous tubules by light microscopic image analysis. Tissue Cell 1998;30:492–501.
[296] De Gendt K, Denolet E, Willems A, Daniels VW, Clinckemalie L, Denayer S, et al. Expression of Tubb3, a beta-tubulin isotype, is regulated by androgens in mouse and rat Sertoli cells. Biol Reprod 2011;85:934–45.
[297] Redenbach DM, Hall ES, Boekelheide K. Distribution of Sertoli cell microtubules, microtubule-dependent motors, and the Golgi apparatus before and after tight junction formation in developing rat testis. Microsc Res Tech 1995;32:504–19.
[298] Grove BD, Pfeiffer DC, Allen S, Vogl AW. Immunofluorescence localization of vinculin in ectoplasmic ("junctional") specializations of rat Sertoli cells. Am J Anat 1990;188:44–56.
[299] Steger K, Schimmel M, Wrobel KH. Immunocytochemical demonstration of cytoskeletal proteins in seminiferous tubules of adult rams and bulls. Arch Histol Cytol 1994;57:17–28.
[300] Lydka M, Kotula-Balak M, Kopera-Sobota I, Tischner M, Bilinska B. Vimentin expression in testes of Arabian stallions. Equine Vet J 2011;43:184–9.
[301] Yu G, Liu Y, Xie L, Wang X. Involvement of Sertoli cells in spermatogenic failure induced by carbendazim. Environ Toxicol Pharmacol 2009;27:287–92.
[302] Suetsugu S, Banzai Y, Kato M, Fukami K, Kataoka Y, Takai Y, et al. Male-specific sterility caused by the loss of CR16. Genes Cells 2007;12:721–33.
[303] Xiao X, Mruk DD, Cheng CY. c-Yes regulates cell adhesion at the apical ectoplasmic specialization-blood–testis barrier axis via its effects on protein recruitment and distribution. Am J Physiol Endocrinol Metab 2013;304:E145–59.
[304] Lee NP, Mruk DD, Conway AM, Cheng CY. Zyxin, axin, and Wiskott–Aldrich syndrome protein are adaptors that link the cadherin/catenin protein complex to the cytoskeleton at adherens junctions in the seminiferous epithelium of the rat testis. J Androl 2004;25:200–15.
[305] Rajpert-De Meyts E, Jorgensen N, Graem N, Muller J, Cate RL, Skakkebaek NE. Expression of anti-Mullerian hormone during normal and pathological gonadal development: association with differentiation of Sertoli and granulosa cells. J Clin Endocrinol Metab 1999;84:3836–44.
[306] Hall ES, Hall SJ, Boekelheide K. Sertoli cells isolated from adult 2,5-hexanedione-exposed rats exhibit atypical morphology and actin distribution [published erratum appears in Toxicol Appl Pharmacol 1993;119(2):310]. Toxicol Appl Pharmacol 1992;117:9–18.
[307] Allard EK, Johnson KJ, Boekelheide K. Colchicine disrupts the cytoskeleton of rat testis seminiferous epithelium in a stage-dependent manner. Biol Reprod 1993;48:143–53.
[308] Hall ES, Hall SJ, Boekelheide K. 2,5-Hexanedione exposure alters microtubule motor distribution in adult rat testis. Fundam Appl Toxicol 1995;24:173–82.

[309] Nakai M, Miller MG, Carnes K, Hess RA. Stage-specific effects of the fungicide carbendazim on Sertoli cell microtubules in rat testis. Tissue Cell 2002;34:73–80.

[310] Beach SF, Vogl AW. Spermatid translocation in the rat seminiferous epithelium: coupling membrane trafficking machinery to a junction plaque. Biol Reprod 1999;60:1036–46.

[311] Russell LD, Malone JP, MacCurdy DS. Effect of the microtubule disrupting agents, colchicine and vinblastine, on seminiferous tubule structure in the rat. Tissue Cell 1981;13:349–67.

[312] Nakai M, Hess RA. Morphological changes in the rat Sertoli cell induced by the microtubule poison carbendazim. Tissue Cell 1994;26:917–27.

[313] Nakai M, Hess RA. Effects of carbendazim (methyl 2-benzimidazole carbamate; MBC) on meiotic spermatocytes and subsequent spermiogenesis in the rat testis. Anat Rec 1997;247:379–87.

[314] Flickinger C, Fawcett DW. The junctional specializations of Sertoli cells in the seminiferous epithelium. Anat Rec 1967;158:207–21.

[315] Stevenson BR, Siliciano JD, Mooseker MS, Goodenough DA. Identification of ZO-1: a high molecular weight polypeptide associated with the tight junction (zonula occludens) in a variety of epithelia. J Cell Biol 1986;103:755–66.

[316] Byers S, Graham R, Dai HN, Hoxter B. Development of Sertoli cell junctional specializations and the distribution of the tight-junction-associated protein ZO-1 in the mouse testis. Am J Anat 1991;191:35–47.

[317] Jensen MB. Vitamin D and male reproduction. Nat Rev Endocrinol 2014;10:175–86.

[318] Schleicher G, Privette TH, Stumpf WE. Distribution of soltriol [1,25(OH)2-vitamin D3] binding sites in male sex organs of the mouse: an autoradiographic study. J Histochem Cytochem 1989;37:1083–6.

[319] Stumpf WE, Sar M, Chen K, Morin J, DeLuca HF. Sertoli cells in the testis and epithelium of the ductuli efferentes are targets for 3H-1,25 (OH)2 vitamin D3. Cell Tissue Res 1987;247:453–5.

[320] Merke J, Hugel U, Ritz E. Nuclear testicular 1,25-dihydroxyvitamin D3 receptors in Sertoli cells and seminiferous tubules of adult rodents. Biochem Biophys Res Commun 1985;127:303–9.

[321] Zanatta L, Zamoner A, Zanatta AP, Bouraima-Lelong H, Delalande C, Bois C, et al. Nongenomic and genomic effects of 1alpha,25(OH)2 vitamin D3 in rat testis. Life Sci 2011;89:515–23.

[322] Carreau S, Hess RA. Oestrogens and spermatogenesis. Philos Trans R Soc Lond B Biol Sci 2010;365:1517–35.

[323] Hess RA, Fernandes SA, Gomes GR, Oliveira CA, Lazari MF, Porto CS. Estrogen and its receptors in efferent ductules and epididymis. J Androl 2011;32:600–13.

[324] King WJ, Greene GL. Monoclonal antibodies localize oestrogen receptor in the nuclei of target cells. Nature 1984;307:745–7.

[325] Fietz D, Bakhaus K, Wapelhorst B, Grosser G, Gunther S, Alber J, et al. Membrane transporters for sulfated steroids in the human testis—cellular localization, expression pattern and functional analysis. PLoS One 2013;8:e62638.

[326] Trezise AE, Linder CC, Grieger D, Thompson EW, Meunier H, Griswold MD, et al. CFTR expression is regulated during both the cycle of the seminiferous epithelium and the oestrous cycle of rodents. Nat Genet 1993;3:157–64.

[327] Hihnala S, Kujala M, Toppari J, Kere J, Holmberg C, Hoglund P. Expression of SLC26A3, CFTR and NHE3 in the human male reproductive tract: role in male subfertility caused by congenital chloride diarrhoea. Mol Hum Reprod 2006;12:107–11.

[328] Mueller S, Rosenquist TA, Takai Y, Bronson RA, Wimmer E. Loss of nectin-2 at Sertoli-spermatid junctions leads to male infertility and correlates with severe spermatozoan head and midpiece malformation, impaired binding to the zona pellucida, and oocyte penetration. Biol Reprod 2003;69:1330–40.

[329] Cheng CY, Silvestrini B, Grima J, Mo MY, Zhu LJ, Johansson E, et al. Two new male contraceptives exert their effects by depleting germ cells prematurely from the testis. Biol Reprod 2001;65:449–61.

[330] Russell L. Observations on rat Sertoli ectoplasmic ("junctional") specializations in their association with germ cells of the rat testis. Tissue Cell 1977;9:475–98.

[331] Grove BD, Vogl AW. Sertoli cell ectoplasmic specializations: a type of actin-associated adhesion junction? J Cell Sci 1989;93(Pt 2):309–23.

[332] Toyama Y. Actin-like filaments in the Sertoli cell junctional specializations in the swine and mouse testis. Anat Rec 1976;186:477–91.

[333] Vogl AW, Grove BD, Lew GJ. Distribution of actin in Sertoli cell ectoplasmic specializations and associated spermatids in the ground squirrel testis. Anat Rec 1986;215:331–41.

[334] Dym M, Fawcett DW. The blood–testis barrier in the rat and the physiological compartmentation of the seminiferous epithelium. Biol Reprod 1970;3:308–26.

[335] Russell LD. Spermiation—the sperm release process: ultrastructural observations and unresolved problems. In: Van Blerkom J, Motta PM, editors. Ultrastructure of reproduction. Boston, MA: Martinus Nijhoff; 1984. p. 46–66.

[336] Russell LD. Role in spermiation. In: Russell LD, Griswold MD, editors. The Sertoli cell. Clearwater, FL: Cache River Press; 1993. p. 269–304.

[337] Russell LD, Goh JC, Rashed RM, Vogl AW. The consequences of actin disruption at Sertoli ectoplasmic specialization sites facing spermatids after *in vivo* exposure of rat testis to cytochalasin D. Biol Reprod 1988;39:105–18.

[338] Cheng CY, Mo M, Grima J, Saso L, Tita B, Mruk D, et al. Indazole carboxylic acids in male contraception. Contraception 2002;65:265–8.

[339] Mruk DD, Cheng CY. Cell–cell interactions at the ectoplasmic specialization in the testis. Trends Endocrinol Metab 2004;15:439–47.

[340] Mruk DD, Cheng CY. Sertoli–Sertoli and Sertoli–germ cell interactions and their significance in germ cell movement in the seminiferous epithelium during spermatogenesis. Endocr Rev 2004;25:747–806.

[341] Weber JE, Turner TT, Tung KS, Russell LD. Effects of cytochalasin D on the integrity of the Sertoli cell (blood–testis) barrier. Am J Anat 1988;182:130–47.

[342] Vogl AW, Soucy LJ, Lew GJ. Distribution of actin in isolated seminiferous epithelia and denuded tubule walls of the rat. Anat Rec 1985;213:63–71.

[343] Vogl AW, Soucy LJ. Arrangement and possible function of actin filament bundles in ectoplasmic specializations of ground squirrel Sertoli cells. J Cell Biol 1985;100:814–25.

[344] Franke WW, Grund C, Fink A, Weber K, Jockusch BM, Zentgraf H, et al. Location of actin in the microfilament bundles associated with the junctional specializations between Sertoli cells and spermatids. Biol Cell 1978;31:7–14.

[345] Lie PP, Mruk DD, Lee WM, Cheng CY. Epidermal growth factor receptor pathway substrate 8 (Eps8) is a novel regulator of cell adhesion and the blood–testis barrier integrity in the seminiferous epithelium. FASEB J 2009;23:2555–67.

[346] Kai M, Irie M, Okutsu T, Inoue K, Ogonuki N, Miki H, et al. The novel dominant mutation Dspd leads to a severe spermiogenesis defect in mice. Biol Reprod 2004;70:1213–21.

[347] Velichkova M, Guttman J, Warren C, Eng L, Kline K, Vogl AW, et al. A human homologue of *Drosophila* kelch associates with myosin-VIIa in specialized adhesion junctions. Cell Motil Cytoskeleton 2002;51:147–64.
[348] Bartles JR, Wierda A, Zheng L. Identification and characterization of espin, an actin-binding protein localized to the F-actin-rich junctional plaques of Sertoli cell ectoplasmic specializations. J Cell Sci 1996;109(Pt 6):1229–39.
[349] Hasson T, Walsh J, Cable J, Mooseker MS, Brown SD, Steel KP. Effects of shaker-1 mutations on myosin-VIIa protein and mRNA expression. Cell Motil Cytoskeleton 1997;37:127–38.
[350] Wang W, Wine RN, Chapin RE. Rat testicular Src: normal distribution and involvement in ethylene glycol monomethyl ether-induced apoptosis. Toxicol Appl Pharmacol 2000;163:125–34.
[351] Chapin RE, Wine RN, Harris MW, Borchers CH, Haseman JK. Structure and control of a cell–cell adhesion complex associated with spermiation in rat seminiferous epithelium. J Androl 2001;22:1030–52.
[352] Zhang J, Mruk DD, Cheng CY. Myotubularin phosphoinositide phosphatases, protein phosphatases, and kinases: their roles in junction dynamics and spermatogenesis. J Cell Physiol 2005;204:470–83.
[353] Palombi F, Salanova M, Tarone G, Farini D, Stefanini M. Distribution of beta 1 integrin subunit in rat seminiferous epithelium. Biol Reprod 1992;47:1173–82.
[354] Salanova M, Stefanini M, De Curtis I, Palombi F. Integrin receptor alpha 6 beta 1 is localized at specific sites of cell-to-cell contact in rat seminiferous epithelium. Biol Reprod 1995;52:79–87.
[355] Siu MK, Cheng CY. Interactions of proteases, protease inhibitors, and the beta1 integrin/laminin gamma3 protein complex in the regulation of ectoplasmic specialization dynamics in the rat testis. Biol Reprod 2004;70:945–64.
[356] Gliki G, Ebnet K, Aurrand-Lions M, Imhof BA, Adams RH. Spermatid differentiation requires the assembly of a cell polarity complex downstream of junctional adhesion molecule-C. Nature 2004;431:320–4.
[357] Russell L. Movement of spermatocytes from the basal to the adluminal compartment of the rat testis. Am J Anat 1977;148:313–28.
[358] Russell LD. Observations on the inter-relationships of Sertoli cells at the level of the blood–testis barrier: evidence for formation and resorption of Sertoli–Sertoli tubulobulbar complexes during the spermatogenic cycle of the rat. Am J Anat 1979;155:259–79.
[359] Russell L, Clermont Y. Anchoring device between Sertoli cells and late spermatids in rat seminiferous tubules. Anat Rec 1976;185:259–78.
[360] Russell LD. Further observations on tubulobulbar complexes formed by late spermatids and Sertoli cells in the rat testis. Anat Rec 1979;194:213–32.
[361] Guttman JA, Obinata T, Shima J, Griswold M, Vogl AW. Non-muscle cofilin is a component of tubulobulbar complexes in the testis. Biol Reprod 2004;70:805–12.
[362] Disanza A, Carlier MF, Stradal TE, Didry D, Frittoli E, Confalonieri S, et al. Eps8 controls actin-based motility by capping the barbed ends of actin filaments. Nat Cell Biol 2004;6:1180–8.
[363] Nicholls PK, Harrison CA, Walton KL, McLachlan RI, O'Donnell L, Stanton PG. Hormonal regulation of Sertoli cell micro-RNAs at spermiation. Endocrinology 2011;152:1670–83.
[364] Weaver AM, Karginov AV, Kinley AW, Weed SA, Li Y, Parsons JT, et al. Cortactin promotes and stabilizes Arp2/3-induced actin filament network formation. Curr Biol 2001;11:370–4.

[365] Russell LD. Spermatid–Sertoli tubulobulbar complexes as devices for elimination of cytoplasm from the head region late spermatids of the rat. Anat Rec 1979;194:233–46.
[366] Russell LD. Deformities in the head region of late spermatids of hypophysectomized-hormone-treated rats. Anat Rec 1980;197:21–31.
[367] Tanii I, Yoshinaga K, Toshimori K. Morphogenesis of the acrosome during the final steps of rat spermiogenesis with special reference to tubulobulbar complexes. Anat Rec 1999;256:195–201.
[368] Kierszenbaum AL, Tres LL. The acrosome–acroplaxome–manchette complex and the shaping of the spermatid head. Arch Histol Cytol 2004;67:271–84.
[369] Taylor MJ, Perrais D, Merrifield CJ. A high precision survey of the molecular dynamics of mammalian clathrin-mediated endocytosis. PLoS Biol 2011;9:e1000604.
[370] Akashi M, Yokota S, Fujita H. Polyubiquitination and proteosome signals in tubulobulbar complexes of rat late spermatids. CellBio 2013;2:173–8.
[371] D'Souza R, Pathak S, Upadhyay R, Gaonkar R, D'Souza S, Sonawane S, et al. Disruption of tubulobulbar complex by high intratesticular estrogens leading to failed spermiation. Endocrinology 2009;150:1861–9.
[372] Neely MD, Boekelheide K. Sertoli cell processes have axoplasmic features: an ordered microtubule distribution and an abundant high molecular weight microtubule-associated protein (cytoplasmic dynein). J Cell Biol 1988;107:1767–76.
[373] Vogl AW, Lin YC, Dym M, Fawcett DW. Sertoli cells of the golden-mantled ground squirrel (*Spermophilus lateralis*): a model system for the study of shape change. Am J Anat 1983;168:83–98.
[374] Vogl AW. Changes in the distribution of microtubules in rat Sertoli cells during spermatogenesis. Anat Rec 1988;222:34–41.
[375] Vogl AW, Soucy LJ, Foo V. Ultrastructure of Sertoli-cell penetrating processes found in germ cells of the golden-mantled ground squirrel (*Spermophilus lateralis*). Am J Anat 1985;172:75–86.
[376] Vogl AW, Weis M, Pfeiffer DC. The perinuclear centriole-containing centrosome is not the major microtubule organizing center in Sertoli cells. Eur J Cell Biol 1995;66:165–79.
[377] Fleming SL, Shank PR, Boekelheide K. gamma-Tubulin overexpression in Sertoli cells *in vivo*: I. Localization to sites of spermatid head attachment and alterations in Sertoli cell microtubule distribution. Biol Reprod 2003;69:310–21.
[378] Hermo L, Oko R, Hecht NB. Differential post-translational modifications of microtubules in cells of the seminiferous epithelium of the rat: a light and electron microscope immunocytochemical study. Anat Rec 1991;229:31–50.
[379] Richburg JH, Redenbach DM, Boekelheide K. Seminiferous tubule fluid secretion is a Sertoli cell microtubule-dependent process inhibited by 2,5-hexanedione exposure. Toxicol Appl Pharmacol 1994;128:302–9.
[380] Vogl AW. Spatially dynamic intercellular adhesion junction is coupled to a microtubule-based motility system: evidence from an *in vitro* binding assay. Cell Motil Cytoskeleton 1996;34:1–12.
[381] Guttman JA, Kimel GH, Vogl AW. Dynein and plus-end microtubule-dependent motors are associated with specialized Sertoli cell junction plaques (ectoplasmic specializations). J Cell Sci 2000;113(Pt 12):2167–76.
[382] Vaid KS, Guttman JA, Singaraja RR, Vogl AW. A kinesin is present at unique Sertoli/spermatid adherens junctions in rat and mouse testes. Biol Reprod 2007;77:1037–48.

[383] Redenbach DM, Boekelheide K, Vogl AW. Binding between mammalian spermatid-ectoplasmic specialization complexes and microtubules. Eur J Cell Biol 1992;59:433–48.

[384] Vogl AW, Linck RW, Dym M. Colchicine-induced changes in the cytoskeleton of the golden-mantled ground squirrel (*Spermophilus lateralis*) Sertoli cells. Am J Anat 1983;168:99–108.

[385] Colucci-Guyon E, Portier MM, Dunia I, Paulin D, Pournin S, Babinet C. Mice lacking vimentin develop and reproduce without an obvious phenotype. Cell 1994;79:679–94.

[386] Franke WW, Grund C, Schmid E. Intermediate-sized filaments present in Sertoli cells are of the vimentin type. Eur J Cell Biol 1979;19:269–75.

[387] Stosiek P, Kasper M, Karsten U. Expression of cytokeratins 8 and 18 in human Sertoli cells of immature and atrophic seminiferous tubules. Differentiation 1990;43:66–70.

[388] Vogl AW, Colucci-Guyon E, Babinet C. Vimentin intermediate filaments are not necessary for the development of a normal differentiated phenotype by mature Sertoli cells. Mol Biol Cell 1996;7(Suppl):555a(#3227).

[389] Russell L. Desmosome-like junctions between Sertoli and germ cells in the rat testis. Am J Anat 1977;148:301–12.

[390] McGinley DM, Posalaky Z, Porvaznik M, Russell L. Gap junctions between Sertoli and germ cells of rat seminiferous tubules. Tissue Cell 1979;11:741–54.

[391] Lie PP, Cheng CY, Mruk DD. The biology of the desmosome-like junction a versatile anchoring junction and signal transducer in the seminiferous epithelium. Int Rev Cell Mol Biol 2011;286:223–69.

[392] Mruk DD, Cheng CY. Desmosomes in the testis: moving into an unchartered territory. Spermatogenesis 2011;1:47–51.

[393] Walker WH. Testosterone signaling and the regulation of spermatogenesis. Spermatogenesis 2011;1:116–20.

[394] Sluka P, O'Donnell L, Bartles JR, Stanton PG. FSH regulates the formation of adherens junctions and ectoplasmic specialisations between rat Sertoli cells *in vitro* and *in vivo*. J Endocrinol 2006;189:381–95.

[395] Shupe J, Cheng J, Puri P, Kostereva N, Walker WH. Regulation of Sertoli–germ cell adhesion and sperm release by FSH and nonclassical testosterone signaling. Mol Endocrinol 2011;25:238–52.

[396] Li SY, Mruk DD, Cheng CY. Focal adhesion kinase is a regulator of F-actin dynamics: new insights from studies in the testis. Spermatogenesis 2013;3:e25385.

[397] Lie PP, Cheng CY, Mruk DD. Signalling pathways regulating the blood–testis barrier. Int J Biochem Cell Biol 2013;45:621–5.

[398] Bremner WJ, Millar MR, Sharpe RM, Saunders PT. Immunohistochemical localization of androgen receptors in the rat testis: evidence for stage-dependent expression and regulation by androgens. Endocrinol 1994;135:1227–34.

[399] Yan HH, Mruk DD, Wong EW, Lee WM, Cheng CY. An autocrine axis in the testis that coordinates spermiation and blood–testis barrier restructuring during spermatogenesis. Proc Natl Acad Sci USA 2008;105:8950–5.

[400] Chung K, Deisseroth K. CLARITY for mapping the nervous system. Nat Methods 2013;10:508–13.

[401] Orth JM, Boehm R. Functional coupling of neonatal rat Sertoli cells and gonocytes in coculture. Endocrinology 1990;127:2812–20.

[402] Kanatsu-Shinohara M, Takehashi M, Takashima S, Lee J, Morimoto H, Chuma S, et al. Homing of mouse spermatogonial stem cells to germline niche depends on beta1-integrin. Cell Stem Cell 2008;3:533–42.

[403] Takashima S, Kanatsu-Shinohara M, Tanaka T, Takehashi M, Morimoto H, Shinohara T. Rac mediates mouse spermatogonial stem cell homing to germline niches by regulating transmigration through the blood–testis barrier. Cell Stem Cell 2011;9:463–75.

[404] Voss C, Lahiri S, Young BP, Loewen CJ, Prinz WA. ER-shaping proteins facilitate lipid exchange between the ER and mitochondria in *S. cerevisiae*. J Cell Sci 2012;125:4791–9.

Establishment of fetal Sertoli cells and their role in testis morphogenesis

2

Humphrey Hung-Chang Yao[a], *Erica Ungewitter*[a], *Heather Franco*[a] and *Blanche Capel*[b]
[a]Reproductive Developmental Biology Group, Laboratory of Reproductive and Developmental Toxicology, NIEHS/NIH, Research Triangle Park, NC, [b]Department of Cell Biology, Duke University Medical Center, Durham, NC

I. Introduction

Sertoli cells, the specialized epithelial cell type that surrounds the male germ line, orchestrate the organization of testicular structures and differentiation of other somatic cell linages in the testis (Leydig cells and peritubular myoid cells). The appearance of Sertoli cells in the gonadal primordium is dependent on the expression of the Y-linked sex-determining gene *Sry*. Differentiation of Sertoli cells marks the onset of dimorphic sexual differentiation of the soma and germ line and the first step of testis morphogenesis. Based on familial mutation studies in humans and genetic models in mice, it has become clear that the path to Sertoli cell fate involves autonomous transcriptional circuitry and paracrine/autocrine regulation. In this chapter, we summarize the current knowledge of the establishment of the Sertoli cell lineage and the molecular and cellular events that follow the appearance of Sertoli cells using the mouse as the model organism.

A. Highlights/milestones since last volume

Technological advances over the past 10 years have fueled the rapid discovery of new regulators of Sertoli cell fate and testis development. In particular, the advent of rapid and inexpensive sequencing technologies and their application to small numbers of cells have resulted in very large datasets that reveal the transcriptional basis of Sertoli fate commitment and early differentiation steps. Improved organ culture and microscopy techniques have made it possible to study the morphogenesis of the testis using live imaging. In addition, the discovery of the antagonistic roles of retinoic acid and CYP26b1 has significantly advanced our understanding of how Sertoli cells regulate germ cell fate.

Sertoli Cell Biology. DOI: http://dx.doi.org/10.1016/B978-0-12-417047-6.00002-8

II. Establishment of the gonadal primordium

The fetal gonads arise as paired structures that are part of the urogenital ridges, which develop on the two medial surfaces of the intermediate mesoderm (Figure 2.1A and B). The adrenal glands develop at the anterior of the urogenital ridges, the kidneys arise in the posterior domain, and the gonads develop in the mid region. The signals that control where the gonads develop are not defined but likely involve *Hox* genes and must be closely linked with the formation of the anterior mesonephric tubules [1], which later give rise to the efferent ductules of the testis. Gene mutations that disrupt the formation of the intermediate mesoderm, such as the transcription factor odd-skipped (*Osr1*), result in a failure to form the gonad and other organs derived from this embryonic region [2].

In the mouse, the gonadal primordium originates as a thickening of the coelomic epithelium (CE) overlaying the mesonephric tubules as a result of cell divisions that occur both within and beneath this layer of cells [3]. Most or all of the somatic cells that form the early gonadal primordium arise from these divisions. At stages prior to embryonic day (E) 11.5 in mice, cells from the CE can give rise to all somatic cell lineages in the gonad (including Sertoli cells), but later, CE cells

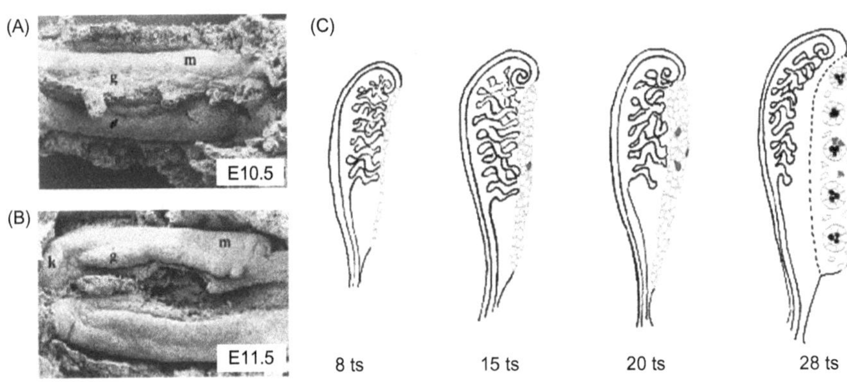

Figure 2.1 Scanning electron micrographs of the urogenital ridge. (A) At E10.5, the urogenital ridge fills most of the body cavity between the limb buds. m, mesonephros; g, genital ridge region; the arrow indicates a pore on the coelomic surface of the genital ridge region. (B) By E11.5, the gonad (g) overlays the mesonephric region of the ridge and the kidney (k) is distinct in the metanephric region. No sexual dimorphisms are apparent. (C) The coelomic epithelium contributes multiple cell types to the XX and XY gonad from the earliest stages of gonadal development. When single coelomic epithelial cells are labeled between 15 and 18 tail somites (t.s.; just prior to E11.5) with DiI (red), they give rise to supporting cell precursors and other lineages within XX and XY gonads. In the XY gonad, some cells labeled at 15–18 t.s. differentiate into Sertoli cells enclosing germ cells (black) inside testis cords by 28 t.s. (E12.5). Other labeled cells reside in the interstitial space. (For interpretation of the references to color in this figure legend, the reader is referred to the web version of this book.)
This figure is adapted from the review by Capel [150] with permission from the publisher.

produce only the non-Sertoli lineages that reside in the interstitial space between the testis cords (Figure 2.1C) [4]. Although the mesonephros was implicated as the source of most somatic cells in the human gonad based on histology [5], this has not been confirmed using modern genetic labeling and live imaging techniques in mouse models.

The outgrowth of the CE requires the coordinated action of transcription factors and secreted signals, including steroidogenic factor 1 (*Nr5a1*, often called *Sf1*) [6]; the Wilms' tumor gene (*Wt1*) [7]; the homeobox genes *Pbx1* [8], *Emx2* [9], and *Lhx9* [10], *Gata4* [11], *Cbx2* [12], *Six1/Six4* [13]; and the insulin family of growth factors and receptors [14,15]. Loss of function of any of these factors leads to failure to form or maintain the gonadal primordium.

Nr5a1, which encodes an orphan nuclear hormone receptor (whose ligand has not yet been identified) is expressed starting during the early stages of urogenital ridge formation, when populations of SF1-positive cells are partitioned to the gonad and the adrenal primordium [16]. Homozygosity for a null allele of *Nr5a1* leads to an arrest and regression of genital ridge development at E11.5 in mice, whereas loss of a single allele does not disrupt gonad development on most genetic backgrounds [6]. However mutation of a single allele of *NR5A1* leads to disruption of gonad development and male sex determination in humans [17–20].

Wt1 encodes a zinc finger transcription factor that is expressed widely in the urogenital ridge. Mutations in this gene affect formation of the mesonephric tubules, the kidney, and the gonad [21,22]. Multiple variants of the WT1 protein have been identified, including two splicoforms that differ in that one includes a tripartite amino acid sequence (+KTS/−KTS) and the other does not. In mice, loss of the −KTS isoform leads to degeneration of the gonad after E11.5 in both sexes. WT1 is likely involved in the activation of many critical genes in the gonad. For example, it directly regulates *Nr5a1* expression in concert with LHX9 [23] and another critical transcription factor, CITED2 [24]. However, loss of the +KTS isoform has a sex-specific effect; XX mice missing the + KTS isoform develop normally, while XY mice lacking the +KTS isoform show sex reversal to female [21]. In humans, reduction of the +KTS: −KTS ratio leads to Fraser syndrome, which is sometimes characterized by gonadal dysgenesis and male-to-female sex reversal [25]. More recent experiments show that Sertoli-cell-specific deletion of *Wt1* after testis development begins also leads to loss of testis cord structures and spermatogenesis in mice [26,27] and humans [28]. Thus, WT1 also plays a maintenance role in Sertoli cell differentiation and testis development after primary sex determination.

Multiple homeobox genes have been implicated in genital ridge development. These genes include the three amino acid loop-extension homeodomain genes *Pbx1* and the LIM homeobox gene *Lhx9*, which are both required for the expansion of SF1-positive cells in the CE. Gonads of *Pbx1*- and *Lhx9*-null mutants are arrested early in development and *Nr5a1* levels are significantly reduced, suggesting that these factors may act upstream of SF1 [10]. The paired-like homeobox gene *Emx2* is required to establish epithelial cell polarity in the CE overlying the gonad. Mutants show an unusual CE with defects in tight junction assembly and subsequent gonadal dysgenesis [9,29]. *Gata4* also plays a critical role in expansion of the CE; loss of *Gata4* results in a failure of the CE to thicken or express *Nr5a1*, *Lhx9*, or other markers associated with early gonadal differentiation [11].

Homozygous mutations in *Cbx2*, a polycomb group gene, or double mutants for *Six1;Six4* (two other homeobox genes) both result in a hypoplastic gonad and cause male-to-female sex reversal [12,13]. In these and other mutants, it has been difficult to determine whether sex reversal was due to a small gonad with a paucity of Sertoli cell progenitors transcriptional effects on expression of *Sry*, and/or the establishment of the Sertoli differentiation program. Two distinct regulatory pathways have been defined for *Six1;Six4*. Hypoplasia of the gonad likely results from the downregulation of the target gene, *Nr5a1*, whereas sex reversal may stem from downregulation of another target gene, *Fog2*, which is directly involved in regulation of *Sry* [13]. In both *Cbx2* and *Six1;Six4* mutants, enforced expression of *Sry* rescues the sex reversal phenotype in XY gonads, even though the gonad remains small [12,13].

As the gonad is forming, proliferation in the CE is detected in both sexes and in cell layers beneath the CE. However, proliferation in the CE increases specifically in XY gonads coincident with expression of *Sry* [30]. Injection of proliferation inhibitors during a narrow window of development, soon after expression of *Sry* is initiated, blocks commitment of progenitor cells to Sertoli fate and leads to male-to-female sex reversal in XY embryos [3]. Analysis of XX<->XY chimeras indicates that a threshold number of Sertoli cells is required to establish testis fate in the gonad. Below this threshold, the gonad forms an ovary [31], which has led to the hypothesis that a block in proliferation reduces the number of Sertoli cells below the threshold required to block female development. However, another possibility is that proliferation *per se* is involved in establishing Sertoli fate.

Insulin, its related growth factors IGF1 and IGF2, and their receptors INSR and IGF1R are also required to establish a normal primordial gonad [14,15]. Double mutants for *Insr;Igf1r* show disruption of proliferation and metabolic pathways. SF1-positive cells in double mutants fail to establish the core gonadal expression program common to bipotential XX and XY cells and, instead, remain in an undifferentiated state. These studies and others cited above suggest a close link between Sertoli cell proliferation and establishment of differentiation pathways [15].

III. Sertoli cell specification and diversion of molecular development toward the testis pathway

At E10.0–11.0 in mice and during the fourth week of development in humans [32], when the gonad is first a distinct primordial organ, it is fully bipotential: regardless of its genotype (XX or XY), it can give rise to a testis or an ovary. Microarray analysis of SF1-GFP-positive cells sorted from the urogenital ridge on the basis of fluorescence [33] or of the isolated gonad at this stage [34] has revealed a virtually identical transcriptome in XX and XY cells. Consistent with the ability of these cells to differentiate into either testis or ovary cells, their transcriptome is "primed" by the expression of genes that will later be associated with either the male or female fate.

Interestingly, transcriptome analysis of the somatic progenitors in the bipotential gonad has revealed a slight bias toward expression of genes that are associated with ovary development [35]. This likely explains why the female fate is the "default" pathway of the gonad and also suggests that the male pathway diverts the female-biased gonad toward a testis fate. The task of diverting gonad development toward the male pathway is accomplished in mammals by the expression of *Sry/SRY*, which is activated in a subset of somatic cells of the XY gonad at E10.5 in mice [36–38] and between 41 and 44 days of development in humans [39]. Expression of *Sry* in an XX gonad leads to development of a testis [40], whereas mouse XY embryos carrying a Y chromosome deleted for *Sry* [41] or XY humans carrying various coding mutations in the gene (for a review, see Wilhelm et al. [42]) develop ovaries. These critical experiments proved that *Sry* is necessary and sufficient to direct gonadal somatic progenitor cells toward a Sertoli cell fate. In XX embryos, or in XY embryos where the *Sry* gene is absent or dysfunctional, these cells differentiate into granulosa cells, and the gonad follows the ovarian pathway.

The induction of *Sry* expression is still incompletely understood; however, several transcription factor complexes are involved in establishing its expression (Figure 2.2). CITED2, WT1, and SF1 act in one regulatory complex upstream of *Sry* [24], while binding between GATA4 and FOG2 is required in a second complex [43]. Several genes in the MAP kinase pathway also have been implicated in induction of *Sry* expression. A nonsense mutation in *Map3k4* leads to a failure to fully induce *Sry* expression and to XY gonadal sex reversal to female on some genetic backgrounds in mice [44]. Interestingly, loss of function of *Gadd45g*, a small acidic protein that interacts with MAP4K4 to phosphorylate and activate p38MAPK and GATA4, also leads to poor induction of *Sry* [45,46]. In humans, a heterozygous mutation in *MAP3K1* is associated with disorders of sexual development, but this allele of *MAP3K1* seems to lead to increased MAPK signaling when tested in patient-derived lymphoblast cell lines [47], which may reflect a difference between mice and humans or differences related to cellular context.

XY embryos with loss-of-function mutations in the polycomb complex member *Cbx2* [12] or the histone demethylase *Jmjd1a* [48] also show disruptions in *Sry* expression, suggesting that chromatin remodeling may be required to induce *Sry*. Loss of sequences far upstream of *Sry* disrupt its expression [49]; however, no mechanistic explanation for this effect has been provided. Transcription of *Sry* from a promoter that lies far upstream gives rise to a circular transcript that is likely not translated [50], whereas a second promoter that is proximate to the coding region of the gene produces a gonad-specific transcript that gives rise to the SRY protein detected in the fetal gonad [37].

Sry is expressed in some of the somatic cells of the early gonad. It is first expressed in the middle of the mouse gonad before spreading anteriorly and then posteriorly [36]. Experiments in mice suggest that *Sry* must be expressed prior to E11.25 to activate the testis pathway [51]. Delays in *Sry* expression typically result

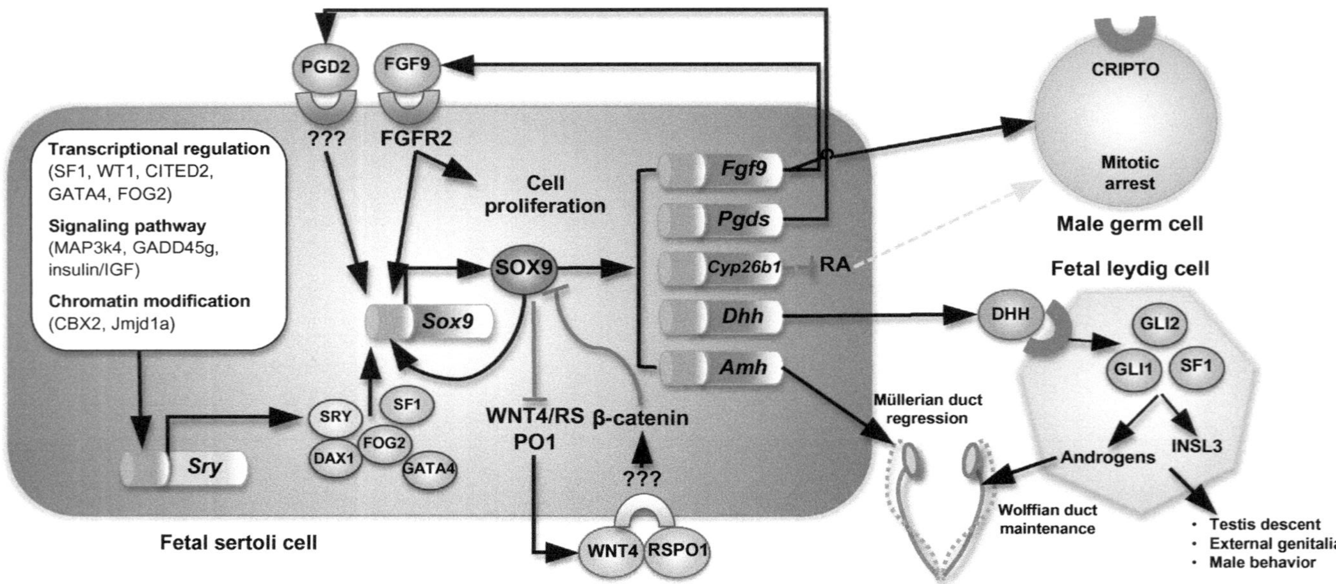

Figure 2.2 Molecular and cellular pathways that connect Sertoli, Leydig, and germ cell development based on mouse genetic models. At the time of sex determination in mouse embryos (E10.5−11.5), *Sry* is expressed in pre-Sertoli cells through coordinate action of transcriptional regulation, signaling pathways, and chromatin modification. SRY along with other transcription regulators (DAX1, FOG2, GATA4, SF1, and other unidentified factors) upregulates the expression of *Sox9*. SOX9 then controls the production of *Fgf9, Pgds, Cyp26b1, Dhh*, and *Amh*. FGF9, via the action of FGFR2, stimulates proliferation of Sertoli cells, expression of *Sox9*, and antagonism of *Wnt4*. PGDS leads to the production of PGD2, which feeds forward and enhances *Sox*9 expression. FGF9 also controls the pluripotency of male germ cells by upregulating the transforming growth factor β/nodal coreceptor *Cripto*. CYP26b1 in Sertoli cells metabolizes and inactivates retinoic acid, therefore preventing male germ cells from entering meiosis. Sertoli-cell-derived DHH, probably via the action of GLI1 and GLI2, induces specification and differentiation of fetal Leydig cells, which are responsible for maintenance of the Wolffian duct, testis descent, and appearance of male secondary sexual characteristics. Sertoli cells also produce AMH, which causes regression of the Müllerian duct.

in ovary or ovotestis development. In the latter case, the central region of the gonad is testicular, while the ends of the gonad are ovarian [52]. In contrast to other mouse strains, C57BL/6 mice show an overall timing delay of the testis program [34], which renders them susceptible to sex reversal under a number of conditions, such as when they are heterozygous for mutations in *Wt1* or *Nr5a1* [53] or when expression of *Sry* is slightly delayed [38]. The molecular explanation for the precise timing requirement for *Sry* has not been determined but is presumed to involve the necessity to silence the ovarian pathway at a critical juncture in gonad development.

The only known target of SRY is *Sox9*, a closely related HMG-box protein. SRY and SF1 bind to multiple sites in a promoter element upstream of *Sox9* called "Tesco" [54]. *Sox9* may be the only requisite target of SRY because overexpression of *Sox9* can substitute for *Sry* in XY mutants in which the native *Sry* gene is disrupted [55] and can induce XX gonads to differentiate as testes [56]. Overexpression of another gene in this family, *Sox3*, can also substitute for *Sry*, [57], suggesting that activation of an HMG-box protein at this specific time in gonad development may be the pivotal event in mammalian sex determination.

In addition to the dominant action of SRY, there is an underlying antagonistic signaling system that directs testis versus ovary development. Two secreted signaling molecules, fibroblast growth factor 9 (FGF9) and WNT4, are both expressed at the bipotential stage of gonad development. In XY gonads, FGF9 suppresses *Wnt4* to establish the male pathway, whereas in XX gonads, *Wnt4* dominates [58]. Expression of the downstream target of WNT4, beta-catenin, in the XY gonad leads to sex reversal to female, suggesting that elevated levels of WNT signaling can override the male pathway [59]. Although XX mutants that are null for either *Wnt4* or beta-catenin show some signs of male development, they do not completely reverse sex [60–62]. In contrast, XY mutants that are null for *Fgf9* show reduced proliferation of Sertoli cell precursors, lose expression of *Sry* and *Sox9*, and completely reverse sex to female [58,63,64]. SOX9 and FGF9 work in a feed-forward loop; SOX9 is likely required to establish FGF9 as the dominant signaling pathway in the male gonad, and FGF9 is required to maintain *Sox9* expression [58]. However, in XY mice that lack both *Fgf9* and *Wnt4*, expression of *Sox9* and the male pathway is recovered, at least during fetal life. This finding has been interpreted to mean that the primary role of *Fgf9* is to repress the female pathway gene *Wnt4*. Other signaling pathways promote and stabilize testis development, including prostaglandin D synthase [65–67], which promotes nuclear localization of the SOX9 protein to establish Sertoli cell fate [68]. These secreted signals may be important in recruiting all cells of the gonad to a testis fate, as in the case of XX<->XY chimeras, where XX cells can be induced to differentiate as Sertoli cells [31].

IV. Organizational functions of fetal Sertoli cells

Once the Sertoli cell lineage is established, these cells begin to orchestrate the organization of testis cords, testis-specific vascular patterning, and the appearance of other somatic cell types (Leydig cells and peritubular myoid cells) (Figure 2.2) [69]. These morphogenetic changes, which depend on a threshold number of Sertoli cells [3,31], eventually lead to compartmentalization of somatic environments into a specialized niche for spermatogenesis (testis cords) and an interstitium that is responsible for androgen production. In addition, Sertoli cells secrete anti-Müllerian hormone (AMH or Müllerian inhibiting substance) that induces regression of Müllerian ducts, the precursors of female reproductive tracts.

A. Formation of testis cords and establishment of testis-specific vasculature

The seminiferous tubules in the adult testis are essential to male fertility because they create the niche that maintains spermatogonial stem cells and directs the progression of spermatogenesis. These structures, which are initially called "testis cords," are organized in the fetal testis soon after *Sry* is expressed in pre-Sertoli cells. The process of fetal cord formation localizes Sertoli and germ cells to the inside of the cords and other somatic cell types in the gonad to the interstitium, the area between cords. A basement membrane and an outside layer of peritubular myoid cells separate the cords from the interstitium [69,70].

The process of fetal cord morphogenesis occurs *de novo* rather than via a branching or invagination process. Between E11.5 and E12.5 in the mouse testis, SOX9-positive pre-Sertoli cells that are originally scattered in the gonadal primordium begin to differentiate. Early in this process, they cluster together and become polarized. Extracellular matrix (ECM) components of the basal membrane are deposited on the basal aspect of the cells, whereas germ cells are sequestered on the apical surface of Sertoli cells. The threshold requirement for the number of Sertoli cells may be related to the mechanics of aggregation dynamics and self-organizing principles or to the accumulation of sufficient levels of secreted signals produced by these cells.

This polarizing/clustering process of Sertoli cells in combination with vascular development (see below and Figure 2.3) eventually leads to formation of multiple tubular loops that are perpendicular to the anterior–posterior axis of the fetal testis. The two ends of each loop fuse to another tubular structure that is assembled along the border of the mesonephros-testis junction, the rete testis. The rete testis connects to the efferent ductules, which are derived from the anterior mesonephric tubules, and connect the testis to the epididymis.

The basal lamina that is deposited between Sertoli cells and peritubular myoid cells is composed of multiple ECM components, including laminin, fibronectin, and collagens [69,70]. While fibronectin is produced by the peritubular myoid cells, laminin is produced by the Sertoli cells [70], and various collagens are produced by both cell types [71]. The importance of ECM components has been demonstrated by two lines of experiments. First, the inclusion of extracellular matrix in Sertoli cell aggregation assays *in vitro* was shown to promote the formation of testis cords

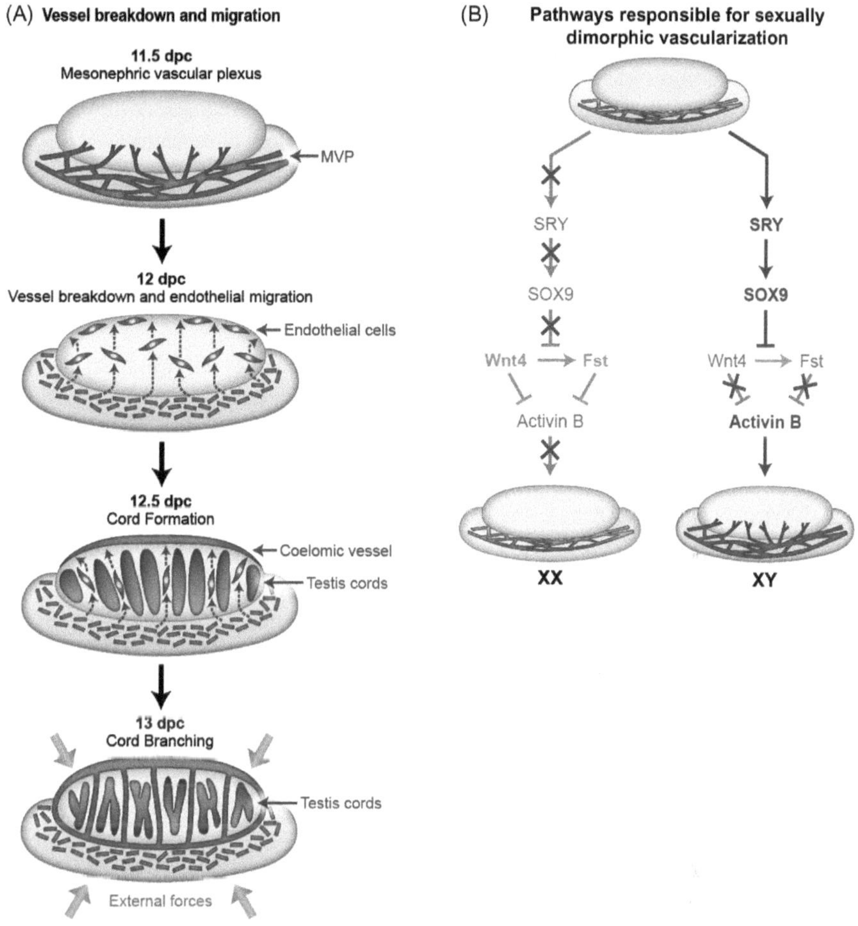

Figure 2.3 (A) At 11.5 days postcoitum (dpc), there is minimal vasculature in the testis, but a prominent vessel structure, the mesonephric vascular plexus (MVP), is present in the neighboring mesonephros. Twelve hours later, the MVP breaks down and individual endothelial cells migrate into the testis, toward the coelomic domain. By 12.5 dpc, the coelomic vessel is clearly visible in the surface domain of the testis and testis cords have formed, followed by consolidation of vessels between cords. (B) At the molecular level, SRY and its downstream effector SOX9 control sexually dimorphic vascularization of the gonad. In the XY gonad (right), SRY/SOX9 suppress *Wnt4* and its downstream target *Fst*. In the absence of WNT4/FST, activin B becomes available and, along with other possible factors, including VEGF, promotes vascularization of the testis. When SRY/SOX9 are absent, such as in the ovary (left), the inhibition of *Wnt4* and *Fst* is relieved. WNT4/FST therefore repress the activity of activin B and subsequently prevent the testis pattern of vascularization. This figure is adapted from the review by Ungewitter et al. [151] with permission from the publisher.

[70,72,73], Second, although antibodies directed against various collagens had no effect on testis cord formation, antibodies or peptides directed against laminin prevented the formation of testis cords [73].

The cellular and molecular mechanisms behind the *de novo* formation of testis cords have not been fully elucidated. However, based on observations using live imaging techniques and genetic mutations [74–76], cord formation involves instructional cues from the vasculature and the interstitium in addition to the cell-sorting mechanisms described above (Figure 2.3A). Most of the major cell-signaling pathways have been linked to testis cord formation or, more generally, to testis morphogenesis. However, in most cases, it is difficult to distinguish primary from secondary effects.

Migration of cells from the mesonephros into the gonad is an XY-specific event downstream of *Sry* [77–79]. The majority of migrating cells are endothelial cells [74,80] as opposed to peritubular or other interstitial cells as originally proposed [71,77,78]. The migration and formation of the testis vascular network is essential for testis cord formation and for establishing blood flow to and from the testis [80–82]. Blocking cell migration into the XY gonad does not affect Sertoli cell differentiation but does inhibit testis cord formation [74,80], indicating that formation of the coelomic vessel and male vasculature is closely linked to cord formation.

Migration of endothelial cells and formation of the testis vasculature are downstream of Sry, presumably via diffusible factors induced by SRY. These factors have not been identified genetically, although many factors, such as FGF9, PDGFA, PDGFB, PDGFC, BMPs, and AMH, can trigger migration of these cells into the XX gonad [63,81,83]. Loss of *Fgf9* eliminates migration but likely acts far upstream, during specification of the male pathway. The fact that loss of any one of the other factors does not eliminate migration completely [83–85] suggests that multiple factors contribute to the induction of cell migration. Although migration is normally specific to the XY gonad, in the absence of *Wnt4*, endothelial migration occurs into the XX gonad, leading to formation of testis-like vasculature [86]. Endothelial migration into the *Wnt4*-null XX gonad appears to be induced by activin B because inactivation of activin B in the *Wnt4* knockout background prevents the occurrence of the testis-like vasculature [87] (Figure 2.3B).

Another factor involved in the early stages of mesonephric cell migration into the gonad is neurotrophin 3 (*Ntf3*). *Ntf3* is expressed in pre-Sertoli cells at the onset of testis cord formation, and its promoter contains a direct binding site for SRY [88]. Exogenous treatment with NTF3 can induce the migration of mesonephric cells into the gonad *in vitro* [89], and null mutants for the receptors of NTF3, *Ntrk1*, or *Ntrk3* show delays and defects in testis cord formation [90]. However, rudimentary seminiferous tubules still form in these mutants, supporting the hypothesis that additional factors are involved in the control of cell migration in the early testis.

Additional factors are involved in Sertoli cell differentiation and testis cord formation, such as members of the transforming growth factor (TGF) β superfamily. TGFβ and activins/inhibins are expressed in Sertoli, interstitial, and germ cells of the fetal testis (reviewed in Fan et al. [91]). Inhibition of intracellular signaling induced by TGFβ or activins in organ culture prevents testis cord formation [92].

Loss of TGFβ2 or the TGFβ coreceptor *Tgfβr3 in vivo* leads to a reduction in the number of testis cords [93] and aberrant testis cord formation [94]. Treatment with TGFβ2 partially rescues the testis cord defect in *Tgfβr3*$^{-/-}$ testes [95]. Sertoli cells in these mutants exhibit defects in their development, which likely contributes to the poor formation of the testis cords. Another TGFβ ligand subunit, inhibin beta B (*Inhbb*), is expressed by Sertoli cells [96] and regulates both testis cord formation and vascularization. In *Inhbb*$^{-/-}$ XY embryos, which are deficient in the protein products activin B, activin AB, and inhibin B, the coelomic vessel forms, but with discontinuities and fewer branches than controls [96]. Treatment of XX gonads with activin B, one of the products of *Inhbb*, is also sufficient to cause the ectopic appearance of the coelomic vessel. As mentioned above, loss of *Wnt4* in XX gonads also results in ectopic formation of testis vasculature. This ectopic testis vasculature disappears when *Inhbb* is also inactivated [87], indicating that Inhbb is one of the factors contributing to the formation of the male vasculature. Thus, formation of the testis cords and testis vasculature is probably facilitated by interaction among multiple signaling pathways.

B. Coordinate development of the interstitium and its resident cell types

The differentiation of the interstitium and its resident cell types, including peritubular myoid and Leydig cells, is very closely intertwined with testis cord formation. Peritubular myoid cells surround the basal lamina of testis cords, while Leydig cells produce androgens and insulin-like growth factor 3, which are necessary for virilization of the male embryo [97]. While there is debate about the cell type of origin for both of these cell types, it is clear that signaling from Sertoli cells is necessary for their development and, reciprocally, their differentiation is involved in the establishment and/or maintenance of testis cords.

For example, Hedgehog signaling has been linked to cord formation, but its effect is likely indirect. The hedgehog ligand desert hedgehog (*Dhh*) is produced by Sertoli cells starting from E11.5 [98–102]. Its receptor, patched 1, is expressed in peritubular myoid and fetal Leydig cells of the testis interstitium. While the phenotype varies depending on the background strain of the mice, ablation of *Dhh* impacts the development and function of testis cords. These effects range from abnormal embryonic testis cord formation to defects in adult spermatogenesis [98,100,102]. In strains in which fetal cord formation is disrupted, testis cords are irregularly shaped and have a discontinuous basal lamina and an absence of peritubular myoid cells [100,102]. Germ cells that are normally localized within testis cords are occasionally observed outside the testis cords in the interstitium. These *in vivo* findings were corroborated by *in vitro* studies in which treatment with the hedgehog inhibitor cyclopamine disrupted testis cord formation and deposition of the basal lamina [103]. Thus, Sertoli cell-derived DHH appears to facilitate testis cord formation or maintenance by controlling the differentiation of peritubular myoid cells in the interstitial compartment.

In addition to this role, the hedgehog signaling pathway has been implicated in fetal Leydig cell development. Ablation of *Dhh* in mice or treatment of cultured fetal testes with the hedgehog inhibitor cyclopamine blocked fetal Leydig cell formation [99,100,102,103]. In addition, inhibition of the downstream hedgehog transcription factors GLI1 and GLI2 impeded fetal Leydig cell development [104]. Activation of the hedgehog signaling pathway is sufficient to drive development of fetal Leydig cells, as shown by the fact that constitutive activation of the hedgehog signaling pathway in the ovary results in the ectopic appearance of fetal Leydig cells [105,106] and overexpression of GLI1 in the testis leads to Leydig cell hyperplasia [107]. Thus, hedgehog signaling originating from the Sertoli cells is both necessary and sufficient for induction of fetal Leydig cells.

Notch signaling is another example of this type of indirect effect. The notch signaling pathway is active in the testis vasculature, Sertoli cells, and testis interstitium [108]. Within the vasculature, the notch ligands delta-like ligand 4, jagged 1 (*Jag1*), jagged 2 and the notch receptors *Notch1* and *Notch4* are expressed [108,109]. The notch receptor *Notch2* and the downstream notch target gene *Hes5* are expressed in Sertoli cells, whereas *Jag1*, the notch receptor *Notch3*, and the downstream notch target gene *Hes1* are expressed in the testis interstitium. A defect in testis cord formation was observed when *Hes1* was ablated or when *Notch1* was overexpressed in the SF1-positive somatic cells [109]. While Sertoli cells differentiated normally in these mice, the testis cords were smaller than those of controls and were abnormally shaped. It is possible that the disruption of cord formation occurred indirectly as the result of changes in the other cell populations in the interstitium. Further work on this pathway showed that inhibition of the notch signaling pathway in the testis *in vitro* using the γ-secretase inhibitor DAPT led to an increase in fetal Leydig cell numbers [109]. These results were corroborated genetically using mouse models in which *Hes1* or *Jag1*, downstream effectors of notch signaling, were ablated [108,109]. Conversely, constitutive activation of the notch signaling pathway led to a decrease in the number of fetal Leydig cells [109]. Together, these data demonstrate that the notch signaling pathway controls maintenance of the fetal Leydig progenitor cell population by restricting these cells' differentiation into steroidogenic cells.

C. Regression of female reproductive tracts by Amh

The involvement of the fetal testes in dimorphic differentiation of the reproductive tracts was first uncovered by the elegant experiments of Jost, who demonstrated that a fetal testicular factor other than testosterone is responsible for Müllerian duct regression in XY rabbit embryos [110]. AMH was later identified as a Sertoli-cell-specific factor [111], and its role in Müllerian duct regression was definitively proven by the retention of Müllerian ducts in XY human patients harboring *AMH* mutations and mouse genetic models in which AMH or its receptor were nonfunctional [112–115]. In both XX and XY embryos, Müllerian ducts are present initially and AMH receptors are not expressed in a sex-specific manner. As soon as testis cords begin to assemble in the XY gonad, SOX9-positive Sertoli cells produce

AMH [116–119]. The expression of *Amh* in Sertoli cells is under the coordinate control of transcription factors, including SOX9, SF1, WT1, and GATA factors. SOX9 initiates *Amh* expression by binding to a specific response element situated at −141 of the *Amh* promoter [120–122]. Mutation of an SF1 binding site in a minimal *Amh* promoter (180 bp) using an *in vitro* transfection system resulted in severely reduced transcriptional activity [123]. However, only a slight decrease of *Amh* expression was observed in transgenic mice in which the same mutation had been introduced [120]. The presence of another SF1 binding site at −218 of the *Amh* promoter suggests that proper expression of *Amh* requires SF1 binding to both sites [124]. In several transient transfection experiments, WT1, GATA4, and other GATA factors enhanced *Amh* promoter activity either by direct binding to the promoter elements or by interacting with SF1 [125–127]. The expression of *Amh* is also subject to negative regulation by DAX1 and FOG-1/FOG-2 [128,129].

D. Regulation of fetal germ cell development by Sertoli cells

Primordial germ cells (PGCs) are the common precursors of spermatozoa and oocytes. They are specified from a small group of cells in the proximal epiblast just prior to gastrulation. PGCs undergo processes of directed migration and passive transport from the base of the allantois, through the extending hindgut, and into the genital ridges. They respond to chemoattractants, including CXCL12 (stem cell derived factor 1) and Kit ligand, expressed by somatic cells in the bipotential gonad [130–132]. Throughout their migratory journey, PGCs proliferate rapidly and continue to expand for a short time after reaching the genital ridge. Upon their arrival at the gonad at around E10, PGCs are rapidly assimilated among somatic cells and, despite their sex chromosome differences, no functional differences between XX and XY germ cells are known prior to this stage. The arrival of germ cells at the genital ridges coincides with the initiation of *Sry* expression in the XY gonad. Sertoli progenitor cells expressing *Sry* and *Sox9* surround PGCs and enclose them within testis cords. Between E13.5 and E15.5, germ cells in the testis undergo mitotic arrest in G0/G1 phase of their cell cycle whereas germ cells in the ovary initiate meiosis soon after E13.5 (for a review, see [133]). This sex-specific developmental decision is dependent on the identity of the neighboring gonadal somatic cells rather than the chromosomal sex of germ cells themselves. XX germ cells that enter a testis environment (e.g., in an XX⇔XY chimera) follow the fate of male germ cells and undergo mitotic arrest. XY germ cells that enter an ovary (e.g., an XY animal whose sex is reversed by deletion of *Sry*) behave similarly to female germ cells and enter meiosis [134–136]. Consistent with this sex-specific entry into mitotic arrest, many genes involved in negative regulation of the cell cycle are upregulated in the XY gonad at this stage [137].

Germ cells in the fetal ovary enter meiosis as the result of the action of retinoic acid (RA) produced in the neighboring mesonephros [138,139] and perhaps additional factors not yet clearly defined [140–142]. In the fetal testis, RA is metabolized and therefore inactivated by the RA-degrading enzyme CYP26b1, which is expressed primarily in Sertoli cells. The ability of Sertoli cells to inhibit the RA pathway is time sensitive; when E11.5 bipotential PGCs were combined with

Sertoli cells isolated from E12.5, E15.5, or E18.5 embryos, only E12.5 Sertoli cells were capable of suppressing meiotic entry and upregulating male germ cell markers [143], suggesting that only Sertoli cells from the earliest stage of testis development are competent to ensure male germ cell differentiation.

The action of RA on sex-specific germ cell meiosis is also counteracted by FGF signals originating from Sertoli cells. FGF9 directly suppresses the female-specific gene *Wnt4* [58] and upregulates male germ cell fate determination genes, including *Nanos2* and DNA methyl transferase 3L, resulting in a block to meiotic entry [144]. FGF9 also controls the pluripotency of male germ cells by upregulating the TGFβ/nodal coreceptor *Cripto* [145]. Fetal germ cells with reduced nodal signaling do not efficiently produce pluripotent embryonic germ cells in culture [146], and it is speculated that high *Cripto* levels protect male germ cells from precocious differentiation and/or meiotic initiation [145].

Germ cells undergo genome-wide DNA demethylation shortly after colonizing the genital ridges. Remethylation of germ cell genomes and the reacquisition of paternal imprints occur late during fetal life, after XY germ cells have committed to the spermatogenic fate [147]. These changes are likely to be closely regulated by Sertoli cells and are critical to stabilize the transition to spermatogonial stem cells. This transition occurs around birth, when germ cells migrate from the center of the testis cords to take up residence in the spermatogonial stem cell niche on the basement membrane of testis cords. The study of germ cell transitions during this period and their regulation by Sertoli cells is an important area of future work.

E. Regulation of testis cord elongation

Once all of the cell lineages of the testis have been established and male germ cells have entered mitotic quiescence, fetal testes undergo their final phase of development, cord elongation and expansion, which leads to the formation of the adult seminiferous tubules. Early Sertoli cell proliferation and testis morphogenesis programs rely heavily on signals from the Sertoli cells themselves (i.e., FGF9); however, the so-called second wave of Sertoli cell proliferation that underlies testis cord elongation is dependent on signals produced by fetal Leydig cells. Conditional ablation of the TGFβ family member activin A in fetal Leydig cells did not affect initial testis cord formation; however, testis cord expansion and elongation were significantly reduced after E15.5 [148]. Sertoli cell proliferation was reduced in mutant embryos at E19.5 and testis cords exhibited abnormal coiling. Conditional ablation of one of the downstream targets of Activin A, *Smad4*, from both fetal Leydig and Sertoli cells also inhibited testis cord elongation during late embryogenesis [149], indicating that communication between fetal Leydig cells and Sertoli cells is essential for late testis cord expansion. Interestingly, the cord defects of conditional *Smad4* knockout fetal testes became more severe as the animals aged and manifested as stromal cell testicular tumors in a high portion of animals by 1 year of age [149]. These findings implicate the TGFβ superfamily as critical regulators of Sertoli cell proliferation during late fetal life and testicular tumor development throughout the life of the animal.

Summary

1. Formation of the bipotential gonad and somatic cell lineages require the coordinated action of transcription factors and secreted signals, including SF1, WT1, GATA4, CBX2, PBX1, EMX2, LHX9, SIX1/SIX4, and the insulin family of growth factors and receptors.
2. The fate decision of the Sertoli cell lineage is dependent on the presence of *Sry*. *Sry* induction is regulated by transcriptional factors, signaling pathways, and chromatin remodeling.
3. Establishment of the Sertoli cell lineage involves autonomous transcriptional circuitry (SOX9) and paracrine/autocrine signaling (FGF9 and PGD2) communication between other cells in the gonad. These signals control both testis morphogenesis and antagonism of the ovarian pathway.
4. Once the Sertoli cell lineage is established, these cells begin to orchestrate the testis-specific vasculature patterning, organization of testis cords, appearance of other somatic cell types (Leydig cells and peritubular myoid cells), and regression of the Müllerian duct.
5. Migration of endothelial cells contributes to the formation of the testis vasculature. This event is downstream of SRY, and presumably occurs via diffusible factors, including FGF9, PDGFA, PDGFB, PDGFC, BMPs, AMH, and activins.
6. Testis cord formation involves instructional cues from the vasculature and the interstitium in addition to cell-sorting mechanisms.
7. Sertoli-cell-derived DHH specifies the Leydig cell lineage. The notch signaling pathway controls maintenance of the fetal Leydig progenitor cell population by restricting these cells' differentiation into steroidogenic cells.
8. CYP26b1 in Sertoli cells metabolizes and inactivates RA, thereby preventing male germ cells from entering meiosis. Sertoli cell-derived FGF9 also controls pluripotency of male germ cells by upregulating the TGFβ/nodal coreceptor *Cripto*.
9. Elongation and coiling of the fetal testis cords occurs as the result of a second wave of Sertoli cell expansion during late fetal development. This second wave of Sertoli proliferation is regulated by signals from fetal Leydig cells and is implicated in testicular tumor development during adulthood.

References

[1] Karl J, Capel B, Ferguson-Smith MA, McLaren A, Ullmann SL, Burgoyne P. Three-dimensional structure of the developing mouse genital ridge [and discussion]. Philos Trans Royal Soc B Biol Sci 1995;350:235–42.

[2] Wang Q, Lan Y, Cho ES, Maltby KM, Jiang R. Odd-skipped related 1 (odd 1) is an essential regulator of heart and urogenital development. Dev Biol 2005;288:582–94.

[3] Schmahl J, Capel B. Cell proliferation is necessary for the determination of male fate in the gonad. Dev Biol 2003;258:264–76.

[4] Karl J, Capel B. Sertoli cells of the mouse testis originate from the coelomic epithelium. Dev Biol 1998;203:323–33.

[5] Satoh M. Histogenesis and organogenesis of the gonad in human embryos. J Anat 1991;177:85–107.

[6] Luo X, Ikeda Y, Parker KL. A cell-specific nuclear receptor is essential for adrenal and gonadal development and sexual differentiation. Cell 1994;77:481–90.

[7] Kreidberg JA, Sariola H, Loring JM, Maeda M, Pelletier J, Housman D, et al. WT-1 is required for early kidney development. Cell 1993;74:679–91.
[8] Schnabel CA, Godin RE, Cleary ML. Pbx1 regulates nephrogenesis and ureteric branching in the developing kidney. Dev Biol 2003;254:262–76.
[9] Miyamoto N, Yoshida M, Kuratani S, Matsuo I, Aizawa S. Defects of urogenital development in mice lacking Emx2. Development 1997;124:1653–64.
[10] Birk OS, Casiano DE, Wassif CA, Cogliati T, Zhao L, Zhao Y, et al. The LIM homeobox gene Lhx9 is essential for mouse gonad formation. Nature 2000;403:909–13.
[11] Hu YC, Okumura LM, Page DC. Gata4 is required for formation of the genital ridge in mice. PLoS Genet 2013;9:e1003629.
[12] Katoh-Fukui Y, Miyabayashi K, Komatsu T, Owaki A, Baba T, Shima Y, et al. Cbx2, a polycomb group gene, is required for Sry gene expression in mice. Endocrinology 2012;153:913–24.
[13] Fujimoto Y, Tanaka SS, Yamaguchi YL, Kobayashi H, Kuroki S, Tachibana M, et al. Homeoproteins six1 and six4 regulate male sex determination and mouse gonadal development. Dev Cell 2013;26:416–30.
[14] Nef S, Verma-Kurvari S, Merenmies J, Vassalli JD, Efstratiadis A, Accili D, et al. Testis determination requires insulin receptor family function in mice. Nature 2003;426:291–5.
[15] Pitetti JL, Calvel P, Romero Y, Conne B, Truong V, Papaioannou MD, et al. Insulin and IGF1 receptors are essential for XX and XY gonadal differentiation and adrenal development in mice. PLoS Genet 2013;9:e1003160.
[16] Morohashi K, Baba T, Tanaka M. Steroid hormones and the development of reproductive organs. Sex Dev 2013;7:61–79.
[17] Achermann JC, Ito M, Ito M, Hindmarsh PC, Jameson JL. A mutation in the gene encoding steroidogenic factor-1 causes XY sex reversal and adrenal failure in humans. Nat Genet 1999;22:125–6.
[18] Achermann JC, Ozisik G, Ito M, Orun UA, Harmanci K, Gurakan B, et al. Gonadal determination and adrenal development are regulated by the orphan nuclear receptor steroidogenic factor-1, in a dose-dependent manner. J Clin Endocrinol Metab 2002;87:1829–33.
[19] Correa RV, Domenice S, Bingham NC, Billerbeck AE, Rainey WE, Parker KL, et al. A microdeletion in the ligand binding domain of human steroidogenic factor 1 causes XY sex reversal without adrenal insufficiency. J Clin Endocrinol Metab 2004;89:1767–72.
[20] Mallet D, Bretones P, Michel-Calemard L, Dijoud F, David M, Morel Y. Gonadal dysgenesis without adrenal insufficiency in a 46, XY patient heterozygous for the nonsense C16X mutation: a case of SF1 haploinsufficiency. J Clin Endocrinol Metab 2004;89: 4829–32.
[21] Hammes A, Guo JK, Lutsch G, Leheste JR, Landrock D, Ziegler U, et al. Two splice variants of the Wilms' tumor 1 gene have distinct functions during sex determination and nephron formation. Cell 2001;106:319–29.
[22] Schedl A, Hastie N. Multiple roles for the Wilms' tumour suppressor gene, WT1 in genitourinary development. Mol Cell Endocrinol 1998;140:65–9.
[23] Wilhelm D, Englert C. The Wilms tumor suppressor WT1 regulates early gonad development by activation of Sf1. Genes Dev 2002;16:1839–51.
[24] Buaas FW, Val P, Swain A. The transcription co-factor CITED2 functions during sex determination and early gonad development. Hum Mol Genet 2009;18:2989–3001.
[25] Klamt B, Koziell A, Poulat F, Wieacker P, Scambler P, Berta P, et al. Frasier syndrome is caused by defective alternative splicing of WT1 leading to an altered ratio of WT1 +/− KTS splice isoforms. Hum Mol Genet 1998;7:709–14.

[26] Chang H, Gao F, Guillou F, Taketo MM, Huff V, Behringer RR. Wt1 negatively regulates beta-catenin signaling during testis development. Development 2008;135:1875–85.
[27] Chen SR, Chen M, Wang XN, Zhang J, Wen Q, Ji SY, et al. The Wilms tumor gene, Wt1, maintains testicular cord integrity by regulating the expression of Col4a1 and Col4a2. Biol Reprod 2013;88:56.
[28] Wang XN, Li ZS, Ren Y, Jiang T, Wang YQ, Chen M, et al. The Wilms tumor gene, Wt1, is critical for mouse spermatogenesis via regulation of Sertoli cell polarity and is associated with non-obstructive Azoospermia in humans. PLoS Genet 2013;9:e1003645.
[29] Kusaka M, Katoh-Fukui Y, Ogawa H, Miyabayashi K, Baba T, Shima Y, et al. Abnormal epithelial cell polarity and ectopic epidermal growth factor receptor (EGFR) expression induced in Emx2 KO embryonic gonads. Endocrinology 2010;151:5893–904.
[30] Schmahl J, Eicher EM, Washburn LL, Capel B. Sry induces cell proliferation in the mouse gonad. Development 2000;127:65–73.
[31] Burgoyne PS, Buehr M, Koopman P, Rossant J, McLaren A. Cell-autonomous action of the testis-determining gene: Sertoli cells are exclusively XY in XX–XY chimaeric mouse testes. Development 1988;102:443–50.
[32] Byskov AG. Differentiation of mammalian embryonic gonad. Physiol Rev 1986;66:71–117.
[33] Nef S, Schaad O, Stallings NR, Cederroth CR, Pitetti JL, Schaer G, et al. Gene expression during sex determination reveals a robust female genetic program at the onset of ovarian development. Dev Biol 2005;287:361–77.
[34] Munger SC, Natarajan A, Looger LL, Ohler U, Capel B. Fine time course expression analysis identifies cascades of activation and repression and maps a putative regulator of mammalian sex determination. PLoS Genet 2013;9:e1003630.
[35] Jameson SA, Natarajan A, Cool J, DeFalco T, Maatouk DM, Mork L, et al. Temporal transcriptional profiling of somatic and germ cells reveals biased lineage priming of sexual fate in the fetal mouse gonad. PLoS Genet 2012;8:e1002575.
[36] Bullejos M, Koopman P. Spatially dynamic expression of Sry in mouse genital ridges. Dev Dynam Off Publ Am Assoc Anatomists 2001;221:201–5.
[37] Hacker A, Capel B, Goodfellow P, Lovell-Badge R. Expression of Sry, the mouse sex determining gene. Development 1995;121:1603–14.
[38] Wilhelm D, Washburn LL, Truong V, Fellous M, Eicher EM, Koopman P. Antagonism of the testis- and ovary-determining pathways during ovotestis development in mice. Mechan Dev 2009;126:324–36.
[39] Hanley NA, Hagan DM, Clement-Jones M, Ball SG, Strachan T, Salas-Cortes L, et al. SRY, SOX9, and DAX1 expression patterns during human sex determination and gonadal development. Mechan Dev 2000;91:403–7.
[40] Koopman P, Gubbay J, Vivian N, Goodfellow P, Lovell-Badge R. Male development of chromosomally female mice transgenic for Sry. Nature 1991;351:117–21.
[41] Gubbay J, Collignon J, Koopman P, Capel B, Economou A, Munsterberg A, et al. A gene mapping to the sex-determining region of the mouse Y chromosome is a member of a novel family of embryonically expressed genes. Nature 1990;346:245–50.
[42] Wilhelm D, Palmer S, Koopman P. Sex determination and gonadal development in mammals. Physiol Rev 2007;87:1–28.
[43] Manuylov NL, Zhou B, Ma Q, Fox SC, Pu WT, Tevosian SG. Conditional ablation of Gata4 and Fog2 genes in mice reveals their distinct roles in mammalian sexual differentiation. Dev Biol 2011;353:229–41.

[44] Bogani D, Siggers P, Brixey R, Warr N, Beddow S, Edwards J, et al. Loss of mitogen-activated protein kinase kinase kinase 4 (MAP3K4) reveals a requirement for MAPK signalling in mouse sex determination. PLoS Biol 2009;7:e1000196.

[45] Gierl MS, Gruhn WH, von Seggern A, Maltry N, Niehrs C. GADD45G functions in male sex determination by promoting p38 signaling and Sry expression. Dev Cell 2012;23:1032–42.

[46] Warr N, Carre GA, Siggers P, Faleato JV, Brixey R, Pope M, et al. Gadd45gamma and Map3k4 interactions regulate mouse testis determination via p38 MAPK-mediated control of Sry expression. Dev Cell 2012;23:1020–31.

[47] Pearlman A, Loke J, Le Caignec C, White S, Chin L, Friedman A, et al. Mutations in MAP3K1 cause 46,XY disorders of sex development and implicate a common signal transduction pathway in human testis determination. Am J Hum Genet 2010;87: 898–904.

[48] Kuroki S, Matoba S, Akiyoshi M, Matsumura Y, Miyachi H, Mise N, et al. Epigenetic regulation of mouse sex determination by the histone demethylase Jmjd1a. Science 2013;341:1106–9.

[49] Capel B, Rasberry C, Dyson J, Bishop CE, Simpson E, Vivian N, et al. Deletion of Y chromosome sequences located outside the testis determining region can cause XY female sex reversal. Nat Genet 1993;5:301–7.

[50] Capel B, Swain A, Nicolis S, Hacker A, Walter M, Koopman P, et al. Circular transcripts of the testis-determining gene Sry in adult mouse testis. Cell 1993;73:1019–30.

[51] Hiramatsu R, Matoba S, Kanai-Azuma M, Tsunekawa N, Katoh-Fukui Y, Kurohmaru M, et al. A critical time window of Sry action in gonadal sex determination in mice. Development 2009;136:129–38.

[52] Eicher EM, Washburn LL, Whitney 3rd JB, Morrow KE. Mus poschiavinus Y chromosome in the C57BL/6J murine genome causes sex reversal. Science 1982;217:535–7.

[53] Correa SM, Washburn LL, Kahlon RS, Musson MC, Bouma GJ, Eicher EM, et al. Sex reversal in C57BL/6J XY mice caused by increased expression of ovarian genes and insufficient activation of the testis determining pathway. PLoS Genet 2012;8:e1002569.

[54] Sekido R, Lovell-Badge R. Sex determination involves synergistic action of SRY and SF1 on a specific Sox9 enhancer. Nature 2008;453:930–4.

[55] Qin Y, Bishop CE. Sox9 is sufficient for functional testis development producing fertile male mice in the absence of Sry. Hum Mol Genet 2005;14:1221–9.

[56] Vidal VP, Chaboissier MC, de Rooij DG, Schedl A. Sox9 induces testis development in XX transgenic mice. Nat Genet 2001;28:216–7.

[57] Moalem S, Babul-Hirji R, Stavropolous DJ, Wherrett D, Bagli DJ, Thomas P, et al. XX male sex reversal with genital abnormalities associated with a de novo SOX3 gene duplication. Am J Med Genet Part A 2012;158A:1759–64.

[58] Kim Y, Kobayashi A, Sekido R, DiNapoli L, Brennan J, Chaboissier MC, et al. Fgf9 and Wnt4 act as antagonistic signals to regulate mammalian sex determination. PLoS Biol 2006;4:e187.

[59] Maatouk DM, DiNapoli L, Alvers A, Parker KL, Taketo MM, Capel B. Stabilization of beta-catenin in XY gonads causes male-to-female sex-reversal. Hum Mol Genet 2008;17:2949–55.

[60] Jeays-Ward K, Dandonneau M, Swain A. Wnt4 is required for proper male as well as female sexual development. Dev Biol 2004;276:431–40.

[61] Liu CF, Bingham N, Parker K, Yao HH. Sex-specific roles of beta-catenin in mouse gonadal development. Hum Mol Genet 2009;18:405–17.

[62] Vainio S, Heikkila M, Kispert A, Chin N, McMahon AP. Female development in mammals is regulated by Wnt-4 signalling. Nature 1999;397:405–9.
[63] Colvin JS, Green RP, Schmahl J, Capel B, Ornitz DM. Male-to-female sex reversal in mice lacking fibroblast growth factor 9. Cell 2001;104:875–89.
[64] Schmahl J, Kim Y, Colvin JS, Ornitz DM, Capel B. Fgf9 induces proliferation and nuclear localization of FGFR2 in Sertoli precursors during male sex determination. Development 2004;131:3627–36.
[65] Moniot B, Declosmenil F, Barrionuevo F, Scherer G, Aritake K, Malki S, et al. The PGD2 pathway, independently of FGF9, amplifies SOX9 activity in Sertoli cells during male sexual differentiation. Development 2009;136:1813–21.
[66] Wilhelm D, Hiramatsu R, Mizusaki H, Widjaja L, Combes AN, Kanai Y, et al. SOX9 regulates prostaglandin D synthase gene transcription *in vivo* to ensure testis development. J Biol Chem 2007;282:10553–60.
[67] Wilhelm D, Martinson F, Bradford S, Wilson MJ, Combes AN, Beverdam A, et al. Sertoli cell differentiation is induced both cell-autonomously and through prostaglandin signaling during mammalian sex determination. Dev Biol 2005;287:111–24.
[68] Malki S, Nef S, Notarnicola C, Thevenet L, Gasca S, Mejean C, et al. Prostaglandin D2 induces nuclear import of the sex-determining factor SOX9 via its cAMP-PKA phosphorylation. EMBO J 2005;24:1798–809.
[69] Cool J, Defalco T, Capel B. Testis formation in the fetal mouse: dynamic and complex de novo tubulogenesis. Wiley Interdiscip Rev Dev Biol 2012;1:847–59.
[70] Tung PS, Skinner MK, Fritz IB. Cooperativity between Sertoli cells and Peritubular Myoid Cells in the Formation of the Basal Lamina in the Seminiferous Tubule. Ann NY Acad Sci 1984;438:435–46.
[71] Skinner MK, Tung PS, Fritz IB. Cooperativity between Sertoli cells and testicular peritubular cells in the production and deposition of extracellular matrix components. J Cell Biol 1985;100:1941–7.
[72] Hadley MA, Byers SW, Suarez-Quian CA, Kleinman HK, Dym M. Extracellular matrix regulates Sertoli cell differentiation, testicular cord formation, and germ cell development *in vitro*. J Cell Biol 1985;101:1511–22.
[73] Hadley MA, Weeks BS, Kleinman HK, Dym M. Lamini promotes formation of cord-like structures by Sertoli cells *in vitro*. Dev Biol 1990;140:318–27.
[74] Combes AN, Lesieur E, Harley VR, Sinclair AH, Little MH, Wilhelm D, et al. Three-dimensional visualization of testis cord morphogenesis, a novel tubulogenic mechanism in development. Dev Dynam Off Publ Am Assoc Anatomists 2009;238:1033–41.
[75] Coveney D, Cool J, Oliver T, Capel B. Four-dimensional analysis of vascularization during primary development of an organ, the gonad. Proc Natl Acad Sci USA 2008; 105:7212–7.
[76] Nel-Themaat L, Vadakkan TJ, Wang Y, Dickinson ME, Akiyama H, Behringer RR. Morphometric analysis of testis cord formation in Sox9-EGFP mice. Dev Dynam Off Publ Am Assoc Anatomists 2009;238:1100–10.
[77] Buehr M, Gu S, McLaren A. Mesonephric contribution to testis differentiation in the fetal mouse. Development 1993;117:273–81.
[78] Martineau J, Nordqvist K, Tilmann C, Lovell-Badge R, Capel B. Male-specific cell migration into the developing gonad. Curr Biol 1997;7:958–68.
[79] Merchant-Larios H, Moreno-Mendoza N, Buehr M. The role of the mesonephros in cell differentiation and morphogenesis of the mouse fetal testis. Int J Dev Biol 1993;37:407–15.

[80] Cool J, DeFalco TJ, Capel B. Vascular-mesenchymal cross-talk through Vegf and Pdgf drives organ patterning. Proc Natl Acad Sci USA 2011;108:167–72.

[81] Brennan J, Karl J, Capel B. Divergent vascular mechanisms downstream of Sry establish the arterial system in the XY gonad. Dev Biol 2002;244:418–28.

[82] Combes AN, Wilhelm D, Davidson T, Dejana E, Harley V, Sinclair A, et al. Endothelial cell migration directs testis cord formation. Dev Biol 2009;326:112–20.

[83] Ross AJ, Tilman C, Yao H, MacLaughlin D, Capel B. AMH induces mesonephric cell migration in XX gonads. Mol Cell Endocrinol 2003;211:1–7.

[84] Brennan J, Tilmann C, Capel B. Pdgfr-alpha mediates testis cord organization and fetal Leydig cell development in the XY gonad. Genes Dev 2003;17:800–10.

[85] Tilmann C, Capel B. Mesonephric cell migration induces testis cord formation and Sertoli cell differentiation in the mammalian gonad. Development 1999;126:2883–90.

[86] Jeays-Ward K, Hoyle C, Brennan J, Dandonneau M, Alldus G, Capel B, et al. Endothelial and steroidogenic cell migration are regulated by WNT4 in the developing mammalian gonad. Development 2003;130:3663–70.

[87] Liu CF, Parker K, Yao HH. WNT4/beta-catenin pathway maintains female germ cell survival by inhibiting activin betaB in the mouse fetal ovary. PLoS One 2010;5:e10382.

[88] Clement TM, Bhandari RK, Sadler-Riggleman I, Skinner MK. SRY directly regulates the neurotrophin 3 promoter during male sex determination and testis development in rats. Biol Reprod 2011;85(2):277–84.

[89] Cupp AS, Uzumcu M, Skinner MK. Chemotactic role of neurotropin 3 in the embryonic testis that facilitates male sex determination. Biol Reprod 2003;68:2033–7.

[90] Cupp AS, Tessarollo L, Skinner MK. Testis developmental phenotypes in neurotropin receptor trkA and trkC null mutations: role in formation of seminiferous cords and germ cell survival. Biol Reprod 2002;66:1838–45.

[91] Fan YS, Hu YJ, Yang WX. TGF-beta superfamily: how does it regulate testis development. Mol Biol Rep 2012;39:4727–41.

[92] Miles DC, Wakeling SI, Stringer JM, van den Bergen JA, Wilhelm D, Sinclair AH, et al. Signaling through the TGF beta-activin receptors ALK4/5/7 regulates testis formation and male germ cell development. PLoS One 2013;8(e54606\dt).

[93] Memon MA, Anway MD, Covert TR, Uzumcu M, Skinner MK. Transforming growth factor beta (TGFbeta1, TGFbeta2 and TGFbeta3) null-mutant phenotypes in embryonic gonadal development. Mol Cell Endocrinol 2008;294:70–80.

[94] Sarraj MA, Escalona RM, Umbers A, Chua HK, Small C, Griswold M, et al. Fetal testis dysgenesis and compromised Leydig cell function in Tgfbr3 (beta glycan) knockout mice. Biol Reprod 2010;82:153–62.

[95] Sarraj MA, Escalona RM, Western P, Findlay JK, Stenvers KL. Effects of TGFbeta2 on wild-type and Tgfbr3 knockout mouse fetal testis. Biol Reprod 2013;88:66.

[96] Yao HH, Aardema J, Holthusen K. Sexually dimorphic regulation of inhibin beta B in establishing gonadal vasculature in mice. Biol Reprod 2006;74:978–83.

[97] Griswold SL, Behringer RR. Fetal Leydig cell origin and development. Sex Dev 2009;3:1–15.

[98] Bitgood MJ, Shen L, McMahon AP. Sertoli cell signaling by Desert hedgehog regulates the male germline. Curr Biol 1996;6:298–304.

[99] Yao HH, Whoriskey W, Capel B. Desert Hedgehog/Patched 1 signaling specifies fetal Leydig cell fate in testis organogenesis. Genes Dev 2002;16:1433–40.

[100] Clark AM, Garland KK, Russell LD. Desert hedgehog (Dhh) gene is required in the mouse testis for formation of adult-type Leydig cells and normal development of peritubular cells and seminiferous tubules. Biol Reprod 2000;63:1825–38.

[101] Bitgood MJ, McMahon AP. Hedgehog and Bmp genes are coexpressed at many diverse sites of cell-cell interaction in the mouse embryo. Dev Biol 1995;172:126–38.

[102] Pierucci-Alves F, Clark AM, Russell LD. A developmental study of the Desert hedgehog-null mouse testis. Biol Reprod 2001;65:1392–402.

[103] Yao HH, Capel B. Disruption of testis cords by cyclopamine or forskolin reveals independent cellular pathways in testis organogenesis. Dev Biol 2002;246:356–65.

[104] Barsoum I, Yao HHC. Redundant and differential roles of transcription factors Gli1 and Gli2 in the development of Mouse Fetal Leydig Cells. Biol Reprod 2011;84:894–9.

[105] Barsoum IB, Bingham NC, Parker KL, Jorgensen JS, Yao HHC. Activation of the Hedgehog pathway in the mouse fetal ovary leads to ectopic appearance of fetal Leydig cells and female pseudohermaphroditism. Dev Biol 2009;329:96–103.

[106] Barsoum IB, Kaur J, Ge RS, Cooke PS, Yao HH. Dynamic changes in fetal Leydig cell populations influence adult Leydig cell populations in mice. FASEB J Off Publ Fed Am Soc Exp Biol 2013;27:2657–66.

[107] Kroft TL, Patterson J, Won Yoon J, Doglio L, Walterhouse DO, Iannaccone PM, et al. GLI1 localization in the germinal epithelial cells alternates between cytoplasm and nucleus: upregulation in transgenic mice blocks spermatogenesis in pachytene. Biol Reprod 2001;65:1663–71.

[108] Defalco T, Saraswathula A, Briot A, Iruela-Arispe ML, Capel B. Testosterone levels influence mouse fetal Leydig cell progenitors through notch signaling. Biol Reprod 2013;88:91.

[109] Tang H, Brennan J, Karl J, Hamada Y, Raetzman L, Capel B. Notch signaling maintains Leydig progenitor cells in the mouse testis. Development 2008;135:3745–53.

[110] Jost A. Recherches sur la differenciation sexuelle de l'embryon de lapin. Arch d'Anatomie Microscopique et de Morpholgie Exp 1947;36:271–315.

[111] Tran D, Muesy-Dessole N, Josso N. Anti-Mullerian hormone is a functional marker of foetal Sertoli cells. Nature 1977;269:411–2.

[112] Behringer RR, Cate RL, Froelick GJ, Palmiter RD, Brinster RL. Abnormal sexual development in transgenic mice chronically expressing mullerian inhibiting substance. Nature 1990;345:167–70.

[113] Behringer RR, Finegold MJ, Cate RL. Müllerian-inhibiting substance function during mammalian sexual development. Cell 1994;79:415–25.

[114] Belville C, Josso N, Picard JY. Persistence of Mullerian derivatives in males. Am J Med Genet 1999;89:218–23.

[115] Mishina Y, Rey R, Finegold MJ, Matzuk MM, Josso N, Cate RL, et al. Genetic analysis of the Mullerian-inhibiting substance signal transduction pathway in mammalian sexual differentiation. Genes Dev 1996;10:2577–87.

[116] Appert A, Fridmacher V, Locquet O, Magre S. Patterns of keratins 8, 18 and 19 during gonadal differentiation in the mouse: sex- and time-dependent expression of keratin 19. Differentiation 1998;63:273–84.

[117] Frojdman K, Pelliniemi LJ, Rey R, Virtanen I. Presence of anti-Mullerian hormone correlates with absence of laminin alpha5 chain in differentiating rat testis and ovary. Histochem Cell Biol 1999;111:367–73.

[118] Hayashi M, Shima H, Hayashi K, Trelstad RL, Donahoe PK. Immunocytochemical localization of Mullerian inhibiting substance in the rough endoplasmic reticulum and Golgi apparatus in Sertoli cells of the neonatal calf testis using a monoclonal antibody. J Histochem Cytochem 1984;32:649–54.

[119] Tran D, Josso N. Localization of anti-Mullerian hormone in the rough endoplasmic reticulum of the developing bovine Sertoli cell using immunocytochemistry with a monoclonal antibody. Endocrinology 1982;111:1562–7.

[120] Arango NA, Lovell-Badge R, Behringer RR. Targeted mutagenesis of the endogenous mouse Mis gene promoter: *in vivo* definition of genetic pathways of vertebrate sexual development. Cell 1999;99:409–19.

[121] De Santa Barbara P, Bonneaud N, Boizet B, Desclozeaux M, Moniot B, Sudbeck P, et al. Direct interaction of SRY-related protein SOX9 and steroidogenic factor 1 regulates transcription of the human anti-Mullerian hormone gene. Mol Cell Biol 1998;18:6653–65.

[122] Morais da Silva S, Hacker A, Harley V, Goodfellow P, Swain A, Lovell-Badge R. Sox9 expression during gonadal development implies a conserved role for the gene in testis differentiation in mammals and birds. Nat Genet 1996;14:62–8.

[123] Shen WH, Moore CC, Ikeda Y, Parker KL, Ingraham HA. Nuclear receptor steroidogenic factor 1 regulates the mullerian inhibiting substance gene: a link to the sex determination cascade. Cell 1994;77:651–61.

[124] Watanabe K, Clarke TR, Lane AH, Wang X, Donahoe PK. Endogenous expression of Mullerian inhibiting substance in early postnatal rat Sertoli cells requires multiple steroidogenic factor-1 and GATA-4-binding sites. Proc Natl Acad Sci USA 2000;97: 1624–9.

[125] Robert NM, Tremblay JJ, Viger RS. Friend of GATA (FOG)-1 and FOG-2 differentially repress the GATA-dependent activity of multiple gonadal promoters. Endocrinology 2002;143:3963–73.

[126] Tremblay JJ, Viger RS. Transcription factor GATA-4 enhances Mullerian inhibiting substance gene transcription through a direct interaction with the nuclear receptor SF-1. Mol Endocrinol 1999;13:1388–401.

[127] Viger RS, Mertineit C, Trasler JM, Nemer M. Transcription factor GATA-4 is expressed in a sexually dimorphic pattern during mouse gonadal development and is a potent activator of the Mullerian inhibiting substance promoter. Development 1998; 125:2665–75.

[128] Nachtigal MW, Hirokawa Y, Enyeart-VanHouten DL, Flanagan JN, Hammer GD, Ingraham HA. Wilms' tumor 1 and Dax-1 modulate the orphan nuclear receptor SF-1 in sex-specific gene expression. Cell 1998;93:445–54.

[129] Tremblay JJ, Viger RS. Nuclear receptor Dax-1 represses the transcriptional cooperation between GATA-4 and SF-1 in Sertoli cells. Biol Reprod 2001;64:1191–9.

[130] Molyneaux K, Wylie C. Primordial germ cell migration. Int J Dev Biol 2004;48: 537–44.

[131] Richardson BE, Lehmann R. Mechanisms guiding primordial germ cell migration: strategies from different organisms. Nat Rev Mol Cell Biol 2010;11:37–49.

[132] Tarbashevich K, Raz E. The nuts and bolts of germ-cell migration. Curr Opin Cell Biol 2010;22:715–21.

[133] McLaren A. Primordial germ cells in the mouse. Dev Biol 2003;262:1–15.

[134] Evans EP, Ford CE, Lyon MF. Direct evidence of the capacity of the XY germ cell in the mouse to become an oocyte. Nature 1977;267:430–1.

[135] Ford CE, Evans EP, Burtenshaw MD, Clegg HM, Tuffrey M, Barnes RD. A functional 'sex-reversed' oocyte in the mouse. Proc R Soc Lond B Biol Sci 1975;190:187–97.

[136] Lovell-Badge R, Robertson E. XY female mice resulting from a heritable mutation in the primary testis-determining gene, Tdy. Development 1990;109:635–46.

[137] Western PS, Miles DC, van den Bergen JA, Burton M, Sinclair AH. Dynamic regulation of mitotic arrest in fetal male germ cells. Stem Cells 2008;26:339−47.
[138] Bowles J, Knight D, Smith C, Wilhelm D, Richman J, Mamiya S, et al. Retinoid signaling determines germ cell fate in mice. Science 2006;312:596−600.
[139] Koubova J, Menke DB, Zhou Q, Capel B, Griswold MD, Page DC. Retinoic acid regulates sex-specific timing of meiotic initiation in mice. Proc Natl Acad Sci USA 2006;103:2474−9.
[140] Chassot AA, Gregoire EP, Lavery R, Taketo MM, de Rooij DG, Adams IR, et al. RSPO1/beta-catenin signaling pathway regulates oogonia differentiation and entry into meiosis in the mouse fetal ovary. PloS One 2011;6:e25641.
[141] Le Bouffant R, Souquet B, Duval N, Duquenne C, Herve R, Frydman N, et al. Msx1 and Msx2 promote meiosis initiation. Development 2011;138:5393−402.
[142] Souquet B, Tourpin S, Messiaen S, Moison D, Habert R, Livera G. Nodal signaling regulates the entry into meiosis in fetal germ cells. Endocrinology 2012;153:2466−73.
[143] Ohta K, Yamamoto M, Lin Y, Hogg N, Akiyama H, Behringer RR, et al. Male differentiation of germ cells induced by embryonic age-specific Sertoli cells in mice. Biol Reprod 2012;86:112.
[144] Bowles J, Feng CW, Spiller C, Davidson TL, Jackson A, Koopman P. FGF9 suppresses meiosis and promotes male germ cell fate in mice. Dev Cell 2010;19:440−9.
[145] Spiller CM, Feng CW, Jackson A, Gillis AJ, Rolland AD, Looijenga LH, et al. Endogenous Nodal signaling regulates germ cell potency during mammalian testis development. Development 2012;139:4123−32.
[146] Matsui Y, Tokitake Y. Primordial germ cells contain subpopulations that have greater ability to develop into pluripotential stem cells. Dev Growth Differ 2009;51:657−67.
[147] Seisenberger S, Andrews S, Krueger F, Arand J, Walter J, Santos F, et al. The dynamics of genome-wide DNA methylation reprogramming in mouse primordial germ cells. Mol Cell 2012;48:849−62.
[148] Archambeault DR, Yao HHC. Activin A, a product of fetal Leydig cells, is a unique paracrine regulator of Sertoli cell proliferation and fetal testis cord expansion. Proc Natl Acad Sci USA 2010;107:10526−31.
[149] Archambeault DR, Yao HH. Loss of smad4 in Sertoli and Leydig cells leads to testicular dysgenesis and hemorrhagic tumor formation in mice. Biol Reprod 2014;90:62.
[150] Capel B. The battle of the sexes. Mechan Dev 2000;92:89−103.
[151] Ungewitter EK, Yao HH. How to make a gonad: cellular mechanisms governing formation of the testes and ovaries. Sex Dev 2013;7:7−20.

Early postnatal interactions between Sertoli and germ cells

3

Qi-En Yang and Jon M. Oatley
Center for Reproductive Biology, School of Molecular Biosciences, College of Veterinary Medicine, Washington State University, Pullman, WA

I. Introduction

The transmission of genetic and epigenetic information to the next generation is achieved via gametes called "spermatozoa" in males. These specialized cells are the terminally differentiated product of a developmental process referred to as "spermatogenesis." Continual production of the millions of spermatozoa daily that are needed for fertility relies on the activities of a robust undifferentiated spermatogonial population that consists of stem cell and progenitor subtypes [1,2]. A primary undifferentiated spermatogonial population is established from gonocyte precursors during early postnatal life in most mammalian species [3]. The initial spermatogonial stem cell (SSC) pool serves as a foundational reservoir from which spermatogenesis originates throughout adult life. Disruption in the formation and maintenance of the primary undifferentiated spermatogonial population can lead to infertility, subfertility, or development of testicular germ-cell cancer [4]. Emerging evidence suggests that miscues in the transition of gonocytes to a postnatal spermatogonial state has dire consequences, including a propensity for transformation to carcinoma *in situ*, which is a common precursor state for many testicular germ-cell tumors [4,5]. Also, failure to form a functional SSC pool leads to loss of the germ cell lineage in adulthood, a condition referred to as a Sertoli-cell-only syndrome.

Establishment of the male germ line during postnatal development requires gonocytes to undergo migration to the basement membrane of seminiferous cords, resume mitotic cell cycle progression, and adopt a spermatogonial lineage commitment. These events are regulated by extrinsic factors of the testicular microenvironment that stimulate molecular pathways in gonocytes. Contributions of Sertoli cells are thought to be the main source of these external factors [6]. Furthermore, emerging evidence suggests that contributions of Sertoli cells are a major component of the niche microenvironment that supports formation of the foundational SSC pool during neonatal development. In this chapter, we address the current understanding of Sertoli cell contributions to the establishment of the primary undifferentiated spermatogonial population from gonocyte precursors during early postnatal development, with an emphasis on knowledge gained from research with mouse models.

Sertoli Cell Biology. DOI: http://dx.doi.org/10.1016/B978-0-12-417047-6.00003-X

II. Neonatal testis development

At birth, seminiferous cords contain centrally located gonocytes and immature Sertoli cells that are anchored to the basement membrane but extend to the center of the cord (Figure 3.1). During neonatal development, gonocytes migrate to the periphery of cords, where physical contact with Sertoli cells and the basement

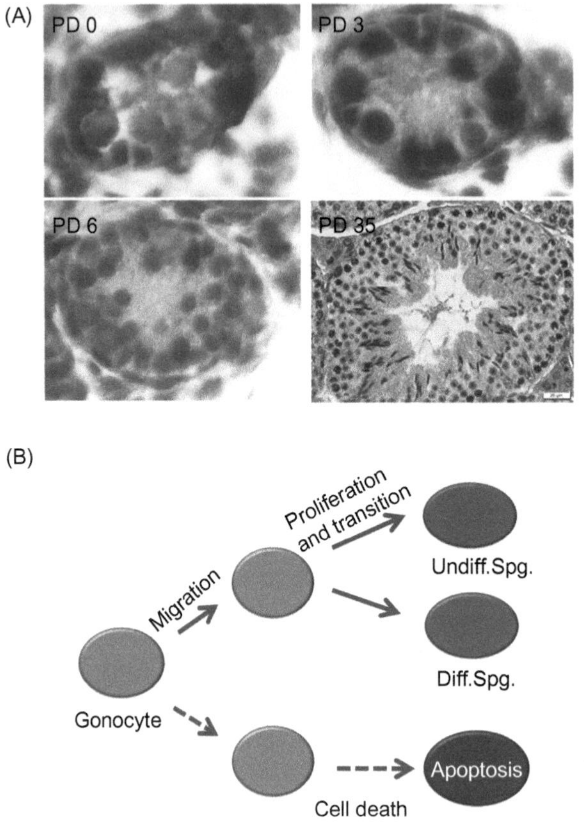

Figure 3.1 Neonatal testis development in mice. (A) Images of cross-sections from testes of mice at postnatal days (PD) 0, 3, 6, and 35. (B) Depiction of different gonocyte fates during neonatal testis development. At PD 0, only gonocytes and Sertoli cells are present within seminiferous cords. Gonocytes are located in the center of cords and migrate to the basement membrane between PD 0–3. Cell cycle progression resumes in gonocytes during migration and the population expands in number while also transitioning to either an undifferentiated or differentiating spermatogonial state. The undifferentiated spermatogonial population consists of spermatogonial stem cell and progenitor pools, and activities of these cells provide the foundation for continual spermatogenesis in adulthood. The initial differentiating spermatogonia are the source of the first round of spermatogenesis. All subsequent rounds of spermatogenesis derive from differentiating spermatogonia that are generated from the progenitor fraction of the undifferentiated spermatogonial population.

membrane is established. Also, this period of transition is marked by a morphological transition from a round shape as gonocytes to an ovoid shape signifying adoption of a spermatogonial state. In addition, nuclear size changes from 10−12 μm in diameter for gonocytes to 9.5−10 μm for spermatogonia [7]. Furthermore, mitotic cell cycle progression is reinitiated during gonocyte migration or shortly after the basement membrane is reached, and the resulting spermatogonial population grows. The current understanding is that a portion of the gonocyte population gives rise to SSCs to form the foundational pool, and another portion transitions to a differentiating spermatogonial state directly (Figure 3.1). The initial gonocyte-derived differentiating spermatogonial population is thought to contribute to the first round of spermatogenesis only, whereas all subsequent rounds of spermatogenesis originate from the SSC pool. The events that define the gonocyte-to-spermatogonial transition occur during a defined period of postnatal life that occurs at 3−6 days of age in mice, after several months in cattle, and after years in humans [8,9]. In addition, evidence from biological studies with mice suggests that contributions from Sertoli cells provide the major influence for establishment of the postnatal spermatogonial lineage.

A. Gonocyte development

In all mammalian species, gonocytes originate from primordial germ cells (PGCs) during embryogenesis. In mice, a small number of PGCs arise from the epiblast around embryonic day (ED) 7.2 and migrate to the genital ridge by ED 11.5, during which time the population grows [10]. Shortly after arriving at the genital ridge, male germ cells are referred to as gonocytes and remain mitotically active, with a division time of approximately 15 h [11,12]. From ED 14.5 to ED 15.5, gonocytes gradually enter the G0/G1 phase of cell cycle that persists for the remainder of fetal life [13]. While in the arrested state, intercellular bridges between gonocytes that formed due to incomplete cytokinesis during the proliferative phase begin to break down, eventually leading to the formation of a single cell population [12]. Gonocytes then follow one of three different fates during early neonatal life (Figure 3.1). First, 30−75% of gonocytes degenerate via necrosis and apoptosis [14−17]. Second, a subset of surviving gonocytes transition to an undifferentiated spermatogonial state to form both a foundational SSC pool and initial progenitor population [7,18]. Third, another subset of the surviving gonocyte pool transitions directly to a differentiating spermatogonial state, and these cells enter meiosis a few days later to initiate the first round of spermatogenesis [19,20]. It is important to note that this third fate has been observed in testes of rodents, but whether a similar pathway occurs in other mammalian species is currently unknown.

The transition of gonocytes to undifferentiated and differentiating spermatogonial subtypes is marked by changes in both cellular morphology and gene expression. In the testes of rats at postnatal days (PDs) 0−4, a subset of gonocytes develop pseudopodia [14,21]. Results of functional transplantation analyses indicate that pseudopodia containing gonocytes possess stem cell capacity for regeneration of spermatogenesis, whereas gonocytes that retain a round cell morphology are cleared by apoptosis and necrosis [14]. Based on nuclear morphology at PD 1, approximately 30% of gonocytes are similar to undifferentiated spermatogonia, whereas the majority of the population resembles differentiating spermatogonia [7,19]. In addition to changes in nuclear morphology, several genes expressed in PGCs that are required for specification and survival of the population are also expressed in gonocytes but become suppressed upon transition to a spermatogonial state. A hallmark example of this paradigm is the expression of the pluripotency factor NANOG, which begins to decline at around ED 16.5 in gonocytes and becomes undetectable in postnatal spermatogonia [22,23]. Similarly, expression of the transcription repressor BLIMP1 becomes suppressed in fetal gonocytes and remains absent in spermatogonia during postnatal life [24,25]. In addition, expression of another gene required for PGC formation, *Dppa3*, begins to decline at around ED 13.5 in gonocytes and becomes undetectable by PD 1 [26,27]. These observations indicate that molecules that influence the establishment of the PGC pool are not required for maintenance of gonocytes or the transition pathway to a postnatal spermatogonial state.

While many genes expressed in PGCs are summarily repressed in gonocytes, the expression of others is maintained even in the postnatal spermatogonial lineage. However, expression of another subset of genes is initiated in gonocytes and then retained in postnatal spermatogonia. The expression of *Oct4* [28], *Sall4* [29], and *Lin28* [30] starts in PGCs and persists in gonocytes, undifferentiated spermatogonia, and subsets of differentiating spermatogonia [29,31]. In contrast, expression of *Nanos2* and *Utf1* occurs in all gonocytes but is restricted to a subset of undifferentiated spermatogonia during postnatal life [32,33]. The transition from a gonocyte to a spermatogonial state is marked by induction of several other genes, including *Plzf* and *Ngn3*. In gonocytes, *Plzf* is expressed at a low-to-undetectable level but is readily detectable in all undifferentiated spermatogonial subtypes [34]. Expression of *Ngn3* is absent in gonocytes and becomes detectable in subsets of A_s, $A_{pr,}$ and A_{al} spermatogonia [30,35]. Collectively, these observations indicate that the retention of *Utf1* and *Nanos2* expression likely marks the transition of a subset of gonocytes to SSCs directly, whereas expression of *Oct4*, *Sall4*, and *Lin28* is a general indication of the premeiotic male germ cell state. Furthermore, attainment of *Plzf* and *Ngn3* expression is indicative of the transition from the gonocyte to the spermatogonial state but likely primarily distinguishes the progenitor spermatogonial state from SSCs. At present, the extrinsic cues that induce differential expression of these genes in gonocytes and postnatal spermatogonial subtypes are undefined.

B. Sertoli cell development

Seminiferous cords of embryonic and fetal testes contain two cell types: gonocytes and Sertoli cells. In rodents, Sertoli cells proliferate throughout fetal life, whereas gonocytes enter cell cycle arrest. The percentage of the Sertoli cell population undergoing cell division is greatest at ED 16–20 and then begins to decline at birth, eventually ceasing at around PD 17–20 [36,37]. In human testes, Sertoli cell proliferation occurs in two waves [38]. The first period of division encompasses late fetal and early neonatal life, when the Sertoli cell population expands more than fivefold. The second period is just prior to puberty, when the Sertoli cell population grows more than twofold [38]. Following sexual maturation, Sertoli cells enter quiescence, a state in which they remain throughout adult life. Evidence from studies with mice suggests that Sertoli cell number is correlated with the number of niches available for SSCs [39]. Thus, the timing and cues governing Sertoli cell mitotic activity in fetal and prepubertal life are crucial for fertility in adulthood. It is likely that communication with gonocytes influences the period of Sertoli cell proliferation, but key signals for this process have yet to be defined.

III. Role of Sertoli cells in gonocyte proliferation and migration

The activities that define the gonocyte stage of male germ-line development are crucial for eventual formation of the spermatogonial lineage during neonatal life. Establishment of a quiescent state during embryonic development with retention of the ability to reenter the cell cycle in neonatal life is essential for gonocyte survival. The majority of gonocyte deaths occur after resumption of cell cycle progression, indicating protection during the quiescent state [15,40,41]. In addition, resumption of cell cycle progression coincides with the migration of gonocytes from the center of seminiferous cords to the basement membrane. Gonocytes that fail to migrate undergo apoptosis and are cleared from the seminiferous epithelium by phagocytic activity of Sertoli cells. Evidence from studies utilizing testis tissue culture systems and conditional knockout mouse models indicate that contributions from Sertoli cells are major influences on the activities of gonocytes during neonatal development (Figure 3.2).

A. *Regulation of gonocyte proliferation*

At present, signals that induce the resumption of mitotic cell cycle progression in gonocytes are undefined. However, a role for retinoblastoma protein 1 (Rb1) has started to emerge. In somatic cells, Rb1 functions as a potent regulator of cell cycle progression by restricting entry to S-phase and thereby inducing quiescence [42,43]. The physical association between Rb1 and members of the E2F transcription factor family abrogates their capacity to promote S-phase progression [44–46]. Complexes of cyclin-dependent kinases (CDKs) and cyclins inactivate Rb1, thus releasing E2F factors [47,48]. The activity of CDKs is regulated by CDK inhibitors (CKIs) [49], including the INK4 proteins (p15, p16, p18, and p19) and the Cip/Kip family members (p21, p27, and p57) [50]. CKIs inactivate CDK/cyclin complexes via physical association to negatively regulate cell cycle progression [50]. In gonocytes of mouse testes, the expression of Rb1, p15, p16, p21, and p27 coincides with the quiescent state [13,51,52], but p27 expression is rapidly downregulated in accordance with the resumption of cell cycle progression and establishment of the postnatal spermatogonial lineage [52].

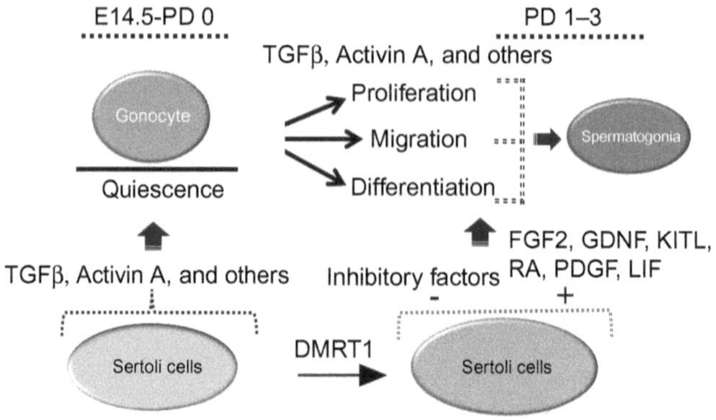

Figure 3.2 Model of the current understanding for the regulation of gonocyte fates during late fetal and early neonatal development in mice. Sertoli cells are thought to be the major contributors of extrinsic signals that influence cell cycle progression in gonocytes, migration of gonocytes to the basement membrane, and transition of gonocytes to an undifferentiated or differentiating spermatogonial state. During the fetal and early neonatal periods, Sertoli cells produce transforming growth factor (TGF) β family proteins (TGFβ1, TGFβ2, and activin A), which induce and maintain cell cycle arrest in gonocytes. After birth, reduced secretion of these ligands and increased secretion of mitogens including fibroblast growth factor 2, leukemia inhibitor factor, GDNF, and others from Sertoli cells promote gonocyte proliferate and adoption of a spermatogonial fate. Concurrently, proliferation by Sertoli cells declines via the actions of DMRT1. Disruption in this process prolongs Sertoli cell proliferation in neonatal testes, leading to impaired gonocyte migration and proliferation.

Despite the correlation between the expression of Rb1 and CKIs in gonocytes, the results of several recent studies indicate that the quiescent state may not be induced and then sustained in a cell-autonomous manner. Hallmark activities of gonocytes—including entry into quiescence, resumption of cell cycle progression, and migration to the basement membrane during neonatal development—are not altered in an Rb1-deficient state [53]. Interestingly, both testis weight and sperm production are increased in male mice that are deficient in p21 and p27, but these phenotypes are attributed to an increased number of Sertoli cells, and proliferation of gonocytes is not altered [52,54]. It is likely that interactions with Sertoli cells direct entry to the quiescent state followed by resumption of cell cycle progression in gonocytes.

The transforming growth factor (TGF) β family of cytokines influences an array of cellular functions, including apoptosis, adhesion, migration, and cell cycle progression [55,56]. In mammals, the TGFβ family consists of three members (TGFβ1, TGFβ2, and TGFβ3), all of which bind the type II receptor that subsequently forms dimers with the type I receptor, leading to downstream activation of the transcription factors SMAD2 and SMAD3 [57]. An activated heterodimer of SMAD2/3 complexes with SMAD4 and DNA-binding cofactors to modulate gene expression [57]. In mice and rats, Sertoli cells secrete TGFβ1 and TGFβ2, and gonocytes express the type I and type II receptors [58,59]. While supplementation of media with recombinant TGFβ1 does not alter gonocyte survival or proliferation in testis organ cultures [59,60], conditional inactivation of the TGFβ type II receptor (TGFβRII) gene in mouse gonocytes leads to precocious resumption of cell cycle progression and increased apoptosis [60]. These findings suggest a major role of TGFβ signaling for control of gonocyte quiescence. Interestingly, some *TgfβrII*-null gonocytes resume proliferation, and an SSC pool appears to arise. A similar phenotype occurs in mice that are deficient in another TGFβ superfamily ligand, activin A, which is able to induce cell cycle arrest by repressing the function of Rb1 and inducing the expression of p21 [61,62]. Activin A is also secreted by Sertoli cells [63–66]. The expression of TGFβ1 and activin A in Sertoli cells is greatest during late fetal development and then decreases significantly after birth [58,67], coincident with the resumption of proliferation by the gonocyte population. Collectively, these findings provide compelling evidence that paracrine signaling from Sertoli cells has a major influence on the induction and subsequent maintenance of cell cycle arrest in gonocytes (Figure 3.2).

In conjunction with the decreasing secretion of inhibitory signals, increased production of mitogenic signals from Sertoli cells to induce resumption of cell cycle progression in gonocytes is thought to occur during early neonatal development. In rat testes, fibroblast grow factor 2 (FGF2) is produced by both Sertoli cells and gonocytes during neonatal development [68]. While gonocyte production can induce an autocrine response, FGF2 secreted by Sertoli cells is likely to be the key mitogenic signal for gonocyte proliferation [68–70]. Supplementation of media with recombinant FGF2 leads to a significant increase in gonocytes in cultured testes from rats at PD 1–3, while neutralization of FGF2 signaling impairs gonocyte proliferation [70]. In addition, expansion of gonocyte number *in vitro* requires supplementation of media with FGF2 [71]. Furthermore, supplementation of culture media with recombinant forms of FGF2 and leukemia inhibitor factor (LIF) appears to enhance gonocyte proliferation *in vitro* [72–74]. Overall, these observations support a model in which FGF2 and LIF secreted by Sertoli cells signal synergistically within gonocytes to regulate proliferation during neonatal development.

B. Control of gonocyte migration

For rodents, the cell cycle activity of gonocytes and Sertoli cells appears to be inversely related and linked to the migration of gonocytes from the center of seminiferous cords to the basement membrane. In testes of mice that are deficient in DMRT1, migration and expansion of the gonocyte population is severely impaired [75]. Interestingly, the initiation of cell cycle arrest in Sertoli cells and subsequent maturation of the population is disrupted in *Dmrt1*-null mice. In testes of wild-type mice, Sertoli cells are distributed evenly along the basement membrane of seminiferous tubules following entry into a quiescent state, and the retention of an active cell cycle status in DMRT1-deficient Sertoli cells leads to abnormal morphology and disrupted distribution [36,75]. A similar phenotype of impaired gonocyte migration has been observed in the testes of mice that are deficient in the transcriptional regulator SIN3A. While Sertoli cell morphology and cell cycle kinetics appear to be normal in *Sin3a*-null mice, the expression of several cytokines, including some factors known to influence migration of somatic cell types, is reduced. Collectively, these findings indicate that the actions of DMRT1 and SIN3A in Sertoli cells influence the activities of gonocytes [76].

The identification of extrinsic signals that induce migration of gonocytes during neonatal development has been challenging because cell cycle progression resumes simultaneously. However, emerging evidence suggests that signaling from a select few cytokines guides the migratory behavior of gonocytes, and Sertoli cells are the source of this signaling. These factors include the cytokines KIT ligand (KITL) and platelet-derived growth factor (PDGF) as well as C-X-C motif chemokine 12 (CXCL12).

Signaling via the transmembrane tyrosine kinase receptor c-KIT influences an array of activities by cells in both the hematopoietic and germ cell lineages [77]. The only known ligand for c-KIT is KITL. Mouse and human PGCs express c-KIT [78,79], and migration is severely impaired in mouse embryos with KITL loss of function [78,80]. Interestingly, expression of c-KIT persists within gonocytes at birth, and activation of a downstream signaling cascade promotes the formation of pseudopodia to mediate migratory activity [21,81–83]. In addition, suppression of c-KIT signaling in cultures of testicular tissue from neonatal mice using a neutralizing antibody leads to a significant decrease in migratory gonocytes, while supplementation of media with recombinant KITL enhances migration [21]. Furthermore, activation of c-KIT signaling does not appear to alter the mitotic activity of either rat or mouse gonocytes [81,83]. Moreover, adhesion of gonocytes to Sertoli cells is impaired in KITL-deficient mice [84]. Taken together, these findings indicate that a paracrine signaling pathway involving Sertoli cell production of KITL that signals via the c-KIT receptor in gonocytes promotes their migration from the center of seminiferous cords to the basement membrane during neonatal development.

PDGF is a potent regulator of cell migration in various types of tissue [85]. The PDGF family contains four members (PDGF-A, PDGF-B, PDGF-C, and PDGF-D), and Sertoli cells express PDGF-B and PDGF-D [86,87]. The receptor complex for PDGF ligands consists of two subunits (PDGF receptor (PDGFR) α and PDGFRβ), which form homodimer or heterodimer complexes upon ligand binding to trigger an intracellular response [88]. In the testes of neonatal mice, PGDFRβ has been detected on the surface of gonocytes [86]. While *Pdgfrβ* null mice die at ED 17–19, defects in PGC specification or migration and formation of the gonocyte pool are not observed. However, pharmacological inhibition of PDGF signaling at PD 1–5 leads to an increased number of centrally located gonocytes and an elevated number of apoptotic cells [86]. These observations suggest that paracrine signaling between Sertoli cells and gonocytes via PDGF ligand activation of PDGFRs may influence the migratory behavior of gonocytes.

The chemokine CXCL12 is a well-described inducer of migration for various somatic cell types, including hematopoietic stem cells [89,90], and the receptor complex includes the transmembrane tyrosine kinases CXCR4 and CXCR7. Signaling via CXCR4 is essential for PGC migration and survival in mice [91,92]. Also, gonocytes and undifferentiated spermatogonia express CXCR4 in both mice and humans [93,94]. Because of its role in guiding the migration of PGCs, CXCL12 secretion from Sertoli cells might direct gonocyte survival and migration during neonatal development, but direct experimental evidence for this process is currently lacking. However, in support of this notion is the finding that CXCL12 expression is reduced significantly in Sertoli cells that are deficient in SIN3a, and gonocyte migration appears to be disrupted in this model [76]. Also, suppression of CXCR4 signaling inhibits the maintenance of the SSC pool in primary cultures of undifferentiated spermatogonia, and migration of SSCs deficient for CXCR4 is impaired following transplantation. While not linked directly to gonocytes, SSCs are direct descendants of gonocytes, and, thus, the influence of CXCR4 signaling may be a common feature.

IV. Role of Sertoli cells in formation of primary undifferentiated and differentiating spermatogonial populations

In mice, gonocytes are precursors of both a primary undifferentiated spermatogonial population that includes the SSC pool and an initial differentiating spermatogonial population. The SSC pool is the source of spermatogenesis throughout adulthood, whereas the first round of spermatogenesis derives from the initial differentiating spermatogonial population [20]. Emerging evidence suggests that contributions from Sertoli cells influence the development of both initial spermatogonial populations. Importantly, many of the factors crucial for the gonocyte to spermatogonia transition also regulate the same spermatogonial populations during steady-state spermatogenesis in adulthood (Figure 3.3).

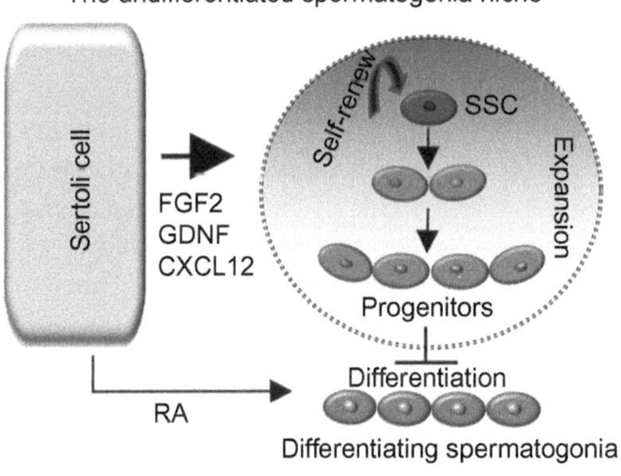

Figure 3.3 Model for the current understanding of Sertoli cell contributions to the undifferentiated spermatogonial niche. It is widely accepted that factors secreted from Sertoli cells influence the behaviors of spermatogonial stem cells and progenitor spermatogonia that make up the undifferentiated spermatogonial population. These Sertoli-cell-borne signals likely influence the formation of the primary undifferentiated and differentiating spermatogonial populations from gonocyte precursors during neonatal development.

A. Transition of gonocyte to stem and progenitor spermatogonia

The formation of an initial, undifferentiated, spermatogonial population from gonocytes is obligatory for the establishment of the spermatogenic lineage in postnatal life. This process involves the formation of initial SSC and progenitor spermatogonial pools. A defining event of the transition from gonocytes to undifferentiated spermatogonia is the repression of c-KIT expression. Gonocytes possess c-KIT on the cell surface [83], but undifferentiated spermatogonia do not [83,95–97]. Because the reestablishment of c-KIT expression defines the transition from an undifferentiated to a differentiating state in postnatal spermatogonia, the repression of expression in gonocytes is likely a requirement for the formation of a primary undifferentiated spermatogonial pool. Studies with mice suggest that factors secreted from Sertoli cells induce a response in undifferentiated spermatogonia that leads to the translational repression of c-KIT expression [98]. Thus, it is tempting to postulate that a similar pathway is present in gonocytes to induce a c-KIT-state during neonatal development.

Results of mouse studies suggest that *c-Kit* gene transcription is active in undifferentiated spermatogonia, but *c-Kit* translation is actively repressed to promote a c-KIT-state [98]. In primary cultures of undifferentiated spermatogonia and *in vivo* counterparts, *c-Kit* transcripts are detectable but c-KIT protein is absent [98]. This finding suggests the existence of a microRNA-mediated, posttranscriptional inhibitory mechanism. In various somatic cell types, *c-Kit* is a known target of the X chromosome microRNA clusters 221 and 222 (*miR221/222*), and undifferentiated spermatogonia express both molecules [98]. Also, a c-KIT+ state is induced by inhibition of *miR221/222* function in primary cultures of undifferentiated spermatogonia at the same time as loss of the SSC pool. Importantly, the expression of *miR221/222* is induced by signaling from cytokines that are known to be expressed by Sertoli cells, including GDNF and FGF2. These findings imply that Sertoli cells actively participate in the induction of an undifferentiated state within spermatogonia that could be initiated during the gonocyte stage of development.

Another critical issue for the establishment of the primary undifferentiated spermatogonial population is the expansion of cells following attainment of an undifferentiated state. Thus, the removal of cell cycle arrest that is initiated in gonocytes during the early neonatal period must be sustained. The mechanism for the promotion of sustained mitotic activity likely includes mitogenic signals from Sertoli cells in the form of secreted cytokines. It is well established that the cytokines GDNF and FGF2 and the chemokine CXCL12 influence the proliferation of undifferentiated spermatogonia [94,99], and Sertoli cells are a source of these factors in the testes of mice [100].

Another component of the mechanism driving mitotic activity in undifferentiated spermatogonia is the repression of cell cycle inhibitors, such as p27 (also known as CDKN1B). In addition to *c-Kit*, *Cdkn1b* is a target of *miR221/222* for translational repression [101,102]. Interestingly, CDKN1B is expressed abundantly in quiescent gonocytes but is undetectable in undifferentiated spermatogonia beginning in the early neonatal period, when expansion of the primary population occurs [52]. Thus, induction of *miR221/222* expression in undifferentiated spermatogonia via paracrine signaling from factors secreted by Sertoli cells is possibly a critical attribute for both the resumption of cell cycle progression in gonocytes and then sustainment in the resulting undifferentiated spermatogonia.

Lastly, expansion of the initial undifferentiated spermatogonial population includes both self-renewal of SSCs to establish a foundational pool and expansion of a progenitor pool. The receptors for GDNF, FGF2, and CXCL12 are expressed by both SSC and progenitor spermatogonial subtypes [29,68,94,103–106]. Thus, these factors are likely general mitogens for undifferentiated spermatogonia, and other factors must provide a fine tuning that promotes self-renewal to establish the SSC pool. Indeed, studies with mice identified CSF-1 as a cytokine that influences self-renewal of SSCs specifically within primary cultures of undifferentiated spermatogonia [107]. In addition, the expression of CSF-1 in mouse testes has been localized to Leydig cells [107]. These findings raise an interesting possibility that cross-communication between Sertoli and Leydig cells influences the expansion of SSCs during neonatal development to establish the foundational pool.

B. Transition of gonocytes to differentiating spermatogonia

After migrating from the center of seminiferous cords to the basement membrane, a subset of gonocytes transitions directly to a differentiating spermatogonial state and is committed to terminal differentiation, thereby providing the basis for the first round of spermatogenesis. This pathway has been observed in mice, but whether a similar mechanism occurs in other mammalian species is not known. Evidence from rodent studies indicates that the transition of gonocytes to differentiating spermatogonia is induced by retinoic acid (RA) secreted by Sertoli cells. In the testes of vitamin-A-deficient (VAD) rats and mice, spermatogenesis is halted at the undifferentiated spermatogonial stage and is reinitiated following the reintroduction of either RA or various retinoids [108–111]. Thus, RA signaling is a primary inducer of the transition from an undifferentiated to a differentiating state in spermatogonia. Importantly, the first round of spermatogenesis is disrupted when a VAD state is initiated prenatally and only undifferentiated spermatogonia are present, demonstrating that RA has a role in the direct transition of gonocytes to a differentiating spermatogonial fate [112]. The major source of RA in rodent testes is thought to be Sertoli cells, which thus establish a paracrine signaling paradigm for influencing gonocyte fate decisions. Indeed, the first round of spermatogenesis is absent in mice with Sertoli cell deficiency of retinol dehydrogenase 10 (RDH10), a key enzyme in the RA biogenesis pathway [113]. However, spermatogenesis is eventually established from the initial undifferentiated spermatogonial population in RDH10-mutant animals later in adulthood. These findings provide compelling evidence that the transition of gonocytes to a differentiating spermatogonial state is regulated by paracrine communication with Sertoli cells via a RA intermediate.

V. Concluding remarks

In the United States alone, approximately 600,000 men are infertile, and many of these cases are attributed to oligospermia (lower than normal number of sperm in the ejaculate) or azoospermia (absence of sperm in the ejaculate). At present, most of these cases are idiopathic. A clinical diagnosis of infertility is almost always made during adulthood. Compromised establishment of the spermatogenic lineage is an overlooked possible cause but could actually be a major underlying reason. The formation of a primary undifferentiated spermatogonial population from gonocyte precursors is the initial event in the postnatal development of the spermatogenic lineage. This process requires gonocyte migration from the center of seminiferous cords to the basement membrane, resumption of cell cycle progression, and transition to an undifferentiated state. Emerging evidence indicates that all of these events are regulated by communication with Sertoli cells. Furthermore, continued communication between Sertoli cells and the newly formed, undifferentiated, spermatogonial population is essential for the formation and maintenance of a foundational SSC pool. As the understanding of the pathways through which Sertoli cells interact with gonocytes and undifferentiated spermatogonia during neonatal development are defined more precisely, new tools for diagnosing and treating male infertility will be developed. Moreover, the advances in understanding will contribute to the design of a methodology that recapitulates the environment within testes *in vitro*, which will lead to enhanced maintenance of gonocytes and SSCs. Given that the gonocyte or SSC pool is often eliminated from the testes of prepubertal boys undergoing chemotherapy or radiotherapy, these advances could provide a means for preserving their germ line that currently unavailable.

References

[1] Clermont Y. Two classes of spermatogonial stem cells in the monkey (*Cercopithecus aethiops*). Am J Anat 1969;126(1):57–71.

[2] Brinster RL, Zimmermann JW. Spermatogenesis following male germ-cell transplantation. Proc Natl Acad Sci USA 1994;91(24):11298–302.

[3] de Rooij DG, Russell LD. All you wanted to know about spermatogonia but were afraid to ask. J Androl 2000;21(6):776–98.

[4] Yang QE, Oatley JM. Spermatogonial stem cell functions in physiological and pathological conditions. Curr Top Dev Biol 2014;107:235–67.

[5] Skakkebaek NE, Rajpert-De Meyts E, Jørgensen N, Carlsen E, Petersen PM, Giwercman A, et al. Germ cell cancer and disorders of spermatogenesis: an environmental connection? APMIS 1998;106(1):3–11.

[6] Griswold MD. Interactions between germ cells and Sertoli cells in the testis. Biol Reprod 1995;52(2):211–6.

[7] Drumond AL, Meistrich ML, Chiarini-Garcia H. Spermatogonial morphology and kinetics during testis development in mice: a high-resolution light microscopy approach. Reproduction 2011;142(1):145–55.

[8] Peters H. Migration of gonocytes into the mammalian gonad and their differentiation. Philos Trans R Soc Lond B Biol Sci 1970;259(828):91–101.
[9] Oatley JM, Brinster RL. Regulation of spermatogonial stem cell self-renewal in mammals. Annu Rev Cell Dev Biol 2008;24:263–86.
[10] Ginsburg M, Snow MH, McLaren A. Primordial germ cells in the mouse embryo during gastrulation. Development 1990;110(2):521–8.
[11] Baillie AH. The histochemistry and ultrastructure of the genocyte. J Anat 1964;98:641–5.
[12] Lei L, Spradling AC. Mouse primordial germ cells produce cysts that partially fragment prior to meiosis. Development 2013;140(10):2075–81.
[13] Western PS, Miles DC, van den Bergen JA, Burton M, Sinclair AH. Dynamic regulation of mitotic arrest in fetal male germ cells. Stem Cells 2008;26(2):339–47.
[14] Orwig KE, Ryu BY, Avarbock MR, Brinster RL. Male germ-line stem cell potential is predicted by morphology of cells in neonatal rat testes. Proc Natl Acad Sci USA 2002;99(18):11706–11.
[15] Wang RA, Nakane PK, Koji T. Autonomous cell death of mouse male germ cells during fetal and postnatal period. Biol Reprod 1998;58(5):1250–6.
[16] Roosen-Runge EC, Leik J. Gonocyte degeneration in the postnatal male rat. Am J Anat 1968;122(2):275–99.
[17] Huff DS, Snyder III HM, Rusnack SL, Zderic SA, Carr MC, Canning DA. Hormonal therapy for the subfertility of cryptorchidism. Horm Res 2001;55(1):38–40.
[18] Clermont Y, Perey B. Quantitative study of the cell population of the seminiferous tubules in immature rats. Am J Anat 1957;100(2):241–67.
[19] Kluin PM, de Rooij DG. A comparison between the morphology and cell kinetics of gonocytes and adult type undifferentiated spermatogonia in the mouse. Int J Androl 1981;4(4):475–93.
[20] Yoshida S, Sukeno M, Nakagawa T, Ohbo K, Nagamatsu G, Suda T, et al. The first round of mouse spermatogenesis is a distinctive program that lacks the self-renewing spermatogonia stage. Development 2006;133(8):1495–505.
[21] Orth JM, Qiu J, Jester Jr WF, Pilder S. Expression of the *c-kit* gene is critical for migration of neonatal rat gonocytes *in vitro*. Biol Reprod 1997;57(3):676–83.
[22] Yamaguchi S, Kimura H, Tada M, Nakatsuji N, Tada T. Nanog expression in mouse germ cell development. Gene Expr Patterns 2005;5(5):639–46.
[23] Nettersheim D, Biermann K, Gillis AJ, Steger K, Looijenga LH, Schorle H. NANOG promoter methylation and expression correlation during normal and malignant human germ cell development. Epigenetics 2011;6(1):114–22.
[24] Ohinata Y, Payer B, O'Carroll D, Ancelin K, Ono Y, Sano M, et al. Blimp1 is a critical determinant of the germ cell lineage in mice. Nature 2005;436(7048):207–13.
[25] Eckert D, Biermann K, Nettersheim D, Gillis AJ, Steger K, Jäck HM, et al. Expression of BLIMP1/PRMT5 and concurrent histone H2A/H4 arginine 3 dimethylation in fetal germ cells, CIS/IGCNU and germ cell tumors. BMC Dev Biol 2008;8:106.
[26] Payer B, Chuva de Sousa Lopes SM, Barton SC, Lee C, Saitou M, Surani MA. Generation of stella-GFP transgenic mice: a novel tool to study germ cell development. Genesis 2006;44(2):75–83.
[27] Sato M, Kimura T, Kurokawa K, Fujita Y, Abe K, Masuhara M, et al. Identification of *PGC7*, a new gene expressed specifically in preimplantation embryos and germ cells. Mech Dev 2002;113(1):91–4.
[28] Ohbo K, Yoshida S, Ohmura M, Ohneda O, Ogawa T, Tsuchiya H, et al. Identification and characterization of stem cells in prepubertal spermatogenesis in mice small star, filled. Dev Biol 2003;258(1):209–25.

[29] Gassei K, Orwig KE. SALL4 expression in gonocytes and spermatogonial clones of postnatal mouse testes. PLoS One 2013;8(1):e53976.
[30] Zheng K, Wu X, Kaestner KH, Wang PJ. The pluripotency factor LIN28 marks undifferentiated spermatogonia in mouse. BMC Dev Biol 2009;9:38.
[31] Hobbs RM, Fagoonee S, Papa A, Webster K, Altruda F, Nishinakamura R, et al. Functional antagonism between Sall4 and Plzf defines germline progenitors. Cell Stem Cell 2012;10(3):284–98.
[32] van Bragt MP, Roepers-Gajadien HL, Korver CM, Bogerd J, Okuda A, Eggen BJ, et al. Expression of the pluripotency marker UTF1 is restricted to a subpopulation of early a spermatogonia in rat testis. Reproduction 2008;136(1):33–40.
[33] Tsuda M, Sasaoka Y, Kiso M, Abe K, Haraguchi S, Kobayashi S, et al. Conserved role of nanos proteins in germ cell development. Science 2003;301(5637):1239–41.
[34] Costoya JA, Hobbs RM, Barna M, Cattoretti G, Manova K, Sukhwani M, et al. Essential role of Plzf in maintenance of spermatogonial stem cells. Nat Genet 2004;36 (6):653–9.
[35] Lecoin L, Gabella G, Le Douarin N. Origin of the c-kit-positive interstitial cells in the avian bowel. Development 1996;122(3):725–33.
[36] Vergouwen RP, Jacobs SG, Huiskamp R, Davids JA, de Rooij DG. Proliferative activity of gonocytes, Sertoli cells and interstitial cells during testicular development in mice. J Reprod Fertil 1991;93(1):233–43.
[37] Hadjiolova KV, Martinova YS, Yankulov KY, Davidov V, Kancheva LS, Hadjiolov AA. An immunocytochemical study of the proliferating cell nuclear matrix antigen p125/6.5 during rat spermatogenesis. J Cell Sci 1989;93(Pt 1):173–7.
[38] Cortes D, Muller J, Skakkebaek NE. Proliferation of Sertoli cells during development of the human testis assessed by stereological methods. Int J Androl 1987;10(4):589–96.
[39] Oatley MJ, Racicot KE, Oatley JM. Sertoli cells dictate spermatogonial stem cell niches in the mouse testis. Biol Reprod 2011;84(4):639–45.
[40] Zogbi C, Tesser RB, Encinas G, Miraglia SM, Stumpp T. Gonocyte development in rats: proliferation, distribution and death revisited. Histochem Cell Biol 2012;138 (2):305–22.
[41] Giampietri C, Petrungaro S, Klinger FG, Coluccia P, Paone A, Vivarelli E, et al. c-Flip expression and function in fetal mouse gonocytes. FASEB J 2006;20(1):124–6.
[42] Spencer SL, Cappell SD, Tsai FC, Overton KW, Wang CL, Meyer T. The proliferation-quiescence decision is controlled by a bifurcation in CDK2 activity at mitotic exit. Cell 2013;155(2):369–83.
[43] Sadasivam S, DeCaprio JA. The DREAM complex: master coordinator of cell cycle-dependent gene expression. Nat Rev Cancer 2013;13(8):585–95.
[44] Weinberg RA. The retinoblastoma protein and cell cycle control. Cell 1995;81 (3):323–30.
[45] Chellappan SP, Hiebert S, Mudryj M, Horowitz JM, Nevins JR. The E2F transcription factor is a cellular target for the RB protein. Cell 1991;65(6):1053–61.
[46] Wu L, Timmers C, Maiti B, Saavedra HI, Sang L, Chong GT, et al. The E2F1-3 transcription factors are essential for cellular proliferation. Nature 2001;414(6862):457–62.
[47] Coller HA. What's taking so long? S-phase entry from quiescence versus proliferation. Nat Rev Mol Cell Biol 2007;8(8):667–70.
[48] Sage J. The retinoblastoma tumor suppressor and stem cell biology. Genes Dev 2012;26(13):1409–20.
[49] Morgan DO. Principles of CDK regulation. Nature 1995;374(6518):131–4.

[50] Denicourt C, Dowdy SF. Cip/Kip proteins: more than just CDKs inhibitors. Genes Dev 2004;18(8):851–5.
[51] Moreno SG, Dutrillaux B, Coffigny H. Status of p53, p21, mdm2, pRb proteins, and DNA methylation in gonocytes of control and gamma-irradiated rats during testicular development. Biol Reprod 2001;64(5):1422–31.
[52] Beumer TL, Kiyokawa H, Roepers-Gajadien HL, van den Bos LA, Lock TM, Gademan IS, et al. Regulatory role of p27kip1 in the mouse and human testis. Endocrinology 1999;140(4):1834–40.
[53] Yang QE, Gwost I, Oatley MJ, Oatley JM. Retinoblastoma protein (RB1) controls fate determination in stem cells and progenitors of the mouse male germline. Biol Reprod 2013;89(5):113.
[54] Holsberger DR, Buchold GM, Leal MC, Kiesewetter SE, O'Brien DA, Hess RA, et al. Cell-cycle inhibitors p27Kip1 and p21Cip1 regulate murine Sertoli cell proliferation. Biol Reprod 2005;72(6):1429–36.
[55] Siegel PM, Massague J. Cytostatic and apoptotic actions of TGF-beta in homeostasis and cancer. Nat Rev Cancer 2003;3(11):807–21.
[56] Fortunel N, Hatzfeld J, Kisselev S, Monier MN, Ducos K, Cardoso A, et al. Release from quiescence of primitive human hematopoietic stem/progenitor cells by blocking their cell-surface TGF-beta type II receptor in a short-term *in vitro* assay. Stem Cells 2000;18(2):102–11.
[57] Massague J. How cells read TGF-beta signals. Nat Rev Mol Cell Biol 2000;1(3): 169–78.
[58] Cupp AS, Kim G, Skinner MK. Expression and action of transforming growth factor beta (TGFbeta1, TGFbeta2, and TGFbeta3) during embryonic rat testis development. Biol Reprod 1999;60(6):1304–13.
[59] Olaso R, Pairault C, Boulogne B, Durand P, Habert R. Transforming growth factor beta1 and beta2 reduce the number of gonocytes by increasing apoptosis. Endocrinology 1998;139(2):733–40.
[60] Moreno SG, Attali M, Allemand I, Messiaen S, Fouchet P, Coffigny H, et al. TGFbeta signaling in male germ cells regulates gonocyte quiescence and fertility in mice. Dev Biol 2010;342(1):74–84.
[61] Yamato K, Koseki T, Ohguchi M, Kizaki M, Ikeda Y, Nishihara T. Activin A induction of cell-cycle arrest involves modulation of cyclin D2 and p21CIP1/WAF1 in plasmacytic cells. Mol Endocrinol 1997;11(8):1044–52.
[62] Sehy DW, Shao LE, Yu AL, Tsai WM, Yu J. Activin A-induced differentiation in K562 cells is associated with a transient hypophosphorylation of RB protein and the concomitant block of cell cycle at G_1 phase. J Cell Biochem 1992;50(3):255–65.
[63] Mather JP, Attie KM, Woodruff TK, Rice GC, Phillips DM. Activin stimulates spermatogonial proliferation in germ-Sertoli cell cocultures from immature rat testis. Endocrinology 1990;127(6):3206–14.
[64] de Winter JP, Vanderstichele HM, Timmerman MA, Blok LJ, Themmen AP, de Jong FH. Activin is produced by rat Sertoli cells *in vitro* and can act as an autocrine regulator of Sertoli cell function. Endocrinology 1993;132(3):975–82.
[65] Wada M, Shintani Y, Kosaka M, Sano T, Hizawa K, Saito S. Immunohistochemical localization of activin A and follistatin in human tissues. Endocr J 1996;43 (4):375–85.
[66] Mendis SH, Meachem SJ, Sarraj MA, Loveland KL. Activin A balances Sertoli and germ cell proliferation in the fetal mouse testis. Biol Reprod 2011;84(2):379–91.

[67] Hasthorpe S, Barbic S, Farmer PJ, Hutson JM. Growth factor and somatic cell regulation of mouse gonocyte-derived colony formation *in vitro*. J Reprod Fertil 2000;119 (1):85–91.
[68] Mullaney BP, Skinner MK. Basic fibroblast growth factor (bFGF) gene expression and protein production during pubertal development of the seminiferous tubule: follicle-stimulating hormone-induced Sertoli cell bFGF expression. Endocrinology 1992;131 (6):2928–34.
[69] Gonzalez-Herrera IG, Prado-Lourenco L, Pileur F, Conte C, Morin A, Cabon F, et al. Testosterone regulates FGF-2 expression during testis maturation by an IRES-dependent translational mechanism. FASEB J 2006;20(3):476–8.
[70] Van Dissel-Emiliani FM, De Boer-Brouwer M, De Rooij DG. Effect of fibroblast growth factor-2 on Sertoli cells and gonocytes in coculture during the perinatal period. Endocrinology 1996;137(2):647–54.
[71] Kanatsu-Shinohara M, Inoue K, Lee J, Yoshimoto M, Ogonuki N, Miki H, et al. Generation of pluripotent stem cells from neonatal mouse testis. Cell 2004;119 (7):1001–12.
[72] De Miguel MP, De Boer-Brouwer M, Paniagua R, van den Hurk R, De Rooij DG, Van Dissel-Emiliani FM. Leukemia inhibitory factor and ciliary neurotropic factor promote the survival of Sertoli cells and gonocytes in coculture system. Endocrinology 1996;137(5):1885–93.
[73] Nikolova DB, Martinova YS, Seidensticker M, Bellvé AR. Leukaemia inhibitory factor stimulates proliferation of prospermatogonial stem cells. Reprod Fertil Dev 1997;9 (7):717–21.
[74] Kanatsu-Shinohara M, Inoue K, Ogonuki N, Miki H, Yoshida S, Toyokuni S, et al. Leukemia inhibitory factor enhances formation of germ cell colonies in neonatal mouse testis culture. Biol Reprod 2007;76(1):55–62.
[75] Raymond CS, Murphy MW, O'Sullivan MG, Bardwell VJ, Zarkower D. *Dmrt1*, a gene related to worm and fly sexual regulators, is required for mammalian testis differentiation. Genes Dev 2000;14(20):2587–95.
[76] Payne CJ, Gallagher SJ, Foreman O, Dannenberg JH, Depinho RA, Braun RE. *Sin3a* is required by Sertoli cells to establish a niche for undifferentiated spermatogonia, germ cell tumors, and spermatid elongation. Stem Cells 2010;28(8):1424–34.
[77] Chabot B, Stephenson DA, Chapman VM, Besmer P, Bernstein A. The proto-oncogene c-kit encoding a transmembrane tyrosine kinase receptor maps to the mouse W locus. Nature 1988;335(6185):88–9.
[78] Runyan C, Schaible K, Molyneaux K, Wang Z, Levin L, Wylie C. Steel factor controls midline cell death of primordial germ cells and is essential for their normal proliferation and migration. Development 2006;133(24):4861–9.
[79] Gkountela S, Li Z, Vincent JJ, Zhang KX, Chen A, Pellegrini M, et al. The ontogeny of cKIT+ human primordial germ cells proves to be a resource for human germ line reprogramming, imprint erasure and *in vitro* differentiation. Nat Cell Biol 2013;15 (1):113–22.
[80] Motro B, van der Kooy D, Rossant J, Reith A, Bernstein A. Contiguous patterns of c-kit and steel expression: analysis of mutations at the W and Sl loci. Development 1991;113(4):1207–21.
[81] Hasthorpe S, Barbic S, Farmer PJ, Hutson JM. Neonatal mouse gonocyte proliferation assayed by an *in vitro* clonogenic method. J Reprod Fertil 1999;116(2):335–44.
[82] Wehrle-Haller B, Weston JA. Altered cell-surface targeting of stem cell factor causes loss of melanocyte precursors in Steel17H mutant mice. Dev Biol 1999;210(1):71–86.

[83] Yoshinaga K, Nishikawa S, Ogawa M, Hayashi S, Kunisada T, Fujimoto T, et al. Role of c-kit in mouse spermatogenesis: identification of spermatogonia as a specific site of c-kit expression and function. Development 1991;113(2):689–99.

[84] Marziali G, Lazzaro D, Sorrentino V. Binding of germ cells to mutant Sld Sertoli cells is defective and is rescued by expression of the transmembrane form of the c-kit ligand. Dev Biol 1993;157(1):182–90.

[85] Clement DL, Mally S, Stock C, Lethan M, Satir P, Schwab A, et al. PDGFRalpha signaling in the primary cilium regulates NHE1-dependent fibroblast migration via coordinated differential activity of MEK1/2-ERK1/2-p90RSK and AKT signaling pathways. J Cell Sci 2013;126(Pt 4):953–65.

[86] Basciani S, De Luca G, Dolci S, Brama M, Arizzi M, Mariani S, et al. Platelet-derived growth factor receptor beta-subtype regulates proliferation and migration of gonocytes. Endocrinology 2008;149(12):6226–35.

[87] Loveland KL, Zlatic K, Stein-Oakley A, Risbridger G, deKretser DM. Platelet-derived growth factor ligand and receptor subunit mRNA in the Sertoli and Leydig cells of the rat testis. Mol Cell Endocrinol 1995;108(1–2):155–9.

[88] Basciani S, Mariani S, Spera G, Gnessi L. Role of platelet-derived growth factors in the testis. Endocr Rev 2010;31(6):916–39.

[89] Aiuti A, Webb IJ, Bleul C, Springer T, Gutierrez-Ramos JC. The chemokine SDF-1 is a chemoattractant for human CD34 + hematopoietic progenitor cells and provides a new mechanism to explain the mobilization of CD34+ progenitors to peripheral blood. J Exp Med 1997;185(1):111–20.

[90] Peled A, Petit I, Kollet O, Magid M, Ponomaryov T, Byk T, et al. Dependence of human stem cell engraftment and repopulation of NOD/SCID mice on CXCR4. Science 1999;283(5403):845–8.

[91] Ara T, Nakamura Y, Egawa T, Sugiyama T, Abe K, Kishimoto T, et al. Impaired colonization of the gonads by primordial germ cells in mice lacking a chemokine, stromal cell-derived factor-1 (SDF-1). Proc Natl Acad Sci USA 2003;100 (9):5319–23.

[92] Molyneaux KA, Zinszner H, Kunwar PS, Schaible K, Stebler J, Sunshine MJ, et al. The chemokine SDF1/CXCL12 and its receptor CXCR4 regulate mouse germ cell migration and survival. Development 2003;130(18):4279–86.

[93] McIver SC, Loveland KL, Roman SD, Nixon B, Kitazawa R, McLaughlin EA. The chemokine CXCL12 and its receptor CXCR4 are implicated in human seminoma metastasis. Andrology 2013;1(3):517–29.

[94] Yang QE, Kim D, Kaucher A, Oatley MJ, Oatley JM. CXCL12-CXCR4 signaling is required for the maintenance of mouse spermatogonial stem cells. J Cell Sci 2013;126 (Pt 4):1009–20.

[95] Shinohara T, Avarbock MR, Brinster RL. Beta1- and alpha6-integrin are surface markers on mouse spermatogonial stem cells. Proc Natl Acad Sci USA 1999;96 (10):5504–9.

[96] Shinohara T, Brinster RL. Enrichment and transplantation of spermatogonial stem cells. Int J Androl 2000;23(Suppl. 2):89–91.

[97] Kubota H, Avarbock MR, Brinster RL. Spermatogonial stem cells share some, but not all, phenotypic and functional characteristics with other stem cells. Proc Natl Acad Sci USA 2003;100(11):6487–92.

[98] Yang QE, Racicot KE, Kaucher AV, Oatley MJ, Oatley JM. MicroRNAs 221 and 222 regulate the undifferentiated state in mammalian male germ cells. Development 2013;140(2):280–90.

[99] Kubota H, Avarbock MR, Brinster RL. Growth factors essential for self-renewal and expansion of mouse spermatogonial stem cells. Proc Natl Acad Sci USA 2004;101 (47):16489–94.
[100] Oatley JM, Brinster RL. The germline stem cell niche unit in mammalian testes. Physiol Rev 2012;92(2):577–95.
[101] Cardinali B, Castellani L, Fasanaro P, Basso A, Alemá S, Martelli F, et al. Microrna-221 and microrna-222 modulate differentiation and maturation of skeletal muscle cells. PLoS One 2009;4(10):e7607.
[102] Felicetti F, Errico MC, Bottero L, Segnalini P, Stoppacciaro A, Biffoni M, et al. The promyelocytic leukemia zinc finger-microRNA-221/-222 pathway controls melanoma progression through multiple oncogenic mechanisms. Cancer Res 2008;68 (8):2745–54.
[103] Suzuki H, Sada A, Yoshida S, Saga Y. The heterogeneity of spermatogonia is revealed by their topology and expression of marker proteins including the germ cell-specific proteins Nanos2 and Nanos3. Dev Biol 2009;336(2):222–31.
[104] Grasso M, Fuso A, Dovere L, de Rooij DG, Stefanini M, Boitani C, et al. Distribution of GFRA1-expressing spermatogonia in adult mouse testis. Reproduction 2012;143 (3):325–32.
[105] Suzuki H, Ahn HW, Chu T, Bowden W, Gassei K, Orwig K, et al. SOHLH1 and SOHLH2 coordinate spermatogonial differentiation. Dev Biol 2012;361(2):301–12.
[106] Cancilla B, Risbridger GP. Differential localization of fibroblast growth factor receptor-1, -2, -3, and -4 in fetal, immature, and adult rat testes. Biol Reprod 1998;58 (5):1138–45.
[107] Oatley JM, Oatley MJ, Avarbock MR, Tobias JW, Brinster RL. Colony stimulating factor 1 is an extrinsic stimulator of mouse spermatogonial stem cell self-renewal. Development 2009;136(7):1191–9.
[108] Coward WA, Howell JM, Thompson JN, Pitt GA. The retinol requirements of rats for spermatogenesis and vision. Br J Nutr 1969;23(3):619–26.
[109] van Pelt AM, de Rooij DG. Synchronization of the seminiferous epithelium after vitamin A replacement in vitamin A-deficient mice. Biol Reprod 1990;43(3):363–7.
[110] van Pelt AM, de Rooij DG. Retinoic acid is able to reinitiate spermatogenesis in vitamin A-deficient rats and high replicate doses support the full development of spermatogenic cells. Endocrinology 1991;128(2):697–704.
[111] Morales C, Griswold MD. Retinol-induced stage synchronization in seminiferous tubules of the rat. Endocrinology 1987;121(1):432–4.
[112] Li H, Palczewski K, Baehr W, Clagett-Dame M. Vitamin A deficiency results in meiotic failure and accumulation of undifferentiated spermatogonia in prepubertal mouse testis. Biol Reprod 2011;84(2):336–41.
[113] Tong MH, Yang QE, Davis JC, Griswold MD. Retinol dehydrogenase 10 is indispensible for spermatogenesis in juvenile males. Proc Natl Acad Sci USA 2013;110 (2):543–8.

The spermatogonial stem cell niche in mammals

4

Dirk G. de Rooij[a,b]
[a]Center for Reproductive Medicine, Academic Medical Center, University of Amsterdam, Amsterdam, The Netherlands, [b]Department of Endocrinology and Metabolism, Faculty of Science, Utrecht University, Utrecht, The Netherlands

I. Research advances related to the mammalian SSC niche since 2003

In 2001, Russell and colleagues were the first to show that in rodent testes, early types of spermatogonia, including spermatogonial stem cells (SSCs), are not distributed randomly over the tubule basal lamina [1]. In 2003, this group suggested that the areas of the tubule basal lamina bordering the interstitial tissue provide a niche for these spermatogonia [2]. In 2007, Yoshida and colleagues narrowed this concept down by suggesting that an SSC niche encompasses those areas of the tubule basal lamina opposite the interstitial arterioles and venules [3]. In addition, since 2003, much progress has been made in understanding the factors that promote SSC self-renewal and differentiation. Two excellent reviews have been published on the factors governing SSC self-renewal and differentiation and the molecular pathways involved [4,5]. Therefore, the emphasis in this chapter is on understanding how the SSC niche in nonprimate mammals is organized, how SSCs behave inside and outside the niche, and how the somatic cells establish the niche environment through the production of particular (growth) factors.

II. Principles of stem cell niches in mammalian tissues

Stem cells are present in all tissues in which functional cells have a finite life-span and need constant replenishment. The viability of a tissue depends on the lifelong presence of a sufficient number of stem cells to keep the tissue at a desired size and functional capacity, indicating that proper maintenance of stem cells must be a top priority in each of these tissues.

Proliferation of stem cells serves two goals: provide differentiating cells that carry out the function of the tissue and replenish the stem cells that are lost to the differentiation pathway. All of these activities need to happen in such a way that the stem cell population does not become depleted or grow, which would lead to

Sertoli Cell Biology. DOI: http://dx.doi.org/10.1016/B978-0-12-417047-6.00004-1

the formation of a tumor. As originally proposed by Schofield in 1978 [6], tissues have reserved a special part of the organ, called the "stem cell niche," to preserve stem cells and to properly manage the balance between self-renewal and differentiation of stem cells in the tissue. Inside the stem cell niche area, other cell types usually secrete growth factors that create an environment that promotes stem cell self-renewal to make sure that these cells are not depleted.

As an illustration of the niche principles described above, in the intestinal epithelium, the stem cells reside near the bottom of the crypts, in the lowest four cell positions [7,8]. Within this niche, Paneth cells in between the stem cells and the mesenchymal cells surrounding the crypt provide factors that stimulate stem cells to self-renew. However, when one or both daughter cells of a dividing stem cell are pushed upward by newly formed neighboring stem cells, these daughter cells leave the niche and enter an environment that stimulates them to differentiate. In adult hemopoiesis, niches for hemopoietic stem cells are present the perivascular or endosteal regions of bone marrow [9,10]. In these niche areas, osteoblastic and osteoclastic cells, macrophages, and endothelial and perivascular stromal cells are all thought to play a role in the stimulation of self-renewal of hemopoietic stem cells. In adipose tissue, the niche for adipose stem cells has been suggested to be in perivascular regions [11]. In the liver, cells called "oval cells" in rodents and hepatic progenitor cells in human, are believed to be stem cells and are localized in the canals of Hering [12] in which the stem cell niche environment stimulating stem cell renewal is established by endothelial cells, macrophages, and myofibroblasts [13]. In the skin, the stem cells are situated in the bulge region of the hair follicles [14]. The stemness of the bulge cells depends on their position within the bulge and the microenvironment supplied by nearby mesenchymal cells.

Tissue stem cells reside in specific areas. In these stem cell niches, various types of surrounding cells promote stem cell self-renewal. In this process, an intimate relationship between the tissue stem cell niche and the vasculature has been noted [15]. During steady-state kinetics, half of the daughter cells of the stem cells must leave the niche to prevent unbridled expansion of these cells within the niche. The cells leaving the niche differentiate into cells that carry out the function of the particular tissue.

In the next part of this review, we discuss the current knowledge about the SSC niche in adult seminiferous epithelium. No data are available on the SSC niche in primates. Therefore, we restrict this discussion to studies in nonprimate mammals, primarily rodents, and compare the SSC niche with stem cell niches in other tissues.

III. Spermatogonial stem cells

In the seminiferous epithelium, single type A spermatogonia, known as A-single (A_s) spermatogonia, are thought to be the SSCs [16–19]. When A_s spermatogonia divide and the division is self-renewing, the daughter cells migrate away from each other, producing two new A_s spermatogonia. When the SSCs go through a differentiating division, the daughter cells stay together and are connected by an intercellular bridge to form a pair of cells, known as A-paired or (A_{pr}) spermatogonia. Further divisions of the A_{pr} spermatogonia yield clones of 4, 8, and, ultimately, 16 A-aligned (A_{al}) spermatogonia. The A_{al} spermatogonia have the ability to differentiate into A1 spermatogonia, the first of six generations of what are known as differentiating spermatogonia. After five subsequent divisions (i.e., from A1 to A2, fromA3 to A4, and from In to B), the B spermatogonia divide into spermatocytes that, after carrying out the S phase, enter the meiotic prophase. Alternatively, it has been recently suggested that most of the A_s spermatogonia are on a differentiation pathway and that only few of the A_s spermatogonia are stem cells. To maintain sufficient numbers of stem cells, clones of A_{al} spermatogonia fragment to form new A_s spermatogonia [20,21]. This possibility is still a matter of debate because the supporting data can be interpreted in other ways [22,23] and some data contradict this idea [19,24,25].

On average, the SSCs divide two to three times per cycle of the seminiferous epithelium, as calculated from H^3-thymidine and cell-count studies [26–28]. It was suggested that some of the A_s spermatogonia have a very long cell cycle and form a separate population of long-cycling stem cells [29]. However, in later studies, the existence of such a population of cells could not be confirmed [30,31]. $A_{s\text{-}pr\text{-}al}$ spermatogonia have a minimal cell cycle time of 60 h in rats [32] and 90 h in Chinese hamsters [26], whereas the duration of the cycle of the seminiferous epithelium is 307 [33] and 408 h [34], respectively. This indicates that during the epithelial cycle, the $A_{s\text{-}pr\text{-}al}$ spermatogonia are not constantly proliferating. Indeed, in Chinese hamsters [35] and mice [28], $A_{s\text{-}pr\text{-}al}$ spermatogonia proliferate most actively during stages X-II of the cycle of the seminiferous epithelium and much less actively in stages III–IX.

In view of the above, a number of tasks can be assigned to the SSC niche. First, the SSC niche stimulates the proliferative activity of the SSCs in such a way that they undergo the required number of divisions. Second, the SSC niche regulates SSC proliferation with respect to the stages of the epithelial cycle. Third, the SSC niche supplies an environment that promotes self-renewal of the SSCs. Fourth, the SSC niche stimulates migration of the SSCs in such a way that about half of their daughter cells always move out of the niche to differentiate into cells that ultimately become spermatozoa. The current knowledge of how the mammalian SSC niche carries out these tasks in the adult seminiferous epithelium is described next.

IV. Location of the SSC

All spermatogonial cell types and the preleptotene spermatocytes reside on the basal lamina of the seminiferous tubules with the Sertoli cells that are the only somatic cells within the seminiferous epithelium. These germ cells are sequestered on the basal lamina by an elaborate system of tight junctions between Sertoli cells that divides the subsequent types of germ cells into those on the basal lamina and those in an adluminal compartment. Because of this Sertoli cell barrier, the SSC niche can only be a two-dimensional space that is one cell layer thick. Similarly in the intestine, the niche also is the one–cell-layer thick epithelium at the bottom of the crypts. Since the beginning of the 21st century, new data have indicated that the SSC niche occupies a particular part of the seminiferous tubule. Chiarini-Garcia and colleagues [1,2] reported that in mice and rats, $A_{s\text{-}pr\text{-}al}$ spermatogonia are not distributed randomly over the basal lamina of the seminiferous tubules but are preferentially localized in the tubule areas that border on patches of interstitial tissue. In tubule areas lying directly against another tubule, fewer $A_{s\text{-}pr\text{-}al}$ spermatogonia are present. The authors concluded that there is an $A_{s\text{-}pr\text{-}al}$ spermatogonial niche [2]. Subsequently, by fluorescently labeling these spermatogonia and studying the location of these cells in whole mounts of seminiferous tubules, Yoshida and colleagues further specified the localization of the $A_{s\text{-}pr\text{-}al}$ spermatogonia [3,36]. This study clearly showed that $A_{s\text{-}pr\text{-}al}$ spermatogonia can preferentially be found in the areas of the tubule basal lamina that are adjacent to interstitial venules and arterioles.

Thus, while differentiating (type A1 to B) spermatogonia are more widespread, $A_{s\text{-}pr\text{-}al}$ spermatogonia are confined to the areas of the circumference of the seminiferous tubules on that border the interstitial tissue and, more specifically, to a location close to blood vessels in the interstitial tissue. Does this mean that such an area that is rich in $A_{s\text{-}pr\text{-}al}$ spermatogonia constitutes an SSC niche? This does not seem likely when one considers that the $A_{s\text{-}pr\text{-}al}$ spermatogonial population contains both stem cells and cells destined to differentiate (i.e., A_{pr} and A_{al} spermatogonia).

By definition, a stem cell niche should protect stem cells from differentiation, and, therefore, the probability of self-renewal should be higher than 50%. When the niche contains a mixture of self-renewing and differentiating cells, the probability of self-renewal must be about 50%, and the cells that support the niche would need to provide a mixture of self-renewal and differentiation promoting factors that would allow a precise balance between renewal and differentiation. The supporting niche cells would need to reliably sense the presence of sufficient numbers of stem cells and differentiating cells.

As yet, there are no indications of the existence of such a mechanism in any tissue. Within the niches, stem cells are generally induced to self-renew, and induction of differentiation takes place in an adjacent area. For this reason, it seems more likely that SSCs and spermatogonia destined to differentiate are separated from each other. As proposed previously, the SSC niche may well be that part of the basal lamina directly opposite to the interstitial blood vessels, and all cells localized further away from the blood vessels are in the differentiation pathway [22]. SSCs in the niche probably divide and the daughter cells move away from each other. As in the intestine, in which daughter cells of stem cells that are pushed out of the niche differentiate [7,8], the SSC daughter cells that happen to migrate to a position outside of the niche area differentiate by dividing into a pair of A_{pr} spermatogonia at their next division, while the ones that stay in the niche remain to become stem cells (Figure 4.1).

A computer program has been developed that simulates the behavior of the SSCs and its descendants inside and outside the niche [22]. A virtual tubule was created with

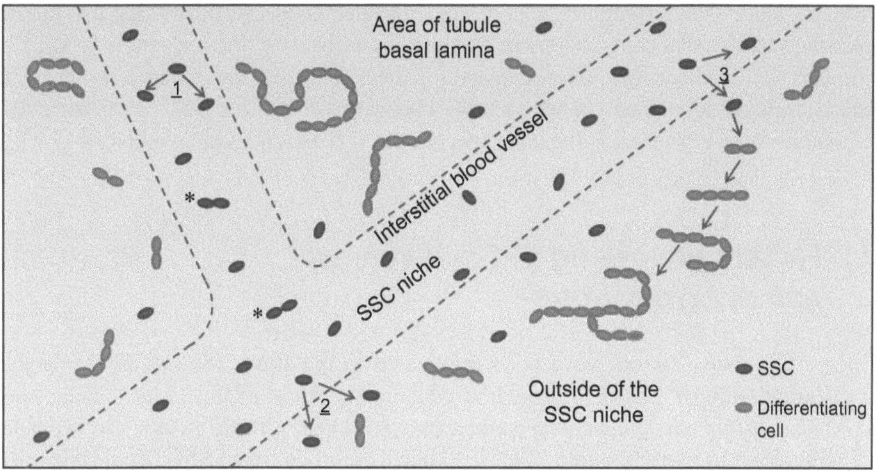

Figure 4.1 Schematic drawing of a flattened surface area of seminiferous tubule basal lamina on which clones of As-pr-al spermatogonia are drawn. The position of a branching interstitial blood vessel running underneath the basal lamina is indicated. It has been supposed that the area directly above the blood vessel constitutes the spermatogonial stem cell (SSC) niche. Single cells representing SSCs are indicated in blue. When SSCs divide there are three possibilities. One, both daughter cells find a place in the niche area (1) in which case the stem cell will have produced two new stem cells. Two, both daughter cells of the SSC migrate out of the niche because of which a stem cell is lost. Three, one of the daughter cells finds a spot in the niche and the other migrates out of the niche. Those SSC that migrate out of the niche will encounter an environment that strongly induces them to enter the differentiation pathway and at their next division form a pair of A_{pr} spermatogonia connected by an intercellular bridge. The A_{pr} spermatogonia will continue to divide and form longer chains of 4, 8, and 16 cells, as indicated by arrows. *SSC daughter cells may not migrate away from each other immediately but do so, much later and form false pairs [26].

a niche area resembling the pattern of branching blood vessels described by Yoshida and colleagues [3]. The probability of self-renewal for SSCs was set to be high within the niche area, while the probability of differentiation was high outside the niche area. Cell kinetic data, which had been previously established for $A_{s\text{-}pr\text{-}al}$ spermatogonia, were used and the authors assumed that after division, SSC daughter cells move in random directions while avoiding other clones of $A_{s\text{-}pr\text{-}al}$ spermatogonial clones localized in the vicinity of these cells. Avoidance of other clones has to be assumed because of the very even distribution of $A_{s\text{-}pr\text{-}al}$ spermatogonial clones over the tubule basal lamina, which has been observed in whole mounts of seminiferous tubules [37]. The computer program rendered a stable, long-term, steady state in which the numbers of $A_{s\text{-}pr\text{-}al}$ spermatogonia remained at the same level as the number counted in whole mounts of seminiferous tubules. Furthermore, the numbers of differentiating A1 spermatogonia produced during the program runs were similar to those *in vivo*. These data support the idea that such an SSC niche, in which SSCs and their descendants behave in accordance with the cell kinetic data gathered in the past, is viable. One of the conclusions from this study is that daughter cells of SSCs migrate after a self-renewing division to areas with few clones of $A_{s\text{-}pr\text{-}al}$ spermatogonia either because the migration of SSCs is inhibited by neighboring clones, preventing them from coming nearby, or because Sertoli cells in low-density areas secrete chemoattractants for SSC daughter cells. Evidence for the existence of the latter mechanism [38–40] is discussed below.

V. Factors governing SSC self-renewal and differentiation

Since 2000, many factors have been implicated in the stimulation of SSC renewal or differentiation of those SSC daughter cells that spill out of the niche.

Glial cell-line derived neurotrophic factor (*GDNF*): GDNF is required for SSC maintenance. In heterozygous GDNF knockout mice, the SSC pool becomes depleted [41]. In contrast, when GDNF is ectopically expressed in spermatogonia, large clusters of single type A spermatogonia appear and, ultimately, seminomatous tumors are formed [41,42]. Furthermore, in mice overexpressing GDNF in Sertoli cells, similar clusters of A spermatogonia and a dramatic increase in the numbers of SSCs per testis have been observed [43].

GDNF is produced by Sertoli cells [41,44] and its synthesis is dependent on follicle stimulating hormone (FSH) levels [44]. GDNF acts by binding to its receptors GFRA1 and c-RET on the plasma membrane of spermatogonia [41,45]. Interestingly, GDNF expression occurs in patches of the seminiferous epithelium, and GFRA1-positive spermatogonia are preferentially localized to these patches of GDNF production [46]. The levels of GDNF seem to vary between the stages of the epithelial cycle, but, unfortunately, conflicting data exist on the timing of the peak in GDNF levels during the cycle of the seminiferous epithelium. In mice, the highest levels of GDNF mRNA and protein have been found in pooled stages II–VI by Grasso and colleagues [24] and stages IX–IV by Caires and colleagues [47]. In rats, stages XII to III show the highest levels of GDNF mRNA and protein [48]; in hamsters, these levels peak at stages V–VII [46]. The cause of these differences may be species differences or, for mice, the combination of pooled stages. Interestingly, Johnston and colleagues [48] suggested that spermatogenic cell density determines GDNF levels, in the sense that these cells inhibit GDNF production by Sertoli cells.

A number of functions of GDNF in the regulation of spermatogonial multiplication, and especially stem cell behavior, have become apparent. First, GDNF promotes self-renewal of SSCs and/or inhibits their differentiation. In addition to the initial findings described above, this notion has been supported by multiple *in vivo* studies. Clearly, normal levels of GDNF and expression of its receptors GFRA1 and RET are required for normal spermatogenesis. Disruption of GDNF signaling in the testis quickly leads to a depletion of SSCs and the appearance of seminiferous tubules without germ cells [49–52]. This aspect of the function of GDNF is, at least partially, mediated by induction of the expression of NANOS2 in SSCs [53].

Second, results from *in vitro* studies indicate that GDNF promotes the proliferation of SSCs and of $A_{pr\text{-}al}$ spermatogonia [54–57]. It is interesting that GDNF levels in rats are highest in stages XII to III [48], which is precisely when $A_{s\text{-}pr\text{-}al}$ spermatogonia proliferate most actively [35]. It is important to note that the stimulatory effect of GDNF on $A_{s\text{-}pr\text{-}al}$ spermatogonial proliferation is not necessarily related to its effect on SSC renewal. There is no evidence that the decision of an SSC to enter the active cell cycle is related to the decision to form an interconnected pair or two single cells. Active proliferation does not necessarily promote self-renewal or differentiation. Indeed, GDNF stimulates the proliferation of $A_{pr\text{-}al}$ spermatogonia as well.

Third, GDNF has an important role in the regulation of the migration of various cells types, such as neuroblasts, Schwann cells, and kidney epithelial cells, and it can also act as a chemoattractant [58–60]. Interestingly, GDNF induces migration of cultured SSCs by acting as a chemoattractant [38–40]. These are important results because cell migration is needed to manage the numbers and the localization of SSCs in their stem cell niche. The SSCs are single cells and after a self-renewing division, the daughter cells must migrate away from each other. In addition, as discussed in more detail below, half of the daughter cells of the SSCs must migrate out of the niche area to produce differentiating cells.

Fibroblast growth factor 2 (FGF2): FGF2 (also known as basic fibroblast growth factor) is also involved in the regulation of SSC behavior. FGF2 is produced by Sertoli, peritubular, Leydig, and germ cells [61,62]. In Sertoli cells, as with GDNF, FGF2 production is dependent on FSH levels [61]. FGF2 has been shown to be an important factor to successfully culture SSCs [55,63,64]. In culture, an increase in the numbers of SSCs, as measured by the SSC transplantation assay [65,66], is observed when FG2 is added [55]. The numbers of colony forming cells also increase in spermatogonial cultures [64]. FGF2 induces MAP2K1 phosphorylation that activates the MAP2K1 pathway, which, in turn, upregulates expression of *Etv5* and *Bcl6b*, both of which play crucial roles in SSC self-renewal [67–69]. In humans, the potential of FGF2 to promote SSC self-renewal has become evident by indications that an activating mutation in FGFR2 in SSCs gives them a proliferative advantage over SSCs lacking this mutation [70].

Chemokine (C-X-C motif) ligand 12 (CXCL12): CXCL12 plays an important role in various stem cell systems, such as hemopoietic stem cells [71], neural stem cells [72], and endothelial precursor cells [73]. In the testis, CXCL12 is produced by Sertoli cells [74,75]. Payne and colleagues [75] generated mice lacking the nuclear corepressor Sin3a in Sertoli cells starting in the fetal gonad. In these mice, Sertoli cells do not produce CXCL12 and a diminished number of germ cells express its receptor. In these mice, the seminiferous epithelium is quickly depleted, suggesting stem cell loss. Kanatsu-Shinohara and colleagues showed that both GDNF and CXCL12 act as essential SSC chemotactic factors *in vitro* and *in vivo* [39]. Cultured early type spermatogonia (germ line stem cells – GS cells), including SSCs, which are deficient in the CXCL12 receptor CXCR4, are less able to colonize a recipient mouse testis in the transplantation assay.

WNT5A: WNT5A is produced by Sertoli cells, and SSCs express receptors for this factor. In culture, WNT5A promotes the maintenance of SSCs though a β-catenin-independent mechanism. WNT5A does not affect SSC proliferation but, more likely, affects the survival of these cells. Fractions of germ cells purified for the presence of WNT5A receptors FZD3 and THY1 contain most of the SSCs as determined by the transplantation assay [76]. The specific role of WNT5A *in vivo* is not yet clear.

CSF1: Spermatogonial populations that are enriched for GFRA1+ and THY1+ cells are also highly enriched for the expression of the CSF1 receptor (CSF1R). CSF1 stimulates the proliferation of freshly isolated type A spermatogonia from prepubertal mice [77]. Oatley and colleagues [78], after isolating Thy1+ cells from testes of 6-day-old mice, concluded that the CSF1 receptor might only be expressed in SSCs and that CSF1 has a special function in regulating the self-renewal of SSCs. In addition, these authors showed that CSF1 is produced by Leydig cells and some peritubular myoid cells. Notably, CSF1 is a special factor in this list because it is not produced by Sertoli cells.

Vascular endothelial growth factor A (VEGFA): VEGFA family members are produced by germ, Sertoli, and interstitial cells, and the corresponding receptors are also expressed by these cells [79–81]. Interestingly, the pro-angiogenic isoform $VEGFA_{164}$ promotes SSC self-renewal, while the antiangiogenic isoform $VEGFA_{165}b$ promotes differentiation of these cells. Administration of these isoforms to mouse pups affects the numbers of SSCs in these animals, as determined by the SSC transplantation assay [81]. In addition, mice in which all isoforms of VEGFA are knocked out in germ and Sertoli cells show decreased numbers of $A_{s\text{-}pr\text{-}al}$ spermatogonia and sperm and become subfertile [79].

Neuregulin 1 (NRG1): The epidermal growth factor family also contributes to the SSC environment. The addition of epidermal growth factor (EGF) to a culture of mouse SSCs does not significantly increase stem cell numbers or the number of A_{al} spermatogonia produced in culture [55,63,82]. However, ERBB2 and ERBB3 receptors are expressed by germ cells. In cultures in the presence of GDNF, the EGF family member NRG1 is able to greatly enhance the formation of A_{al} spermatogonia while decreasing SSC numbers [82]. In the testis, NRG1 is expressed by peritubular cells [83], which suggests a role for peritubular cells and its secreted NRG1 in the differentiation of SSCs.

Bone morphogenetic protein 4 (BMP4): BMP4 induces differentiation of cultured $A_{s\text{-}pr\text{-}al}$ spermatogonia and SSCs [84,85], In a rat stem cell line, BMP4 induces the expression of c-KIT, which is involved in spermatogonial differentiation [86]. It is still a matter of debate whether BMP4 is produced in the seminiferous epithelium by spermatogonia and spermatocytes and/or Sertoli cells [84,87]. When BMP4 is primarily produced by spermatogonia, it is difficult to understand how this factor could play a decisive role in the decision of A_s spermatogonia whether to self-renew or to form a pair of A_{pr} spermatogonia, unless BMP4 is simply a downstream link in a pathway governed by another factor. It is interesting that BMP4 expression is regulated by vitamin A levels [87] and that levels of the vitamin A precursor retinoic acid are known to fluctuate during the epithelial cycle [88]. More data will be needed on this subject to obtain a clear picture of the role of BMP4 in spermatogonial differentiation.

Inhibin beta A (*INHBA*): INHBA (activin A) is produced by Sertoli cells and, possibly, some germ cells. INHBA is present during the entire cycle of the seminiferous epithelium but is at peak levels in epithelial stage VIII [89]. This member of the transforming growth factor β family induces differentiation of SSCs in culture [85]. However, INHBA is not absolutely required for SSC differentiation because mice that are deficient for its receptor (ACTRIIA) do show differentiating germ cells [90].

NOTCH: For stem cells of several tissues (e.g., intestine [7], liver [91]) and neuronal stem cells, proper NOTCH signaling is essential [92,93]. In the testis, spermatogonia have NOTCH receptors, and Sertoli cells produce NOTCH ligands [94]. Notch signaling is essential during fetal and neonatal spermatogenesis [95,96]. However, in adult mice, Notch signaling in Sertoli cells is redundant at best because inactivation of NOTCH signaling in these cells does not affect spermatogenesis [97]. Nevertheless, activation of NOTCH1 signaling in spermatogonia leads to a decrease in sperm production, and Huang and colleagues suggest that in this case, SSC differentiation is inhibited because of the observed decreased expression of neurogenin 3 (NGN3) [98]. NGN3 has been implicated in SCC differentiation [99]. Additional studies will be needed to confirm this hypothesis.

Stem cell factor (*SCF*): The c-KIT receptor and its ligand SCF produced by Sertoli cells play an important role in spermatogenesis. Many mutations in both *c-kit* and *scf* are known to cause defects in the spermatogenic process. These defects range from a total lack of germ cells due to effects in primordial germ cells to almost normal spermatogenesis [100,101]. Without a functioning SCF/c-KIT pathway, $A_{s\text{-}pr\text{-}al}$ spermatogonia can proliferate and form clones of A_{al} spermatogonia, but they cannot differentiate into A1 spermatogonia [102,103]. The c-Kit/SCF pathway does not seem to play a role in the regulation of SSC behavior.

Reactive oxygen species (*ROS*): ROS have been shown to play a role in the regulation of hemopoietic and neural stem cells [104–107]. Interestingly, in the testis, ROS may also have an important role. Depletion of ROS in culture stopped SSC proliferation, while the addition of ROS enhanced self-renewal [108]. Furthermore, in mice deficient in NADPH oxidase1, which catalyzes the formation of ROS, SSCs show a reduced self-renewal capacity upon serial transplantation. The amount of ROS necessary for normal spermatogenesis is, at least in part, supplied via GDNF and FGF2 because both these factors increase the formation of ROS by SSCs [108]. In addition, ROS generated by peritubular cells and interstitial cells may also contribute to SSC maintenance.

Extracellular matrix (*ECM*): The ECM is thought to be of crucial importance in the regulation of the fate of stem cells in several tissues [109]. With respect to spermatogenesis, most of the research has focused on processes other than SSC behavior, such as the passage of spermatocytes through the Sertoli cell barrier [110]. It is also possible to enrich testis cell populations for SSCs by purifying these populations for cells expressing α6-integrin or β1-integrin (ITGB1) [111]. Importantly, when ITGB1 is knocked out in Sertoli cells, transplanted SSCs have difficulty finding their way to the basal lamina [112,113]. It is not known yet whether the expression of ITGB1 by Sertoli cells is specific to the SSC niche. ITGB1 may be expressed by all Sertoli cells, including those outside the niche. One finding suggesting a relation between the ECM and SSC behavior is that in a rat SSC line, differentiation is accompanied by changes in the expression of adhesion molecules by SSCs [86]. Nevertheless, there are as yet no data showing differential expression of ECM proteins inside and outside the SSC niche, and a possible role for these proteins in SSC fate decisions has yet to be established.

VI. The environment inside and outside the niche

As described above, the factors promoting SSC self-renewal are GDNF, FGF2, CXCL2, CSF1, $VEGFA_{164}$, and ROS. In contrast, BMP4, INHBA, and $VEGFA_{165}b$ are involved in SSC differentiation (Figure 4.2 and Table 4.1). All of these factors are produced by Sertoli cells, except for CSF1, which is produced by peritubular cells and Leydig cells, and ROS, which are induced in SSCs by GDNF and FGF2 and can also be brought into the niche environment by peritubular and interstitial cells. Clearly, a large number of factors are involved in providing the optimal environment for SSC renewal inside and differentiation outside the niche, and there is no reason to assume that no additional factors will be discovered.

Will the multitude of factors governing SSC self-renewal require extensive regulatory fine-tuning to ensure proper SSC maintenance? When the tubule basal lamina can be subdivided into a SSC niche area with high chances of self-renewal and the rest of the basal lamina where SSCs differentiate, the answer is "probably not".

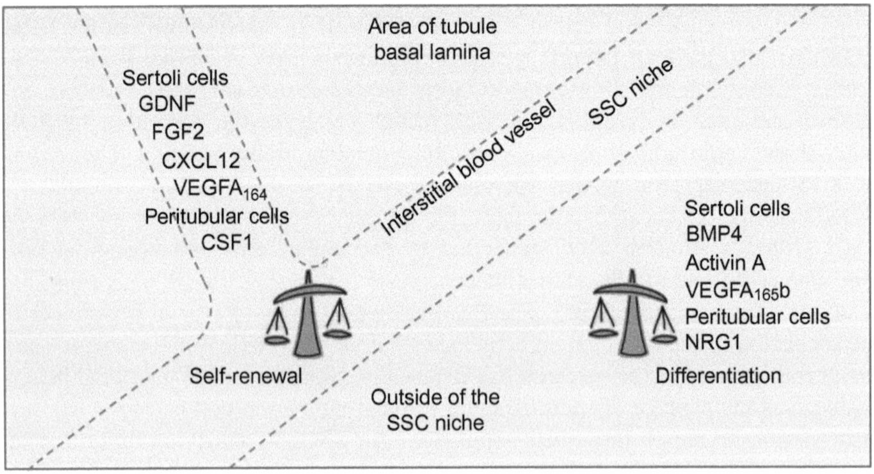

Figure 4.2 This drawing depicts the balance between the factors, known so far, that clearly play a role in establishing an environment that stimulates self-renewal in the niche area and differentiation outside of the niche.

Table 4.1 Overview of the factors that regulate SSC self-renewal and differentiation

Factor	Self-renewal or differentiation	Production site	References
GDNF	Self-renewal	Sertoli cells	24,41–60
FGF2	Self-renewal	Sertoli cells	55,61–70
CXCL12	Self-renewal	Sertoli cells	39,69,70
CSF1	Self-renewal	Peritubular and Leydig cells	74,75
$VEGFA_{164}$	Self-renewal	Sertoli cells, germ cells, interstitial cells	79–81
$VEGFA_{165}b$	Differentiation	Sertoli cells, germ cells, interstitial cells	79–81
NRG1	Differentiation	Peritubular cells	82,83
BMP4	Differentiation	Sertoli cells, spermatogonia, spermatocytes	84,85,87
INHBA	Differentiation	Sertoli cells	85,89,90
ROS	Self-renewal	Induced in SSCs by GDNF and FGF2, Other cells in the niche	108

As long as the stimulation of self-renewal of SSCs in their niche is very strong and the niche is sufficiently large, the SSC niche functions properly in safeguarding sufficient numbers of SSCs. Outside the niche, these cells generally produce a pair of A_{pr} spermatogonia that proceed to form larger clones of 4, 8, and 16 cells that further differentiate into A_{al} spermatogonia. In such a system, the contribution of each factor involved is likely not essential as long as the total result is induction of self-renewal in the niche and of differentiation outside of the niche.

VII. The role of cell migration in SSC self-renewal and differentiation

As described above, FSH stimulates the secretion of GDNF [44], and, in turn, GDNF stimulates the expression of CXCR4 on GS cells [39]. Because both GDNF and CXCL12 are involved in migration, migration of cells is greatly stimulated in the tubule areas bordering on the interstitial tissue. What might be the reason for promoting the migration of cells within and just outside the niche? A possible answer is that a tissue area that constantly produces cells for self-renewal and differentiation needs a substantial amount of cell migration for cells to move to places within and outside the niche. Unlike in the intestinal crypts, the cells in the seminiferous tubules destined to differentiate are not physically pushed away to a differentiation area by newly formed adjacent cells. Therefore, promotion of their migration may be necessary to enable some of the SSCs to move out of the niche.

In addition to promoting the motility of cells, GDNF has been shown to act as a chemoattractant for SSCs [38–40]. This might seem odd at first because, in the steady state, a niche area must always get rid of half of the cells that are produced because these cells are needed to replenish cells that are lost to differentiation outside the niche area. It does not make sense then to attract SSCs because the niche area would continuously expand or become overcrowded, and too few differentiating cells would be formed. A combination of the recent finding that germ cells inhibit GDNF production by Sertoli cells and the finding that clones of $A_{s\text{-}pr\text{-}al}$ spermatogonia are very evenly distributed over the tubule basal lamina [37] may provide the answer. As discussed by de Rooij and colleagues [22], the even distribution of clones of $A_{s\text{-}pr\text{-}al}$ spermatogonia can either be achieved by inhibition of the migration of daughter cells of SSCs by neighboring clones, forcing them to migrate to areas with few other clones, or by attraction of SSCs to areas harboring only a few clones [22]. When GDNF is preferentially produced by Sertoli cells that have only a few $A_{s\text{-}pr\text{-}al}$ spermatogonia in their immediate surroundings, as suggested by Johnston and colleagues [48], it is conceivable that these Sertoli cells will attract SSC daughter cells "looking" for a place to migrate to by enhanced production of GDNF. The area marked by these Sertoli cells can be situated inside or outside the SSC niche, and, in a steady state, half of the SSC daughter cells migrate out of the niche and the other half stay in the niche.

VIII. Spermatogonial differentiation

Spermatogonial differentiation occurs in two steps. First, A_s spermatogonia that move out of the niche start expressing NGN3 and, at their next division, have a substantial probability of producing a pair of A_{pr} spermatogonia that will continue to divide into clones of 4, 8, and 16 A_{al} spermatogonia. As discussed elsewhere [23], some A_{pr} split up and become SSCs again, but it is unclear whether clones of A_{al} spermatogonia can do this in the normal epithelium and subsequently regain SSC capacity because A_{al} spermatogonia are mainly, if not exclusively, formed outside the niche in an environment that promotes differentiation. Second, in stage VI/VII, most clones of A_{al} spermatogonia differentiate into A1 spermatogonia. This is a clear step of differentiation that requires retinoic acid and c-KIT expression in the differentiating spermatogonia [88,102,114]. After this step, the spermatogonia drastically change their pattern of proliferative activity as additional divisions and development of these cells take place in specific stages of the seminiferous epithelial cycle [115]. Moreover, these differentiating spermatogonia undergo visible changes in their chromatin structure [116,117] and start to express the structural maintenance of chromosomes (SMC) protein 6 [118].

The two growth factors that have been shown to induce SSC differentiation are BMP4 and INHBA. SCF is the Sertoli-cell-produced ligand for the c-KIT receptor that is known to be essential for the transition from A_{al} to A1 spermatogonia, which is the second step in the differentiation of spermatogonia.

IX. Niche localization: what controls the controller?

The important question that arises next is, what determines the localization of the niche? Another question is why is one part of the tubule basal lamina subject to high levels of self-renewal factors while differentiation promoting factors have the upper hand in another part of the tubule basal lamina? One obvious reason may be that FSH levels are higher in the interstitial tissue and consequently also at the basal lamina of the parts of the tubule that border the interstitial tissue. FSH stimulates Sertoli cells to produce GDNF and FGF2, which both promote self-renewal. This reasoning can address the presence of the niche in areas bordering the interstitial tissue. However, Yoshida and colleagues also showed that the niche is situated in places where arterioles and venules are close to the seminiferous tubules and not the entire tubule area that borders the interstitial tissue [3]. One would expect FSH levels to be highest close to capillaries and not near the somewhat larger arterioles and venules. Can the localization of the niche be related to the clumps of Leydig cells that can often be seen near the blood vessels in the interstitial tissue? In this context, a study of collared peccaries (*Tayassu tajacu*) is very important because it shows that Leydig cells probably do not play this role [119]. In these peccaries, Leydig cells and seminiferous tubules are not intermixed as in rodents. Large groups of Leydig cells surround lobes with seminiferous tubules. Importantly, GFRA1-positive spermatogonia (i.e., A_s, A_{pr}, and a few A_{al} spermatogonia) are preferentially located in the parts of the seminiferous tubules that do not border interstitial tissue regions filled with Leydig cells. The authors suggest that Leydig cells may, instead, play a role in differentiation.

As discussed above, in stem cell niches in other tissues, there often appears to be a connection between the niche and the vasculature [9,11,13,120]. In the testis, blood-borne factors and/or factors produced by perivascular cells might determine the place of the SSC niche. Clearly, this must be studied in further detail.

X. The SSC niche and the cycle of the seminiferous epithelium

As described above, it has become abundantly clear that factors secreted by Sertoli cells play a major role in the regulation of SSC behavior. Furthermore, Sertoli cells themselves are also subject to regulatory factors. First, there are factors that determine the location of the niche and whether Sertoli cells produce factors that induce self-renewal or differentiation. Second, Sertoli cell behavior is strongly dependent on the stages of the cycle of the seminiferous epithelium. A major cause of this stage dependency is epithelial-stage-dependent expression of the FSH receptor in Sertoli cells [121,122]. A cyclical stimulation of Sertoli cells by FSH has a major impact on the environment inside and outside the SSC niche. As discussed above, GDNF and FGF2 secretion by Sertoli cells depends on FSH levels [44,61], and FSH affects the production of inhibin A, which has a role in the regulation of the proliferation of $A_{s\text{-}pr\text{-}al}$ spermatogonia [123], and of INHBA [89]. Therefore, the environment inside and outside the SSC niche is not a static one, but levels of factors that promote self-renewal and differentiation as well as proliferation-stimulating factors fluctuate with the stages of the cycle of the seminiferous epithelium. How this impacts SSC behavior is not clear. Interestingly, Ventela and colleagues found that the $A_{s\text{-}pr\text{-}al}$ spermatogonial population in stages IX–XII contains the most cells that are capable of forming a colony in culture [124]. Clearly, an SSC niche that is dependent on the stages of the epithelial cycle adds a new level of complexity to the understanding of the mechanisms involved in regulating SSC maintenance and the production of differentiating germ cells.

XI. The SSC niche during cell loss

During (stem) cell loss, for example because of irradiation or administration of the alkylating agent busulphan, the seminiferous epithelium has been shown to have a substantial ability to recover [125–127]. How can this be achieved? Among spermatogonial cell types, SSCs are the most radioresistant [128,129]; therefore, the cells that survive a dose of radiation are likely to be SCCs. In addition, the surviving SSCs are primarily localized in niche areas because most A_s spermatogonia are localized in these areas. Interestingly, after irradiation during the first five divisions of surviving SSCs, the average probability of self-renewal is much higher than that of differentiation [130].

How is this shift accomplished from 50% differentiation and 50% self-renewal of stem cell daughter cells to virtually only self-renewal? To answer this question, it is important to realize what actually happens in the testis when germ cell numbers decline. Spermatogonia that are lethally damaged by the DNA damage inflicted by radiation enter apoptosis when trying to undergo mitosis. As a result, most of these cells disappear within 1 or 2 days and a progressive depletion of the seminiferous epithelium and a concomitant shrinkage of the tubules start to occur. In contrast, there are no actively proliferating somatic cells in the adult interstitial tissue. Therefore, no early response of this tissue to irradiation can be expected, and no such response has ever been reported. The result is an overall decrease of tubule volume and no change in the interstitial tissue volume (Figure 4.3). It follows that as germ cell depletion occurs and the tubules shrink, more and more of the tubule circumference borders the interstitial tissue. SSCs that were just outside the niche at the time of irradiation move into the niche because of the tubule shrinkage, and those that were in the niche move to a more central position within the niche. Clearly, this process leads to enhanced self-renewal of SSCs after irradiation.

After irradiation or busulphan administration, FSH levels increase [44,131–133]. This increases suggests that, as has been reported for busulphan [134], after cell loss, GDNF and FGF2 levels increase and a larger area of the tubule basal lamina is exposed to high enough levels of these growth factors to result in the promotion of self-renewing divisions of SSCs.

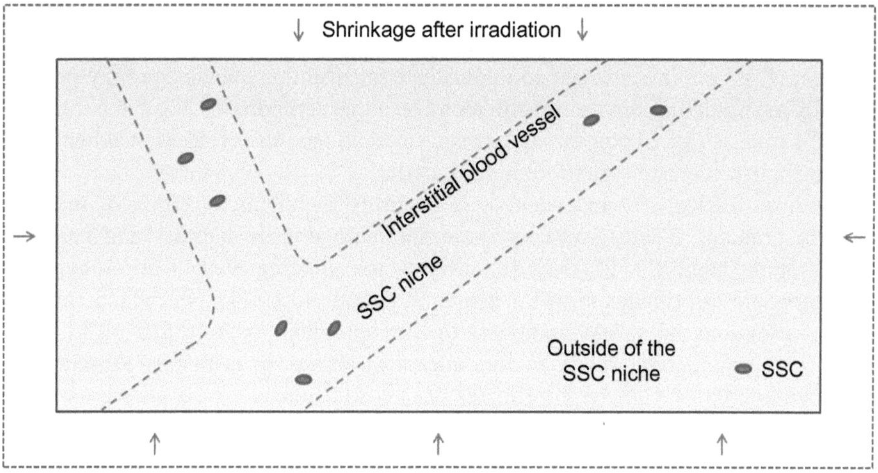

Figure 4.3 Seminiferous tubules shrink in case of SSC loss, for example after irradiation, while no immediate changes in the interstitial tissue can be expected. Because of this the percentage of tubule basal lamina constituting the niche area becomes larger. SSCs outside of the niche at the time of irradiation may get back into the niche area as the tubules shrink and SSCs that were already in the niche will get into a more central position. Both phenomena will promote SSC renewal after cell loss.

Thus, when SSC numbers become diminished and depletion of the epithelium starts, the SSC niche grows and the secretion of SSC self-renewal promoting factors is stimulated. Interestingly, in the transplantation assay of SSCs, the recipient seminiferous epithelium is depleted first to facilitate the homing of the transplanted SSCs toward the basal lamina. Another good reason for depleting the recipient epithelium first is therefore to stimulate the formation of repopulating colonies by the transplanted SSCs through an increase in the probability that these cells will home in to a niche area and that their daughter cells will thereby self-renew. In addition, increased levels of FSH are present in the recipient mice.

XII. Perspectives

The picture that emerges from this review is that the seminiferous epithelium has a stem cell niche that shares many characteristics with stem cell niches of other mammalian tissues. The stem cells are confined to a specific part of the tissue, and the cells surrounding the stem cells secrete growth factors that regulate the proliferative activity and the probability of self-renewal during stem cell proliferation. Daughter cells of stem cells that migrate out of the niche enter an environment that strongly promotes their differentiation and the formation of A_{pr} spermatogonia at their next division.

There are at least two important gaps in our understanding of the SSC niche. First, we do not know what factors determine the location of the niche near the arterioles and venules in the interstitial tissue. Second, the role of the epithelial cycle in SSC behavior has yet to be studied with respect to self-renewal and differentiation. This role is important to understand because this knowledge may provide tools for manipulating the size of the niche to either temporarily boost the numbers of SSCs in cases of hypospermatogenesis or to induce differentiation when there are insufficient numbers of differentiating cells.

The most difficult future endeavor is probably the characterization of the SSC niche in primates. There are many more spermatogonia in humans and monkeys than in other mammals [28,135]. In addition, the function of the various types of spermatogonia in primates is still a matter of debate. It is unknown which cells act as stem cells, and the massive numbers of spermatogonia make it difficult to imagine how spermatogonia relate to the interstitial tissue in primates. Research on understanding the SSC niche in primates has yet to begin.

Acknowledgments

The author is grateful to Drs. Elena Vicini (Rome, Italy) and Geert Hamer (Amsterdam, Netherlands) for critical reading of the manuscript.

References

[1] Chiarini-Garcia H, Hornick JR, Griswold MD, Russell LD. Distribution of type A spermatogonia in the mouse is not random. Biol Reprod 2001;65:1179–85.

[2] Chiarini-Garcia H, Raymer AM, Russell LD. Non-random distribution of spermatogonia in rats: evidence of niches in the seminiferous tubules. Reproduction 2003;126: 669–80.

[3] Yoshida S, Sukeno M, Nabeshima YI. A vasculature-associated niche for undifferentiated spermatogonia in the mouse testis. Science 2007;317:1722–6.

[4] Oatley JM, Brinster RL. The germline stem cell niche unit in mammalian testes. Physiol Rev 2012;92:577–95.

[5] Kanatsu-Shinohara M, Shinohara T. Spermatogonial stem cell self-renewal and development. Annu Rev Cell Dev Biol 2013;29:163–87.

[6] Schofield R. The relationship between the spleen colony-forming cell and the haemopoietic stem cell. Blood Cells 1978;4:7–25.

[7] Stange DE, Clevers H. Concise review: the Yin and Yang of intestinal (cancer) stem cells and their progenitors. Stem Cells 2013;31:2287–95.

[8] Snippert HJ, van der Flier LG, Sato T, van Es JH, van den Born M, Kroon-Veenboer C, et al. Intestinal crypt homeostasis results from neutral competition between symmetrically dividing Lgr5 stem cells. Cell 2010;143:134–44.

[9] Ugarte F, Forsberg EC. Haematopoietic stem cell niches: new insights inspire new questions. EMBO J 2013;32:2535–47.

[10] Smith JN, Calvi LM. Concise review: current concepts in bone marrow microenvironmental regulation of hematopoietic stem and progenitor cells. Stem Cells 2013;31: 1044–50.

[11] Berry DC, Stenesen D, Zeve D, Graff JM. The developmental origins of adipose tissue. Development 2013;140:3939–49.
[12] Kordes C, Haussinger D. Hepatic stem cell niches. J Clin Invest 2013;123:1874–80.
[13] Lorenzini S, Bird TG, Boulter L, Bellamy C, Samuel K, Aucott R, et al. Characterisation of a stereotypical cellular and extracellular adult liver progenitor cell niche in rodents and diseased human liver. Gut 2010;59:645–54.
[14] Goldstein J, Horsley V. Home sweet home: skin stem cell niches. Cell Mol Life Sci 2012;69:2573–82.
[15] Gomez-Gaviro MV, Lovell-Badge R, Fernandez-Aviles F, Lara-Pezzi E. The vascular stem cell niche. J Cardiovasc Transl Res 2012;5:618–30.
[16] Huckins C. The spermatogonial stem cell population in adult rats. I. Their morphology, proliferation and maturation. Anat Rec 1971;169:533–57.
[17] Oakberg EF. Spermatogonial stem-cell renewal in the mouse. Anat Rec 1971;169: 515–31.
[18] de Rooij DG. Spermatogonial stem cell renewal in the mouse. I. Normal situation. Cell Tissue Kinet 1973;6:281–7.
[19] Oatley MJ, Kaucher AV, Racicot KE, Oatley JM. Inhibitor of DNA binding 4 is expressed selectively by single spermatogonia in the male germline and regulates the self-renewal of spermatogonial stem cells in mice. Biol Reprod 2011;85:347–56.
[20] Nakagawa T, Nabeshima YI, Yoshida S. Functional identification of the actual and potential stem cell compartments in mouse spermatogenesis. Dev Cell 2007;12: 195–206.
[21] Nakagawa T, Sharma M, Nabeshima Y, Braun RE, Yoshida S. Functional hierarchy and reversibility within the murine spermatogenic stem cell compartment. Science 2010;328:62–7.
[22] de Rooij DG, van Beek ME. Computer simulation of the rodent spermatogonial stem cell niche. Biol Reprod 2013;88:131.
[23] de Rooij DG, Griswold MD. Questions about spermatogonia posed and answered since 2000. J Androl 2012;33:1085–95.
[24] Grasso M, Fuso A, Dovere L, de Rooij DG, Stefanini M, Boitani C, et al. Distribution of GFRA1-expressing spermatogonia in adult mouse testis. Reproduction 2012;143:325–32.
[25] Chakraborty P, Buaas FW, Sharma M, Snyder E, de Rooij DG, Braun RE. LIN28A marks the spermatogonial progenitor population and regulates its cyclic expansion. Stem Cells 2014.
[26] Lok D, Jansen MT, de Rooij DG. Spermatogonial multiplication in the Chinese hamster. II. Cell cycle properties of undifferentiated spermatogonia. Cell Tissue Kinet 1983;16:19–29.
[27] Lok D, Weenk D, de Rooij DG. Morphology, proliferation, and differentiation of undifferentiated spermatogonia in the Chinese hamster and the ram. Anat Rec 1982;203: 83–99.
[28] Tegelenbosch RA, de Rooij DG. A quantitative study of spermatogonial multiplication and stem cell renewal in the C3H/101 F1 hybrid mouse. Mutat Res 1993;290:193–200.
[29] Huckins C. The spermatogonial stem cell population in adult rats. 3. Evidence for a long-cycling population. Cell Tissue Kinet 1971;4:335–49.
[30] Lok D, Jansen MT, de Rooij DG. Spermatogonial multiplication in the Chinese hamster. IV. Search for long cycling stem cells. Cell Tissue Kinet 1984;17:135–43.
[31] Grisanti L, Falciatori I, Grasso M, Dovere L, Fera S, Muciaccia B, et al. Identification of spermatogonial stem cell subsets by morphological analysis and prospective isolation. Stem Cells 2009;27:3043–52.

[32] Huckins C. The spermatogonial stem cell population in adult rats. II. A radioautographic analysis of their cell cycle properties. Cell Tissue Kinet 1971;4:313–34.
[33] Huckins C. Cell cycle properties of differentiating spermatogonia in adult Sprague-Dawley rats. Cell Tissue Kinet 1971;4:139–54.
[34] Oud JL, de Rooij DG. Spermatogenesis in the Chinese hamster. Anat Rec 1977;187:113–24.
[35] Lok D, de Rooij DG. Spermatogonial multiplication in the Chinese hamster. III. Labelling indices of undifferentiated spermatogonia throughout the cycle of the seminiferous epithelium. Cell Tissue Kinet 1983;16:31–40.
[36] Shetty G, Meistrich ML. The missing niche for spermatogonial stem cells: do blood vessels point the way? Cell Stem Cell 2007;1:361–3.
[37] de Rooij DG, Janssen JM. Regulation of the density of spermatogonia in the seminiferous epithelium of the Chinese hamster: I. Undifferentiated spermatogonia. Anat Rec 1987;217:124–30.
[38] Dovere L, Fera S, Grasso M, Lamberti D, Gargioli C, Muciaccia B, et al. The niche-derived glial cell line-derived neurotrophic factor (GDNF) induces migration of mouse spermatogonial stem/progenitor cells. PLoS One 2013;8:e59431.
[39] Kanatsu-Shinohara M, Inoue K, Takashima S, Takehashi M, Ogonuki N, Morimoto H, et al. Reconstitution of mouse spermatogonial stem cell niches in culture. Cell Stem Cell 2012;11:567–78.
[40] Huleihel M, Fadlon E, Abuelhija A, Piltcher Haber E, Lunenfeld E. Glial cell line-derived neurotrophic factor (GDNF) induced migration of spermatogonial cells *in vitro* via MEK and NF-kB pathways. Differentiation 2013;86:38–47.
[41] Meng X, Lindahl M, Hyvonen ME, Parvinen M, de Rooij DG, Hess MW, et al. Regulation of cell fate decision of undifferentiated spermatogonia by GDNF. Science 2000;287:1489–93.
[42] Meng X, de Rooij DG, Westerdahl K, Saarma M, Sariola H. Promotion of seminomatous tumors by targeted overexpression of glial cell line-derived neurotrophic factor in mouse testis. Cancer Res 2001;61:3267–71.
[43] Yomogida K, Yagura Y, Tadokoro Y, Nishimune Y. Dramatic expansion of germinal stem cells by ectopically expressed human glial cell line-derived neurotrophic factor in mouse sertoli cells. Biol Reprod 2003;69:1303–7.
[44] Tadokoro Y, Yomogida K, Ohta H, Tohda A, Nishimune Y. Homeostatic regulation of germinal stem cell proliferation by the GDNF/FSH pathway. Mech Dev 2002;113:29–39.
[45] Airaksinen MS, Saarma M. The GDNF family: signalling, biological functions and therapeutic value. Nat Rev Neurosci 2002;3:383–94.
[46] Sato T, Aiyama Y, Ishii-Inagaki M, Hara K, Tsunekawa N, Harikae K, et al. Cyclical and patch-like GDNF distribution along the basal surface of Sertoli cells in mouse and hamster testes. Plos One 2011;6:e28367.
[47] Caires KC, de Avila J, McLean DJ. Endocrine regulation of spermatogonial stem cells in the seminiferous epithelium of adult mice. Biores Open Access 2012;1:222–30.
[48] Johnston DS, Olivas E, DiCandeloro P, Wright WW. Stage-specific changes in GDNF expression by rat Sertoli cells: a possible regulator of the replication and differentiation of stem spermatogonia. Biol Reprod 2011;85:763–9.
[49] He ZP, Jiang JJ, Hofmann MC, Dym M. Gfra1 silencing in mouse spermatogonial stem cells results in their differentiation via the inactivation of RET tyrosine kinase. Biol Reprod 2007;77:723–33.

[50] Naughton CK, Jain S, Strickland AM, Gupta A, Milbrandt J. Glial cell-line derived neurotrophic factor-mediated RET signaling regulates spermatogonial stem cell fate. Biol Reprod 2006;74:314–21.

[51] Savitt J, Singh D, Zhang C, Chen LC, Folmer J, Shokat KM, et al. The *in vivo* response of stem and other undifferentiated spermatogonia to the reversible inhibition of glial cell line-derived neurotrophic factor signaling in the adult. Stem Cells 2012;30: 732–40.

[52] Jijiwa M, Kawai K, Fukihara J, Nakamura A, Hasegawa M, Suzuki C, et al. GDNF-mediated signaling via RET tyrosine 1062 is essential for maintenance of spermatogonial stem cells. Genes Cells 2008;13:365–74.

[53] Sada A, Suzuki A, Suzuki H, Saga Y. The RNA-binding protein NANOS2 is required to maintain murine spermatogonial stem cells. Science 2009;325:1394–8.

[54] Ebata KT, Yeh JR, Zhang XF, Nagano MC. Soluble growth factors stimulate spermatogonial stem cell divisions that maintain a stem cell pool and produce progenitors *in vitro*. Exp Cell Res 2011;317:1319–29.

[55] Kubota H, Avarbock MR, Brinster RL. Growth factors essential for self-renewal and expansion of mouse spermatogonial stem cells. Proc Natl Acad Sci USA 2004;101: 16489–94.

[56] Hofmann MC, Braydich-Stolle L, Dym M. Isolation of male germ-line stem cells; influence of GDNF. Dev Biol 2005;279:114–24.

[57] Aponte PM, Soda T, van de Kant HJ, de Rooij DG. Basic features of bovine spermatogonial culture and effects of glial cell line-derived neurotrophic factor. Theriogenology 2006;65:1828–47.

[58] Paratcha G, Ledda F. GDNF and GFRalpha: a versatile molecular complex for developing neurons. Trends Neurosci 2008;31:384–91.

[59] Cornejo M, Nambi D, Walheim C, Somerville M, Walker J, Kim L, et al. Effect of NRG1, GDNF, EGF and NGF in the migration of a Schwann cell precursor line. Neurochem Res 2010;35:1643–51.

[60] Tang MJ, Worley D, Sanicola M, Dressler GR. The RET-glial cell-derived neurotrophic factor (GDNF) pathway stimulates migration and chemoattraction of epithelial cells. J Cell Biol 1998;142:1337–45.

[61] Mullaney BP, Skinner MK. Basic fibroblast growth-factor (bFGF) gene-expression and protein-production during pubertal development of the seminiferous tubule - follicle-stimulating hormone-induced Sertoli-cell bFGF expression. Endocrinology 1992;131: 2928–34.

[62] Han IS, Sylvester SR, Kim KH, Schelling ME, Venkateswaran S, Blanckaert VD, et al. Basic fibroblast growth-factor is a testicular germ-cell product which may regulate Sertoli-cell function. Mol Endo 1993;7:889–97.

[63] Kanatsu-Shinohara M, Ogonuki N, Inoue K, Miki H, Ogura A, Toyokuni S, et al. Long-term proliferation in culture and germline transmission of mouse male germline stem cells. Biol Reprod 2003;69:612–6.

[64] Aponte PM, Soda T, Teerds KJ, Mizrak SC, van de Kant HJ, de Rooij DG. Propagation of bovine spermatogonial stem cells *in vitro*. Reproduction 2008;136:543–57.

[65] Brinster RL, Avarbock MR. Germline transmission of donor haplotype following spermatogonial transplantation. Proc Natl Acad Sci USA 1994;91:11303–7.

[66] Brinster RL, Zimmermann JW. Spermatogenesis following male germ-cell transplantation. Proc Natl Acad Sci USA 1994;91:11298–302.

[67] Chen C, Ouyang W, Grigura V, Zhou Q, Carnes K, Lim H, et al. ERM is required for transcriptional control of the spermatogonial stem cell niche. Nature 2005;436: 1030–4.

[68] Ishii K, Kanatsu-Shinohara M, Toyokuni S, Shinohara T. FGF2 mediates mouse spermatogonial stem cell self-renewal via upregulation of Etv5 and Bcl6b through MAP2K1 activation. Development 2012;139:1734–43.

[69] Oatley JM, Avarbock MR, Telaranta AI, Fearon DT, Brinster RL. Identifying genes important for spermatogonial stem cell self-renewal and survival. Proc Natl Acad Sci USA 2006;103:9524–9.

[70] Goriely A, McVean GA, van Pelt AM, O'Rourke AW, Wall SA, de Rooij DG, et al. Gain-of-function amino acid substitutions drive positive selection of FGFR2 mutations in human spermatogonia. Proc Natl Acad Sci USA 2005;102:6051–6.

[71] Anthony BA, Link DC. Regulation of hematopoietic stem cells by bone marrow stromal cells. Trends Immunol 2013.

[72] Li M, Hale JS, Rich JN, Ransohoff RM, Lathia JD. Chemokine CXCL12 in neurodegenerative diseases: an SOS signal for stem cell-based repair. Trends Neurosci 2012; 35:619–28.

[73] Salvatore P, Pagliarulo C, Colicchio R, Napoli C. CXCR4-CXCL12-dependent inflammatory network and endothelial progenitors. Curr Med Chem 2010;17:3019–29.

[74] Yang QE, Kim D, Kaucher A, Oatley MJ, Oatley JM. CXCL12-CXCR4 signaling is required for the maintenance of mouse spermatogonial stem cells. J Cell Sci 2013;126: 1009–20.

[75] Payne CJ, Gallagher SJ, Foreman O, Dannenberg JH, Depinho RA, Braun RE. Sin3a is required by Sertoli cells to establish a niche for undifferentiated spermatogonia, germ cell tumors, and spermatid elongation. Stem Cells 2010;28:1424–34.

[76] Yeh JR, Zhang XF, Nagano MC. Wnt5a is a cell-extrinsic factor that supports self-renewal of mouse spermatogonial stem cells. J Cell Sci 2011;124:2357–66.

[77] Kokkinaki M, Lee TL, He ZP, Jiang JJ, Golestaneh N, Hofmann MC, et al. The molecular signature of spermatogonial stem/progenitor cells in the 6-day-old mouse testis. Biol Reprod 2009;80:707–17.

[78] Oatley JM, Oatley MJ, Avarbock MR, Tobias JW, Brinster RL. Colony stimulating factor 1 is an extrinsic stimulator of mouse spermatogonial stem cell self-renewal. Development 2009;136:1191–9.

[79] Lu N, Sargent KM, Clopton DT, Pohlmeier WE, Brauer VM, McFee RM, et al. Loss of vascular endothelial growth factor A (VEGFA) isoforms in the testes of male mice causes subfertility, reduces sperm numbers, and alters expression of genes that regulate undifferentiated spermatogonia. Endocrinology 2013;154:4790–802.

[80] Bott RC, McFee RM, Clopton DT, Toombs C, Cupp AS. Vascular endothelial growth factor and kinase domain region receptor are involved in both seminiferous cord formation and vascular development during testis morphogenesis in the rat. Biol Reprod 2006;75:56–67.

[81] Caires KC, de Avila JM, Cupp AS, McLean DJ. VEGFA family isoforms regulate spermatogonial stem cell homeostasis *in vivo*. Endocrinology 2012;153:887–900.

[82] Hamra FK, Chapman KM, Nguyen D, Garbers DL. Identification of neuregulin as a factor required for formation of aligned spermatogonia. J Biol Chem 2007;282: 721–30.

[83] Hoeben E, Swinnen JV, Heyns W, Verhoeven G. Heregulins or neu differentiation factors and the interactions between peritubular myoid cells and Sertoli cells. Endocrinology 1999;140:2216–23.

[84] Pellegrini M, Grimaldi P, Rossi P, Geremia R, Dolci S. Developmental expression of BMP4/ALK3/SMAD5 signaling pathway in the mouse testis: a potential role of BMP4 in spermatogonia differentiation. J Cell Sci 2003;116:3363–72.
[85] Nagano M, Ryu BY, Brinster CJ, Avarbock MR, Brinster RL. Maintenance of mouse male germ line stem cells *in vitro*. Biol Reprod 2003;68:2207–14.
[86] Carlomagno G, van Bragt MP, Korver CM, Repping S, de Rooij DG, van Pelt AM. BMP4-induced differentiation of a rat spermatogonial stem cell line causes changes in its cell adhesion properties. Biol Reprod 2010;83:742–9.
[87] Baleato RM, Aitken RJ, Roman SD. Vitamin A regulation of BMP4 expression in the male germ line. Dev Biol 2005;286:78–90.
[88] Hogarth CA, Griswold MD. The key role of vitamin A in spermatogenesis. J Clin Invest 2010;120:956–62.
[89] Okuma Y, O'Connor AE, Hayashi T, Loveland KL, de Kretser DM, Hedger MP. Regulated production of activin A and inhibin B throughout the cycle of the seminiferous epithelium in the rat. J Endocrinol 2006;190:331–40.
[90] Wreford NG, Kumar TR, Matzuk MM, De Kretser DM. Analysis of the testicular phenotype of the follicle-stimulating hormone beta-subunit knockout and the activin type II receptor knockout mice by stereological analysis. Endocrinology 2001;142: 2916–20.
[91] Boulter L, Govaere O, Bird TG, Radulescu S, Ramachandran P, Pellicoro A, et al. Macrophage-derived Wnt opposes Notch signaling to specify hepatic progenitor cell fate in chronic liver disease. Nat Med 2012;18:572–9.
[92] Nyfeler Y, Kirch RD, Mantei N, Leone DP, Radtke F, Suter U, et al. Jagged1 signals in the postnatal subventricular zone are required for neural stem cell self-renewal. EMBO J 2005;24:3504–15.
[93] Basak O, Giachino C, Fiorini E, Macdonald HR, Taylor V. Neurogenic subventricular zone stem/progenitor cells are Notch1-dependent in their active but not quiescent state. J Neurosci 2012;32:5654–66.
[94] Dirami G, Ravindranath N, Achi MV, Dym M. Expression of Notch pathway components in spermatogonia and Sertoli cells of neonatal mice. J Androl 2001;22:944–52.
[95] Garcia TX, Hofmann MC. NOTCH signaling in Sertoli cells regulates gonocyte fate. Cell Cycle 2013;12:2538–45.
[96] Garcia TX, DeFalco T, Capel B, Hofmann MC. Constitutive activation of NOTCH1 signaling in Sertoli cells causes gonocyte exit from quiescence. Dev Biol 2013;377: 188–201.
[97] Hasegawa K, Okamura Y, Saga Y. Notch signaling in Sertoli cells regulates cyclical gene expression of Hes1 but is dispensable for mouse spermatogenesis. Mol Cell Biol 2012;32:206–15.
[98] Huang Z, Rivas B, Agoulnik AI. NOTCH1 gain of function in germ cells causes failure of spermatogenesis in male mice. PLoS One 2013;8:e71213.
[99] Kaucher AV, Oatley MJ, Oatley JM. NEUROG3 is a critical downstream effector for STAT3-regulated differentiation of mammalian stem and progenitor spermatogonia. Biol Reprod 2012;86(164):1–11.
[100] Russell ES. Analysis of pleiotropism at the W-locus in the mouse; relationship between the effects of W and Wv substitution on hair pigmentation and on erythrocytes. Genetics 1949;34:708–23.
[101] Besmer P, Manova K, Duttlinger R, Huang EJ, Packer A, Gyssler C, et al. The Kit-ligand (steel factor) and its receptor c-Kit/W - pleiotropic roles in gametogenesis and melanogenesis. Development 1993;125–37.

[102] de Rooij DG, Okabe M, Nishimune Y. Arrest of spermatogonial differentiation in jsd/jsd, Sl17H/Sl17H, and cryptorchid mice. Biol Reprod 1999;61:842–7.
[103] Ohta H, Tohda A, Nishimune Y. Proliferation and differentiation of spermatogonial stem cells in the W/W-v mutant mouse testis. Biol Reprod 2003;69:1815–21.
[104] Ito K, Hirao A, Arai F, Takubo K, Matsuoka S, Miyamoto K, et al. Reactive oxygen species act through p38 MAPK to limit the lifespan of hematopoietic stem cells. Nat Med 2006;12:446–51.
[105] Piccoli C, D'Aprile A, Ripoli M, Scrima R, Lecce L, Boffoli D, et al. Bone-marrow derived hematopoietic stem/progenitor cells express multiple isoforms of NADPH oxidase and produce constitutively reactive oxygen species. Biochem Biophys Res Commun 2007;353:965–72.
[106] Le Belle JE, Orozco NM, Paucar AA, Saxe JP, Mottahedeh J, Pyle AD, et al. Proliferative neural stem cells have high endogenous ROS levels that regulate self-renewal and neurogenesis in a PI3K/Akt-dependant manner. Cell Stem Cell 2011;8:59–71.
[107] Kim J, Wong PK. Loss of ATM impairs proliferation of neural stem cells through oxidative stress-mediated p38 MAPK signaling. Stem Cells 2009;27:1987–98.
[108] Morimoto H, Iwata K, Ogonuki N, Inoue K, Atsuo O, Kanatsu-Shinohara M, et al. ROS are required for mouse spermatogonial stem cell self-renewal. Cell Stem Cell 2013;12:774–86.
[109] Watt FM, Huck WT. Role of the extracellular matrix in regulating stem cell fate. Nat Rev Mol Cell Biol 2013;14:467–73.
[110] Xiao X, Mruk DD, Cheng CY. Intercellular adhesion molecules (ICAMs) and spermatogenesis. Hum Reprod Update 2013;19:167–86.
[111] Shinohara T, Avarbock MR, Brinster RL. beta(1)- and alpha(6)-integrin are surface markers on mouse spermatogonial stem cells. Proc Natl Acad Sci USA 1999;96:5504–9.
[112] Kanatsu-Shinohara M, Takehashi M, Takashima S, Lee J, Morimoto H, Chuma S, et al. Homing of mouse spermatogonial stem cells to germline niche depends on beta1-integrin. Cell Stem Cell 2008;3:533–42.
[113] de Rooij DG, Repping S, van Pelt AM. Role for adhesion molecules in the spermatogonial stem cell niche. Cell Stem Cell 2008;3:467–8.
[114] van Pelt AM, de Rooij DG. The origin of the synchronization of the seminiferous epithelium in vitamin A-deficient rats after vitamin A replacement. Biol Reprod 1990;42:677–82.
[115] de Rooij DG, Russell LD. All you wanted to know about spermatogonia but were afraid to ask. J Androl 2000;21:776–98.
[116] Chiarini-Garcia H, Russell LD. High-resolution light microscopic characterization of mouse spermatogonia. Biol Reprod 2001;65:1170–8.
[117] Chiarini-Garcia H, Russell LD. Characterization of mouse spermatogonia by transmission electron microscopy. Reproduction 2002;123:567–77.
[118] Verver DE, van Pelt AM, Repping S, Hamer G. Role for rodent Smc6 in pericentromeric heterochromatin domains during spermatogonial differentiation and meiosis. Cell Death Dis 2013;4:e749.
[119] Campos-Junior PH, Costa GM, Lacerda SM, Rezende-Neto JV, de Paula AM, Hofmann MC, et al. The spermatogonial stem cell niche in the collared peccary (Tayassu tajacu). Biol Reprod 2012;86(155):1–10.
[120] Nicolas M, Wolfer A, Raj K, Kummer JA, Mill P, van Noort M, et al. Notch1 functions as a tumor suppressor in mouse skin. Nat Genet 2003;33:416–21.

[121] Heckert LL, Griswold MD. Expression of Follicle-Stimulating-Hormone receptor messenger-Rna in rat testes and Sertoli cells. Mol Endocrinol 1991;5:670–7.
[122] Kliesch S, Penttila TL, Gromoll J, Saunders PTK, Nieschlag E, Parvinen M. FSH receptor messenger-RNA is expressed stage-dependently during rat spermatogenesis. Mol Cell Endocrinol 1992;84:R45–9.
[123] van Dissel-Emiliani FM, Grootenhuis AJ, de Jong FH, de Rooij DG. Inhibin reduces spermatogonial numbers in testes of adult mice and Chinese hamsters. Endocrinology 1989;125:1899–903.
[124] Ventela S, Makela JA, Kulmala J, Westermarck J, Toppari J. Identification and regulation of a stage-specific stem cell niche enriched by Nanog-positive spermatogonial stem cells in the mouse testis. Stem Cells 2012;30:1008–20.
[125] van Keulen CJ, de Rooij DG. The recovery from various gradations of cell loss in the mouse seminiferous epithelium and its implications for the spermatogonial stem cell renewal theory. Cell Tissue Kinet 1974;7:549–58.
[126] van Keulen CJ, de Rooij DG. Spermatogonial stem cell renewal in the mouse. II. After cell loss. Cell Tissue Kinet 1973;6:337–45.
[127] van den Aardweg GJ, de Ruiter-Bootsma AL, Kramer MF, Davids JA. Growth and differentiation of spermatogenetic colonies in the mouse testis after irradiation with fission neutrons. Radiat Res 1983;94:447–63.
[128] van der Meer Y, Cattanach BM, de Rooij DG. The radiosensitivity of spermatogonial stem cells in C3H/101 F1 hybrid mice. Mutat Res 1993;290:201–10.
[129] van der Meer Y, Huiskamp R, Davids JA, van der Tweel I, de Rooij DG. The sensitivity to X rays of mouse spermatogonia that are committed to differentiate and of differentiating spermatogonia. Radiat Res 1992;130:296–302.
[130] van Beek MEAB, Meistrich ML, de Rooij DG. Probability of self-renewing divisions of spermatogonial stem cells in colonies, formed after fission neutron irradiation. Cell Tissue Kinet 1990;23:1–16.
[131] van Alphen MM, van de Kant HJ, de Rooij DG. Depletion of the spermatogonia from the seminiferous epithelium of the rhesus monkey after X irradiation. Radiat Res 1988;113:473–86.
[132] Meistrich ML, Kangasniemi M. Hormone treatment after irradiation stimulates recovery of rat spermatogenesis from surviving spermatogonia. J Androl 1997;18:80–7.
[133] Shetty G, Wilson G, Huhtaniemi I, Shuttlesworth GA, Reissmann T, Meistrich ML. Gonadotropin-releasing hormone analogs stimulate and testosterone inhibits the recovery of spermatogenesis in irradiated rats. Endocrinology 2000;141:1735–45.
[134] Zohni K, Zhang X, Tan SL, Chan P, Nagano MC. The efficiency of male fertility restoration is dependent on the recovery kinetics of spermatogonial stem cells after cytotoxic treatment with busulfan in mice. Hum Reprod 2012;27:44–53.
[135] Rowley MJ. Quantitation of cells of seminiferous epithelium of human testis employing Sertoli cell as a constant. Z Zellforsch Mikrosk Anat 1971;115:461–72.

DMRT1 and the road to masculinity

5

Leslie L. Heckert and Valentine A. Agbor
Department of Molecular and Integrative Physiology, University of Kansas Medical Center, Kansas City, KS

I. Introduction

Doublesex (dsx) and male abnormal-3 (mab-3) related transcription factor and the DM domain

In 1998, a study by Raymond and colleagues identified the doublesex (dsx) and male abnormal-3 (mab-3) related transcription factor (TF) 1 (DMRT1) [1]. Mab-3 and dsx are distantly related TFs that regulate male sexual differentiation in *Caenorhabditis elegans* and *Drosophila melanogaster*, respectively. The functional and structural comparison identified several conserved features; both dsx and mab-3 regulate sex specific neuroblast differentiation and yolk protein gene transcription and contain a shared DNA binding motif that was called the "DM domain" [1]. The finding that the male-specific dsx isoform (dsx^M) regulates male-specific neuroblast differentiation in *C. elegans* further illustrated the functional similarity between these two phylogenetically diverse proteins [1]. Given that molecular mechanisms of sex determination and differentiation are highly variable across species, this finding revealed a unique situation in which orthologous genes from different phyla retained similar functions in sex determination and differentiation and suggested a common ancient origin for at least some features of sexual regulation. A search for other genes that encode DM-domain-containing proteins identified a cDNA for human *DMRT1* (originally known as *DMT1*) from a testis library, and fluorescence *in situ* hybridization localized *DMRT1* to the distal short arm of human chromosome 9 (9p24.3), a region linked to XY sex reversal and testis dysgenesis [1]. Importantly, this initial report was the first to suggest that DM-domain-containing proteins are an important class of evolutionarily conserved proteins that regulate male sexual development across different phyla and that *DMRT1*, the vertebrate counterpart of dsx and mab-3, might exhibit similar conserved reproductive functions.

These captivating findings and their implications prompted many new studies that ultimately defined a family of vertebrate DM-domain-containing TFs, or DMRT proteins, and demonstrating their varied developmental roles. Consequently, since the discovery of *DMRT1*, many vertebrate DMRT genes, both orthologs and paralogs of *DMRT1*, have been identified and characterized that have distinct

Sertoli Cell Biology. DOI: http://dx.doi.org/10.1016/B978-0-12-417047-6.00005-3

expression profiles and functions associated with somitogenesis, nervous system development, and reproduction (recently reviewed in Refs. [2,3]). Among the many different DMRT genes, several *DMRT1* paralogs and *DMRT1* and its orthologs, in particular, are known to manage various reproductive functions, such as gonad differentiation, gametogenesis, and sex determination [3]. By early 2005, DMRT1 was recognized as a conserved vertebrate TF with an expression profile that implicated it in sex determination and/or testis development. At this time, the first report on *Dmrt1*$^{-/-}$ mice had been published; this report described the essential role of DMRT1 in postnatal testis development. However, because DMRT1 was associated with sex determination in some species, the phenotype of *Dmrt1*$^{-/-}$ mice also suggested that its role might differ among vertebrates. In the early 2000s, there was also some mechanistic insight on *Dmrt1* transcriptional regulation and elements within its promoter region that contribute to testis expression.

The goal of this chapter is to summarize the research forming our current understanding of *DMRT1*, with a particular focus on its regulation and role in male sexual development and the progress made since 2005. To help convey its unique evolutionary position, the chapter includes discussion on DMRT1 in a variety of vertebrate species and both its divergent and conserved features. However, the greatest emphasis is on mammalian DMRT1 and studies in humans and mice. To limit confusion about the variety of gene and protein nomenclatures of different species, we employ the following nomenclature for genes and proteins, respectively, in different species: humans and birds, *DMRT1*/DMRT1; pig *DMRT-1*/DMRT-1; rodents, *Dmrt1*/DMRT1; reptiles, *dmrt1*/DMRT1; and amphibians and fishes, *dmrt1*/Dmrt1. In discussions involving more than one species, we use only one set of terms according to the following hierarchy: humans/birds > rodents > reptiles > amphibians/fishes.

II. DMRT1 expression

Although gonad structure and development differ dramatically among vertebrates, *DMRT1* expression is specific to the gonads across all vertebrate species, including mammals, birds, turtles, frogs, and fish [4–16]. *DMRT1* also exhibits a dynamic, sexually dimorphic expression pattern during gonad development that signifies a dominant role in the testis. That is, at some point during gonad development, *DMRT1* expression predominates in the testis compared to the ovary. However, the timing of this dimorphic expression differs among species and suggests phylogenetic differences in *DMRT1* function, such as sex determination versus sexual differentiation. Thus, *DMRT1* is expressed at higher levels in males than females prior to and/or during sex determination in some vertebrates, while male-dominant expression develops later, primarily during testis differentiation, in others [10,17–24]. While *DMRT1* expression analysis has been performed in many different species, studies in mice have provided the most detail and, therefore, these are emphasized in the sections below. Note that many of the studies, both expression and functional, employed mouse embryos that were staged based on either days postcoitum (dpc) or embryonic day. For consistency, however, dpc is used to designate all embryonic stages in this chapter.

As in other vertebrates, *Dmrt1* expression in mice is gonad specific and, during development, becomes sexually dimorphic (testis or ovary) as the gonads differentiate [4]. Reverse transcription-polymerase chain reaction (RT-PCR) of whole mouse embryos indicated that *Dmrt1* is expressed as early as 9.5 dpc, and *in situ* hybridization and immunohistochemistry (IHC) studies have documented its unique expression characteristics [4,20,25]. Because *Dmrt1* mRNA and protein are expressed similarly in mice, the patterns described here emphasize IHC data because these data best portray DMRT1's distinct cellular profiles and functional correlations with sex determination, testis/Sertoli cell differentiation, and germ cell proliferation and development [4,20,25].

IHC of DMRT1 and germ cell nuclear antigen 1 (GCNA1), a germ-cell-specific protein, was instrumental in revealing the cellular expression patterns of DMRT1 in XX and XY embryonic gonads and postnatal testes [20,26]. During the early stages of gonadogenesis, before gonads are sexually distinct (approximately 11.5 dpc), DMRT1 expression in indifferent XX and XY gonads was indistinguishable and observed in both germ cells and somatic cells (Figure 5.1A and B) [10,20,25]. At that time, the signal for DMRT1 was robust in most somatic cells and weaker or absent in germ cells (Figure 5.1A and B) [20]. With the onset of gonad differentiation at approximately 12.5 dpc, sex-specific profiles were observed for DMRT1 mRNA and protein that revealed differences in both the location and identity of DMRT1-positivecells [4,20,25]. In testes at 12.5 dpc, DMRT1 localized to Sertoli cells and germ cells within testis cords, and levels in Sertoli cells were notably higher than in germ cells (Figure 5.1C). In ovaries at this time, DMRT1-positive cells showed no obvious organization, and DMRT1 levels were greatly diminished in somatic (pregranulosa) cells, while levels in germ cells were higher than in male germ cells (Figure 5.1D) [20]. By 13.5 dpc, DMRT1-positive somatic cells were nearly absent from the ovary, while, in the testis, DMRT1-positive Sertoli cells were prominent and localized to the tubule periphery (Figure 5.1E and F) [20]. Regardless of gonadal sex, germ cells showed variable DMRT1 expression at 13.5 dpc and, by 15.5 dpc, they no longer expressed DMRT1 (Figure 5.1E–H) [10].

DMRT1 never returns to the ovary; however, in the testis, it is continuously expressed in Sertoli cells and eventually returns to the germ cells shortly after birth [4,20]. Hence, in testes from mice ranging in age from postnatal day 0.5 (P0.5) to adulthood, IHC of DMRT1 and GCNA1 showed that DMRT1 was always expressed in Sertoli cells and, once again, in germ cells beginning at P0.5 (Figure 5.2A, top left). By P7, nearly all germ cells were DMRT1 positive, but, by P8, with the appearance of differentiated germ cells, increasing numbers of DMRT1-negative germ cells were observed within the tubules (Figure 5.2A). Given that P8 mouse testes reportedly have only type A and B spermatogonia, the emergence of DMRT1-negative germ cells at P8 suggested that *Dmrt1* silencing is nearly complete in type B spermatogonia and that its expression is largely restricted to type A spermatogonia (Figure 5.2A) [20,27]. In addition, P8 testes showed distinct patterns of DMRT1-positive germ cells that were indicative of its cyclic profile observed in the adult seminiferous epithelium (Figure 5.2A) [20,28].

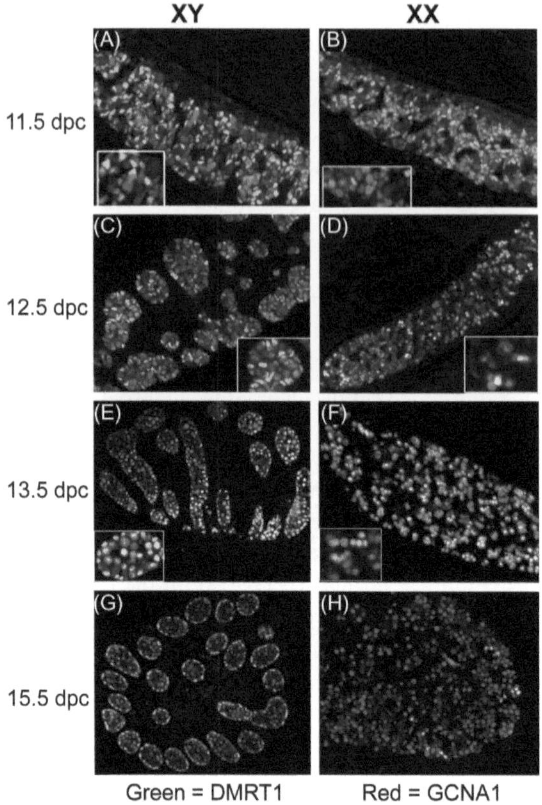

Figure 5.1 Gonad expression of *doublesex* and *male abnormal-3* related transcription factor 1 (DMRT1) and germ cell nuclear antigen 1 (GCNA1) in developing mouse embryos. Dual immunofluorescent staining of DMRT1 (green) and GCNA1 (red) in XY (A, C, E, G) and XX (B, D, F, H) gonads of developing mouse embryos. Embryonic time points are 11.5 (A, B), 12.5 (C, D), 13.5 (E, F), and 15.5 (G, H) days postcoitum (dpc). Insets show higher magnification. Yellow/orange cells indicate DMRT1-positive germ cells. Red cells indicate DMRT1-negative germ cells. Green cells indicate DMRT1-positive somatic cells. (For interpretation of the references to color in this figure legend, the reader is referred to the web version of this book.)
Modified from Lei et al. [20].

Notably, *Dmrt1* expression in both male and female germ cells changes in accordance with differentiation and, in male germ cells, with mitotic activity. Hence, in embryos, DMRT1 is lost from male germ cells coincident with their mitotic arrest and reemerges just after birth, at around the time that gonocytes reenter the cell cycle, marking their transition to type A spermatogonia [20,29]. Furthermore, DMRT1 is lost from both male and female germ cells upon their differentiation. In females, this occurs in the embryo, when the cells enter meiosis between 13.5 and 16.5 dpc while, in male germ cells, DMRT1 expression is lost during the transition

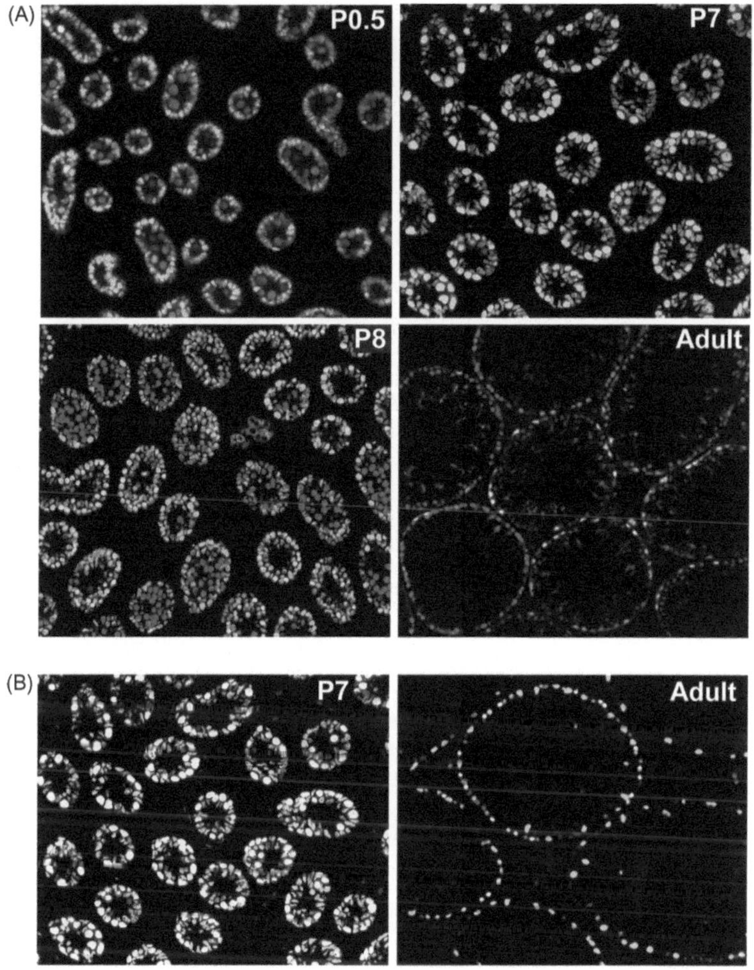

Figure 5.2 *Doublesex* and *male abnormal-3* related transcription factor 1 (DMRT1) expression and cell proliferation in postnatal mouse testes. (A) Dual immunofluorescent staining of DMRT1 (green) and germ cell nuclear antigen 1 (GCNA1) (red) in testes from postnatal day 0.5 (P0.5), P7, P8, and adult mice. Sertoli cells (green), DMRT1-positive germ cells (yellow), and DMRT1-negative germ cells (red) are shown. (B) *Left*: Triple immunofluorescent staining for DMRT1 (green), GCNA1 (red), and PCNA (blue) in testes isolated from P7 mice. *Right*: Immunofluorescence of DMRT1 (green) and ki67 (red) and adult mouse testis. (For interpretation of the references to color in this figure legend, the reader is referred to the web version of this book.)
Modified from Lei et al. [20].

from type A to type B spermatogonia and does not reappear in germ cells at later stages of spermatogenesis (Figure 5.2A) [20,28,30,31].

Colocalization of DMRT1 and other proteins provided important clues to its potential functions. The association between DMRT1 and mitotic activity/proliferation in germ cells was further examined by IHC studies that colocalized DMRT1 with proliferation markers Ki67 and proliferating cell nuclear antigen (PCNA). At P0.5, DMRT1 was observed in a small number of Ki67-negative gonocytes, suggesting that DMRT1 precedes the onset of mitosis [20]. Colocalization of PCNA, DMRT1, and GCNA1 revealed that, by P2, numerous mitotically active, DMRT1-positive germ cells were located at the tubule periphery [20]. The presence of DMRT1 in proliferating germ cells continued and, by P7, most or all germ cells expressed both DMRT1 and PCNA and were positioned at the tubule periphery (Figure 5.2B, left) [20]. With the start of germ cell differentiation at P8, DMRT1-negative germ cells appeared, and the DMRT1-positive germ cells continued to proliferate and were still confined to the tubule periphery [20]. In adult testes, DMRT1 still coexpressed with Ki67, indicating its presence in mitotically active germ cells and the retention of its associated functions after the seminiferous epithelium was fully established (Figure 5.2B, right) [20]. Complementary studies also supported a role for DMRT1 in germ cell expansion during the first week of life. Thus, unlike in wild-type (wt) mice, the weight of *Dmrt1*-null (*Dmrt1*$^{-/-}$) mouse testes declined significantly between P2 and P7 and, at P7, showed little staining for P-H3, a mitotic metaphase marker [11,32].

In addition to mitotic markers, DMRT1 has been colocalized with proteins that mark specific populations of germ cells. In particular, studies showed that it coexpressed with POU domain, class 5, TF 1 (POU5F1, also known as OCT4) and neurogenin 3 (NGN3) in P7 testes and, in adult testes, with POU5F1, glial cell line-derived neurotrophic factor (GDNF) family receptor α-1 (GFRA1), zinc finger and blood-testis barrier (BTB) domain-containing protein 16 (ZBTB16, also known as PLZF), and proto-oncogene c-Kit (c-KIT) [20,28]. DMRT1's coexpression with POU5F1 and NGN3 at P7 and with POU5F1, ZBTB16, and GFRA1 in adults suggests that DMRT1-positive cells contribute to the spermatogonial stem cell (SSC) pool at around the time of niche formation and in adults. The coexpression of DMRT1 with c-KIT indicated that DMRT1 is present in differentiating spermatogonia in adult testes, which is different from what was indicated by an analysis of P7 and P8 testes [20,27]. Thus, it appears that during the first wave of spermatogenesis, DMRT1 is restricted to type A spermatogonia, but in subsequent waves, its expression extends through type B spermatogonia [20,28].

In summary, a male preference for DMRT1 is first apparent in the somatic cells of embryonic gonads, in accordance with sexual differentiation. This preference remains throughout embryonic and postnatal development because DMRT1-positive Sertoli cells are present throughout adulthood, while DMRT1 is lost from ovarian somatic cells by approximately 13.5 dpc. In germ cells, DMRT1 is silenced in both males and females by 15.5 dpc, coinciding with mitotic arrest and meiotic entry, respectively. At this point, DMRT1 is lost permanently from the ovary and restricted to Sertoli cells of the testis. In postnatal testes, Sertoli cell expression is maintained and germ cell expression returns shortly after birth, at a time when germ cells become mitotically active and transition into A undifferentiated spermatogonia and SSCs [4,20]. In adult testes, DMRT1 expression is robust in undifferentiated spermatogonia and low in differentiating spermatogonia, and it is absent once germ cells initiate meiosis.

III. Regulation of DMRT1

A. Endocrine

Several hormones and signaling pathways influence *Dmrt1* expression. The cell's differentiation status and culture environment, at least for Sertoli cells, are also important because *Dmrt1* expression was silent in the mouse MSC-1 and TM4 Sertoli cell lines and was dramatically reduced in primary rat Sertoli cell cultures after 24 h [33]. This dramatic reduction suggested that an important growth factor or hormone was missing from the cultures. In support of this theory, subsequent supplementation of the media with follicle-stimulating hormone (FSH) showed a dramatic rise in *Dmrt1* mRNA levels [33]. Because FSH emits signals primarily through the second messenger cAMP, Sertoli cells were treated with 8-bromo-cAMP, a cAMP analogue, which also rapidly increased *Dmrt1* mRNA levels. This nearly 18-fold increase resulted from increased transcription and depended on protein kinase (PK) A activation [33]. In contrast, phorbol 12-myristate 13-acetate, a known activator of PKC, reduced *Dmrt1* mRNA levels in Sertoli cells and blocked the effects of cAMP via a mechanism inhibited by PD98059, a MAP kinase inhibitor [33]. Together, these findings suggested that PKC activation of MAP kinase inhibits *Dmrt1* transcription and interferes with the stimulatory effects of PKA on *Dmrt1* transcription. Interestingly, FSH was also shown to stimulate *dmrt1* expression in rainbow trout [34]. However, in trout, FSH also enhanced the production of androgen 11-ketotestosterone (11KT), which, when blocked by an inhibitor of Δ4 steroidogenesis (trilostane), strongly reduced but did not eliminate the response to FSH [34]. Furthermore, a second study showed that 11KT alone could significantly increase *dmrt1* expression [34,35]. Hence, the mechanisms of FSH regulation appear to differ in rats and fish. In rats, FSH signaling in Sertoli cells activates *Dmrt1* transcription directly, whereas in rainbow trout, FSH primarily stimulates *dmrt1* expression indirectly via induction of 11KT.

Sex steroids also influence *DMRT1* expression in many species of fishes in addition to rainbow trout as well as in turtles, frogs, chickens, and rats (Table 5.1). Notably, in nonmammalian species, sex steroids also are known to influence gonadal sex differentiation, an event in which *DMRT1* is differentially expressed in males and females [12,54–56]. Except in rats, the effects of sex steroids on *DMRT1* expression and sexual phenotype have been consistent across species. In females, androgenic compounds enhanced *DMRT1* expression in ovaries, where its expression is normally low or absent, and masculinize the animals (Table 5.1, top half). In males, estrogenic compounds reduced expression in the testis, where expression is normally high, and feminized the animals (Table 5.1, bottom half). The effects of flutamide, an androgen antagonist, and aromatase inhibitors were also evaluated. In juvenile male Asian catfish and adult male fathead minnows, flutamide treatment resulted in feminization and reduced *dmrt1* expression [43,44]. In chickens and rainbow trout, inhibition of aromatase in females increased *DMRT1* expression and masculinized the animals [36,52]. The synthetic estrogen diethylstilbestrol (DES) had no effect on *DMRT1* expression in chick embryos [51]. Although this finding appears to contrast with the

Table 5.1 **Effects of sex steroids on *Dmrt1* expression**

Species	[a]Compound	Activity	*Dmrt1* expression	Comments/References
Rainbow trout	11OHA4	Androgenic	Increased	All-female population masculinized by androgen [36]
Rainbow trout	11OHA4	Androgenic	Increased	All-female population masculinized by androgen [37]
Rainbow trout	11OHA4	Androgenic	Increased	Androgen-induced masculinization [38]
Rainbow trout	11OHA4	Androgenic	None	All-female population masculinized by androgen during gonad differentiation [39]
Rainbow trout	11KT	Androgenic	Increased	Immature males treated with 11KT [35]
Rainbow trout	T	Androgenic	Increased	Immature males treated with T [35]
Tilapia	MT	Androgenic	Increased	XX fish sex reversal with androgen [40]
Catfish	MT	Androgenic	Increased	Testicular differentiation initiated by exogenous MT [41]
Self-fertilizing fish	MT	Androgenic	Increased	MT treated embryos developed mostly into males and expressed dmrt1 [42]
Frog	T	Androgenic	Increased	All-female (XX) tadpoles (w/ovaries) sex-reversed by T injection [22]
Catfish	Fl	Anti-androgenic	Decreased	All-male population exposed to flutamide [43]
Fathead minnow	Fl	Anti-androgenic	Decreased	Adult males feminized by flutamide treatment [44]
Rainbow trout	EE2	Estrogenic	Decreased	Male-to-female gonad reversal. Expression change after initial feminization [45]
Rainbow trout	E2	Estrogenic	Decreased	All-male population treated with E2 during gonad differentiation [39]
Tilapia	EE2	Estrogenic	Decreased	XY sex reversal induced by EE2 [40]
Catfish	Es	Estrogenic	Decreased	All-male population exposed to Es [43]
Self-fertilizing fish	E2	Estrogenic	Decreased	E2 decreased dmrt1 mRNA in hermaphrodite and secondary male gonads [46]
Self-fertilizing fish	BPA	Estrogenic	Decreased	BPA decreased dmrt1 mRNA in hermaphrodite and secondary male gonads [46]
Fathead minnow	EE2	Estrogenic	Decreased	Adult males feminized by EE2 [44]

(*Continued*)

Table 5.1 (Continued)

Species	[a]Compound	Activity	*Dmrt1* expression	Comments/References
Rare minnow	4-NP	Estrogenic	Decreased	Males exposed to 4-NP [47]
Zebra fish	MCP	Estrogenic	Decreased	MCP exposure increased proportion of female fish and reduced dmrt1 [48]
Zebra fish	EE2	Estrogenic	Decreased	EE2 exposure retarded/ceased male gonadal sex development [49]
Turtle	E2	Estrogenic	Decreased	Male-to-female sex reversal. Embryos treated with E2 at male-producing temperature [50]
Chicken	DES	Estrogenic	None	DES injected on day 0. Both sex embryos evaluated days 8, 12, and 16 [51].
Chicken	E2	Estrogenic	Decreased	E2 injected at E4.5. Gonads evaluated at E9.5. Male gonads feminized [52]
Rat	E2	Estrogenic	Increased	Rat Sertoli cell cultures treated with E2 [53]
Rat	DPN	Estrogenic	Increased	Rat Sertoli cell cultures treated with DPN [53]
Rat	PPT	Estrogenic	No change	Rat Sertoli cell cultures treated with PPT [53]
Chicken	F	Aromatase inhibitor	Increased	Female gonads masculinized by Fadrozole injected at E3.5 evaluated at E9.5 [52]
Rainbow trout	ATD	Aromatase inhibitor	Increased	All-female population masculinized by ATD [36]

[a]*Compound abbreviations*: 11KT, 11-ketotestosterone; 11OHA4, 11β-hydroxyandrostenedione; 4-NP, 4-nonylphenol; ATD, 1,4,6-androstatriene-3,17-dione; BPA, bisphenol A; DES, diethylstilbestrol; DPN, 2,3-*bis*(4-hydroxyphenyl)-propionitrile (ESR2-selective agonist); E2, 17β-estradiol; EE2, 17α-ethynylestradiol; Es, endosulfane; F, fadrozole; Fl, flutamide; MCP, monocrotophos; MT, 17α-methyltestosterone; PPT, 4,40,400-(4-propyl-(1H)-pyrazole-1,3,5-triyl)trisphenol (ESR1-selective agonist); T, testosterone.

observed effects of 17β-estradiol (E2) and fadrozole (an aromatase inhibitor) on *DMRT1* expression in chick embryos, the properties of DES are variable and the treatment protocols of the two studies differed [51,52]. In rats, the effects of estrogens differed from those observed in other species. In these rat studies, E2 and 2,3-bis(4-hydroxyphenyl)-propionitrile (DPN), a selective agonist for estrogen receptor 2 (ESR2), increased DMRT1 protein levels in cultured Sertoli cells, while no effect was observed with the ESR1-selective agonist 4,40,400-(4-propyl-(1H)-pyrazole-1,3,5-triyl)trisphenol [53]. Interestingly, inhibitors of PI3K (wortmannin) and cAMP response element binding protein (KG-501) blocked the E2-stimulated increase in DMRT1. While steroids clearly impact *DMRT1* expression, the mechanism for this effect remains unclear and is complicated by the fact that both steroids and *DMRT1* expression are closely tied to gonad development/differentiation. Therefore, steroids may elicit their effects on *DMRT1* expression directly through *DMRT1* or indirectly, via their effects on sexual development [12,55,57]. A better understanding of how sex steroids regulate *DMRT1* and/or other sexual differentiation genes is likely to provide important insights into mechanisms associated with environmental estrogens and their detrimental effects on reproductive health.

B. Temperature

In many organisms, gonadal sex is determined by environmental factors, such as temperature, a property that allows for a certain amount of plasticity with respect to sex phenotype [54]. Many reptiles employ temperature-dependent sex determination (TSD), whereby the temperature of egg incubation determines whether the bipotential gonads develop into testes or ovaries. Expression of *dmrt1* has been evaluated in a number of species with TSD, typically during the period in which temperature dictates gonad development (i.e., the temperature-sensitive period (TSP)). In red-eared slider turtles (*Trachemys scripta*), *dmrt1* mRNA was readily detected at male-promoting temperatures but not at female-promoting temperatures [6,58,59]. Detailed *in situ* hybridization studies showed that *dmrt1* transcripts were present at the beginning of the TSP in all embryos at the male-promoting temperature and in about half of them at the female-promoting temperature, although expression in females was more diffuse and weaker than in males [59]. Subsequently, *dmrt1* expression increased at male-promoting temperatures and decreased at female-promoting temperatures. Also in *T. scripta*, gonad cultures were used to show that temperature regulation occurred within cells of the bipotential gonad [60]. Similar dimorphic expression patterns were noted in sea turtles (*Lepidochelys olivacea*), snapping turtles (*Chelydra serpentina*), and salamanders (*Hynobius retardatus*). During the TSP, *dmrt1* expression in these species was consistently higher at male-promoting temperatures than at female-promoting temperatures [61–63]. Temperature regulation of *dmrt1* is also implicated in several species of fish with TSD and, as in reptiles, its expression is favored at male-promoting temperatures [64–67]. However, as with steroid regulation, the association of temperature with gonad development obscures the mechanism associated with *dmrt1* regulation because changes in its expression may occur after the developmental events initiated by temperature.

IV. DMRT1 locus and gene expression

Studies indicate that DMRT1 expression is regulated at both the transcriptional and posttranscriptional levels and, based on expression and functional data, is directed by mechanisms that are highly important to gonad function because they direct and/or maintain cell-specific expression of *Dmrt1* in gonads as well as manage transcriptional changes that occur with meiotic entry, mitotic activity, and stage specificity. Unfortunately, our current understanding of these regulatory mechanisms is still quite modest. Thus, we need greater insight into the elements and proteins that control *DMRT1* expression before we can fully appreciate the associated components and their contributions to germ cell and Sertoli cell function.

Fortunately, technical advances in genomics have greatly accelerated our understanding of genome structure, function, and evolution and have provided many new databases and tools to facilitate research on gene function and regulation. Sequence data are now available for the genomes of hundreds of species and are used extensively to guide research on transcriptional mechanisms. These studies use the data either *in silico* to help identify important regulatory regions or use them directly in genome-wide or gene-specific experiments to decipher mechanisms [68–71]. With respect to *DMRT1* expression, both *in silico* approaches and functional assays have been used to predict potential mechanisms and to determine the transcriptional activity of specific sequences, respectively. The following section describes the *DMRT1* locus and the insights gained from sequence analysis and experiments on transcriptional mechanisms.

A. DMRT1 locus

The *DMRT1* locus has now been sequenced for at least 100 different vertebrate species and has been characterized for many others [72–79]. There are numerous similarities between the *DMRT1* genes and their loci in different vertebrates. Therefore, to help illustrate some of these similarities, a schematic representation of the human *DMRT1* locus depicted together with its evolutionary conservation profile generated from 100 vertebrate species was captured from the University of California, Santa Cruz (UCSC) Genome Browser (Figure 5.3A, top; http://genome.ucsc.edu) [75,81]. The five exons of human *DMRT1* are in the same range as the five to seven exons noted for all other species, with the exception of mugger crocodile, which reportedly has only three exons [72–79]. In addition, human *DMRT1*, like the DMRT1 of most species, is located between *KANK1*, its nearest 5' neighbor, and *DMRT3* and *DMRT2* on its 3' side (Figure 5.3A, top) [75–77,82–86]. The syntenic relationship with DMRT3 and DMRT2 is conserved in all, or nearly all, vertebrates, including several species of fish, and most have *KANK1* positioned upstream as well [75,77,87–90]. Such conserved synteny suggests a functional advantage to maintaining their positional relationship and/or the surrounding regulatory environment; the latter being particularly relevant for genes regulated by distal sequences and coregulated syntenic genes [91–93].

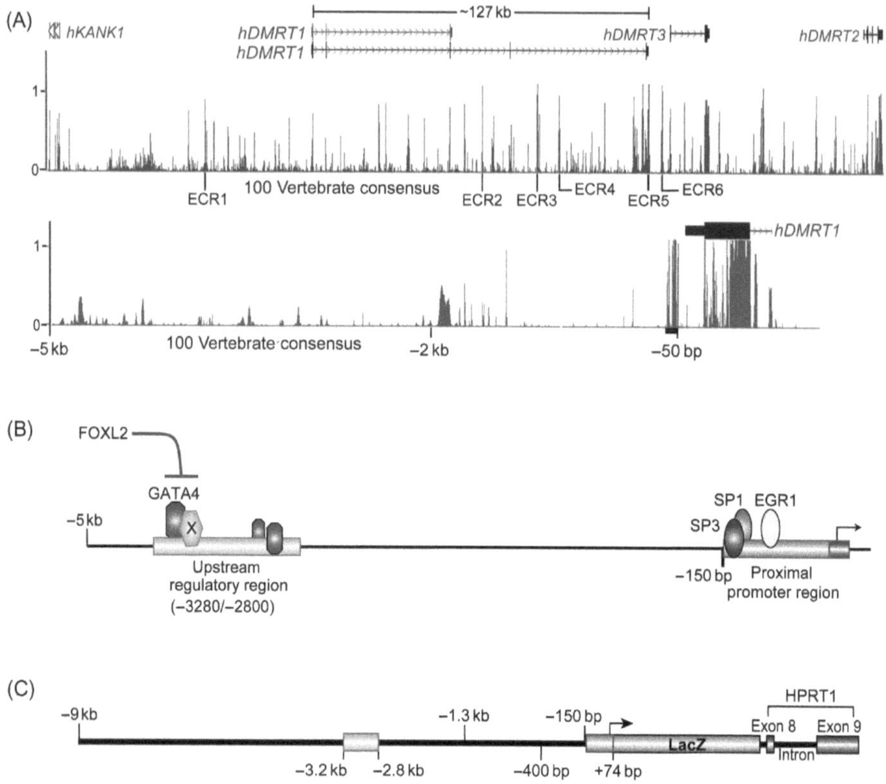

Figure 5.3 *DMRT1* and transcriptional regulation. (A) *Top*: Human *doublesex* and *male abnormal-3* related transcription factor 1 (*DMRT1*) locus viewed on the University of California, Santa Cruz (UCSC), Genome Browser (http://genome.ucsc.edu/). Horizontal lines show the genes *KANK1* (3' end only), *DMRT1*, DMRT3, and DMRT2, with vertical lines indicating exons and arrows indicating direction of transcription. The 100-vertebrate consensus (bottom) shows the measurement of evolutionary conservation determined by phastCons using the multiple alignments of 100 vertebrate species. The viewing range on the y-axis is set from 0 to 1. Previously reported noncoding evolutionarily conserved regions 1−6 (ECR1−ECR6) are denoted below the 100-vertebrate consensus [72,76,77,80]. *Bottom*: 5 kb of DMRT1 5' flanking sequence from top. (B) Diagrammatic representation of the rat *Dmrt1* 5'-flanking region showing the transcriptionally active regions as determined by transient transfection of primary rat Sertoli cells. Transcriptional activity was associated with two major regions. The upstream regulatory region, located between −3.2 and −2.8 kb, enhanced *Dmrt1* transcriptional activity only in Sertoli cells. Within the upstream regulatory region, four active sequences were identified, three of which bound the transcription factor GATA4 and a fourth, containing a YY1-like motif, bound an unknown protein "X." Cotransfection studies showed that the transcription factor forkhead box L2 (FOXL2) repressed *Dmrt1* transcriptional activity through the upstream regulatory region. The proximal promoter region (the core promoter plus approximately 150 bp upstream) was active in all transfected cells tested and contained active sequences that bound the transcription factors SP1, SP3, and EGR1. (C) Schematic representation of the reporter constructs used to generate transgenic mice. Rectangles denote regulatory regions, spheres and hexagons denote proteins, and bent arrows denote several *Dmrt1* transcriptional start sites identified in Sertoli cells.

B. *In silico* sequence analysis

With the enormous growth in genome sequence data and bioinformatics resources, investigators can readily view features of the genome either globally or in specific regions. Comparative genomics and/or identification of specific TF motifs were used to reveal potential regulatory sequences in *DMRT1*'s 5'-flanking region, the home of its core/proximal promoter, and in distal regions associated with the locus (i.e., distal enhancers). Several studies compared *DMRT1* sequences from different species of varying evolutionary distance and identified evolutionarily conserved regions (ECRs) in the noncoding sequences. These noncoding ECRs are predicted to have an important regulatory function because of the imposed constraints to retain them [68–70,94]. *DMRT1* sequences have also been scanned for specific TF motifs or DNA sequence patterns recognized by known TFs to help identify TFs that regulate *DMRT1*. However, both of these approaches only implicate sequences and TFs, and their actual importance must be determined experimentally, which, unfortunately, has not been done in many studies on *DMRT1*.

To help elaborate the features of the *DMRT1* locus and to provide a point of reference for regions identified in the studies described below, the chapter includes a schematic of the human *DMRT1* locus viewed via the UCSC Genome Browser (Figure 5.3A) [75,81]. The 100-vertebrate conservation profile displayed on the UCSC Genome Browser is similar to the published VISTA plot produced previously by El-Mogharbel and colleagues, who evaluated sequence conservation across the entire *DMRT1* locus from approximately 30 kb 5' to *DMRT1* through *DMRT2* (Figure 5.3A) [77]. In the El-Mogharbel study, the VISTA Web server was used to align *DMRT1* sequences from platypuses, humans, mice, rats, dogs, opossums, and chickens and the resulting VISTA alignment showed considerable sequence homology in coding and noncoding sequences across all mammals and declining sequence homology with platypuses and chickens [77,95]. A review of the multiz alignments of the 100 available vertebrate sequences along a similar region of the human genome on the UCSC Genome Browser supported the VISTA analysis and also revealed some interesting evolutionary features of the noncoding conserved regions. In particular, the species contributions to noncoding ECRs were as follows: most noncoding conserved ECRs were present in nearly all or all mammals, birds, and reptiles; some were restricted to mammals; and only a few were found in fishes (unpublished observation) [75,81]. Intriguingly, the noncoding ECRs that were conserved across all vertebrates, including multiple species of fishes, were located either in the last intron or in the 3' intergenic region of *hDMRT1*.

C. *DMRT1* 5'-flanking region

Comparative sequence analysis of the 5'-flanking region of *DMRT1* revealed little sequence similarity between evolutionarily distant species, such as between mammals (i.e., humans or mice) and chickens, frogs, or fish. However, some sequence similarity was observed between more similar species, such as among different mammals [5,72,76,96]. Thus, among mammals, comparisons of the 5'-flanking sequences of human, mouse, and pig *DMRT1* identified several ordered, conserved sequence blocks with greater than 60% homology [5]. The limited sequence conservation of the 5'-flanking region can be appreciated by viewing the 100-vertebrate conservation profile of *hDMRT1* from 5 kb upstream (−5 kb) through exon 1 (Figure 5.3A, bottom). While most of the region was shown to be poorly or modestly conserved, there was one highly-conserved area between −138 and −50 bp. (Figure 5.3A, black rectangle, bottom). Interestingly, the multiz alignment track associated with this sequence showed that it was conserved in all mammals, except big brown bats (absent) and platypuses (less conserved) and was absent or poorly conserved in other species (unpublished observation).

Even in the absence of overt sequence conservation, different or divergent species may still employ conserved regulatory mechanisms if the relevant TFs and their binding sites are retained. Thus, the presence of similar TF motifs in sequences of different species, particularly if they are arranged in distinct clusters, is also indicative of conserved regulation [68,71,97]. Sequence analysis has identified many TF motifs in the 5'-flanking regions of *DMRT1* from various species. In one study, TF motifs were identified across the first 500 bp of 5'-flanking sequence from human, mouse, chicken, frog, and medaka *DMRT1* [72]. However, no TF motif clusters were conserved between diverse species and only one—containing binding sites for AML-1α, C/EBPβ, and GATA binding protein 2 (GATA2)—was conserved between humans and mice [72]. While clusters of TF motifs were not conserved between diverse species, all 5'-flanking sequences contained multiple motifs for GATA binding factors and CdxA and at least one motif for AML-1α, except for chickens, which had only a single CdxA site [72].

TF binding site motifs were also identified in 5'-flanking regions of tilapia *dmrt1*. These regions included Gata1, CdxA, C/ebp, Stat, Sry, Aml-1α, and wrasse *dmrt1*, which revealed numerous motifs for Gata binding factors (e.g., Gata1, Gata2, Gata3, and GATAx) and for activator proteins, such as Ap1 and Ap4 [96,98]. In addition, Sox5 or Sry binding sites were observed in the 5' regions of tilapia, zebra fish, wrasse, mouse, and human *DMRT1* [72,96,98]. Unfortunately, few putative binding sites have been functionally tested, and, thus, their contribution to *DMRT1* transcriptional activity is unclear. However, as discussed below, functional evaluation of the rat 5'-flanking region identified several functional sequences corresponding to GATA1 and GATA3 motifs, which were shown to bound the TF GATA4 [99]. Thus, it appears that GATA factors may play a conserved role in regulating *DMRT1*, at least via its 5'-flanking region.

D. Distal ECRs

Comparative sequence analysis of *DMRT1* has identified six noncoding ECRs, which, because of their potential for housing important regulatory sequences, are regions of considerable interest and referred to here as "ECRs 1–6" (Fig. 5.3A). Mawaribuchi and colleagues identified a highly conserved region of approximately 400 bp that is present in *DMRT1* sequences from humans, mice, dogs, opossums, platypuses, chickens, and lizards but not frogs [72]. This sequence, ECR1, is located about 40 kb upstream of *DMRT1* in humans, while its position varied in other species, ranging from 6 (chickens) to 68 kb (platypuses) upstream [72]. In the multiz alignment track of the UCSC Genome Browser, which contains 100 vertebrate genomes, ECR1 was conserved in all of the available genomes from mammals, birds, and reptiles, except for aardvarks and frogs, and was absent from the genomes of fishes (unpublished observation) [75]. Inspection of ECR1 for common TF motifs identified potential binding sites for CdxA, GATA-X, Pbx-1, AP-1, and Nkx-2.5, but transcriptional activity of the region was not evaluated [72].

Shetty and colleagues identified ECR2 by comparing sequences of emu and human *DMRT1*, and El-Mogharbel and colleagues identified ECR4 by comparing *DMRT1* sequences from platypuses and humans (Figure 5.3A) [77,80]. ECR2 is a 270 bp region located in intron 3 and has 90% sequence homology between emus and humans [80]. Importantly, RT-PCR analysis failed to detect novel *DMRT1* transcripts containing ECR2, making it unlikely that ECR2 represents an undisclosed exon. Inspection of ECR2 in the multiz alignment track of the UCSC Genome Browser showed that it was conserved in all available genomes from mammals, birds, and reptiles and was absent in all fishes, except spotted gar (unpublished observation) [75,80]. ECR4 is a 370 bp region located in intron 4 with 79% sequence homology between platypuses and humans [77]. It was also identified in the genomes of all other mammals evaluated by the investigators, as well as in snakes, chickens, frogs, and turtles, and a sequence similar to ECR4 was identified in fish. Inspection of ECR4 in 100 vertebrate genomes on the multiz alignment track showed that it was present in all mammals, birds, and reptiles and was missing from many but not all species of fishes (UCSC Genome Browser, unpublished observation) [75,81].

Lastly, Brunner and colleagues identified the conserved regions known as ECR3, ECR5, and ECR6 by comparing the sequence of *Fugu rubripes dmrt1* to that of human *DMRT1* [76]. Use of the more distant fish genome proved beneficial because it identified the three most conserved noncoding ECRs of the six ECRs so far reported. PCR analysis of these regions using genomic DNA from different species identified these ECRs in most of the 20 additional vertebrate genomes tested, including those of fish. With two to three exceptions each, these three ECRs were present in all of the 100 vertebrate genomes present in the multiz alignment track of the UCSC Genome Browser (unpublished observation) [75,81]. ECR3 (region A in the report by Brunner and colleagues) is a 229 bp region with 82% sequence similarity and is located in intron 4 (Figure 5.3A) [76]. ECR5 (region B) is a 385 bp region with 89% sequence similarity located 66 bp 3' to the putative poly(A) signal of *hDMRT1*. ECR6 (region C) is 220 bp with 78% sequence similarity and is positioned approximately 5 kb 3' to *DMRT1* and 2.5 kb 5' to *DMRT3* (Figure 5.3A) [76].

E. Transcriptional activity

Although sequence analysis of *DMRT1*, either published or publicly available, reveals numerous sites of potential importance to its regulation, few sequences have been tested experimentally for activity. In fact, functional data on *DMRT1* transcriptional regulation are derived almost entirely from studies of its 5'-flanking region gathered either by transient transfection or transgenic analyses [5,96,99–104]. Transcriptional start site (TSS) mapping has been performed for only a few species but is important to ensure accurate identification of the core promoter region, the sequence surrounding the site of transcriptional initiation [72,101,105]. Using RNA from rat Sertoli cells, RNAse protection and 5'RACE identified one major TSS, located 102 bp 5' to the translational start codon, and several minor TSSs [101]. No discernible TATA motif was identified near the rat core promoter, but this region did contain a motif for an initiator element that similarly facilitates accurate basal transcription [105,106]. TSSs identified in mice and chickens were positioned 236 and 54 bp upstream of the translational start codon, respectively [72].

F. Transient transfection analysis

The most extensive information on *Dmrt1* transcription comes from the activity of the rat 5'-flanking region determined by transient transfection of primary rat Sertoli cells [99,101]. Performed with various amounts of 5' flanking sequence, ranging from 5000 to 150 bp, the transfection studies identified two main regions of transcriptional activity: an upstream regulatory region located between approximately 3.2 and −2.8 kb and a proximal promoter region containing the core promoter that was 150 bp of upstream sequence (Figure 5.3B) [99–101]. Site-directed mutagenesis and transient transfection further refined the sequences required for transcriptional activity, and *in vitro* DNA–protein binding assays helped characterize the elements' binding proteins. The results showed that the proximal promoter features two discrete active sequences: a 5' sequence (GGCGTG), which bound to TFs SP1 and SP3, and a 3' sequence (CCCACG), which bound early growth response 1 (EGR1) [101]. The 5' sequence is highly conserved in mammals, and the 3' region is partly conserved. Identified binding sites for EGR1, EGR2, and EGR3 are conserved in the aligned sequences for human, mouse, and rat *DMRT1* [75,81]. The data suggest that the transcriptional mechanisms regulating the *DMRT1* proximal promoter are conserved in mammals and employ the TFs SP1, SP3, and EGR1 or their paralogs (Figure 5.3B).

Unlike the proximal promoter, which was active in various cell lines and primary Sertoli cells, the upstream regulatory region was active only in primary Sertoli cells. *DMRT1* expression, and thus transcription, was limited to cultures of primary testis cells because *DMRT1* transcripts were not observed in any of the cell lines examined [99]. Together, these findings suggest that the proximal promoter region is associated with general or basal transcription, while transcriptional activity of the upstream region is more restricted or cell specific.

Within the upstream regulatory region, four active sequences were identified. Two active sequences contained consensus motifs for the TF GATA1, while the other two sequences contained GATA3 and YY1-like motifs. The two GATA1 consensus sites, both of which bound GATA4 when they were assayed with Sertoli cell nuclear extracts, were in the 3' portion of the upstream regulatory region and contributed modestly to the overall activity of the region [99]. In contrast, the GATA3 and YY1-like sequences within the 5' portion of the region contributed robustly to the activity of the upstream regulatory region because their mutation completely eliminated its activity. The closely positioned GATA3 and YY1-like sites also function synergistically because mutation of either site alone reduced transcriptional activity to the same extent as their simultaneous mutation. Like the GATA1 consensus sites, the GATA3 site, when assayed with Sertoli cell nuclear extracts, bound GATA4, while the specific protein complexes bound to the YY1-like motif have not been identified [99]. Interestingly, YY1 regulates many genes that, like *Dmrt1*, are involved in cell proliferation and differentiation, and it is known to cooperate with GATA3 and GATA4 [107–113]. Furthermore, a role for GATA4 in DMRT1 expression is supported by *in vivo* data, which showed markedly reduced DMRT1 levels in GATA4-deficient testes [114].

As discussed below, studies clearly indicate that forkhead box L2 (FOXL2) silences *Dmrt1* expression in the ovaries to maintain its female sexual identity. To elucidate the potential mechanism behind the FOXL2 silencing, the effects of FOXL2 on the transcription activity of the *Dmrt1* 5'-flanking region were measured by transient transfection analysis, which showed that FOXL2 directly repressed *Dmrt1* transcriptional activity in a manner that depended on the upstream regulatory region (Figure 5.3B) [100]. Together, these studies suggest a mechanism whereby *Dmrt1* transcription in Sertoli cells depends on the cooperation between GATA4 and the protein(s) bound to the YY1-like motif and that FOXL2 represses *Dmrt1* expression by interfering with the activity of these proteins (Figure 5.3B). The finding that FOXL2 and GATA4 interact directly further supports this idea [115].

Transient transfection analysis was employed to evaluate the 5' flanking regions of *dmrt1* in two species of fishes. In one study, 1721 bp of *dmrt1* flanking sequence from the protogynous wrasse (*Halichoeres tenuispinis*) was employed [96]. Activity was not observed in the intact 1721 bp region when transfected into either TM4 or Cos-1 cell lines, which represent Sertoli and fibroblast-like cells, respectively. However, when placed downstream of the SV40 enhancer, the 1721 bp region showed considerable activity in both cell types, and further analysis associated the activity with three main regions, each of approximately 200 bp. The study also found putative binding sites within these regions for Sry (proximal element B), Gata1, Ap4, Gata2 (distal element H), and Gatax (distal element I), but none of the sites was tested for activity or TF binding [96]. In a second study, the 5'-flanking region of zebra fish *dmrt1* was evaluated [104]. In this study, up to 1828 bp of flanking sequence was transfected into a fibroblast-like cell line, Cos-7, and assayed for transcriptional activity. The activity of the region was about twice that of a control lacking the promoter, suggesting that either the cells or the DNA employed were insufficient to support *dmrt1* transcription.

G. Transgenic analysis

In vivo transcriptional activity of the *Dmrt1* 5'-flanking regions from rats, pigs, and frogs has been evaluated [5,100,102]. Yoshimoto and colleagues examined the activity of a 3 kb region of *Xenopus tropicalis dmrt1* 5'-flanking sequence, which they used to direct green fluorescent protein (GFP) expression in transgenic frogs [102].

While the analysis was limited, transgenic tadpoles did express GFP in the gonads, but the authors also observed ectopic expression and did not identify expressing cell types. Thus, the study suggested that the 3 kb region can contribute to gonad expression but does not fully restrict its expression to this site.

Boyer and colleagues used 2.6 kb of porcine *DMRT-1* 5'-flanking sequence to express either GFP (pDMRT1-GFP) or Cre-recombinase (Cre, pDMRT1-Cre) in transgenic mice [5]. Three of 11 *pDMRT1-GFP* lines showed identical expression in the genital ridge at 11.5 dpc. While GFP fluorescence was observed in only a few cells at 11.5 dpc, further analysis of one of the three lines showed robust fluorescence in testes at 12.5, 14.5, and 18.5 dpc. Notably, transgene expression was not observed outside the genital ridge. In developing ovaries, GFP expression was not observed by fluorescence but was detected by RT-PCR at 13.5 and 15.5 dpc but not at 17.5 dpc. After cell sorting, *DMRT-1* transcripts were present only in GFP-positive cells, indicating correspondence endogenous *DMRT1* expression; in the majority of *MIS*-positive Sertoli cells; and in roughly half of the *OCT4*-positive germinal cells. In adult testes, GFP detection was difficult, suggesting that the transgene lacked sequences needed to maintain its expression in adults. In addition, *pDMRT1-Cre*, which directed Cre-mediated expression of a second fluorescent reporter (EYFP) from the R26R-EYFP mouse line, showed similar results, with the exception of sites of ectopic expression, which likely reflects the greater sensitivity of this assay [5]. The authors also noted that the transgene contains the proximal promoter elements identified in rats (i.e., SP1, SP3, and EGR1) and that, while the upstream region was less clearly defined because of sequence alignment shifts, the 2.6 kb porcine sequence contained a GATA4 binding motif.

In vivo activity of the rat *Dmrt1* 5'-flanking region was evaluated using transgenic mice to measure the activity of different lengths of 5'-flanking sequence that ranged from 9 kb to 150 bp [100]. The flanking sequences directed expression of a LacZ reporter, and expression was determined by several methods: tissue specificity by RT-PCR and cellular expression by enzymatic and immunological detection of β-galactosidase (β-gal). Unlike the previous studies, which focused predominantly on prenatal time points, this study examined the activity of rat *Dmrt1* transgenes in prepubertal mice at P15. Thirty-two distinct lines, each containing one of six different transgenes, were evaluated by RT-PCR to determine tissue specificity (Figure 5.3C). The results indicated that different regions are involved in directing *Dmrt1* expression to the testis and in limiting transcriptional influences of the surrounding chromatin. Thus, many transgenes were expressed in the testis, but those having 2.8 kb or less flanking sequence were significantly prone to ectopic expression or complete silencing. IHC and RT-PCR after busulfan treatment, which eliminates germ cells, was used to examine transgene expression in Sertoli cells and germ cells. Both IHC and RT-PCR indicated that the 9 and 3.2 kb flanking sequences could direct expression to both Sertoli cells and germ cells, whereas reporters driven by smaller regions were largely restricted to germ cells. The results indicated that the upstream regulatory region (− 3.2 to −2.8 kb) is important for Sertoli-cell-specific expression and for restricting *Dmrt1* expression to the testis. The results are also consistent with the transient transfection results and thereby implicate the GATA4 binding site and the YY1-like motif in these roles [99–101]. As was observed for the porcine transgenes, postnatal expression levels, particularly in adult mice, were very difficult to detect using histological techniques and were significantly lower than that of endogenous DMRT1, particularly in later generations ([100] and unpublished observation).

The transgenic studies suggest that 2.6 kb of pig flanking sequence and 3.2 kb of rat flanking sequence function similarly with respect to restricting expression to the testis and directing it to the germ cells and Sertoli cells. The studies in rat *Dmrt1* suggest that distinct flanking sequences regulate *Dmrt1* in Sertoli cells and germ cells; the −3.2/−2.8 kb region regulates the former, and sequences within −1.3 kb regulate the latter. Furthermore, both the porcine and rat transgenic studies indicated that a more extensive sequence is required to properly maintain *DMRT1/Dmrt1* expression in testes [5,99–101]. While a good comparison could not be made with the information provided in the frog study, this study too suggested that more sequence is needed for appropriate levels and restricted expression of *dmrt1* in the gonads [102]. Notably, the frog sequence showed no similarity in the regions used for the rat and pig *Dmrt1/DMRT-1* transgenes.

More recently, the TF TFAP2C, which is a target of the germ cell specification factor BLIMP1 (aka PRDM1), was implicated in direct regulation of *Dmrt1* [116,117]. Thus, *Dmrt1* expression was decreased in *Tfap2c*$^{-/-}$ mouse primordial germ cell (PGC)-like cells compared to control PGC-like cells and with reduced *TFAP2C* levels in a human seminoma-like cell line [116]. Chromatin immunoprecipitation (ChIP) suggested that TFAP2C bound to sequences within ECR4 (Figure 5.3A), but the relevance of this finding is uncertain because there was a very large standard deviation in TFAP2C binding enrichment and specific sequences were not tested for binding or transcriptional activity [116]. With respect to some of the studies discussed below, it is noteworthy that, in response to TFAP2C deficiency, the expression changes of other genes were similar to that observed in studies of pluripotency and teratoma formation related to *Dmrt1* deficiency [116,117].

Lastly, two of the ECRs described above, ECR5 and ECR6, were functionally evaluated independently of their identification described above, in a transgenic mouse enhancer assay (Figure 5.3A) [118]. In this study, human genomic sequences were amplified by PCR and cloned into an Hsp68-LacZ reporter vector used to generate transgenic mice. Enhancer activity was determined by β-gal staining of 11.5 dpc embryos, which can be viewed on either the VISTA Enhancer Browser (http://enhancer.lbl.gov) or the UCSC Genome Browser under the VISTA enhancers track [118]. ECR5 was present within a 1296 bp region (element_196) that had no activity in 11.5 dpc embryos, while ECR6, included in a region of 1853 bp (element_112), showed considerable forebrain activity in transgenic embryos. Notably, *DMRT3* is known to be expressed in the forebrains of mice and chickens, suggesting that ECR6 is important for *DMRT3* expression in the forebrain [119]. While ECR5 and ECR6 appear not to regulate *DMRT1* at this time point, they may gain the ability at another time point, and, because *DMRT3* expression is also reported in the gonads, the element(s) may coordinate the expression of both genes [120].

H. Posttranscriptional regulation

Several studies and annotations from public databases have revealed that, in many or all vertebrates, *DMRT1* produces several different transcripts via alternative splicing [19,23,73–75,78,82,84,121–123]. The transcripts typically produce different types if encoded proteins and have different 3'UTR, but their functional significance remains uncertain. Quantification of the different splice variants from zebra fish, mice, and humans showed that transcripts encoding the full-length DMRT1 protein were much more abundant than other variants [74,82,84]. Variation in the 3'UTR may, as suggested by one study, reflect differences in mRNA stability. Hence, in medaka, *dmrt1* mRNA stability was regulated by a conserved sequence in its 3'UTR [13]. The sequence was identified in the 3'-UTRs of both medaka *dmrt1* genes: the autosomal *dmrt1a* and its duplicate *dmrt1bY* [13]. The two *dmrt1* 3'UTRs, when fused to GFP and injected into one-cell-stage embryos, caused an initial robust, uniform fluorescence throughout the embryo that eventually disappeared everywhere except in the primordial gonads [13]. Additional analysis identified an 11-base core sequence (CUGCUGCAGGU) that stabilized chimeric transcripts and, when mutated in the 3'UTR of *dmrt1bY*, dramatically reduced gonad-specific expression [13]. Interestingly, this sequence is present in 3'UTRs of many other vertebrates, including humans and mice, and evaluation of the human 3'UTR fused to GFP showed that it acted in a similar way. While too few studies currently exist to generalize the impact of posttranscriptional regulation on *Dmrt1* expression, the findings that 3' splice variants of *DMRT1* are common in vertebrates and, at least in medaka, *dmrt1* mRNA stability is controlled via the 3'UTR offer the intriguing possibility that *DMRT1* expression may, in part, depend on which 3'UTR splice variant is produced and its stability.

V. DMRT1 function

A. Sexual identity and gonad differentiation

Vertebrate sex determination, while equated with the differentiation of gonads into either ovaries or testes, is directed by a variety of diverse mechanisms typified by two main categories, genetic and nongenetic sex determination (see reviews by Bull and Matson and Zarkower [3,124]). Genetic or chromosomal sex determination is observed in species that have heterogametic sex chromosomes, such as birds and mammals, in which sex is established at fertilization by the nature of the inherited sex chromosomes [125]. Vertebrates employ two types of genetic sex determination systems, the heterogametic male (XY) and the heterogametic female (ZW) system, and it is the differential inheritance of a dominant gene on these chromosomes that determines the developmental pathway taken by the gonad and, consequently, whether the animal will be male or female [72,126–128]. In nongenetic or environmental sex determination, sex is determined by specific environmental conditions (e.g., temperature, nutrient availability, or visual cues) that are present after fertilization during a discrete sensitive period of development [54,72,125]. In general, the initial sex-determining event occurs or stimulates within the gonads the recruitment of other autosomal "sex-determining" genes that direct gonad sex differentiation and typically expressed in a sexually dimorphic manner [54,125,129–132]. In mammals and other vertebrates, the gonad subsequently produces hormones that instruct the rest of the body to develop as either male or female (i.e., sexual differentiation or secondary sex determination).

The conserved features and expression profiles of *DMRT1* have implicated it in various reproductive functions, including the determination and maintenance of sexual identity, testis differentiation, and spermatogenesis. DMRT1 functions differ among vertebrates; in some species, DMRT1 appears to act as a master regulator of sex determination, while in others, it operates at later stages to instruct and/or maintain testis differentiation, regulate gametogenesis, and preserve germ cell unipotency. The following section focuses primarily on the knowledge of DMRT1 function gained from studies in mice and humans.

Studies in mice have contributed significantly to our understanding of DMRT1 and its importance in male fertility. In 2000, Raymond and colleagues first reported the phenotype of *Dmrt1*-null mice and showed that, despite its expression in both male and female embryonic gonads, *Dmrt1* is required only for male fertility and its associated functions are restricted to postnatal testes [10]. Since that time, our understanding of DMRT1 has grown significantly, aided, in particular, by further characterization of $Dmrt1^{-/-}$ mice and the creation and characterization of new models of *Dmrt1* deficiency. These phenotypic analyses have revealed the subtle but biologically relevant effects of *Dmrt1* in embryos and emphasized its importance for postnatal testicular function and its cell-specific roles in Sertoli cells and germ cells.

B. Sertoli cell functions

Sertoli cells are specialized testicular somatic cells with distinct prenatal and postnatal roles. In mice and other eutherian mammals, the indifferent gonads differentiate into testes in response to the sex-determining region of Chr Y (SRY) [133,134]. This initial sex-determining signal occurs within the gonads' somatic cells and induces their differentiation into Sertoli cells. In the embryo, Sertoli cells orchestrate testis development and, in postnatal testes, establish and maintain the required environment for germ cell development [135–139]. As revealed through various *Dmrt1* knockout, transgenic, and rescue mice, DMRT1 loss affects only postnatal Sertoli cells, in which it plays important autonomous and nonautonomous roles in their maturation and maintenance of sexual identity [8,10,11,32,140].

C. Sertoli cell maturation: neonatal and prepubertal gene expression changes

DMRT1 was first implicated in Sertoli cell maturation when proliferation and/or differentiation defects were suggested by the overabundance of immature Sertoli cells in seminiferous tubules of $Dmrt1^{-/-}$ mice [10]. In these mice, increased

Sertoli cell numbers were noted from approximately P14 onward, and the morphology and expression profiles for GATA1 and GATA4 indicated an immature phenotype [10]. In a subsequent study, Sertoli cell numbers were measured in testes from *Dmrt1*$^{-/-}$ mice of different ages (P2 to P20) [11]. Between P2 and P7, the number of Sertoli cells per seminiferous tubule was similar and stable in testes from both wt and *Dmrt1*$^{-/-}$ mice. From P7 onward, Sertoli cell numbers remained relatively stable in wt mice, with only a modest increase at P10 followed by a gradual decline. In *Dmrt1*$^{-/-}$ mice, however, Sertoli cell numbers rose more sharply, particularly between P10 and P11, and then declined slightly at P12 and P14 and rose again at P20, at which time the Sertoli cell number per tubule was more than twice that observed in wt mice [11]. In the absence of altered apoptosis, the results indicated that the increased Sertoli cell numbers resulted from a defect in proliferation [11]. Normally, mouse Sertoli cell proliferation begins prenatally, decreases after birth, and ceases by about P15 [138,141,142]. Thus, in *Dmrt1*$^{-/-}$ mice, Sertoli cell proliferation began and progressed normally until around P10, when proliferation continued or increased instead of coming to a stop [11,142]. The timing of Sertoli cell divergence from normalcy is significant because it corresponds to an important maturation step for Sertoli cells, which gain their adult-like characteristics between P10 and P14 [138,142]. The findings suggest that DMRT1 is important for Sertoli cell maturation and the increased cell number occurred because the cells did not cease proliferating as a consequence of their failed terminal differentiation.

Studies that evaluated gene expression differences in the testes of control (wt and *Dmrt1*$^{+/-}$) and *Dmrt1* mutant mice further supported a role of DMRT1 in Sertoli cell maturation [138]. Although variability exists between species, there are several useful markers that help characterize the differentiation status of Sertoli cells and distinguish between mature and immature Sertoli cells [138]. In humans and rodents, expression of the androgen receptor (AR) has been used to distinguish immature and mature Sertoli cells [138,142]. In humans, AR expression is associated with mature Sertoli cells, and loss or weak expression of AR indicates maturation failure [138,143]. AR expression in rat Sertoli cells is associated with maturing Sertoli cells, and its loss or reduction indicates maturation failure [138,144]. In addition to AR, other markers of mature Sertoli cells include GATA-1, sulfated glycoprotein 2, and laminin alpha 5, while makers of immature Sertoli cells include cytokeratin-18 (*Krt18*) and anti-Müllerian hormone [138,145].

The initial characterization of the *Dmrt1*$^{-/-}$ mice first suggested abnormal Sertoli cell differentiation when IHC revealed that GATA-1 expression was delayed and GATA-4 expression persisted in mutant testes compared to control animals [10]. Similar changes were observed in mice carrying a conditional deletion of *Dmrt1* in Sertoli cells (*SCDmrt1KO* mice), indicating that the changes were due to the specific effects of DMRT1 in Sertoli cells [8]. Quantification of *Gata-1* and *Gata-4* transcripts in P7 testes showed that *Gata-1* mRNA was reduced (approximately two-fold) and *Gata-4* mRNA was elevated (approximately threefold) in *Dmrt1*$^{-/-}$ mice compared to control animals [32]. *Gata-1* and *Gata-4* transcripts were also measured in P7 testes from *Dmrt1*$^{-/-}$ mice carrying a transgene that expressed DMRT1 only in Sertoli cells of the testis (also known as *Dmrt1*$^{-/-;tg}$ mice) [32]. In *Dmrt1*$^{-/-;tg}$

mice, *Gata-1* mRNA levels matched that of *Dmrt1*$^{-/-}$ mice, while *Gata-4* mRNA levels were partially rescued, suggesting that, at P7, testicular transcript levels of *Gata-1* were insensitive to DMRT1 in Sertoli cells and of *Gata-4* were partly sensitive [32]. Together, these results suggest that DMRT1 in Sertoli cells controls the timing of GATA-1 and GATA-4 expression and that DMRT1 in both Sertoli cells and germ cells influences the magnitude of their expression.

The expression of *Ar* and *Krt18* was also measured in testes of wt, *Dmrt1*$^{-/-}$, and *Dmrt1*$^{-/-;tg}$ mice. Quantification of P7 mRNA levels showed that *Ar* mRNA was similar in the three genotypes and that *Krt18* mRNA levels were dramatically higher (approximately sixfold) in *Dmrt1*$^{-/-}$ and *Dmrt1*$^{-/-;tg}$ testes compared to wt testes [32]. IHC supported the AR expression results at P7 but, at P15, when Sertoli cells had completed their differentiation, Sertoli cell AR immunoreactivity was noticeably lower in *Dmrt1*$^{-/-}$ mice compared to wt mice and, in *Dmrt1*$^{-/-;tg}$ mice, it was markedly improved relative to *Dmrt1*$^{-/-}$ mice but did not reach the levels observed in wt mice [32]. Likewise, IHC showed similar AR levels in Sertoli cells of *SCDmrt1KO* and control mice at P5 and reduced AR levels in Sertoli cells of *SCDmrt1KO* testes at P9 and P21 [8]. Together, these studies indicate that DMRT1 impacts Sertoli cell maturation as early as P7 and continues to induce changes that ultimately thwart their terminal differentiation.

D. Sertoli cell morphology: integrity, polarity, and nuclear structure

In addition to the expression profiles described above, mutant and wt testes were examined for morphological features associated with Sertoli cell maturation [138,32]. After birth, Sertoli cells undergo significant morphological changes. Initially, Sertoli cells are irregularly oriented, cuboidal, or columnar cells. However, with age, Sertoli cells evolve into complex, elongated cells that spread from the basement membrane to the lumen that are oriented with their long axes perpendicular to the basal lamina and form numerous cytoplasmic extensions with developing germ cells [146–148]. During maturation, the apical–lateral surfaces of adjacent Sertoli cells develop junctional complexes, which provide them with the polarity and structural integrity that support germ cell development. From birth to puberty, Sertoli cell nuclei also undergo significant changes, increasing in both size and complexity condensation patterns [146–148]. The nuclei of neonatal Sertoli cells are round or oval and have few nuclear membrane infoldings, finely granular chromatin, and unremarkable nucleoli [148]. As the cells mature, their nuclei become oblong or elongated with borders that contain prominent infoldings and reticular nucleoli that have a central area of condensation [148].

E. Nuclear structure

Sertoli cell nuclear structure was evaluated in electron micrographs of testis sections from wt, *Dmrt1*$^{-/-}$, and *Dmrt1*$^{-/-;tg}$ mice [32]. At P7, Sertoli cell nuclei of wt mice had a columnar shape and few if any membrane infoldings, while those of *Dmrt1*$^{-/-}$ and *Dmrt1*$^{-/-;tg}$ mice were reminiscent of mature nuclei, with their irregular shape and prominent infoldings [32]. Sertoli cell nuclei in the two mutant mouse models also had a larger average cross-sectional area than those from wt mice. By P42, Sertoli cell nuclei in wt mice had grown (approximately twofold) and gained the invaginations and chromatin characteristic (central condensation area) of mature Sertoli cells. However, in the *Dmrt1*$^{-/-}$ mice, the nuclei were of similar in size to or smaller than at P7, no longer contained the infoldings seen at P7, and did not acquire the typical chromatin characteristics seen in wt mice [32]. The Sertoli cell nuclei from P42 *Dmrt1*$^{-/-;tg}$ mice shared features with both wt and *Dmrt1*$^{-/-}$ mice, displaying a combination of their nuclear membrane topographies and nucleolar structure, which suggested that Sertoli cells in *Dmrt1*$^{-/-;tg}$ mice maintained an intermediate maturation state between the mature Sertoli cells of wt mice and the immature Sertoli cells of *Dmrt1*$^{-/-}$ mice [32]. The data also indicated that returning DMRT1 to Sertoli cells only partially rescued their maturation defect, implying that DMRT1 in germ cells or the presence of germ cells is important to Sertoli cell maturation. Interestingly, the nuclear features of Sertoli cells from *Dmrt1*$^{-/-}$ mice were similar to those observed in men who have undergone long-term estrogen treatment, suggesting that a hormonal imbalance might accompany the DMRT1 deficiency [149].

F. Structural integrity and polarity

In further support of DMRT1 as an important maturation factor for postnatal Sertoli cells, it was found that their epithelial structure and polarity were compromised in *Dmrt1*-mutant mice. Other than the observed nuclear changes at P7, no structural changes in the seminiferous epithelium were reported for *Dmrt1*$^{-/-}$, *Dmrt1*$^{-/-;tg}$, or *SCDmrt1KO* mice at either P1 (*Dmrt1*$^{-/-}$) or P7 (*Dmrt1*$^{-/-;tg}$ and *SCDmrt1KO*) [8,10,32]. However, disorganized Sertoli cells were noted in the *SCDmrt1KO* mice at P9 and in *Dmrt1*$^{-/-}$ mice at P10. By P14, the seminiferous tubules of both mice lacked perceptible lumens, indicating that the epithelial structure normally formed by Sertoli cells had failed [8,10]. At the later time points of P28 in *SCDmrt1KO* mice and P42 in *Dmrt1*$^{-/-}$ mice, the testes had deteriorated significantly, with little evidence of any tubule-like structures [8,10,32]. However, testes from P42 *Dmrt1*$^{-/-;tg}$ mice showed significant improvement compared to *Dmrt1*$^{-/-}$ mice because the former contained clearly defined tubules with lumens [32]. Consistent with a role in Sertoli cell terminal differentiation, the results from *SCDmrt1KO* mice and comparison of *Dmrt1*$^{-/-}$ and *Dmrt1*$^{-/-;tg}$ mice demonstrated that, during the second postnatal week, DMRT1 is needed in Sertoli cells to establish an intact seminiferous epithelium.

Further investigation of *Dmrt1* mutant mice linked their effects on seminiferous epithelium integrity to changes in expression and localization of proteins that establish junctional complexes and polarity of the Sertoli cells. In vertebrates, there are three major types of multiprotein complexes referred to as cell junctions; anchoring junctions (adhesion junctions, desmosomes, hemidesmosomes), communicating junctions (gap junctions), and occluding junctions (tight junctions) [150–155]. Testes of *Dmrt1* mutant mice were evaluated for expression changes in proteins that contribute to occluding and adhesion junctions. Expression profiling of *Dmrt1*$^{-/-}$ testes showed that the mRNA for *occludin* (*Ocln*), which encodes a major component of tight junctions and consequently of the BTB, was the most highly reduced transcript at P2. IHC revealed that OCLN levels were greatly reduced in *Dmrt1*$^{-/-}$ testes at P14 and undetectable by P20 [11]. Other tight junction proteins, claudin-11 (CLDN11) and tight junction protein ZO-1 (ZO-1), were examined by IHC, which revealed that levels of CLDN11, like OCLN, were greatly reduced in *Dmrt1*$^{-/-}$ testes at P14, while ZO-1, with levels similar to controls, was mislocalized and confined to the lumens of the seminiferous tubules [11]. *Ocln* and *Cldn11* transcripts were also measured in testes from P7 *Dmrt1*$^{+/+}$, *Dmrt1*$^{-/-}$, and *Dmrt1*$^{-/-;tg}$ mice [32]. As with the IHC results, the expression of *Ocln* and *Cldn11* mRNA was significantly reduced in *Dmrt1*$^{-/-}$ mice relative to *Dmrt1*$^{+/+}$ mice [11,32]. However, the reduced expression of *Ocln* and *Cldn11* in *Dmrt1*$^{-/-}$ mice was not rescued in the *Dmrt1*$^{-/-;tg}$ mice, indicating that DMRT1 in Sertoli cells is not sufficient to maintain the expression of these genes.

Expression of the proteins associated with ectoplasmic specializations (ESs) was also evaluated in *Dmrt1* mutant mice. ESs are testis-specific adherens junctions that are required for both seminiferous epithelium integrity and spermatid maturation. ESs are classified as either basal ESs, which join adjacent Sertoli cells and form part of the BTB, or apical ESs, which join Sertoli cells to developing spermatids [153,154,156,157]. Transcript levels were measured for *Espn*, which encodes ESPIN, an actin-binding/bundling and adapter protein of basal and apical ESs [158–161]. At P7, *Espn* mRNA levels were approximately 2.5-fold higher in testes from *Dmrt1*$^{-/-}$ mice than in testes from *Dmrt1*$^{+/+}$ or *Dmrt1*$^{-/-;tg}$ mice [32]. Thus, *Dmrt1*$^{-/-}$ transgenic mice (*Dmrt1*$^{-/-;tg}$), which produce DMRT1 only in Sertoli cells, expressed normal levels of *Espn*, indicating that the gene is regulated autonomously by DMRT1 in Sertoli cells. ESPIN, ACTIN, and poliovirus receptor-related protein 2 (also known as NECTIN-2), another ES adapter protein, were also localized in control and *Dmrt1* mutant testes by IHC [162–168]. In postpubertal (P42) mice, IHC showed that ESPIN staining in *Dmrt1*$^{+/+}$ testes was localized with actin at the basal and apical ESs, while, in *Dmrt1*$^{-/-}$ testes, the staining was reduced and found in irregular, random patches without the polarity noted in *Dmrt1*$^{+/+}$ testes [32]. In *Dmrt1*$^{-/-;tg}$ testes, ESPIN staining localized with ACTIN only in the adluminal area, along clearly oriented projections. Similar results were observed for NECTIN-2, which, in *Dmrt1*$^{-/-}$ testes, was barely detectable and, in *Dmrt1*$^{-/-;tg}$ testes, was distributed similarly to ESPIN. The findings revealed that DMRT1 is needed for proper expression and localization of the ES proteins and, consequently, junction integrity and cell polarity. More specifically, in Sertoli cells, DMRT1 controls the levels of ESPIN and NECTIN-2 and directs their localization to the adluminal compartment and apical ESs [32]. In brief, the findings indicate that DMRT1 regulates Sertoli cell structural functions and polarity via autonomous and nonautonomous actions that control the expression and distribution of proteins that contribute to the BTB and apical and basal ESs.

G. Antagonizing forkhead box L2 (FOXL2) and Sertoli cell differentiation

FOXL2 is a winged-helix/forkhead TF that is highly conserved in vertebrates and, in the gonad, displays female-specific expression (recently reviewed in Ref. [169]). As with DMRT1 activities in the testis, FOXL2 is a key factor in ovarian differentiation, and both expression profiles and functional analyses indicate that *Dmrt1* and *Foxl2* have antagonistic roles in the gonads [169–174]. This antagonism was first implicated by expression studies performed in developing gonads of various species of fishes, mammals, reptiles, and birds, which revealed that *Foxl2* was expressed in an inverse manner to *Dmrt1*, with *Foxl2* expression coinciding with ovarian development and *Dmrt1* expression with testis development [54,62,67,122,170,175–185]. In developing mouse ovaries, *Foxl2* expression was first observed at 12.5 dpc and then increased at 13.5 dpc, while little or no *Foxl2* expression was observed in developing mouse testes [100,172,173,181,186]. Notably, in mice, the silencing of *Dmrt1* observed in ovarian somatic cells corresponded to the induction of *Foxl2* because DMRT1 expression was clearly evident in the somatic cells prior to the emergence of *Foxl2* (i.e., at 11.5 dpc) but dropped dramatically with the induction of *Foxl2* expression at 12.5 and 13.5 dpc (Figure 5.1F) [20]. Furthermore, in species in which gonad sexual identity can be manipulated by temperature and/or steroids, numerous studies have shown that increased *Dmrt1* expression and suppressed *Foxl2* expression favors testis development and, conversely, increased *Foxl2* expression and suppressed *Dmrt1* expression favor ovarian development [36,45,52,54,55,60,62,65,67,182,187].

In addition to the numerous expression studies, functional studies in mice and fish support the antagonistic relationship between FOXL2 and DMRT1. In mice, deletion of *Foxl2* caused female-specific infertility and ovarian defects that indicated that *Foxl2* is needed to secure female sexual identity by maintaining the ovarian somatic program and, consequently, preventing the transformation of granulosa cells into their male counterparts, the Sertoli cells [169,171,173]. Changes in *Foxl2*$^{-/-}$ ovaries were not evident until after birth, when female-to-male sex reversal was indicated by the continued presence of fetal sex cords and gene expression changes that showed a loss of female-specific somatic cell markers and a gain of male-specific markers, including *Dmrt1* and the TF SRY-box 9 (SOX9), which is expressed in Sertoli cells and required for testis development [171,173,174,188,189]. As implicated by the many expression studies, DMRT1 in mouse Sertoli cells appears to be a male counterpart of FOXL2 in granulosa cells because *SCDmrt1KO* mice also lose their gender-specific programming, indicated by the loss of male-specific markers and gain of female-specific markers in Sertoli cells [140]. FOXL2, which is normally absent from testes, was expressed in testes of P28 *Dmrt1*$^{-/-}$ and *SCDmrt1KO* mice, but not in testes of mice with a conditional deletion of *Dmrt1* in germ cells (*GCDmrt1KO*) [140]. In *SCDmrt1KO* testes, FOXL2 was not observed until around P14, when it was observed in a limited number of cells. However, between P14 and P28, the number of FOXL2-expressing cells increased significantly and, proportionately as the number of

SOX9-expressing cells decreased [3]. Consequently, by P28, most Sertoli cells had lost SOX9 and gained FOXL2 expression as well as other feminizing genes, including *Esr1*, *Esr2*, *Wnt4*, and *Rspo1* [140]. Furthermore, when *Dmrt1* was deleted from adult mice using a ubiquitous, tamoxifen-inducible Cre, some FOXL2-positive cells were noted in the testis 28 days after *Dmrt1* deletion [140].

Interestingly, antagonistic effects of FOXL2 and DMRT1 were recently noted in the Nile tilapia (*Oreochromis niloticus*) [190]. In this study, transcription activator-like effector nucleases were used to produce deleterious mutations in *foxl2* and *dmrt1* [190]. Similarly to the observations in mice, *dmrt1* deficiency induced *foxl2* expression in the gonads, and *foxl2* deficiency induced *dmrt1* expression [190]. Thus, the antagonistic functions of DMRT1 and FOXL2 are not limited to mammals, and, when considered with the expression data from various species, it appears that their antagonism and ability to preserve gonad sexual identity may be conserved across all vertebrates. Interestingly, in mice, the sexual transformations caused by DMRT1 and FOXL2 deficiency occurred postnatally and, therefore, considerable time passed between gene deletion and phenotype. Why was the delay so long between *Foxl2* and *Dmrt1* deletion and the resulting cell fate changes? Currently, the answer is unknown, but the available evidence does suggest a model in which DMRT1 and FOXL2 help preserve established genomic states that dictate the sexual identity of Sertoli cells and granulosa cells, respectively, and their loss destabilizes this state, resulting in its slow but inevitable deterioration.

H. Germ cell functions

Germ cell development begins with the establishment of unipotent PGCs, which colonize the gonads after migrating from their position outside the embryo proper along the hindgut to the indifferent gonads [191–193]. In mice, PGCs proliferate in the gonads until approximately 13.5 dpc and, depending on the sex of the gonadal somatic cells, either become oocytes or gonocytes (also known as prospermatogonia) [191,192,194]. In mouse embryos, DMRT1 was present in the PGCs of XX and XY gonads, indicated by IHC of DMRT1 and GCNA1 at 11.5 dpc, but not in germ cells that had transitioned into oocytes and gonocytes [20]. DMRT1 expression remained silent in the female germ line, while it reemerged in the male germ line just after birth, a time when gonocytes migrate to the basement membrane of the seminiferous cords, become A undifferentiated spermatogonia, and resume proliferation (Figure 5.2A, top left) [20,192]. Not all gonocytes, however, become undifferentiated spermatogonia because the first round of mouse spermatogenesis is initiated directly from gonocytes [195]. During this neonatal period, the number of DMRT1-positive germ cells continues to increase, and, by P7, nearly all express DMRT1. With further maturation, DMRT1 is lost from the germ cells as they differentiate and enter meiosis. The expression of DMRT1 in PGCs, early spermatogenic precursors, proliferating cells, and undifferentiated and differentiation spermatogonia suggested that, in germ cells, DMRT1 helps to sustain an undifferentiated state and proliferative capacity.

I. Gonocyte development

In mice, DMRT1 is important in the early postnatal period (i.e., within the first week) for the transition of gonocytes to spermatogonia. In *Dmrt1*$^{-/-}$ mice, both germ cell mitosis and radial migration to the tubule periphery were compromised [10,11]. Hence, most (approximately 75%) gonocytes in neonatal *Dmrt1*$^{-/-}$ mice remained in the center of the seminiferous tubules, while, in wt mice, most (approximately 75%) migrated to the tubule periphery [11]. Conditional deletions of *Dmrt1* from germ cells (*Tnap*-directed) and Sertoli cells (*Dhh*-directed) showed that loss of *Dmrt1* in germ cells caused many but not all germ cells to remain in the tubule center, while loss of *Dmrt1* in Sertoli cells indicated delayed migration because, unlike wt testes, germ cells in the mutant mice were evident in tubule centers at P5 but most had migrated to the tubule periphery by P7 [8]. A loss of germ cell migration was also observed in the *Dmrt1*$^{-/-;tg}$ mice [32]. In all, it appears that DMRT1 in Sertoli cells influences the timing of germ cell migration, and, in germ cells, it is essential to their migration.

In addition to migration, DMRT1 is needed for germ cell expansion in the neonatal testis [11,32]. At P2, ratios of testis to body weight were similar among wt, *Dmrt1*$^{-/-}$, and *Dmrt1*$^{-/-;tg}$ mice, but, by P7, differences suggested growth deficits in the two mutant models [32]. Histological evaluation of P7 testes revealed reduced germ cell numbers in *Dmrt1*$^{-/-}$ and *Dmrt1*$^{-/-;tg}$ mice compared to wt mice [32]. In addition, the number of cells expressing phosphohistone H3, a mitotic marker, was reduced at P7 (below the level attributable to cell loss), and lower phosphohistone H3 levels were indicated at P2–P6 [11]. Thus, in contrast to its effects on Sertoli cells, DMRT1 deficiency reduced proliferation and germ cell numbers. Furthermore, staining for γH2AX, which is expressed in intermediate and B spermatogonia and preleptotene to zygotene spermatocytes, was greatly reduced in P7 *Dmrt1*$^{-/-}$ testes, indicating that germ cell maturation was defective [11]. Overall, the studies show that DMRT1 is needed during the neonatal period to support germ cell migration, proliferation, and differentiation, all of which accompany the transformation of gonocytes into A undifferentiated spermatogonia. In a consideration of its role in Sertoli cells, the findings suggest that, during the neonatal/prepubertal period, DMRT1 helps establish a balanced cellular environment in the seminiferous epithelium that can continue to support ongoing spermatogenesis.

J. Mitotic–meiotic transition

In both male and female germ cells, DMRT1 is implicated in the mitotic-to-meiotic transition. This transition requires retinoic acid (RA) signaling and the gene stimulated by RA 8 (*Stra8*), which, as it name suggests, is transcriptionally activated by RA [196–199]. In embryonic ovaries, altered expression of STRA8, SYCP3, and γH2AX indicated that meiotic prophase was atypical in female germ cells of *Dmrt1*$^{-/-}$ mice [200]. At 13.5 dpc, quantitative RT-PCR showed that *Stra8* mRNA levels were significantly reduced in *Dmrt1*$^{-/-}$ ovaries compared to controls, and IHC confirmed this finding in ovaries from 13.5 to 15.5 dpc *Dmrt1*$^{-/-}$ and wt mice [200]. In addition, qChIP showed that DMRT1 was bound to the *Stra8* promoter region in 13.5 dpc ovaries but not in 13.5 dpc testes, suggesting that DMRT1 directly activates *Stra8* transcription in mouse embryonic ovaries but not embryonic testes [200]. IHC of SYCP3, a protein that forms part of the synaptonemal complex, and γH2AX, a phosphorylated histone H2AX associated with double-strand breaks and homologous recombination, showed significantly reduced expression of both proteins in *Dmrt1*$^{-/-}$ embryonic ovaries [200]. Yet, at P4, expression of the diplotene-specific protein MSY2 was similar between *Dmrt1*$^{-/-}$ and wt ovaries, indicating that, despite the preceding expression changes, the mutant germ cells progressed through the meiotic prophase to diplonema. Notably, *Dmrt1*$^{-/-}$ ovaries at P8 were smaller than those of wt mice, with fewer primordial (approximately 50% reduced), primary (approximately 25% reduced) and secondary (approximately 25% reduced) follicles [200]. Inexplicably, however, the significantly reduced follicle numbers in *Dmrt1*$^{-/-}$ females did not diminish fertility, even under continuous breeding conditions [200].

In male germ cells, the meiotic transition occurs after birth. *GCDmrt1KO* using a *Neurog3-cre* transgene resulted in defects that indicated that DMRT1 manages the entry of male germ cells into meiosis [28]. In these mice, there were obvious signs of disrupted spermatogenesis in adult testes, and, remarkably, some tubules contained spermatogonia, spermatocytes, and spermatids, indicating that the mutant germ cells had completed meiosis. However, the interpretation of these results was complicated by the fact that *Neurog3* (and the *Neurog3*-directed Cre transgene) was initially expressed in germ cells at ~P3–P4 and *Neurog3*-positive germ cells did not contribute to the first wave of spermatogenesis [195,201–203]. Nevertheless, the continual presence of STRA8-positive germ cells, which were undergoing DNA replication (determined by 5-bromo-2-deoxyuridine, BrdU), indicated that they were regularly being produced from undifferentiated spermatogonia that lacked DMRT1. This finding led the authors to conclude that deletion of *Dmrt1* in undifferentiated spermatogonia caused their uncontrolled meiotic entry [28]. Entry into meiosis was supported by the presence of SYPC3 in germ cells of mutant testes [28]. Like normal meiosis, the precocious entry into meiosis was RA dependent.

In all, this study indicated that one of DMRT1's key germ cell functions is to prevent spermatogonia from differentiating by maintaining their undifferentiated state. This conclusion was further supported by a qChIP study, which suggested that DMRT1 directly regulates transcription of *Stra8* and spermatogenesis- and oogenesis-specific basic helix–loop–helix 1 (*Sohlh1*), which encodes a TF expressed in undifferentiated spermatogonia that is needed for spermatogonial development [200,204]. Loss of SOHLH1 in *Neurog3*-GC*Dmrt1KO* testes is consistent with a role for DMRT1 activation of *Sohlh1* transcription. Although STRA8 protein levels appeared to be elevated in *Neurog3*-GC*Dmrt1KO* mutant testes, *Stra8* mRNA levels at P9 were unchanged compared to wt mice [28,205]. However, this finding may reflect the late activity of the *Neurog3-cre* transgene rather than an absence of silencing by DMRT1. Interestingly, despite use of the same binding sites in the *Stra8* promoter region, DMRT1 functions very differently in the ovary, where it activates *Stra8* transcription, than in the testis, where it silences *Stra8* [28,200].

K. Pluripotency and cancer

The importance of DMRT1 to pluripotency and formation of germ cell tumors (GCTs) has been emphasized in studies that employed *Dmrt1* mutant mice and *Dmrt1* knockdown in mouse SSC cultures [9,206,207]. One of the earliest associations between *Dmrt1* and pluripotency was in the reported expression profiles of neonatal *Dmrt1*$^{-/-}$ mice, which showed that mRNA levels of developmental pluripotency-associated 3 (*Dppa3*) increased 3.6-fold and 17-fold in testes from P1 and P2 mutant mice, respectively [11]. *Dppa3* (also known as *Stella* and *Pgc7*) is expressed in PGCs, undifferentiated embryonic cells, and GCTs and is a marker of pluripotency [208,209]. Its induction in *Dmrt1*$^{-/-}$ mice suggests that DMRT1 opposes pluripotency and/or regulates the transition of PGCs to gonocytes. Yet evidence for this, such as the formation of GCTs, was absent in these mice and other *Dmrt1* mutant mice on genetic backgrounds that were C57BL/6J (B6) or mixed [9,207].

The link between DMRT1 and GCTs, however, was revealed when the *Dmrt1* deletions were in the 129Sv genetic background. In this background, 90% of the *Dmrt1*$^{-/-}$ mice formed teratomas, compared to 4% of heterozygous mice and none in wt [9]. Furthermore, 63% of the 129Sv mice with a conditional deletion of *Dmrt1* in germ cells (*Tnap-cre* driven) developed teratomas, compared to 42% with a Sertoli cell conditional deletion (*Dhh-cre* driven) and 9% with only the floxed alleles. Germ cell proliferation, as revealed by Ki67 IHC, was 10-fold higher in *Dmrt1*$^{-/-}$ 129Sv embryos at 16.5 dpc than in wt embryos, indicating that DMRT1 facilitates the PGCs' mitotic arrest [9]. Yet studies in neonatal mice of mixed backgrounds indicated, by the reduced phosphohistone H3 levels in *Dmrt1*$^{-/-}$ germ cells, that DMRT1 facilitated entry into or progression through the mitotic cell cycle [11]. While it is difficult to judge to what extent genetic background differences contributed to the conflicting outcomes, an interpretation consistent with both observations and its role in meiotic entry is that DMRT1 helps facilitate key germ cell transitions—that is, the transitions from proliferating PGC to quiescent gonocyte, from quiescent gonocyte to proliferating undifferentiated spermatogonia, and from spermatogonia to spermatocyte. How, mechanistically, DMRT1 manages the transitions or whether its activity transcends from one transition to the next has yet to be determined.

The connection between GCTs and DMRT1 was further supported by additional gene expression changes observed in testes of wt and *Dmrt1* mutant mice on 129Sv and C57BL/6J backgrounds [9,207]. In 129Sv mice, IHC of newborn testes showed that *Dmrt1* deletion caused elevated levels of SOX2 as well as other proteins associated with pluripotency (e.g., NANOG, OCT3/4, and E-cadherin) [9]. Quantitative RT-PCR corroborated the finding by showing increased levels of their corresponding mRNAs in mutant testes from 15.5 dpc embryos and revealed decreased levels of *Ret*, a receptor for GDNF [9]. Moreover, *Sox2* was implicated as a direct target of DMRT1 by ChIP, which showed DMRT1 bound to sequences in its promoter region, suggesting that DMRT1 silences *Sox2* expression directly [9].

Gene expression profiles of wt and *Dmrt1*-mutant testes from 15.5 dpc 129Sv and C57BL/6J embryos were used to perform pathway analysis. This analysis revealed that DMRT1 regulated pluripotency genes regardless of the genetic background. However, in the embryos on the 129Sv background, the deletion altered the expression of more pluripotency-associated genes and showed greater induction for some of these genes (e.g., *Sox2*, *E-cad*, *Foxd3*), indicating that DMRT1 had greater silencing potential in the 129Sv mouse strain [207]. Expression changes in the *Dmrt1* mutants also revealed that DMRT1 regulates NODAL, NOTCH, and GDNF signaling pathways in both strains, but, again, more significant changes were noted in the 129Sv mutant mice [207]. Further genetic analysis suggested that *Nanos3*, a gene encoding an RNA binding protein, and *Dmrt1* have related functions in regulating pluripotency and proliferation because their combined heterozygosity increased tumor susceptibility [207].

Experiments in SSC cultures, also known as germ-line stem (GS) cells, emphasized the importance of DMRT1 in regulating germ cell pluripotency and revealed important mechanistic details, including its silencing of *Sox2*, that support its role in maintaining germ cell identity [206]. In GS cells, knockdown of *Dmrt1* or the DNA methyltransferase (cytosine-5) 1 (*Dnmt1*), a maintenance methylase, resulted in programmed cell death [206]. However, when programmed cell death was inhib ited in the GS cells by knockdown or deletion of transformation related protein 53 (p53), knockdown of either *Dmrt1* or *Dmnt1* converted the unipotent GS cells into embryonic stem-like cells or multipotent GS (mGS) cells, which are referred to as *Dmrt1*-mGS and *Dmnt1*-mGS cells, respectively [206]. Notably, *Dmrt1* expression was reduced in the *Dmnt1*-mGS cells, and the expression of *Dmrt1* cDNA during *Dnmt1*-mGS cell formation significantly reduced the number of mGS cells formed, indicating that *Dmrt1* acted downstream of *Dnmt1* in regulating pluripotency [206]. Both *Dmrt1*- and *Dnmt1*-derived mGS cells showed reduced DNA methylation at the H19 locus, but the reduction was greater with *Dmrt1* than with *Dnmt1* knockdown. Further comparisons showed that, relative to *Dnmt1* knockdown, *Dmrt1* knockdown produced mGS cells at a higher frequency and within a shorter period of time [206]. Comparisons of wt GS cells to *Dmrt1*-mGS cells indicated that *Dmrt1*-mGS cells had acquired an expression profile with enhanced features of pluripotency and decreased features of germ cell identity. In favor of pluripotency, *Dmrt1*-mGS cells began to express *Nanog* and *Cripto* and increased the expression of *Sox2*, *Oct4*, *Utf1*, *Eras*, and *Etv5* [206]. The cells also showed decreased expression of cyclin-dependent kinase inhibitor 2C (aka $p18^{INK4c}$). Declining germ cell identity was indicated by the loss of *Nanos3* and the reduction in *Sohlh1* and *Mvh* [206]. The continued expression of *Sohlh1* and *Mvh*, albeit at lower levels, as well as the presence of other genes associated with germ cell functions (i.e., *Neurog3*) suggested that *Dnmt1*-mGS cells still retained some spermatogonial characteristics. The *Dmrt1*-mGS cells could also form teratomas when transplanted under the subcutaneous tissues of nude mice and, in culture, differentiated into cells that expressed a neuronal marker [206]. Furthermore, when embryoid bodies generated from *Dmrt1*-mGS cells were evaluated for differentiation, cells were identified that expressed α-smooth muscle actin, a mesodermal marker, and α-fetoprotein, suggestive of endodermal lineage. When injected into blastocysts, the cells also produced chimeric offspring, although with low efficiency [206]. The study suggested that

DMRT1 is required for SSC survival and, by suppressing pluripotency genes and maintaining germ cell identity genes, for maintenance of their unipotent characteristics. The concept that DMRT1 is important to germ cell unipotency was introduced in an earlier study on human PGCs and various pluripotent cell types. After comparing the expression profiles of these cells, the authors concluded that upregulation of DMRT1 helped determined the unipotent state of PGCs [210].

Although *Dmrt1*-mutant gonads from 129Sv and C57BL/6J embryos and spermatogonia with combined *Dmrt1* and *p53* deficiency represent very different biological models, their results have several important similarities that underscore some important functions of DMRT1 with respect to pluripotency [9,28,206,207]. In particular, each identified increased *Sox2* expression as an important component of the mechanism linking *Dmrt1* loss to pluripotency and tumor formation, and each model implicated induction of *Oct4* and loss of *Nanos3* as contributing factors [9,206,207]. The studies in *Dmrt1*-mGS cells provided important functional evidence that supported roles for *Oct4* and *Sox2* in the regulation of pluripotency by DMRT1. Accordingly, double knockdown of *Dmrt1* and *Sox2* in p53-deficient GS cells inhibited the formation of *Dmrt1*-mGS cells, indicating that *Sox2* is required to promote pluripotency in response to DMRT1 loss [206]. Furthermore, overexpression of *Sox2* in p53-deficient GS cells induced *Oct4* and produced mGS (*Sox2*-mGS) cells, and formation of *Sox2*-mGS cells was blocked by *Oct4* knockdown [206]. Therefore, *Oct4* is required for *Sox2*-induced pluripotency. With the mouse studies, the data support a model in which DMRT1 prevents pluripotency, at least in part, by directly silencing *Sox2*, which, in turn, inhibits *Oct4* expression.

The authors of the analyses of *Dmrt1*-mGS cells and *Dmrt1* mutant mice reported other similarities that are indicative of pluripotency in SSCs and GCTs. These similarities include increases in other genes associated with pluripotency, in particular *Nanog* and *Eras*, and perturbations in DNA methylation suggested by *H19* hypomethylation in *Dmrt1*-mGS cells and misregulated *Dmnt3l* and *Dmnt3b* in *Dmrt1* mutant mice [9,28,206,207]. The effects on DNA methylation are intriguing because, if they are globally applied, may be an important component of reprogramming the germ cells. Thus, either through an unrecognized function or its regulation of genes that control DNA methylation, DMRT1 may help establish and/or maintain a unique DNA methylation state that suppresses pluripotency genes and accommodates the germ cell identity genes that determine unipotency. Regardless of the mechanisms involved, the studies clearly demonstrate that DMRT1 deficiency induces pluripotency but only does so in combination with other genetic changes, such as loss of p53 or switching to the 129Sv genetic background.

Interestingly, data from *Dmrt1* knockdown in wt GS cells (i.e., with functional p53) and *Dmrt1* mutant mice suggested that *Dmrt1* deficiency induces apoptosis only of neonatal germ cells. Wt GS cells are cultured germ cells isolated from newborn testes [206,211–215]. Therefore, at the time of isolation, the germ cells were gonocytes and, at the time of *Dmrt1* knockdown, likely represented neonatal undifferentiated spermatogonia and/or gonocytes [206]. In these cells, *Dmrt1* knockdown caused apoptosis [206]. Unfortunately, the effects on germ cell apoptosis in *Dmrt1* mutant mice are more difficult to determine. *Dmrt1*$^{-/-}$ embryos (all strains) and P10 mice showed no change in germ cell apoptosis relative to control mice [10,200,207]. In addition, no change in apoptosis was observed in mice with a conditional deletion of *Dmrt1* in germ cells at 1, 2, and 6 months of age [28]. In these mice, deletion of *Dmrt1* was directed by *Neurog3-cre*, which does not efficiently delete genes in neonatal germ cells. In fact, expression studies of *Neurog3* and *Neurog3-cre* indicate that efficient germ cell deletion does not occur until several weeks after birth, and some SSCs may escape deletion [28,202,203]. However, an indication that *Dmrt1* deficiency does influence germ cell apoptosis in mice was identified in a report on neonatal *Dmrt1*$^{-/-}$ mice at ages P1–P7 [11]. In these mice, IHC indicated that germ cells did not progress beyond the gonocyte stage, and the authors noted that there were some cells with a nuclear morphology that was characteristic of apoptosis but provided no definitive evidence for this comment [11]. Without a closer evaluation of germ cell apoptosis in *Dmrt1*$^{-/-}$ mice, it is uncertain whether apoptosis occurs in the mice or whether it is a phenomenon that is related to cell culture. However, if apoptosis does occur in the *Dmrt1*$^{-/-}$ mice, the reports, thus far, suggest that it is restricted to some but not all neonatal germ cells because germ cells are present and apoptosis normal in *Dmrt1*$^{-/-}$ mice at P10 and some germ cells in *Dmrt1*$^{-/-}$ mice are mitotically active at P3 [10,11]. If apoptosis is restricted to a subset of germ cells, then, perhaps, these are selected for when the SSC cultures are generated.

L. DMRT1 and GCTs in humans

Consistent with the studies in mice, *DMRT1* in humans has been linked to testicular GCTs (TGCTs) [9,206,207,216–218]. Human TGCTs originate from carcinoma *in situ* cells, which are thought to be developmentally arrested, transformed fetal germ cells that survived throughout childhood [219]. As with mouse mGS cells and GCTs, human TGCTs often express genes affiliated with pluripotency, such as *OCT4* and *NANOG*, and exhibit changes in DNA methylation, particularly hypomethylation at the imprinted *H19* locus [206,220–224]. Notably, several genome-wide association (GWA) studies of TGCTs identified *DMRT1* as a susceptibility locus. In 2010, Turnbull and colleagues identified the TGCT-associated single nucleotide polymorphism (SNP) rs745383, which was located in the second intron of *DMRT1*, and was subsequently identified in three independent GWA studies [216,217,225,226]. In contrast, one 52-case study found no correlation between rs745383 and pediatric GCTs [227]. However further support for *DMRT1* as a susceptibility allele in human TGCTs was provided by a fifth GWA study that identified a different TGCT-associated SNP, rs7040024, located in the first intron of DMRT1 [217].

Expression profiling of various types of human TGCTs showed that *DMRT1* expression was increased in spermatocytic seminomas and decreased in embryonal carcinomas, while expression in seminomas and dysgerminomas was inconsistent [9,218]. Notably, spermatocytic seminomas are derived from postnatal germ cells (spermatogonia or spermatocytes), while other TGCTs arise from fetal germ cells (PGCs or gonocytes) [228]. Thus, in tumors derived from the more differentiated germ cells, expression/overexpression of *DMRT1* correlated with tumor formation. In TGCTs derived from undifferentiated fetal germ cells, reduced *DMRT1* expression correlated with tumor formation. Interestingly, expression profiling also revealed a strong correlation between the expression of *DMRT1* expression and that of *RET* (a GDNF receptor), *E-CADHERIN*, *NANOG*, and *CDKN2C*, such that tumors with higher DMRT1 levels (i.e., spermatocytic seminomas) had higher levels of *RET* and $P18^{INK4C}$ and lower levels of *E-CADHERIN* and *NANOG*, while tumors with lower *DMRT1* levels (i.e., embryonal carcinomas) also had lower *RET* and $P18^{INK4C}$ and higher *E-CADHERIN* and *NANOG* levels [207]. While $P19^{INK4D}$ expression was also measured, the correlation was less consistent. Thus, the changes in gene expression and DNA methylation observed in human TGCTs are similar to those observed in GCTs in mice and *Dmrt1*-mGS cells, suggesting that the latter are good models for investigating the mechanisms responsible for TGCT formation in humans and identifying possible treatments for these tumors.

M. Testis development and infertility in humans

In addition to TGCTs, *DMRT1* in humans is associated with male disorders related to sexual development and infertility [1,9,216–218,229–231]. Evidence for the importance of *DMRT1* to male sexual development derives from studies that link its position on chromosome 9 (9p24.3) to disorders of sexual development (DSD; i.e., testis dysgenesis and XY sex reversal) [1,85,232–240]. Partial deletions in the short arm of chromosome 9 are responsible for the syndrome known as "chromosome 9p deletion syndrome" or "monosomy 9p syndrome" (online Mendelian inheritance in man 158170), which is characterized by mental retardation, craniofacial abnormalities, and delayed motor development [235,241–244]. However, some XY patients with monosomy 9p syndrome also present with a spectrum of genital and/or gonadal abnormalities, including external genitalia phenotypes ranging from complete female (XY sex reversal) to male with hypospadias and gonadal phenotypes ranging from complete gonadal dysgenesis and ovotestes, cryptorchid, and/or hypoplastic testes [234,237,241,243–246]. In addition, cases of DSD without other features of monosomy 9p syndrome have been identified in connection with smaller 9p24.3 deletions [240,247,248]. Chromosome mapping results of such deletions revealed that combined hemizygosity of *DMRT1*, *DMRT2*, and *DMRT3* were strongly associated with testis dysgenesis and XY sex reversal in patients with 9p deletions and strongly implicated haploinsufficiency of one or more of these *DMRT* genes as causative in 9p-associated DSD [85,232–236,238,242,247–251]. There are, however, a few reports that apparently contradict this interpretation because these authors mapped the 9p deletion breakpoints of two patients with 46XY sex reversal to the intergenic region between *KANK1* and *DMRT1*, and, hence, the patients were homozygous for all three *DMRT* genes [235,240,252]. Notably, the breakpoints were positioned approximately 6.5–25 kb and 40 kb 5' to *DMRT1* and, therefore, reasonably within the range that is likely to contain an important regulatory sequence for the proper expression of *DMRT1* and possibly *DMRT2* and

DMRT3. In fact, ECR1, a proposed distal regulatory region of *DMRT1*, is located approximately 40 kb upstream of human *DMRT1* (Figure 5.3A).

Even though multiple cases of nonsyndromic 46XY DSD show that point mutations in the *DMRT1* coding region are rare, the functional ties of *DMRT1* orthologs to reproduction and recent analysis of smaller 9p deletions place the evidence in favor of *DMRT1* haploinsufficiency as the primary or sole contributing factor to DSD in XY individuals with 9p deletions [85,232–234,249]. Of the three *DMRT* genes implicated in 9p DSD, only *DMRT1* has a defined role in testis development and function, and the known roles of DMRT2 and DMRT3 are unrelated to reproduction [10,253–255]. *Dmrt2*$^{-/-}$ mice showed defects in somite development, but the gonads at 18.5 dpc were normal [253]. However, early lethality prevented an evaluation of the effects on the gonads after birth. In horses and mice, DMRT3 is important for the circuitry of spinal interneurons [254,255].

Although the phenotype observed in *Dmrt1*$^{-/-}$ mice differs in several ways from testis disorders of patients with 9p deletions, this phenotype revealed that *Dmrt1* is needed in mice to maintain testis differentiation [10,28,140]. Therefore, aside from the number of alleles required (i.e., two in humans and one in mice) and the window of activity, the outcomes largely agree with respect to testis differentiation. In mice, the timing or window of *Dmrt1* activity is postnatal with respect to sexual development; therefore, its deletion does not influence masculinization, and XY pups are all born male. In humans with DSD and 9p deletions, the active window is *in utero*, but the specific timing appears to differ between individuals, resulting in the varied outcomes of the syndrome. If the defect arises early in testis development, androgen production is compromized and the individual is feminized. This effect offers a better explanation than a direct influence of DMRT1 on sex determination of cases of 9p deletions with XY sex reversal in which complete feminization occurred. In fact, histological evaluation of the gonads from a 46, XY fetus with 9p deletion and ambiguous genitalia led the authors to suggest that the gene(s) on 9p were required for maintenance of testis differentiation rather than for sex determination [239]. However, a role for DMRT1 in sex determination cannot be emphatically ruled out because there exists at least one report of a patient with XY sex reversal and ovarian tissue that showed association to *DMRT1* [248]. The less severe phenotypes of the syndrome (e.g. hypospadias) are consistent with later windows of activity, in which testis development is less disrupted and androgen production is normal or only moderately reduced.

In addition to evidence for its functional ties to testis differentiation, support for *DMRT1* acting in human testis development was provided by reports of patients with XY gonadal dysgenesis and small 9p deletions that implicated only *DMRT1*. In one study, high-resolution, array-comparative genomic hybridization identified chromosome abnormalities in a 46, XY female patient with testicular tissue on one side and an ovary on the other [248]. This study revealed a 35 kb deletion that affected only exons 3 and 4 of *DMRT1*. While the patient also carried 11q24.2 and 19p13.3 duplications, the genes within these regions have no known connection to gonadogenesis [248]. A second small deletion restricted to the coding sequence of *DMRT1* was identified in a 46, XY patient with gonadal dysgenesis [256].

The deletion was 103 kb in length and spanned approximately 88 kb upstream of *DMRT1* into its second intron, thereby deleting its first two exons and a considerable portion of its 5'-flanking sequence. In addition, deletions were identified in a genome-wide SNP and copy number variant analysis of idiopathic male infertility that linked *DMRT1* to human spermatogenic failure [231]. In this study, five *DMRT1* deletions were uncovered from 1302 infertile men, while none were found in the 7353 male controls [231].

Although *DMRT1* is most frequently included in the 9p deletions associated with testicular defects, variability in 9p deletions and phenotypic outcomes emphasize the uncertainty regarding the specific contributions of *DMRT1* haploinsufficiency as well as temporal plasticity with respect to the activity of the deleted gene(s). Such variable outcomes may result from *DMRT1* hemizygosity alone and reflect variable expression from the undeleted *DMRT1* allele or the status of other factors (i.e., modifier genes) that influence testis development and function [231]. These variable outcomes also might result from differences in the deleted sequences that affect the function or expression of other genes associated with *DMRT1*. Even the smallest deletions (i.e., the ones that effect only *DMRT1*) may remove important regulatory sequences that control other genes that influence the phenotype. Regardless of these possibilities, there is significant evidence that *DMRT1* has an important role in human testis development and/or function, the specifics of which await the identification and characterization of more refined *DMRT1* mutations in patients with testicular abnormalities.

VI. Conclusions

Since 1998, studies of *DMRT1* in a variety of vertebrate species have contributed greatly to our knowledge of its conserved importance in reproduction and support its pivotal position in pathways that control testis differentiation and germ-line potential. Although many features of DMRT1 are still poorly understood, there is considerable promise that, through its continued evaluation, we will gain greater knowledge of the mechanistic processes that maintain Sertoli cell differentiation and control germ cell unipotency and development. Some of the more pressing areas needing investigation, described below, reflect deficits in information on the factors that regulate DMRT1 and the mechanistic features of its transcriptional activity.

Knowledge of how DMRT1 functions as a TF, both with respect to its biochemical properties and mechanism of action, is significantly lacking. Studies have identified *in vitro* binding site motifs (t/c t/g G n G t/a A C A t/a T G T t/a g/t c/t) and sequences enriched at *in vivo* ChIP binding sites (similar to the *in vitro* motif as well as C-rich motifs) and shown that it preferably binds DNA as a dimer *in vitro* [200,207,257]. In addition, more than 4000 DMRT1 binding sites, and, consequently, putative target genes have been identified, although the actual number of individual sites is likely to be smaller because there was correction for redundancy between studies [9,28,140,200,205,207,258–260]. Regardless of these strides, we still know little about DMRT1's functional domains, interacting factors, or post-translational modifications that alter its capacity to regulate transcription. Furthermore, associated functional data supporting putative target genes as true direct targets of DMRT1 are available for only a small percentage of these genes. Quantitative ChIP has confirmed that DMRT1 binds to a number of identified sites, but only a small proportion of the target genes showed reported expression changes in response to DMRT1. With few exceptions, the expression changes were reported using *Dmrt1* mutant models in which close temporal correlation between changes in DMRT1 and target gene expression was not possible and primary (direct) effects could not be distinguished from indirect (secondary) effects. Thus, to date, there are but a few examples of studies showing that DMRT1 elicited a direct transcriptional effect on a target gene element or promoter, and even the results of those studies may have occurred not via DNA binding but through interaction with a bound TF [258–260].

Unfortunately, the studies are complicated by the lack of a good cell system for studying DMRT1 and the lack of appropriate function of some regulatory elements in standard transcriptional assays. However, given that target gene identity is fundamental to our understanding of DMRT1 and its activities in Sertoli cells and germ cells, greater emphasis needs to be placed on confirming binding sequences and establishing that DMRT1 controls the expression of proposed target gene(s). At least for germ cells, studies suggest that GS cells will be an excellent model system for mechanistic studies of *Dmrt1*.

There are also many lapses in our understanding of what controls the expression of DMRT1. Studies in several species indicate the importance of GATA factors in the regulation of *Dmrt1*, and advances have been made in understanding its specific expression in Sertoli cells, in which evidence indicates that GATA4 plays an important role. However, evidence also indicates that DMRT1 expression in Sertoli cells is established and maintained by several unknown factors that bind regulatory sequences in proximity to the GATA4 binding sites and at distal sites. Genome analysis also implicated distal regulatory elements in the regulation of DMRT1 expression, and transient transfection analysis indicated that transcriptional regulation of *Dmrt1* by FSH requires sequences outside its 5'-flanking sequence (unpublished observation). Defining a genomic region that recapitulates *Dmrt1* expression *in vivo* will facilitate the search for distal elements because defining this region will limit the amount of genomic sequence data to be queried and provides a platform for genomics data to help guide enhancer identification.

In germ cells, almost nothing is known about what regulates DMRT1 expression. In particular, there is no knowledge of how *Dmrt1* expression is restricted to PGCs and then later to male germ cells, nor of how *Dmrt1* is silenced when germ cells differentiate (i.e., enter meiosis) or enter mitotic arrest or what reactivates *Dmrt1* in gonocytes shortly after birth. These events are implicated in the control of germ cell pluripotency and cancer as well as the regulation of spermatogenesis. Therefore, their disclosure is certain to shed new light on these events.

Lastly, posttranscriptional regulation of DMRT1 is another important understudied area. In almost every species examined, various transcripts are produced for

DMRT1. Yet little or nothing is known about how these transcripts are regulated or what, if any, functional significance they have. The apparently greater variability at the 3' end suggests that some variants may play an important role in transcript stability, which, in the medaka *dmrt1* genes, is controlled by a conserved regulatory sequence in the 3'-UTR. Given that the sequence is found in many vertebrates, investigation of this sequence and transcript stability in other vertebrates may reveal an important and conserved mechanism for controlling DMRT1 levels.

In closing, there remains much more to learn about DMRT1 and its structure, function, and regulation. It is hoped that its intriguing properties and significance for male reproduction will continue to stimulate interest from the research community that will advance our understanding of DMRT1 and the pathway to masculinity.

References

[1] Raymond CS, Shamu CE, Shen MM, Seifert KJ, Hirsch B, Hodgkin J, et al. Evidence for evolutionary conservation of sex-determining genes. Nature 1998;391:691–5.

[2] Zarkower D. DMRT genes in vertebrate gametogenesis. Curr Top Dev Biol 2013;102:327–56.

[3] Matson CK, Zarkower D. Sex and the singular DM domain: insights into sexual regulation, evolution and plasticity. Nat Rev Genet 2012;13:163–74.

[4] Raymond CS, Kettlewell JR, Hirsch B, Bardwell VJ, Zarkower D. Expression of Dmrt1 in the genital ridge of mouse and chicken embryos suggests a role in vertebrate sexual development. Dev Biol 1999;215:208–20.

[5] Boyer A, Dornan S, Daneau I, Lussier J, Silversides DW. Conservation of the function of DMRT1 regulatory sequences in mammalian sex differentiation. Genesis 2002;34:236–43.

[6] Kettlewell JR, Raymond CS, Zarkower D. Temperature-dependent expression of turtle Dmrt1 prior to sexual differentiation. Genesis 2000;26:174–8.

[7] Moniot B, Berta P, Scherer G, Sudbeck P, Poulat F. Male specific expression suggests role of DMRT1 in human sex determination. Mech Dev 2000;91:323–5.

[8] Kim S, Bardwell VJ, Zarkower D. Cell type-autonomous and non-autonomous requirements for Dmrt1 in postnatal testis differentiation. Dev Biol 2007;307:314–27.

[9] Krentz AD, Murphy MW, Kim S, Cook MS, Capel B, Zhu R, et al. The DM domain protein DMRT1 is a dose-sensitive regulator of fetal germ cell proliferation and pluripotency. Proc Natl Acad Sci USA 2009;106:22323–8.

[10] Raymond CS, Murphy MW, O'Sullivan MG, Bardwell VJ, Zarkower D. *Dmrt1*, a gene related to worm and fly sexual regulators, is required for mammalian testis differentiation. Genes Dev 2000;14:2587–95.

[11] Fahrioglu U, Murphy MW, Zarkower D, Bardwell VJ. mRNA expression analysis and the molecular basis of neonatal testis defects in Dmrt1 mutant mice. Sex Dev 2007;1:42–58.

[12] Rhen T, Schroeder A. Molecular mechanisms of sex determination in reptiles. Sex Dev 2010;4:16–28.

[13] Herpin A, Nakamura S, Wagner TU, Tanaka M, Schartl M. A highly conserved cis-regulatory motif directs differential gonadal synexpression of Dmrt1 transcripts during gonad development. Nucleic Acids Res 2009;37:1510–20.

[14] Ferguson-Smith M. The evolution of sex chromosomes and sex determination in vertebrates and the key role of DMRT1. Sex Dev 2007;1:2–11.

[15] Smith CA, Roeszler KN, Ohnesorg T, Cummins DM, Farlie PG, Doran TJ, et al. The avian Z-linked gene DMRT1 is required for male sex determination in the chicken. Nature 2009;461:267–71.

[16] Koopman P. Sex determination: the power of DMRT1. Trends Genet 2009;25:479–81.

[17] Aoyama S, Shibata K, Tokunaga S, Takase M, Matsui K, Nakamura M. Expression of Dmrt1 protein in developing and in sex-reversed gonads of amphibians. Cytogenet Genome Res 2003;101:295–301.

[18] Kobayashi T, Matsuda M, Kajiura-Kobayashi H, Suzuki A, Saito N, Nakamoto M, et al. Two DM domain genes, DMY and DMRT1, involved in testicular differentiation and development in the medaka, *Oryzias latipes*. Dev Dyn 2004; 231:518–26.

[19] Sreenivasulu K, Ganesh S, Raman R. Evolutionarily conserved, DMRT1, encodes alternatively spliced transcripts and shows dimorphic expression during gonadal differentiation in the lizard, *Calotes versicolor*. Mech Dev 2002;119(Suppl. 1):S55–64.

[20] Lei N, Hornbaker KI, Rice DA, Karpova T, Agbor VA, Heckert LL. Sex-specific differences in mouse DMRT1 expression are both cell type- and stage-dependent during gonad development. Biol Reprod 2007;77:466–75.

[21] Osawa N, Oshima Y, Nakamura M. Molecular cloning of Dmrt1 and its expression in the gonad of *Xenopus*. Zoolog Sci 2005;22:681–7.

[22] Shibata K, Takase M, Nakamura M. The Dmrt1 expression in sex-reversed gonads of amphibians. Gen Comp Endocrinol 2002;127:232–41.

[23] Raghuveer K, Senthilkumaran B. Identification of multiple dmrt1s in catfish: localization, dimorphic expression pattern, changes during testicular cycle and after methyltestosterone treatment. J Mol Endocrinol 2009;42:437–48.

[24] Jorgensen A, Morthorst JE, Andersen O, Rasmussen LJ, Bjerregaard P. Expression profiles for six zebrafish genes during gonadal sex differentiation. Reprod Biol Endocrinol 2008;6:25.

[25] De Grandi A, Calvari V, Bertini V, Bulfone A, Peverali G, Camerino G, et al. The expression pattern of a mouse doublesex-related gene is consistent with a role in gonadal differentiation. Mech Dev 2000;90:323–6.

[26] Enders GC, May 2nd JJ. Developmentally regulated expression of a mouse germ cell nuclear antigen examined from embryonic day 11 to adult in male and female mice. Dev Biol 1994;163:331–40.

[27] Bellve AR, Cavicchia JC, Millette CF, O'Brien DA, Bhatnagar YM, Dym M. Spermatogenic cells of the prepuberal mouse. Isolation and morphological characterization. J Cell Biol 1977;74:68–85.

[28] Matson CK, Murphy MW, Griswold MD, Yoshida S, Bardwell VJ, Zarkower D. The mammalian doublesex homolog DMRT1 is a transcriptional gatekeeper that controls the mitosis versus meiosis decision in male germ cells. Dev Cell 2010;19:612–24.

[29] de Rooij DG. Stem cells in the testis. Int J Exp Pathol 1998;79:67–80.

[30] Bullejos M, Koopman P. Germ cells enter meiosis in a rostro-caudal wave during development of the mouse ovary. Mol Reprod Dev 2004;68:422–8.

[31] Menke DB, Koubova J, Page DC. Sexual differentiation of germ cells in XX mouse gonads occurs in an anterior-to-posterior wave. Dev Biol 2003;262:303–12.

[32] Agbor VA, Tao S, Lei N, Heckert LL. A *Wt1-Dmrt1* transgene restores DMRT1 to sertoli cells of *Dmrt1(−/−)* testes: a novel model of DMRT1-deficient germ cells. Biol Reprod 2013;88:51.

[33] Chen JK, Heckert LL. Dmrt1 expression is regulated by follicle-stimulating hormone and phorbol esters in postnatal Sertoli cells. Endocrinology 2001;142:1167–78.

[34] Sambroni E, Lareyre JJ, Le Gac F. FSH controls gene expression in fish both independently of and through steroid mediation. PLoS One 2013;8:e76684.

[35] Rolland AD, Lardenois A, Goupil AS, Lareyre JJ, Houlgatte R, Chalmel F, et al. Profiling of androgen response in rainbow trout pubertal testis: relevance to male gonad development and spermatogenesis. PLoS One 2013;8:e53302.

[36] Vizziano D, Baron D, Randuineau G, Mahe S, Cauty C, Guiguen Y. Rainbow trout gonadal masculinization induced by inhibition of estrogen synthesis is more physiological than masculinization induced by androgen supplementation. Biol Reprod 2008;78:939–46.

[37] Baron D, Montfort J, Houlgatte R, Fostier A, Guiguen Y. Androgen-induced masculinization in rainbow trout results in a marked dysregulation of early gonadal gene expression profiles. BMC Genomics 2007;8:357.

[38] Baron D, Houlgatte R, Fostier A, Guiguen Y. Expression profiling of candidate genes during ovary-to-testis trans-differentiation in rainbow trout masculinized by androgens. Gen Comp Endocrinol 2008;156:369–78.

[39] Marchand O, Govoroun M, D'Cotta H, McMeel O, Lareyre J, Bernot A, et al. DMRT1 expression during gonadal differentiation and spermatogenesis in the rainbow trout, *Oncorhynchus mykiss*. Biochim Biophys Acta 2000;1493:180–7.

[40] Kobayashi T, Kajiura-Kobayashi H, Guan G, Nagahama Y. Sexual dimorphic expression of DMRT1 and Sox9a during gonadal differentiation and hormone-induced sex reversal in the teleost fish Nile tilapia (*Oreochromis niloticus*). Dev Dyn 2008;237:297–306.

[41] Raghuveer K, Garhwal R, Wang DS, Bogerd J, Kirubagaran R, Rasheeda MK, et al. Effect of methyl testosterone- and ethynyl estradiol-induced sex differentiation on catfish, *Clarias gariepinus*: expression profiles of DMRT1, cytochrome P450aromatases and 3 beta-hydroxysteroid dehydrogenase. Fish Physiol Biochem 2005;31:143–7.

[42] Kanamori A, Yamamura A, Koshiba S, Lee JS, Orlando EF, Hori H. Methyltestosterone efficiently induces male development in the self-fertilizing hermaphrodite fish, *Kryptolebias marmoratus*. Genesis 2006;44:495–503.

[43] Rajakumar A, Singh R, Chakrabarty S, Murugananthkumar R, Laldinsangi C, Prathibha Y, et al. Endosulfan and flutamide impair testicular development in the juvenile Asian catfish, *Clarias batrachus*. Aquat Toxicol 2012;110-111:123–32.

[44] Filby AL, Thorpe KL, Maack G, Tyler CR. Gene expression profiles revealing the mechanisms of anti-androgen- and estrogen-induced feminization in fish. Aquat Toxicol 2007;81:219–31.

[45] Vizziano-Cantonnet D, Baron D, Mahe S, Cauty C, Fostier A, Guiguen Y. Estrogen treatment up-regulates female genes but does not suppress all early testicular markers during rainbow trout male-to-female gonadal transdifferentiation. J Mol Endocrinol 2008;41:277–88.

[46] Rhee JS, Kim BM, Lee CJ, Yoon YD, Lee YM, Lee JS. Bisphenol a modulates expression of sex differentiation genes in the self-fertilizing fish, *Kryptolebias marmoratus*. Aquat Toxicol 2011;104:218–29.

[47] Zhang X, Zha J, Wang Z. Influences of 4-nonylphenol on doublesex- and mab-3-related transcription factor 1 gene expression and vitellogenin mRNA induction of adult rare minnow (*Gobiocypris rarus*). Environ Toxicol Chem 2008;27:196–205.

[48] Zhang X, Gao L, Yang K, Tian H, Wang W, Ru S. Monocrotophos pesticide modulates the expression of sexual differentiation genes and causes phenotypic feminization in

zebrafish (*Danio rerio*). Comp Biochem Physiol C Toxicol Pharmacol 2013;157:33–40.

[49] Schulz RW, Bogerd J, Male R, Ball J, Fenske M, Olsen LC, et al. Estrogen-induced alterations in amh and dmrt1 expression signal for disruption in male sexual development in the zebrafish. Environ Sci Technol 2007;41:6305–10.

[50] Murdock C, Wibbels T. Dmrt1 expression in response to estrogen treatment in a reptile with temperature-dependent sex determination. J Exp Zool B Mol Dev Evol 2006;306:134–9.

[51] Kamata R, Takahashi S, Morita M. Gene expression of sex-determining factors and steroidogenic enzymes in the chicken embryo: influence of xenoestrogens. Gen Comp Endocrinol 2004;138:148–56.

[52] Bannister SC, Smith CA, Roeszler KN, Doran TJ, Sinclair AH, Tizard ML. Manipulation of estrogen synthesis alters MIR202* expression in embryonic chicken gonads. Biol Reprod 2011;85(1):22–30.

[53] Lucas TF, Lazari MF, Porto CS. Differential role of the estrogen receptors ESR1 and ESR2 on the regulation of proteins involved with proliferation and differentiation of Sertoli cells from 15-day-old rats. Mol Cell Endocrinol 2014;382:84–96.

[54] Matsumoto Y, Crews D. Molecular mechanisms of temperature-dependent sex determination in the context of ecological developmental biology. Mol Cell Endocrinol 2012;354:103–10.

[55] Leet JK, Gall HE, Sepulveda MS. A review of studies on androgen and estrogen exposure in fish early life stages: effects on gene and hormonal control of sexual differentiation. J Appl Toxicol 2011;31:379–98.

[56] Nakamura M. Sex determination in amphibians. Semin Cell Dev Biol 2009;20:271–82.

[57] Nakamura M. Is a sex-determining gene(s) necessary for sex-determination in amphibians? Steroid hormones may be the key factor. Sex Dev 2013;7:104–14.

[58] Murdock C, Wibbels T. Expression of Dmrt1 in a turtle with temperature-dependent sex determination. Cytogenet Genome Res 2003;101:302–8.

[59] Shoemaker C, Ramsey M, Queen J, Crews D. Expression of Sox9, Mis, and Dmrt1 in the gonad of a species with temperature-dependent sex determination. Dev Dyn 2007;236:1055–63.

[60] Shoemaker-Daly CM, Jackson K, Yatsu R, Matsumoto Y, Crews D. Genetic network underlying temperature-dependent sex determination is endogenously regulated by temperature in isolated cultured *Trachemys scripta* gonads. Dev Dyn 2010;239:1061–75.

[61] Torres Maldonado LC, Landa Piedra A, Moreno Mendoza N, Marmolejo Valencia A, Meza Martinez A, Merchant Larios H. Expression profiles of *Dax1*, *Dmrt1*, and *Sox9* during temperature sex determination in gonads of the sea turtle *Lepidochelys olivacea*. Gen Comp Endocrinol 2002;129:20–6.

[62] Rhen T, Metzger K, Schroeder A, Woodward R. Expression of putative sex-determining genes during the thermosensitive period of gonad development in the snapping turtle, *Chelydra serpentina*. Sex Dev 2007;1:255–70.

[63] Sakata N, Miyazaki K, Wakahara M. Up-regulation of *P450arom* and down-regulation of *Dmrt-1* genes in the temperature-dependent sex reversal from genetic males to phenotypic females in a salamander. Dev Genes Evol 2006;216:224–8.

[64] Fernandino JI, Hattori RS, Shinoda T, Kimura H, Strobl-Mazzulla PH, Strussmann CA, et al. Dimorphic expression of dmrt1 and cyp19a1 (ovarian aromatase) during early gonadal development in pejerrey, *Odontesthes bonariensis*. Sex Dev 2008;2:316–24.

[65] Baroiller JF, D'Cotta H, Saillant E. Environmental effects on fish sex determination and differentiation. Sex Dev 2009;3:118–35.

[66] Blasco M, Fernandino JI, Guilgur LG, Strussmann CA, Somoza GM, Vizziano-Cantonnet D. Molecular characterization of cyp11a1 and cyp11b1 and their gene expression profile in pejerrey (*Odontesthes bonariensis*) during early gonadal development. Comp Biochem Physiol A Mol Integr Physiol 2010;156:110–8.

[67] Poonlaphdecha S, Pepey E, Canonne M, de Verdal H, Baroiller JF, D'Cotta H. Temperature induced-masculinisation in the Nile tilapia causes rapid up-regulation of both dmrt1 and amh expressions. Gen Comp Endocrinol 2013;193:234–42.

[68] Yanez-Cuna JO, Kvon EZ, Stark A. Deciphering the transcriptional cis-regulatory code. Trends Genet 2013;29:11–22.

[69] Alfoldi J, Lindblad-Toh K. Comparative genomics as a tool to understand evolution and disease. Genome Res 2013;23:1063–8.

[70] Wittkopp PJ, Kalay G. Cis-regulatory elements: molecular mechanisms and evolutionary processes underlying divergence. Nat Rev Genet 2012;13:59–69.

[71] Hardison RC, Taylor J. Genomic approaches towards finding cis-regulatory modules in animals. Nat Rev Genet 2012;13:469–83.

[72] Mawaribuchi S, Yoshimoto S, Ohashi S, Takamatsu N, Ito M. Molecular evolution of vertebrate sex-determining genes. Chromosome Res 2012;20:139–51.

[73] Anand A, Patel M, Lalremruata A, Singh AP, Agrawal R, Singh L, et al. Multiple alternative splicing of Dmrt1 during gonadogenesis in Indian mugger, a species exhibiting temperature-dependent sex determination. Gene 2008;425:56–63.

[74] Guo Y, Cheng H, Huang X, Gao S, Yu H, Zhou R. Gene structure, multiple alternative splicing, and expression in gonads of zebrafish Dmrt1. Biochem Biophys Res Commun 2005;330:950–7.

[75] Karolchik D, Barber GP, Casper J, Clawson H, Cline MS, Diekhans M, et al. The UCSC Genome Browser database: 2014 update. Nucleic Acids Res 2014;42:D764–70.

[76] Brunner B, Hornung U, Shan Z, Nanda I, Kondo M, Zend-Ajusch E, et al. Genomic organization and expression of the doublesex-related gene cluster in vertebrates and detection of putative regulatory regions for DMRT1. Genomics 2001;77:8–17.

[77] El-Mogharbel N, Wakefield M, Deakin JE, Tsend-Ayush E, Grutzner F, Alsop A, et al. DMRT gene cluster analysis in the platypus: new insights into genomic organization and regulatory regions. Genomics 2007;89:10–21.

[78] Zhao Y, Lu H, Yu H, Cheng H, Zhou R. Multiple alternative splicing in gonads of chicken DMRT1. Dev Genes Evol 2007;217:119–26.

[79] Wan QH, Pan SK, Hu L, Zhu Y, Xu PW, Xia JQ, et al. Genome analysis and signature discovery for diving and sensory properties of the endangered Chinese alligator. Cell Res 2013;23:1091–105.

[80] Shetty S, Kirby P, Zarkower D, Graves JA. DMRT1 in a ratite bird: evidence for a role in sex determination and discovery of a putative regulatory element. Cytogenet Genome Res 2002;99:245–51.

[81] Kent WJ, Sugnet CW, Furey TS, Roskin KM, Pringle TH, Zahler AM, et al. The human genome browser at UCSC. Genome Res 2002;12:996–1006.

[82] Cheng HH, Ying M, Tian YH, Guo Y, McElreavey K, Zhou RJ. Transcriptional diversity of DMRT1 (dsx- and mab3-related transcription factor 1) in human testis. Cell Res 2006;16:389–93.

[83] Jang S, Zhou F, Xia L, Zhao W, Cheng H, Zhou R. Construction of a BAC library and identification of *Dmrt1* gene of the rice field eel, *Monopterus albus*. Biochem Biophys Res Commun 2006;348:775–80.

[84] Lu H, Huang X, Zhang L, Guo Y, Cheng H, Zhou R. Multiple alternative splicing of mouse Dmrt1 during gonadal differentiation. Biochem Biophys Res Commun 2007;352:630–4.

[85] Raymond CS, Parker ED, Kettlewell JR, Brown LG, Page DC, Kusz K, et al. A region of human chromosome 9p required for testis development contains two genes related to known sexual regulators. Hum Mol Genet 1999;8:989–96.
[86] Ottolenghi C, Fellous M, Barbieri M, McElreavey K. Novel paralogy relations among human chromosomes support a link between the phylogeny of doublesex-related genes and the evolution of sex determination. Genomics 2002;79:333–43.
[87] Johnsen H, Andersen O. Sex dimorphic expression of five dmrt genes identified in the Atlantic cod genome. The fish-specific dmrt2b diverged from dmrt2a before the fish whole-genome duplication. Gene 2012;505:221–32.
[88] Nanda I, Zend-Ajusch E, Shan Z, Grutzner F, Schartl M, Burt DW, et al. Conserved synteny between the chicken Z sex chromosome and human chromosome 9 includes the male regulatory gene DMRT1: a comparative (re)view on avian sex determination. Cytogenet Cell Genet 2000;89:67–78.
[89] Kondo M, Froschauer A, Kitano A, Nanda I, Hornung U, Volff JN, et al. Molecular cloning and characterization of DMRT genes from the medaka *Oryzias latipes* and the platyfish *Xiphophorus maculatus*. Gene 2002;295:213–22.
[90] Kersey PJ, Allen JE, Christensen M, Davis P, Falin LJ, Grabmueller C, et al. Ensembl Genomes 2013: scaling up access to genome-wide data. Nucleic Acids Res 2014;42:D546–52.
[91] Akalin A, Fredman D, Arner E, Dong X, Bryne JC, Suzuki H, et al. Transcriptional features of genomic regulatory blocks. Genome Biol 2009;10:R38.
[92] Engstrom PG, Ho Sui SJ, Drivenes O, Becker TS, Lenhard B. Genomic regulatory blocks underlie extensive microsynteny conservation in insects. Genome Res 2007;17:1898–908.
[93] Kikuta H, Laplante M, Navratilova P, Komisarczuk AZ, Engstrom PG, Fredman D, et al. Genomic regulatory blocks encompass multiple neighboring genes and maintain conserved synteny in vertebrates. Genome Res 2007;17:545–55.
[94] Parveen N, Masood A, Iftikhar N, Minhas BF, Minhas R, Nawaz U, et al. Comparative genomics using teleost fish helps to systematically identify target gene bodies of functionally defined human enhancers. BMC Genomics 2013;14:122.
[95] Frazer KA, Pachter L, Poliakov A, Rubin EM, Dubchak I. VISTA: computational tools for comparative genomics. Nucleic Acids Res 2004;32:W273–9.
[96] Jeong HB, Park JG, Park YJ, Takemura A, Hur SP, Lee YD, et al. Isolation and characterization of DMRT1 and its putative regulatory region in the protogynous wrasse, *Halichoeres tenuispinis*. Gene 2009;438:8–16.
[97] Nelson AC, Wardle FC. Conserved non-coding elements and cis regulation: actions speak louder than words. Development 2013;140:1385–95.
[98] Guan G, Kobayashi T, Nagahama Y. Sexually dimorphic expression of two types of DM (*Doublesex/Mab-3*)-domain genes in a teleost fish, the Tilapia (*Oreochromis niloticus*). Biochem Biophys Res Commun 2000;272:662–6.
[99] Lei N, Heckert LL. Gata4 regulates testis expression of Dmrt1. Mol Cell Biol 2004;24:377–88.
[100] Lei N, Karpova T, Hornbaker KI, Rice DA, Heckert LL. Distinct transcriptional mechanisms direct expression of the rat Dmrt1 promoter in sertoli cells and germ cells of transgenic mice. Biol Reprod 2009;81:118–25.
[101] Lei N, Heckert LL. *Sp1* and *Egr1* regulate transcription of the *Dmrt1* gene in Sertoli cells. Biol Reprod 2002;66:675–84.
[102] Yoshimoto S, Okada E, Oishi T, Numagami R, Umemoto H, Tamura K, et al. Expression and promoter analysis of *Xenopus DMRT1* and functional characterization of the transactivation property of its protein. Dev Growth Differ 2006;48:597–603.

[103] Yao K, Zhao Y, Zhang T, Cheng HH, Zhou RJ. Analysis of regulatory regions of mammalian *DMRT1* genes. Yi Chuan 2006;28:825–30.

[104] Gao S, Zhang T, Zhou X, Zhao Y, Li Q, Guo Y, et al. Molecular cloning, expression of Sox5 and its down-regulation of Dmrt1 transcription in zebrafish. J Exp Zool B Mol Dev Evol 2005;304:476–83.

[105] Lenhard B, Sandelin A, Carninci P. Metazoan promoters: emerging characteristics and insights into transcriptional regulation. Nat Rev Genet 2012;13:233–45.

[106] Weis L, Reinberg D. Transcription by RNA polymerase II: initiator-directed formation of transcription-competent complexes. FASEB J 1992;6:3300–9.

[107] Hwang SS, Kim YU, Lee S, Jang SW, Kim MK, Koh BH, et al. Transcription factor YY1 is essential for regulation of the Th2 cytokine locus and for Th2 cell differentiation. Proc Natl Acad Sci USA 2013;110:276–81.

[108] Kassardjian A, Rizkallah R, Riman S, Renfro SH, Alexander KE, Hurt MM. The transcription factor YY1 is a novel substrate for Aurora B kinase at G2/M transition of the cell cycle. PLoS One 2012;7:e50645.

[109] Pan X, Jones M, Jiang J, Zaprazna K, Yu D, Pear W, et al. Increased expression of PcG protein YY1 negatively regulates B cell development while allowing accumulation of myeloid cells and LT-HSC cells. PLoS One 2012;7:e30656.

[110] Potula HH, Morel L. Genetic variation at a Yin-Yang 1 response site regulates the transcription of cyclin-dependent kinase inhibitor p18INK4C transcript in lupus-prone mice. J Immunol 2012;188:4992–5002.

[111] Zhang S, Jiang T, Feng L, Sun J, Lu H, Wang Q, et al. Yin Yang-1 suppresses differentiation of hepatocellular carcinoma cells through the downregulation of CCAAT/enhancer-binding protein alpha. J Mol Med (Berl) 2012;90:1069–77.

[112] Bhalla SS, Robitaille L, Nemer M. Cooperative activation by GATA-4 and YY1 of the cardiac B-type natriuretic peptide promoter. J Biol Chem 2001;276:11439–45.

[113] Gregoire S, Karra R, Passer D, Deutsch MA, Krane M, Feistritzer R, et al. Essential and unexpected role of YY1 to promote mesodermal cardiac differentiation. Circ Res 2013;112:900–10.

[114] Manuylov NL, Zhou B, Ma Q, Fox SC, Pu WT, Tevosian SG. Conditional ablation of *Gata4* and *Fog2* genes in mice reveals their distinct roles in mammalian sexual differentiation. Dev Biol 2011;353:229–41.

[115] Anttonen M, Pihlajoki M, Andersson N, Georges A, L'Hote D, Vattulainen S, et al. FOXL2, GATA4, and SMAD3 co-operatively modulate gene expression, cell viability and apoptosis in ovarian granulosa cell tumor cells. PLoS One 2014;9:e85545.

[116] Schemmer J, Arauzo-Bravo MJ, Haas N, Schafer S, Weber SN, Becker A, et al. Transcription factor TFAP2C regulates major programs required for murine fetal germ cell maintenance and haploinsufficiency predisposes to teratomas in male mice. PLoS One 2013;8:e71113.

[117] Schafer S, Anschlag J, Nettersheim D, Haas N, Pawig L, Schorle H. The role of BLIMP1 and its putative downstream target TFAP2C in germ cell development and germ cell tumours. Int J Androl 2011;34:e152–8.

[118] Pennacchio LA, Ahituv N, Moses AM, Prabhakar S, Nobrega MA, Shoukry M, et al. *In vivo* enhancer analysis of human conserved non-coding sequences. Nature 2006;444:499–502.

[119] Smith CA, Hurley TM, McClive PJ, Sinclair AH. Restricted expression of DMRT3 in chicken and mouse embryos. Mech Dev 2002;119(Suppl. 1):S73–6.

[120] Kim S, Kettlewell JR, Anderson RC, Bardwell VJ, Zarkower D. Sexually dimorphic expression of multiple doublesex-related genes in the embryonic mouse gonad. Gene Expr Patterns 2003;3:77–82.
[121] Miyake Y, Sakai Y, Kuniyoshi H. Molecular cloning and expression profile of sex-specific genes, *Figla* and *Dmrt1*, in the protogynous hermaphroditic fish, *Halichoeres poecilopterus*. Zoolog Sci 2012;29:690–701.
[122] Alam MA, Kobayashi Y, Horiguchi R, Hirai T, Nakamura M. Molecular cloning and quantitative expression of sexually dimorphic markers Dmrt1 and Foxl2 during female-to-male sex change in *Epinephelus merra*. Gen Comp Endocrinol 2008;157:75–85.
[123] Huang X, Guo Y, Shui Y, Gao S, Yu H, Cheng H, et al. Multiple alternative splicing and differential expression of dmrt1 during gonad transformation of the rice field eel. Biol Reprod 2005;73:1017–24.
[124] Bull JJ. Evolution of sex determining mechanisms. Menlo Park, CA: Evolution, Benjamin/Cummings Pub. Co., Advanced Book Program; 1983.
[125] Cutting A, Chue J, Smith CA. Just how conserved is vertebrate sex determination? Dev Dyn 2013;242:380–7.
[126] Marshall Graves JA. Weird animal genomes and the evolution of vertebrate sex and sex chromosomes. Annu Rev Genet 2008;42:565–86.
[127] Livernois AM, Graves JA, Waters PD. The origin and evolution of vertebrate sex chromosomes and dosage compensation. Heredity (Edinb) 2012;108:50–8.
[128] Bellott DW, Page DC. Reconstructing the evolution of vertebrate sex chromosomes. Cold Spring Harb Symp Quant Biol 2009;74:345–53.
[129] Capel B. Sex in the 90s: SRY and the switch to the male pathway. Annu Rev Physiol 1998;60.497–523.
[130] Smith CA, Sinclair AH. Sex determination: insights from the chicken. Bioessays 2004;26:120–32.
[131] Gamble T, Zarkower D. Sex determination. Curr Biol 2012;22:R257–62.
[132] Angelopoulou R, Lavranos G, Manolakou P. Sex determination strategies in 2012: towards a common regulatory model? Reprod Biol Endocrinol 2012;10:13.
[133] Hacker A, Capel B, Goodfellow P, Lovell-Badge R. Expression of Sry, the mouse sex determining gene. Development 1995;121:1603–14.
[134] Koopman P, Munsterberg A, Capel B, Vivian N, Lovell-Badge R. Expression of a candidate sex-determining gene during mouse testis differentiation. Nature 1990;348:450–2.
[135] Capel B. The battle of the sexes. Mech Dev 2000;92:89–103.
[136] Svingen T, Koopman P. Building the mammalian testis: origins, differentiation, and assembly of the component cell populations. Genes Dev 2013;27:2409–26.
[137] Burgoyne PS, Buehr M, Koopman P, Rossant J, McLaren A. Cell-autonomous action of the testis-determining gene: Sertoli cells are exclusively XY in XX—XY chimaeric mouse testes. Development 1988;102:443–50.
[138] Sharpe RM, McKinnell C, Kivlin C, Fisher JS. Proliferation and functional maturation of Sertoli cells, and their relevance to disorders of testis function in adulthood. Reproduction 2003;125:769–84.
[139] Palmer SJ, Burgoyne PS. *In situ* analysis of fetal, prepuberal and adult XX—XY chimaeric mouse testes: Sertoli cells are predominantly, but not exclusively, XY. Development 1991;112:265–8.

[140] Matson CK, Murphy MW, Sarver AL, Griswold MD, Bardwell VJ, Zarkower D. DMRT1 prevents female reprogramming in the postnatal mammalian testis. Nature 2011;476:101–4.
[141] Walker WH. Molecular mechanisms controlling Sertoli cell proliferation and differentiation. Endocrinology 2003;144:3719–21.
[142] Vergouwen RP, Jacobs SG, Huiskamp R, Davids JA, de Rooij DG. Proliferative activity of gonocytes, Sertoli cells and interstitial cells during testicular development in mice. J Reprod Fertil 1991;93:233–43.
[143] Regadera J, Martinez-Garcia F, Gonzalez-Peramato P, Serrano A, Nistal M, Suarez-Quian C. Androgen receptor expression in Sertoli cells as a function of seminiferous tubule maturation in the human cryptorchid testis. J Clin Endocrinol Metab 2001;86:413–21.
[144] Bremner WJ, Millar MR, Sharpe RM, Saunders PT. Immunohistochemical localization of androgen receptors in the rat testis: evidence for stage-dependent expression and regulation by androgens. Endocrinology 1994;135:1227–34.
[145] Yomogida K, Ohtani H, Harigae H, Ito E, Nishimune Y, Engel JD, et al. Developmental stage- and spermatogenic cycle-specific expression of transcription factor GATA-1 in mouse Sertoli cells. Development 1994;120:1759–66.
[146] Ramos Jr. AS, Dym M. Ultrastructural differentiation of rat Sertoli cells. Biol Reprod 1979;21:909–22.
[147] Russell LD, Ren HP, Sinha Hikim I, Schulze W, Sinha Hikim AP. A comparative study in twelve mammalian species of volume densities, volumes, and numerical densities of selected testis components, emphasizing those related to the Sertoli cell. Am J Anat 1990;188:21–30.
[148] Gondos B, Berndstaon WE. Postnatal and pubertal development. In: Russell LD, Griswold MD, editors. The Sertoli cell. 1st ed. Clearwater, FL: Cache River Press; 1993. p. 116–54.
[149] Schulze C. Response of the human testis to long-term estrogen treatment: morphology of Sertoli cells, Leydig cells and spermatogonial stem cells. Cell Tissue Res 1988;251:31–43.
[150] Mruk DD, Cheng CY. Tight junctions in the testis: new perspectives. Philos Trans R Soc Lond B Biol Sci 2010;365:1621–35.
[151] Morrow CM, Mruk D, Cheng CY, Hess RA. Claudin and occludin expression and function in the seminiferous epithelium. Philos Trans R Soc Lond B Biol Sci 2010;365:1679–96.
[152] Kopera IA, Bilinska B, Cheng CY, Mruk DD. Sertoli-germ cell junctions in the testis: a review of recent data. Philos Trans R Soc Lond B Biol Sci 2010;365:1593–605.
[153] Yan HH, Mruk DD, Lee WM, Cheng CY. Ectoplasmic specialization: a friend or a foe of spermatogenesis? Bioessays 2007;29:36–48.
[154] Russell L. Observations on rat Sertoli ectoplasmic ("junctional") specializations in their association with germ cells of the rat testis. Tissue Cell 1977;9:475–98.
[155] Chihara M, Otsuka S, Ichii O, Hashimoto Y, Kon Y. Molecular dynamics of the blood-testis barrier components during murine spermatogenesis. Mol Reprod Dev 2010;77:630–9.
[156] Vogl AW. Distribution and function of organized concentrations of actin filaments in mammalian spermatogenic cells and Sertoli cells. Int Rev Cytol 1989;119:1–56.
[157] Wong EW, Mruk DD, Cheng CY. Biology and regulation of ectoplasmic specialization, an atypical adherens junction type, in the testis. Biochim Biophys Acta 2008;1778:692–708.

[158] Bartles JR, Wierda A, Zheng L. Identification and characterization of espin, an actin-binding protein localized to the F-actin-rich junctional plaques of Sertoli cell ectoplasmic specializations. J Cell Sci 1996;109(Pt 6):1229–39.

[159] Chen B, Li A, Wang D, Wang M, Zheng L, Bartles JR. Espin contains an additional actin-binding site in its N terminus and is a major actin-bundling protein of the Sertoli cell-spermatid ectoplasmic specialization junctional plaque. Mol Biol Cell 1999;10:4327–39.

[160] Toyama Y, Maekawa M, Yuasa S. Ectoplasmic specializations in the Sertoli cell: new vistas based on genetic defects and testicular toxicology. Anat Sci Int 2003;78:1–16.

[161] Sekerkova G, Zheng L, Loomis PA, Changyaleket B, Whitlon DS, Mugnaini E, et al. Espins are multifunctional actin cytoskeletal regulatory proteins in the microvilli of chemosensory and mechanosensory cells. J Neurosci 2004;24:5445–56.

[162] Cheng CY, Mruk DD. Cell junction dynamics in the testis: Sertoli-germ cell interactions and male contraceptive development. Physiol Rev 2002;82:825–74.

[163] Ozaki-Kuroda K, Nakanishi H, Ohta H, Tanaka H, Kurihara H, Mueller S, et al. Nectin couples cell-cell adhesion and the actin scaffold at heterotypic testicular junctions. Curr Biol 2002;12:1145–50.

[164] Palombi F, Salanova M, Tarone G, Farini D, Stefanini M. Distribution of beta 1 integrin subunit in rat seminiferous epithelium. Biol Reprod 1992;47:1173–82.

[165] Salanova M, Stefanini M, De Curtis I, Palombi F. Integrin receptor alpha 6 beta 1 is localized at specific sites of cell-to-cell contact in rat seminiferous epithelium. Biol Reprod 1995;52:79–87.

[166] Takahashi K, Nakanishi H, Miyahara M, Mandai K, Satoh K, Satoh A, et al. Nectin/PRR: an immunoglobulin-like cell adhesion molecule recruited to cadherin-based adherens junctions through interaction with Afadin, a PDZ domain-containing protein. J Cell Biol 1999;145:539–49.

[167] Bouchard MJ, Dong Y, McDermott Jr. BM, Lam DH, Brown KR, Shelanski M, et al. Defects in nuclear and cytoskeletal morphology and mitochondrial localization in spermatozoa of mice lacking nectin-2, a component of cell-cell adherens junctions. Mol Cell Biol 2000;20:2865–73.

[168] Ikeda W, Nakanishi H, Miyoshi J, Mandai K, Ishizaki H, Tanaka M, et al. Afadin: a key molecule essential for structural organization of cell-cell junctions of polarized epithelia during embryogenesis. J Cell Biol 1999;146:1117–32.

[169] Georges A, Auguste A, Bessiere L, Vanet A, Todeschini AL, Veitia RA. FOXL2: a central transcription factor of the ovary. J Mol Endocrinol 2014;52:R17–33.

[170] Wu GC, Tomy S, Lee MF, Lee YH, Yueh WS, Lin CJ, et al. Sex differentiation and sex change in the protandrous black porgy, *Acanthopagrus schlegeli*. Gen Comp Endocrinol 2010;167:417–21.

[171] Uda M, Ottolenghi C, Crisponi L, Garcia JE, Deiana M, Kimber W, et al. Foxl2 disruption causes mouse ovarian failure by pervasive blockage of follicle development. Hum Mol Genet 2004;13:1171–81.

[172] Schmidt D, Ovitt CE, Anlag K, Fehsenfeld S, Gredsted L, Treier AC, et al. The murine winged-helix transcription factor Foxl2 is required for granulosa cell differentiation and ovary maintenance. Development 2004;131:933–42.

[173] Ottolenghi C, Omari S, Garcia-Ortiz JE, Uda M, Crisponi L, Forabosco A, et al. Foxl2 is required for commitment to ovary differentiation. Hum Mol Genet 2005;14:2053–62.

[174] Ottolenghi C, Pelosi E, Tran J, Colombino M, Douglass E, Nedorezov T, et al. Loss of Wnt4 and Foxl2 leads to female-to-male sex reversal extending to germ cells. Hum Mol Genet 2007;16:2795–804.

[175] Smith EK, Guzman JM, Luckenbach JA. Molecular cloning, characterization, and sexually dimorphic expression of five major sex differentiation-related genes in a Scorpaeniform fish, sablefish (*Anoplopoma fimbria*). Comp Biochem Physiol B Biochem Mol Biol 2013;165:125–37.
[176] Ashida H, Ueyama N, Kinoshita M, Kobayashi T. Molecular identification and expression of FOXL2 and DMRT1 genes from willow minnow *Gnathopogon caerulescens*. Reprod Biol 2013;13:317–24.
[177] Raghuveer K, Senthilkumaran B, Sudhakumari CC, Sridevi P, Rajakumar A, Singh R, et al. Dimorphic expression of various transcription factor and steroidogenic enzyme genes during gonadal ontogeny in the air-breathing catfish, *Clarias gariepinus*. Sex Dev 2011;5:213–23.
[178] Hale MC, Xu P, Scardina J, Wheeler PA, Thorgaard GH, Nichols KM. Differential gene expression in male and female rainbow trout embryos prior to the onset of gross morphological differentiation of the gonads. BMC Genomics 2011;12:404.
[179] Yang X, Zheng J, Qu L, Chen S, Li J, Xu G, et al. Methylation status of cMHM and expression of sex-specific genes in adult sex-reversed female chickens. Sex Dev 2011;5:147–54.
[180] Chue J, Smith CA. Sex determination and sexual differentiation in the avian model. FEBS J 2011;278:1027–34.
[181] Loffler KA, Zarkower D, Koopman P. Etiology of ovarian failure in blepharophimosis ptosis epicanthus inversus syndrome: FOXL2 is a conserved, early-acting gene in vertebrate ovarian development. Endocrinology 2003;144:3237–43.
[182] Shoemaker CM, Queen J, Crews D. Response of candidate sex-determining genes to changes in temperature reveals their involvement in the molecular network underlying temperature-dependent sex determination. Mol Endocrinol 2007;21:2750–63.
[183] Baron D, Houlgatte R, Fostier A, Guiguen Y. Large-scale temporal gene expression profiling during gonadal differentiation and early gametogenesis in rainbow trout. Biol Reprod 2005;73:959–66.
[184] Janes DE, Elsey RM, Langan EM, Valenzuela N, Edwards SV. Sex-biased expression of sex-differentiating genes FOXL2 and FGF9 in American alligators, *Alligator mississippiensis*. Sex Dev 2013;7:253–60.
[185] Yan H, Ijiri S, Wu Q, Kobayashi T, Li S, Nakaseko T, et al. Expression patterns of gonadotropin hormones and their receptors during early sexual differentiation in Nile tilapia *Oreochromis niloticus*. Biol Reprod 2012;87:116.
[186] Ottolenghi C, Uda M, Crisponi L, Omari S, Cao A, Forabosco A, et al. Determination and stability of sex. Bioessays 2007;29:15–25.
[187] Liu ZH, Zhang YG, Wang DS. Studies on feminization, sex determination, and differentiation of the Southern catfish, *Silurus meridionalis*—a review. Fish Physiol Biochem 2010;36:223–35.
[188] Vidal VP, Chaboissier MC, de Rooij DG, Schedl A. Sox9 induces testis development in XX transgenic mice. Nat Genet 2001;28:216–7.
[189] Chaboissier MC, Kobayashi A, Vidal VI, Lutzkendorf S, van de Kant HJ, Wegner M, et al. Functional analysis of Sox8 and Sox9 during sex determination in the mouse. Development 2004;131:1891–901.
[190] Li MH, Yang HH, Li MR, Sun YL, Jiang XL, Xie QP, et al. Antagonistic roles of Dmrt1 and Foxl2 in sex differentiation via estrogen production in tilapia as demonstrated by TALENs. Endocrinology 2013;154:4814–25.
[191] McLaren A. Primordial germ cells in the mouse. Dev Biol 2003;262:1–15.
[192] Griswold MD, Oatley JM. Concise review: defining characteristics of mammalian spermatogenic stem cells. Stem Cells 2013;31:8–11.

[193] Surani MA, Hajkova P. Epigenetic reprogramming of mouse germ cells toward totipotency. Cold Spring Harb Symp Quant Biol 2010;75:211–8.
[194] Donovan PJ, De Miguel MP. Gonocyte development and differentiation. In: Skinner MK, Griswold MD, editors. Sertoli cell biology. San Diego, CA: Elsevier Academic Press; 2005. p. 417–35.
[195] Yoshida S, Sukeno M, Nakagawa T, Ohbo K, Nagamatsu G, Suda T, et al. The first round of mouse spermatogenesis is a distinctive program that lacks the self-renewing spermatogonia stage. Development 2006;133:1495–505.
[196] Mc CP, Cerecedo LR. Vitamin A deficiency in the mouse. J Nutr 1952;46:361–76.
[197] Mark M, Jacobs H, Oulad-Abdelghani M, Dennefeld C, Feret B, Vernet N, et al. STRA8-deficient spermatocytes initiate, but fail to complete, meiosis and undergo premature chromosome condensation. J Cell Sci 2008;121:3233–42.
[198] Anderson EL, Baltus AE, Roepers-Gajadien HL, Hassold TJ, de Rooij DG, van Pelt AM, et al. Stra8 and its inducer, retinoic acid, regulate meiotic initiation in both spermatogenesis and oogenesis in mice. Proc Natl Acad Sci USA 2008;105:14976–80.
[199] Bowles J, Koopman P. Retinoic acid, meiosis and germ cell fate in mammals. Development 2007;134:3401–11.
[200] Krentz AD, Murphy MW, Sarver AL, Griswold MD, Bardwell VJ, Zarkower D. DMRT1 promotes oogenesis by transcriptional activation of Stra8 in the mammalian fetal ovary. Dev Biol 2011;356:63–70.
[201] Nakagawa T, Sharma M, Nabeshima Y, Braun RE, Yoshida S. Functional hierarchy and reversibility within the murine spermatogenic stem cell compartment. Science 2010;328:62–7.
[202] Yoshida S. Stem cells in mammalian spermatogenesis. Dev Growth Differ 2010;52:311–7.
[203] Yoshida S, Takakura A, Ohbo K, Abe K, Wakabayashi J, Yamamoto M, et al. Neurogenin3 delineates the earliest stages of spermatogenesis in the mouse testis. Dev Biol 2004;269:447–58.
[204] Ballow D, Meistrich ML, Matzuk M, Rajkovic A. *Sohlh1* is essential for spermatogonial differentiation. Dev Biol 2006;294:161–7.
[205] Murphy MW, Sarver AL, Rice D, Hatzi K, Ye K, Melnick A, et al. Genome-wide analysis of DNA binding and transcriptional regulation by the mammalian Doublesex homolog DMRT1 in the juvenile testis. Proc Natl Acad Sci USA 2010;107:13360–5.
[206] Takashima S, Hirose M, Ogonuki N, Ebisuya M, Inoue K, Kanatsu-Shinohara M, et al. Regulation of pluripotency in male germline stem cells by Dmrt1. Genes Dev 2013;27:1949–58.
[207] Krentz AD, Murphy MW, Zhang T, Sarver AL, Jain S, Griswold MD, et al. Interaction between DMRT1 function and genetic background modulates signaling and pluripotency to control tumor susceptibility in the fetal germ line. Dev Biol 2013;377:67–78.
[208] Bowles J, Teasdale RP, James K, Koopman P. Dppa3 is a marker of pluripotency and has a human homologue that is expressed in germ cell tumours. Cytogenet Genome Res 2003;101:261–5.
[209] Payer B, Saitou M, Barton SC, Thresher R, Dixon JP, Zahn D, et al. Stella is a maternal effect gene required for normal early development in mice. Curr Biol 2003;13:2110–7.
[210] Pashai N, Hao H, All A, Gupta S, Chaerkady R, De Los Angeles A, et al. Genome-wide profiling of pluripotent cells reveals a unique molecular signature of human embryonic germ cells. PLoS One 2012;7:e39088.
[211] Kanatsu-Shinohara M, Inoue K, Lee J, Yoshimoto M, Ogonuki N, Miki H, et al. Generation of pluripotent stem cells from neonatal mouse testis. Cell 2004;119:1001–12.

[212] Kanatsu-Shinohara M, Inoue K, Ogonuki N, Morimoto H, Ogura A, Shinohara T. Serum- and feeder-free culture of mouse germline stem cells. Biol Reprod 2011;84:97–105.

[213] Kanatsu-Shinohara M, Lee J, Inoue K, Ogonuki N, Miki H, Toyokuni S, et al. Pluripotency of a single spermatogonial stem cell in mice. Biol Reprod 2008;78:681–7.

[214] Kanatsu-Shinohara M, Muneto T, Lee J, Takenaka M, Chuma S, Nakatsuji N, et al. Long-term culture of male germline stem cells from hamster testes. Biol Reprod 2008;78:611–7.

[215] Kanatsu-Shinohara M, Ogonuki N, Inoue K, Miki H, Ogura A, Toyokuni S, et al. Long-term proliferation in culture and germline transmission of mouse male germline stem cells. Biol Reprod 2003;69:612–6.

[216] Turnbull C, Rapley EA, Seal S, Pernet D, Renwick A, Hughes D, et al. Variants near DMRT1, TERT and ATF7IP are associated with testicular germ cell cancer. Nat Genet 2010;42:604–7.

[217] Kanetsky PA, Mitra N, Vardhanabhuti S, Vaughn DJ, Li M, Ciosek SL, et al. A second independent locus within DMRT1 is associated with testicular germ cell tumor susceptibility. Human Mol Genet 2011;20:3109–17.

[218] Looijenga LH, Hersmus R, Gillis AJ, Pfundt R, Stoop HJ, van Gurp RJ, et al. Genomic and expression profiling of human spermatocytic seminomas: primary spermatocyte as tumorigenic precursor and DMRT1 as candidate chromosome 9 gene. Cancer Res 2006;66:290–302.

[219] Jorgensen A, Nielsen JE, Almstrup K, Toft BG, Petersen BL, Rajpert-De Meyts E. Dysregulation of the mitosis-meiosis switch in testicular carcinoma *in situ*. J Pathol 2013;229:588–98.

[220] Hoei-Hansen CE, Almstrup K, Nielsen JE, Brask Sonne S, Graem N, Skakkebaek NE, et al. Stem cell pluripotency factor NANOG is expressed in human fetal gonocytes, testicular carcinoma *in situ* and germ cell tumours. Histopathology 2005;47:48–56.

[221] Jones TD, Ulbright TM, Eble JN, Cheng L. OCT4: a sensitive and specific biomarker for intratubular germ cell neoplasia of the testis. Clin Cancer Res 2004;10:8544–7.

[222] Jones TD, Ulbright TM, Eble JN, Baldridge LA, Cheng L. OCT4 staining in testicular tumors: a sensitive and specific marker for seminoma and embryonal carcinoma. Am J Surg Pathol 2004;28:935–40.

[223] Rajpert-De Meyts E, Hanstein R, Jorgensen N, Graem N, Vogt PH, Skakkebaek NE. Developmental expression of POU5F1 (OCT-3/4) in normal and dysgenetic human gonads. Hum Reprod 2004;19:1338–44.

[224] Looijenga LH, Verkerk AJ, Dekker MC, van Gurp RJ, Gillis AJ, Oosterhuis JW. Genomic imprinting in testicular germ cell tumours. APMIS 1998;106:187–95 discussion 196–7.

[225] Ruark E, Seal S, McDonald H, Zhang F, Elliot A, Lau K, et al. Identification of nine new susceptibility loci for testicular cancer, including variants near DAZL and PRDM14. Nat Genet 2013;45:686–9.

[226] Chung CC, Kanetsky PA, Wang Z, Hildebrandt MA, Koster R, Skotheim RI, et al. Meta-analysis identifies four new loci associated with testicular germ cell tumor. Nat Genet 2013;45:680–5.

[227] Poynter JN, Hooten AJ, Frazier AL, Ross JA. Associations between variants in KITLG, SPRY4, BAK1, and DMRT1 and pediatric germ cell tumors. Genes Chromosomes Cancer 2012;51:266–71.

[228] Boublikova L, Buchler T, Stary J, Abrahamova J, Trka J. Molecular biology of testicular germ cell tumors: unique features awaiting clinical application. Crit Rev Oncol Hematol 2014;89:366–85.

[229] Tannour-Louet M, Han S, Corbett ST, Louet JF, Yatsenko S, Meyers L, et al. Identification of *de novo* copy number variants associated with human disorders of sexual development. PLoS One 2010;5:e15392.

[230] Kratz CP, Han SS, Rosenberg PS, Berndt SI, Burdett L, Yeager M, et al. Variants in or near KITLG, BAK1, DMRT1, and TERT-CLPTM1L predispose to familial testicular germ cell tumour. J Med Genet 2011;48:473–6.

[231] Lopes AM, Aston KI, Thompson E, Carvalho F, Goncalves J, Huang N, et al. Human spermatogenic failure purges deleterious mutation load from the autosomes and both sex chromosomes, including the gene DMRT1. PLoS Genet 2013;9:e1003349.

[232] Repetto GM, Wagstaff J, Korf BR, Knoll JH. Complex familial rearrangement of chromosome 9p24.3 detected by FISH. Am J Med Genet 1998;76:306–9.

[233] Christ LA, Crowe CA, Micale MA, Conroy JM, Schwartz S. Chromosome breakage hotspots and delineation of the critical region for the 9p-deletion syndrome. Am J Hum Genet 1999;65:1387–95.

[234] Swinkels ME, Simons A, Smeets DF, Vissers LE, Veltman JA, Pfundt R, et al. Clinical and cytogenetic characterization of 13 Dutch patients with deletion 9p syndrome: delineation of the critical region for a consensus phenotype. Am J Med Genet A 2008;146A:1430–8.

[235] Barbaro M, Balsamo A, Anderlid BM, Myhre AG, Gennari M, Nicoletti A, et al. Characterization of deletions at 9p affecting the candidate regions for sex reversal and deletion 9p syndrome by MLPA. Eur J Hum Genet 2009;17:1439–47.

[236] Vinci G, Chantot-Bastaraud S, El Houate B, Lortat-Jacob S, Brauner R, McElreavey K. Association of deletion 9p, 46,XY gonadal dysgenesis and autistic spectrum disorder. Mol Hum Reprod 2007;13:685–9.

[237] Vasquez-Velasquez AI, Arnaud-Lopez L, Figuera LE, Padilla-Gutierrez JR, Rivas F, Rivera H. Ambiguous genitalia by 9p deletion inherent to a dic(Y;9)(q12;p24). J Appl Genet 2005;46:415–8.

[238] Ounap K, Uibo O, Zordania R, Kiho L, Ilus T, Oiglane-Shlik E, et al. Three patients with 9p deletions including DMRT1 and DMRT2: a girl with XY complement, bilateral ovotestes, and extreme growth retardation, and two XX females with normal pubertal development. Am J Med Genet A 2004;130A:415–23.

[239] Vialard F, Ottolenghi C, Gonzales M, Choiset A, Girard S, Siffroi JP, et al. Deletion of 9p associated with gonadal dysfunction in 46,XY but not in 46,XX human fetuses. J Med Genet 2002;39:514–8.

[240] Calvari V, Bertini V, De Grandi A, Peverali G, Zuffardi O, Ferguson-Smith M, et al. A new submicroscopic deletion that refines the 9p region for sex reversal. Genomics 2000;65:203–12.

[241] Alfi O, Donnell GN, Crandall BF, Derencsenyi A, Menon R. Deletion of the short arm of chromosome no.9 (46,9p-): a new deletion syndrome. Ann Genet 1973;16:17–22.

[242] Quinonez SC, Park JM, Rabah R, Owens KM, Yashar BM, Glover TW, et al. 9p partial monosomy and disorders of sex development: review and postulation of a pathogenetic mechanism. Am J Med Genet A 2013;161:1882–96.

[243] Flejter WL, Fergestad J, Gorski J, Varvill T, Chandrasekharappa S. A gene involved in XY sex reversal is located on chromosome 9, distal to marker D9S1779. Am J Hum Genet 1998;63:794–802.

[244] Guioli S, Schmitt K, Critcher R, Bouzyk M, Spurr NK, Ogata T, et al. Molecular analysis of 9p deletions associated with XY sex reversal: refining the localization of a

sex-determining gene to the tip of the chromosome. Am J Hum Genet 1998;63:905–8.

[245] McDonald MT, Flejter W, Sheldon S, Putzi MJ, Gorski JL. XY sex reversal and gonadal dysgenesis due to 9p24 monosomy. Am J Med Genet 1997;73:321–6.

[246] Bennett CP, Docherty Z, Robb SA, Ramani P, Hawkins JR, Grant D. Deletion 9p and sex reversal. J Med Genet 1993;30:518–20.

[247] Muroya K, Okuyama T, Goishi K, Ogiso Y, Fukuda S, Kameyama J, et al. Sex-determining gene(s) on distal 9p: clinical and molecular studies in six cases. J Clin Endocrinol Metab 2000;85:3094–100.

[248] Ledig S, Hiort O, Wunsch L, Wieacker P. Partial deletion of DMRT1 causes 46,XY ovotesticular disorder of sexual development. Eur J Endocrinol 2012;167:119–24.

[249] Machado AZ, da Silva TE, Frade Costa EM, Dos Santos MG, Nishi MY, Brito VN, et al. Absence of inactivating mutations and deletions in the DMRT1 and FGF9 genes in a large cohort of 46,XY patients with gonadal dysgenesis. Eur J Med Genet 2012;55:690–4.

[250] Mello MP, Coeli FB, Assumpcao JG, Castro TM, Maciel-Guerra AT, Marques-de-Faria AP, et al. Novel DMRT1 3′UTR + 11insT mutation associated to XY partial gonadal dysgenesis. Arq Bras Endocrinol Metabol 2010;54:749–53.

[251] Privitera O, Vessecchia G, Bernasconi B, Bettio D, Stioui S, Giordano G. Prenatal diagnosis of del(9)(p24): a sex reverse case. Prenat Diagn 2005;25:945–8.

[252] Lerer I, Sagi M, Meiner V, Cohen T, Zlotogora J, Abeliovich D. Deletion of the ANKRD15 gene at 9p24.3 causes parent-of-origin-dependent inheritance of familial cerebral palsy. Hum Mol Genet 2005;14:3911–20.

[253] Seo KW, Wang Y, Kokubo H, Kettlewell JR, Zarkower DA, Johnson RL. Targeted disruption of the DM domain containing transcription factor Dmrt2 reveals an essential role in somite patterning. Dev Biol 2006;290:200–10.

[254] Andersson LS, Larhammar M, Memic F, Wootz H, Schwochow D, Rubin CJ, et al. Mutations in DMRT3 affect locomotion in horses and spinal circuit function in mice. Nature 2012;488:642–6.

[255] Promerova M, Andersson LS, Juras R, Penedo MC, Reissmann M, Tozaki T, et al. Worldwide frequency distribution of the "Gait keeper" mutation in the *DMRT3* gene. Anim Genet 2014;45:274–82.

[256] Ledig S, Hiort O, Scherer G, Hoffmann M, Wolff G, Morlot S, et al. Array-CGH analysis in patients with syndromic and non-syndromic XY gonadal dysgenesis: evaluation of array CGH as diagnostic tool and search for new candidate loci. Hum Reprod 2010;25:2637–46.

[257] Murphy MW, Zarkower D, Bardwell VJ. Vertebrate DM domain proteins bind similar DNA sequences and can heterodimerize on DNA. BMC Mol Biol 2007;8:58.

[258] Wang DS, Zhou LY, Kobayashi T, Matsuda M, Shibata Y, Sakai F, et al. Doublesex- and Mab-3-related transcription factor-1 repression of aromatase transcription, a possible mechanism favoring the male pathway in tilapia. Endocrinology 2010;151:1331–40.

[259] Herpin A, Braasch I, Kraeussling M, Schmidt C, Thoma EC, Nakamura S, et al. Transcriptional rewiring of the sex determining *dmrt1* gene duplicate by transposable elements. PLoS Genet 2010;6:e1000844.

[260] Li M, Wu F, Gu Y, Wang T, Wang H, Yang S, et al. Insulin-like growth factor 3 regulates expression of genes encoding steroidogenic enzymes and key transcription factors in the Nile tilapia gonad. Biol Reprod 2012;86(163):1–10.

Hormonal regulation of spermatogenesis through Sertoli cells by androgens and estrogens

6

Lee B. Smith[a], *William H. Walker*[b] *and Liza O'Donnell*[c]
[a]MRC Centre for Reproductive Health, University of Edinburgh, The Queen's Medical Research Institute, Edinburgh, UK, [b]Department of Obstetrics, Gynecology and Reproductive Sciences, Magee Womens Research Institute, University of Pittsburgh, School of Medicine, Pittsburgh, PA, [c]Monash Institute of Medical Research-Prince Henry's Institute, Clayton, Victoria, Australia

I. Introduction

In the context of the hormonal regulation of spermatogenesis, androgen signaling is fundamental because it is both necessary and sufficient to support sperm development. Testosterone can drive spermatogenesis in the absence of gonadotropin or other steroid hormone support, but no other hormone that is present at physiological concentrations can drive spermatogenesis in the absence of testosterone. For this reason, androgens are the preeminent factor that controls normal spermatogenesis. Androgens are also a substrate for local estrogen production, and, as such, overall androgen influence over spermatogenesis extends beyond a direct influence to encompass a complex and multifactorial network. Despite decades of research, many of the mechanisms underpinning androgen control of spermatogenesis still remain to be established.

Since the previous volume of *Sertoli Cell Biology* was published, the development and application of a multitude of previously unavailable genetic models has revolutionized our understanding of the cell–cell interactions that are essential for the transmission of the androgen signal and identified a wealth of molecules and mechanisms impacting Sertoli cell function; yet, there remain many questions to answer. While the requirement for androgen signaling in Sertoli cells is beyond doubt, the minutiae of how this signal is propagated remain one of the major challenges ahead in the field. For example, the elucidation of the nonclassical androgen receptor (AR) signaling cascade has been one of the major steps forward, but this advance highlights how much more there is to learn.

This chapter summarizes our current understanding of how androgens, acting via Sertoli cells, support spermatogenesis and wider testis function. The chapter also highlights areas of research requiring significant further investigation.

Sertoli Cell Biology. DOI: http://dx.doi.org/10.1016/B978-0-12-417047-6.00006-5

II. Androgen signaling

A. Classical testosterone signaling

Testosterone is a steroid hormone that is produced by the testis and is vital for masculinization and the initiation and maintenance of male fertility [1,2]. Therefore, testosterone is the major androgen acting in the testis to regulate spermatogenesis. Androgen's effects are largely mediated by the cognate AR, a 110-kDa member of the ligand-activated transcription factor superfamily that is encoded by a single X-linked gene. AR is a modular protein with four structurally and functionally distinct domains, including a variable and intrinsically disordered N-terminal domain encoded by exon 1, a highly conserved zing-finger DNA-binding domain (DBD) encoded by exons 2 and 3, and a C-terminal ligand-binding domain encoded by exons 4–8 that is connected to the DBD by a short hinge region that also contains a bipartite ligand-dependent nuclear localization signal necessary for AR nuclear localization [3,4]. In the absence of ligand, AR is bound to heat shock chaperone proteins in the cell cytoplasm.

Genomic (also known as classical) testosterone signaling begins with testosterone diffusing through the plasma membrane and binding AR (Figure 6.1). Ligand binding results in the release of AR from the heat shock proteins, homodimerization of AR molecules, and nuclear transport. The androgen–AR complex binds specific androgen response elements (AREs) in promoters of target genes and recruits coactivators or corepressor proteins and protein complexes that modulate gene expression together. This classical pathway of testosterone action requires at least 30–45 min to alter gene expression and at least 1 h to produce nascent proteins [5].

B. Nonclassical testosterone signaling

In addition to the classical pathway of testosterone signaling, there is evidence in Sertoli cells of more rapid, nonclassical (also called nongenomic) pathways of testosterone action (Figure 6.1). Elucidation of this mechanism of testosterone action represents one of the most important advances in Sertoli cell androgen research. These testosterone effects can be detected within seconds, and cellular responses can be assayed within minutes.

Two nonclassical testosterone signaling pathways have been identified. In one pathway, AR-mediated activation of Src kinase causes the activation of kinase signaling pathways plus downstream gene expression (reviewed in [6]). In the second pathway, testosterone causes the Sertoli cell to depolarize, resulting in Ca^{2+} influx and the activation of intracellular signaling factors (reviewed in [7]). The two pathways are discussed in greater detail below.

In the first nonclassical testosterone-mediated pathway, treatment of cultured Sertoli cells with physiologic levels of testosterone transiently increases the localization of AR to the plasma membrane and results in AR interacting with and activating Src tyrosine kinase [8] (Figure 6.1). It was found that AR and not a novel receptor for androgens was responsible for initiating the intracellular signaling

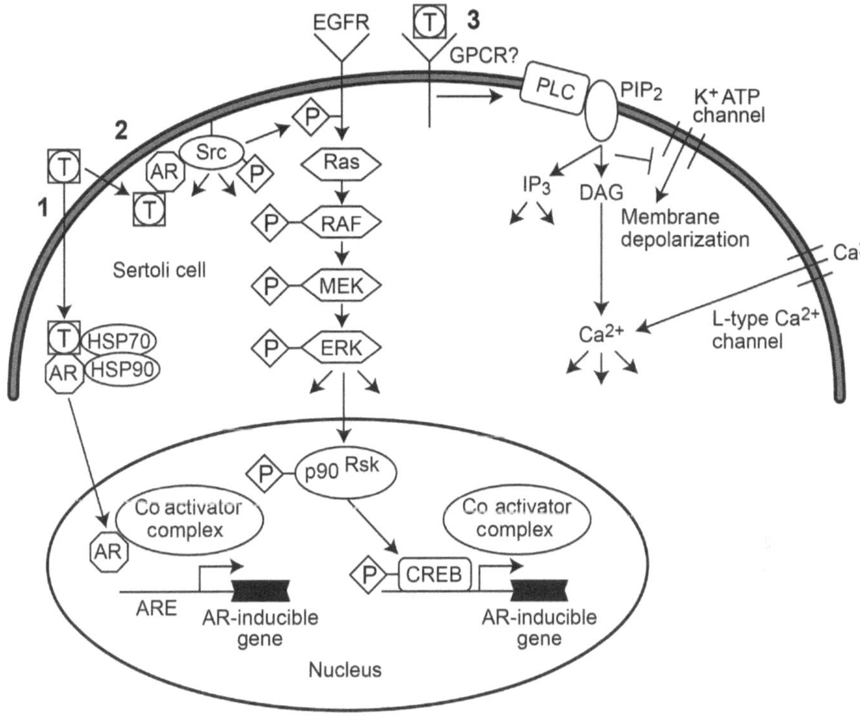

Figure 6.1 Classical and nonclassical androgen signaling cascades. In classical AR signaling (1), testosterone, diffusing through the cell membrane, binds AR, promoting AR dimerization and releasing AR from chaperone proteins. The testosterone AR complex localizes to the cell nucleus and binds AREs in the promoters of target genes. Recruitment of cofactors then promotes transcription of downstream target genes. In nonclassical androgen signaling, the androgen–AR complex activates a Src-dependent signaling cascade, leading to the transcription of downstream genes (2). A second method of nonclassical signaling involves androgen-mediated influx of calcium ions into Sertoli cells (3), which induces rapid changes in cellular processes (see Section II for details).

because kinase activation was blocked in AR-defective cells and cells treated with siRNA against AR [9]. In response to androgen stimulation, the proline-rich region of AR (amino acids 352–359) interacts with the SH3 domain of Src to activate the kinase [10]. Activated Src then causes the phosphorylation and stimulation of the epidermal growth factor receptor (EGFR) via an intracellular pathway that then stimulates the MAP kinase cascade (Raf, MEK, ERK). One result of increasing ERK activity is that $p90^{Rsk}$ kinase phosphorylates and activates the CREB transcription factor [8] and the expression of CREB-regulated genes [9]. The testosterone-mediated phosphorylation of ERK and CREB phosphorylation is rapid (within 1 min) and can be sustained for at least 12 h [9].

In the second pathway, there is rapid $[Ca^{2+}]$ influx into cultured Sertoli cells (within 20–40 s) through L-type $[Ca^{2+}]$ channels in response to testosterone stimulation [11,12] (Figure 6.1). This pathway begins with testosterone activating an as yet-unidentified G_q-type GPCR that subsequently stimulates phospholipase C that then hydrolyzes PIP2 in the plasma membrane to produce IP_3 and diacylglycerol. In response to lower levels of PIP2, the closing of the K^+_{ATP} channels is favored, causing an increase in membrane resistance and depolarization of the cell. These events cause voltage-dependent L-type Ca^{2+} channels to open and allow the influx of Ca^{2+}, which may alter many cellular processes, including the cytoskeleton structure of Sertoli cells [13,14]. Thus far, it has not been established whether AR or another receptor is required to initiate this pathway. Furthermore, the downstream targets of the testosterone-mediated [Ca^{2+}] influx pathway and its effect on spermatogenesis remain to be determined [15].

III. Testosterone production and action

Testosterone signaling via both classical and nonclassical pathways is influenced by local testosterone availability. Interestingly, there is some evidence that Sertoli cells play a key role in testosterone production by the fetal testis because they are the only testicular cell type that expresses 17βHsd3, an essential enzyme in the conversion of the precursor androgen androstenedione produced by fetal Leydig cells to testosterone [16,17]. For this reason, Sertoli cells appear to play a significant role in testosterone-driven masculinization that is not usually ascribed to these cells.

In adulthood, testosterone is produced almost exclusively by adult Leydig cells, and the amount of testosterone in the testis far exceeds that required for function [18], with testosterone concentrations in testes of both men and rodents approximately 25- to 125-fold greater than circulating levels [19–23]. The biology underlying the necessity for the high levels of testosterone in the testis is poorly understood, although a functional gradient is evident even at these high concentrations. Sperm counts decrease at a logarithmic rate as testosterone concentrations drop below 70 nM in rodents [24], and completion of meiosis and spermiogenesis requires different levels of testosterone support (reviewed in [25,26]). In adult rats, testicular testosterone concentrations of around 3% of normal levels maintain meiosis, whereas levels that are 12% of normal levels are required for elongated spermatid production [27,28].

Thus, the completion of meiosis requires less androgen action than the completion of spermiogenesis. This concept is supported by studies in transgenic mouse models. Reduced, but not ablated, androgen production or signaling can support meiosis but not spermiogenesis [29–31], whereas the ablation of androgen signaling causes meiotic arrest and significantly restricts spermiogenesis [32,33]. A dose-dependent effect at the level of androgen signaling within Sertoli cells was demonstrated in mice with targeted Sertoli cell-specific disruption of the DBD of AR [30]. Haploid spermatid production is evident in mice that are hemizygous for the Sertoli cell-specific ABP-Cre allele, whereas meiotic arrest is seen in mice that are homozygous for ABP-Cre.

These findings not only indicate that the genomic pathway of androgen action in Sertoli cells is required for spermatogenesis [30], but also demonstrate that Sertoli cells can respond to different levels of androgen signaling to support either meiosis or spermiogenesis, depending on the level of androgen support available. A further intriguing finding is that the higher concentrations of testosterone may also be required to promote novel, nonclassical mechanisms of testosterone action that are important for supporting spermatogenesis [6].

IV. Androgen receptor

A. AR expression in the testis

In most physiological scenarios, regardless of the concentration of testosterone, the ability to effect androgen signaling requires the presence of AR in target cells. In the testis, AR is expressed in Sertoli, Leydig, peritubular, arteriole smooth muscle, and vascular endothelial cells (Figure 6.2). In contrast, AR is not expressed in germ cells (reviewed in [34]) and is not required in germ cells for the completion of spermatogenesis, an observation supported by three complementary studies. First, chimeric mice carrying both AR-defective and wild-type germ cells produce offspring from AR-defective germ cells [35]. Second, germ cells carrying a nonfunctional AR gene transplanted into the testes of recipient wild-type mice following busulphan treatment are able to undergo spermatogenesis [36]. Third, cell-specific ablation of AR in germ cells using Cre/loxP methodologies has no impact on spermatogenesis or fertility [37]. Together, these results indicate that testosterone must support spermatogenesis indirectly by acting through AR-expressing somatic cells, including the Sertoli cells, in the testis.

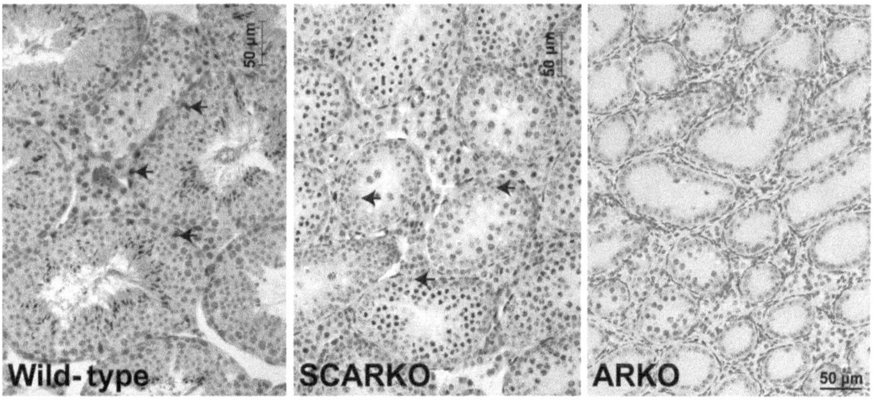

Figure 6.2 AR expression in wild-type, SCARKO and ARKO mice. In wild-type mice, AR is expressed in the Leydig, peritubular myoid, vascular smooth muscle, and vascular endothelial cells in addition to Sertoli cells (arrows). Conversely, in SCARKO mice, while AR expression is retained in other somatic cell types. AR expression is ablated in Sertoli cells (arrows), which is associated with an apparent arrest of spermatogenesis during meiosis. ARKO mice do not express AR in any cell type and display a more prominent disruption to spermatogenesis (see Section IV for details).

B. Sertoli cell-specific ablation of AR

In humans, AR expression is activated in Sertoli cells at the age of 5 months but is not significantly upregulated until 4 years of age [38–41], while in mice and rats, Sertoli cell AR expression is first detected 3–5 days after birth [42–45]. One of the major advances in our understanding of androgen's influence on Sertoli cell function has developed from the application of the Cre-loxP system to ablate AR in a cell-specific manner. At least five "floxed" alleles of AR have been generated by different groups [29,33,46–48] (reviewed in [25,49]); the subtle differences between alleles have proved very informative for producing a comprehensive picture of the role of AR signaling in the testis, including providing a greater understanding of the role of androgens in Sertoli cells (reviewed in [25,34,49–52]).

Prior to 2004, our understanding of the role of AR signaling in Sertoli cells derived primarily from analysis of testicular feminized (Tfm) and AR knockout (ARKO) mice [48,53] (Figure 6.2). These models have several limitations that impact the ability to tie specific observations to AR action within Sertoli cells. First, AR is ablated in all cell types in ARKO mice, making it impossible to ascribe any phenotypic change to loss of androgen signaling from a specific cell type. Second, Tfm and ARKO mice have only approximately 10% of normal levels of circulating androgens (and, because androgens are a substrate for estrogen production, low estrogen levels). Finally, the testes in Tfm and ARKO mice are cryptorchid; development of the testis at a temperature that is significantly higher than normal confounds analyses of the role of androgens in adult testis function, including any role in spermatogenesis.

In 2004, three separate teams published descriptions of independently generated Sertoli cell-specific AR knockout (SCARKO) models [29,32,33] (Figure 6.2). Each group utilized a Cre recombinase driven by the Amh/Mis promoter, and, thus, induced AR ablation of AR gene function from fetal life. Because Sertoli cell expression of AR begins at approximately postnatal day 4 in mice [54], Sertoli cells were never exposed to androgen–AR stimulation in all three SCARKO models. While this has subsequently proven to be very informative, the timing of AR ablation presents challenges in attempts to distinguish between developmental effects impacting later events and functional requirements of Sertoli cell AR action in adulthood.

One of the most important observations arising from the SCARKO mouse studies versus those of total ARKO or Tfm mice was that when ablation of AR is restricted to Sertoli cells, androgen action at the level of the gubernaculum is retained [55], and the testes descend normally. Retention of this critical event demonstrated the significance of being able to restrict the location in which AR is ablated and, for the first time, permitted investigation of the role of AR signaling in Sertoli cells without the confounding factor provided by cryptorchidism.

While researchers were well aware of the limitations of the Tfm and ARKO models, analysis of these mice significantly advanced our understanding of testis biology. In hindsight and with the benefit of having analyzed cell-specific ARKO mice, however, one thing that is apparent from studies of the Tfm and ARKO mice is that the interpretation of the role of AR in different cell types tended to follow autocrine lines. That is, any observed impact on a specific cell type was attributed to loss of AR from within that cell type. It was impossible to determine whether AR signaling in one cell type could determine the development and function of other cell types. For example, the significant reduction in Sertoli cell numbers observed in the Tfm and ARKO mice [48,53] in adulthood suggested that Sertoli cell AR signaling influenced final Sertoli cell numbers. For this reason, a similar reduction in Sertoli cell number was expected to arise in SCARKO mice. However, surprisingly, this was not the case. SCARKO mice, which lack AR starting in fetal life, develop appropriate numbers of Sertoli cells that mature normally (albeit with a slight delay [56]) during puberty, demonstrating that Sertoli cell AR is in fact dispensable for final Sertoli cell number and maturation [57].

By inference, this observation suggests that androgens must act via another cell type to determine the final Sertoli cell number; however, despite analyses of several models with specific AR ablation from other somatic cell types, this cell type still remains to be identified. Counterintuitively, transgenically driven expression of AR in Sertoli cells starting in fetal life (as opposed to postnatal day 4) also leads to a reduction in the numbers of Sertoli cell in adulthood, although in this case, this reduction is associated with premature maturation of seminiferous tubules. This premature maturation occurs despite increased expression of follicle-stimulating hormone (FSH) and FSH receptor, which normally promote Sertoli cell proliferation. This finding suggests that increased AR expression leads to the premature termination of Sertoli cell proliferation and the initiation of Sertoli cell differentiation regardless of FSH signaling [58]. Therefore, premature Sertoli cell AR expression may cause premature Sertoli cell differentiation, thus reducing the pool of undifferentiated Sertoli cells that are capable of proliferation. The reduction in numbers of Sertoli cells is reflected in the reduction of germ cell numbers in the testis. Together, these data demonstrate the complexity of androgen-mediated control of Sertoli cell number and shows, for the first time, that a wider network of androgen control of testis development exists. In this network, Sertoli cells are influenced by androgen signaling in other cell types and, in turn, are able to influence other cell types through AR signaling in Sertoli cells.

Even though Sertoli cell numbers are unaffected by the loss of Sertoli cell AR, all three models of Sertoli cell-specific AR ablation are infertile and display very similar phenotypes of spermatogenic arrest during meiosis. While numbers of spermatogonia are unaffected, ablation of Sertoli cell AR leads to a small reduction in the number of spermatocytes and a significant reduction in or absence of postmeiotic spermatids. These observations have been confirmed in a fourth model of Sertoli cell AR ablation [30], and are similar to the arrest in spermatogenesis observed following pharmacological suppression of testosterone production [59]. However, the etiology is different in that SCARKO mice display normal or near-normal concentration of androgens, and this phenotype cannot be rescued in hypogonadal.SCARKO (hpg.SCARKO) mice treated acutely with exogenous androgens [60].

In one model of Sertoli cell AR ablation, loss of AR resulted in a significant reduction in numbers of Leydig cells when they were examined in adulthood. This finding suggests that a Sertoli cell AR-dependent paracrine signal may exist and prove important for supporting adult Leydig cell development. However, this apparent reduction in Leydig cell number is compensated for by increased testosterone production per Leydig cell to maintain normal serum levels [61]. Furthermore, while Leydig cell number was determined using 3βHsD as a Leydig cell marker, employment of a second counting methodology, the optical disector method, revealed no reduction in Leydig cell number in the same mouse line [62]. Thus, whether Sertoli cell AR signaling does indeed impact final numbers of adult Leydig cells remains unresolved, although some evidence in support of such a relationship has been provided by a novel transgenic model of induced Sertoli cell AR expression. Premature activation of Sertoli cell AR impacts Leydig cell development, leading to reductions in total numbers of fetal and adult-type Leydig cells, despite normal circulating concentrations of LH. Adult-type Leydig cells from transgenic animals show advanced maturation and increased androgen production per Leydig cell to maintain normal circulating testosterone concentrations, which is consistent with a role for Sertoli cell AR signaling in controlling the numbers, maturation, and function of the Leydig cell population [63].

V. The role of androgens in Sertoli cells

A. Androgens and the blood–testis barrier

The first impact arising from loss of Sertoli cell AR is observable around postnatal day 10.5, when changes in the gene expression of functional components of the Sertoli cell cytoskeleton and basement membrane components become phenotypically apparent [64]. Furthermore, maturation of Sertoli cells and Sertoli cell polarity are impacted [56].

An important aspect of Sertoli cell maturation is the development of inter-Sertoli cell tight and occluding junctions, which contribute to the so-called blood–testis barrier (BTB). The BTB provides the environment of meiotic and postmeiotic germ cells and maintains immune privilege within the seminiferous epithelium (Figure 6.3A). Development of the BTB begins in SCARKO mice but is inhibited by changes in the expression of key BTB components, leading to permeabilization of the BTB, demonstrated by use of a biotin tracer [65]. Defective development of Sertoli cell tight junctions exposes postmeiotic germ cells to the immune system, resulting in detectable antibodies to germ cell antigens in the bloodstream and an influx of immune cells into the testicular interstitium [66]. As germ cells "pass through" the BTB, the proteins above the cell in the original BTB are relocated to below the cell to form a new BTB. Testosterone stimulation of *hpg* mice causes the rapid redistribution of claudin 11 to the newly forming BTB prior to the activation of the transcription of genes encoding BTB components [67]. Thus, nonclassical testosterone signaling may regulate the maintenance of the BTB. In addition, kinases that are known to be regulated by nonclassical testosterone signaling [8,68] contribute to the formation and dissolution of tight and adherens junctions that make up the BTB [69–71].

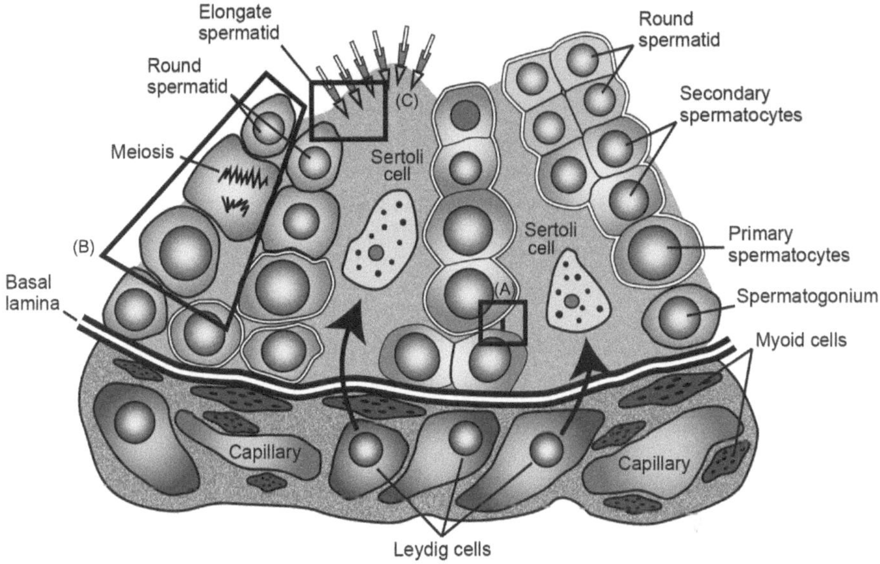

Figure 6.3 Schematic overview of impacts of Sertoli cell AR signaling on spermatogenesis. Androgen–AR signaling impacts Sertoli cell function at several key points. (A) Androgen–AR signaling is important in the development and support of the BTB, which retains the immune privilege of the adluminal compartment. (B) Androgen–AR signaling within Sertoli cells is essential for the completion of meiosis. (C) Sertoli cell androgen signaling plays a key role in the process of spermiation. Each of these roles varies across the stages of the spermatogenic cycle, and, together, they reflect a dynamic system of regulation that functions cohesively to promote spermatogenesis.

B. Androgens in meiosis

Androgens are absolutely required for the ability of spermatocytes to proceed through meiotic division and to produce haploid spermatids, as has been demonstrated in a number of mouse models (e.g., those described in [30,32,33,72] and reviewed in [25,51]; Figure 6.3B). In the absence of AR in Sertoli cells, spermatogonia can enter meiosis; however, cell survival through the long meiotic prophase is impaired and there is a gradual loss of spermatocytes [33]. At the end of the first meiotic prophase, spermatocytes normally enter two rapid and successive meiotic divisions to produce haploid round spermatids. However, in the absence of AR in Sertoli cells, this meiotic division and the subsequent production of spermatids cannot be completed [32,33], perhaps due to a failure of diplotene spermatocytes to enter meiosis I [32]. Meiotic completion can be accelerated by the premature induction of Sertoli cell AR during early postnatal development [58], highlighting the importance of Sertoli cell AR as a driver of meiosis.

The precise androgen-dependent signals from the Sertoli cells and how they regulate the survival and division of spermatocytes remain unknown. Proteomic analysis of pachytene spermatocytes isolated after androgen suppression and replacement reveals that the expression and posttranslational modifications of proteins involved in meiosis and cell survival are androgen responsive [73], despite a lack of AR within these cells. Thus, Sertoli cells must produce paracrine or extracellular factors, which, in turn, modulate protein expression and posttranslational modifications in spermatocytes. Another mechanism by which androgen action on Sertoli cells may facilitate the completion of meiosis is the establishment of a functional BTB and the resulting immune-privileged environment within the epithelium [65,74,75]. This environment is normally formed at the onset of meiosis during the first wave of spermatogenesis (see section V.A).

C. Androgens in spermiogenesis and sperm release

Androgen action is also essential for the completion of spermiogenesis and, hence, the production of elongated spermatids (reviewed in [26,51,76]). This requirement was noted in early studies that administered testosterone to hypophysectomized rats [77–79] and was further proven by the androgen-dependent stimulation of sperm production in *hpg* mice (reviewed in [80]). An arrest of spermiogenesis was also seen in several genetically modified mouse models of reduced androgen signaling [29,30,81]. When testicular androgen levels are reduced below a certain threshold (around 3% of normal levels), elongated spermatid production ceases due to the premature detachment of step 8 round spermatids from Sertoli cells, whereas higher dose testosterone replacement can reverse this detachment and allow elongated spermatid production to continue [27,82]. Androgen action likely stimulates the expression of the cell adhesion molecules that are necessary for round spermatids to adhere to Sertoli cells and to complete their maturation [83] (reviewed in [2]).

The final stage of spermiogenesis, known as spermiation, is the process by which elongated spermatids undergo their final remodeling and ultimate release by Sertoli cells prior to their transit to the epididymis (Figure 6.3C). Spermiation is well known to be acutely sensitive to reduced testicular androgen production or signaling (reviewed in [84]). In this setting, elongated spermatids are retained by the Sertoli cells at the end of spermiation and, instead of being released, are phagocytosed by the Sertoli cells [85]. Failure of spermiation is observed in rats with reduced testicular androgen levels [86] and is exacerbated by the addition of an AR antagonist to block residual androgen action [87]. Consistent with this finding, spermiation failure is also observed in mice with reduced levels of androgen signaling due to a hypomorphic allele of AR [29] and can be induced *in vitro* by the treatment of seminiferous tubules with an AR antagonist [68].

There is indirect evidence that nonclassical testosterone signaling contributes to the release of mature sperm from Sertoli cells because Src kinase, which is regulated by nonclassical signaling, regulates sperm release. Specifically, treatment of cultured seminiferous tubule fragments with a Src kinase inhibitor decreased the relative numbers of sperm released. Consistent with this finding was the increased localization of activated Src to the region around the Sertoli-elongated spermatid adhesion complex during stages VII–VIII, when sperm are released [88–90]. In addition, protein members of the adhesion complexes between Sertoli cells and elongated spermatids, including focal adhesion kinase, are known to be phosphorylated by Src [91–93].

Spermiation failure is an early feature of androgen suppression [86] and likely underlies the rapid suppression of sperm counts seen in monkeys and men during androgen-based contraceptive administration [94,95]. Androgen stimulation is unlikely to be the sole regulator of spermiation because many other endocrine and pharmacological disturbances, including elevated testicular estrogen [96], cause spermiation failure [84,88]. It seems more likely that spermiation failure is a relatively nonspecific response of Sertoli cells to various onslaughts, including endocrine disturbances (reviewed in [84]).

D. Stage-specific effects of androgens

Germ cell development in the seminiferous epithelium is arranged in strictly defined cellular associations referred to as "stages," and the number of cellular associations that can be distinguished from each other dictates the number of stages in each species. Spermatogenesis is initiated at puberty at different points along the seminiferous tubules when a spermatogonial stem cell becomes committed to the process of differentiation and undergoes multiple mitoses to provide a large pool of spermatogonia that enter meiosis and spermiogenesis. Thus, a stem cell eventually populates a large area of tubule, and spermatogenesis continues to occur at this site throughout adulthood [97].

Germ cell development proceeds in a highly ordered fashion along the tubule, resulting in the appearance of particular cohorts in tubule cross-sections. In other words, a particular spermatogonial subtype is always seen in the epithelium along with a particular subtype of meiotic cell and with spermatids at particular phases of development. As germ cells proceed through development, the stages follow one another in a consecutive wave along the tubule; in mice, germ cells progress from stage I to stage XII of spermatogenesis and then again into stage I during one complete spermatogenic cycle [98]. These "waves" of spermatogenesis allow continual "bursts" of sperm production along the tubules.

Because of the highly ordered nature of spermatogenesis and its organization into stages, it is important to note that many genes are expressed, and cellular functions are modulated, in a stage-dependent manner. The timing of each stage and of the completion of one cycle of stages is species specific and highly consistent within species [98]. The timing of the spermatogenic stages appears to be controlled by the interactions between Sertoli cells and the developing germ cells. Sertoli cells apparently have an internal "clock" because some proteins show stage-related changes even during early embryonic development, when only the most immature germ cells are present [99]. However, germ cells can set the timing of the clock because rat germ cells transplanted into mouse seminiferous tubule cause the spermatogenic stages to proceed according to rat timing [100].

Both androgen action and receptor expression show stage specificity during spermatogenesis. The mid-spermatogenic stages (VII–VIII in mice and rats and II–III in humans) are well known to be the most androgen responsive (reviewed in [101,102]).

In addition, AR protein is maximally expressed in Sertoli cells in these stages in rats [103], monkeys [104], and men [105]. Acute ablation of androgen action in adult rats modulates the expression of various cohorts of genes in Sertoli cells and germ cells in these stages [87]. Androgens act on AR to modulate the expression of various genes and proteins within Sertoli cells, which, in turn, mediate Sertoli cells' ability to support germ cell development (reviewed in [51]). The transcription of some androgen-responsive genes appears to be directly stimulated by AR binding to AREs within their promoter region, such as *Rhox5* [106] and *Tubb3* [107] (reviewed in [51]), whereas others are likely to be affected indirectly, perhaps downstream of early response genes (see Section VI).

Various studies have indicated that the expression of a surprisingly large number of genes is changed in response to androgen suppression (reviewed in [51]). Sertoli cell mRNA transcripts with stage-specific expression patterns have been identified in rats [108], and the effects of androgen suppression on these transcripts in stages VII and VIII were subsequently examined [87]. This analysis revealed that the expression of genes enriched in Sertoli cells and maximally expressed in the earlier stages of spermatogenesis (I–VI) were mainly increased during androgen suppression in stages VII and VIII. This finding suggests that, in the mid-spermatogenic stages, androgens normally facilitate the decrease in Sertoli cell mRNA transcripts that are maximally expressed earlier in the cycle [87] (an observation consistent with the results of microarray studies described later). Conversely, mRNA transcripts that are maximally expressed in Sertoli cells in the later stages of the cycle [108] are either increased or decreased in the mid-spermatogenic stages by androgen suppression. Thus, Sertoli cells in the mid-spermatogenic stages appear to respond to androgens by coordinating the expression of the genes needed later in the spermatogenic cycle [87]. These findings suggest that androgens can facilitate the stage-specific expression of cohorts of Sertoli cell genes.

Androgens can also modulate the expression of various microRNA (miRNA) species in stage VIII seminiferous tubules *in vivo* and in Sertoli cells *in vitro*, which may underpin stage-dependent changes [109,110]. miRNAs are an evolutionarily conserved class of small (18- to 24-nucleotide), noncoding RNAs that participate in posttranscriptional gene regulation. Acute androgen suppression causes changes in miRNA species that are predicted to be involved in key Sertoli cell processes [109,110]. Interestingly, a greater proportion of miRNAs are increased by androgen suppression, which would be predicted to decrease target protein translation. Closer investigation of a few miRNAs in Sertoli cells reveals that androgen-mediated modulation of miRNA expression is rapid, within 6 h [109]. These findings suggest that androgens modulate the expression of key miRNA species in Sertoli cells that target the translation of groups of proteins involved in spermatogenesis and point to miRNAs as acute molecular switches that androgens utilize to modulate Sertoli cell function [109].

Thus, it seems that the peak of Sertoli cell AR expression in the mid-spermatogenic stages receives the androgen signal and modulates both the transcription and translation of a cohort of genes. These transcriptional changes can be direct, occurring via AR-dependent gene expression and effects on Sertoli cell miRNAs, which then have downstream impacts on Sertoli cell gene and protein expression. Changes in androgen-responsive proteins in Sertoli cells ultimately impact germ cell gene [87] and protein expression [73] and function. Thus, androgens act on Sertoli cells, particularly in the mid-spermatogenic stages, to coordinate germ cell development and, thus, the timing of the spermatogenesis.

VI. AR-dependent gene expression in Sertoli cells

Androgens may mediate gene expression changes via classical and nonclassical signaling pathways, although the specific details of this process remain underexplored. As described above (section V.D), the surprising finding is that many testosterone-regulated genes identified in recent microarray studies are apparently downregulated by the steroid. However, the genes identified by these studies show little overlap, and fold change in gene expression is relatively low, with very few genes showing a greater than twofold change in gene expression. Furthermore, few of the genes identified are associated with known fertility phenotypes (reviewed in [51]).

The development of cell-specific knockouts of AR in Sertoli cells has provided a significant opportunity to dissect AR-responsive pathways within Sertoli cells and to identify androgen-responsive genes underpinning male fertility. The challenge has been identifying genes that change because they are direct targets of AR in Sertoli cells versus genes whose expression changes because of changes in cellular composition or testicular architecture. Thus, experimental timing and temporal distance between genetic insult and gene expression analysis have well-recognized confounding effects on results.

As described above (section IV.B), ablation of Sertoli cell AR prior to the onset of expression means that Sertoli cells never develop with AR activity. In the first analysis of gene expression using SCARKO mice, a time point of 10 days after birth was chosen because this precedes any overt histological differences in testis architecture. Forty genes exceeded a twofold cutoff in gene expression difference between SCARKO and control animals. Twenty-eight of these genes were revealed to be downregulated by testosterone and 12 were upregulated [111]. Of the 40 genes that displayed significant changes in expression, three (*Galgt1* [112,113], *PCI* [114], and *epididymal protease inhibitor [Eppin]* [115]) are known to be required for fertility. This initial study was later extended to cover a wider time scale. This continuation study revealed that differences in gene expression in SCARKO mice are first detectable 8 days after birth [54]. The results also showed that genes encoding extracellular matrix components, cell adhesion proteins, proteases, and cytoskeletal factors are regulated by androgen–AR signaling during the initiation of spermatogenesis, which is consistent with a role for Sertoli cell AR signaling in this process [56,111]. A more targeted qPCR assay approach identified similar results in SCARKO mice [57] to those obtained by a second group using a similar Sertoli cell ARKO mouse model [64].

Another study comparing AR hypomorphic mice and AR hypomorphs with Sertoli cell-specific deletion of AR identified 46 and 57 genes regulated by AR-dependent signaling (more than twofold), respectively, with, again, significantly more genes suppressed than upregulated by testosterone [116]. Furthermore, 65% of these genes contain a conserved ARE within 6 kb of their transcription start sites, and a third of these genes are known to be regulated by validated AREs [116]. The presence of known AREs in the promoters of some target genes supports the role of classical AR signaling in the control of Sertoli cell function. Conversely, the absence of AREs in other androgen-regulated genes suggests either indirect or downstream regulation, or perhaps regulation by nonclassical AR signaling.

In a refinement to the basic Sertoli cell ablation model, SCARKO mice have been generated on a juvenile spermatogonial depletion mutant (*jsd*) background. The *jsd* mutation leads to infertility after the initial waves of spermatogenesis, with only undifferentiated type A spermatogonia remaining from the germ cell complement [117]. Comparison of SCARKO-*jsd* to the *jsd* model revealed that the expression of 157 genes is suppressed and of 197 genes is upregulated in the *SCARKO-jsd* mice. Genes associated with signal transduction, cell adhesion, nuclear localization, and meiosis are differentially regulated in the absence of AR. Interestingly, while suppression of testosterone in *jsd* mice is known to support spermatogonial differentiation, ablation of Sertoli cell AR has no impact on this process, demonstrating that this process occurs via a Sertoli cell AR-independent mechanism [118].

In summary, the trends identified from these several gene expression studies allow us to make several statements regarding the overarching mechanism of Sertoli cell AR signaling in the control of testis function. The first general observation is that the number of genes downregulated by testosterone far exceeds the number that is upregulated. This observation is perhaps surprising, given AR's role as a transcription factor, but perhaps reflects the timing of the analyses, which in several cases were conducted days or weeks after the initial genetic insult. For this reason, many identified genes represent secondary androgen responses rather than specific targets of AR.

Second, there is a wide discrepancy between datasets, with only the four highly testosterone-responsive genes *Rhox5*, *Lrp4*, *Drd4*, and *Fhod3* quantitatively regulated in the same direction in studies employing the SCARKO, SCARKO-jsd, and Tfm models [119]. This divergence probably reflects differences in the ages of the mice, duration of testosterone stimulation or deprivation, hormone levels, and germ cell complement of the testes and highlights both the complexity of the system and the need to restrict analyses to specific temporal and functional windows.

Third, surprisingly few gene expression changes rise above a twofold difference in expression levels, suggesting that Sertoli cells' AR control of spermatogenesis is likely to result from small changes in the expression of a large number of genes and not simply the regulation of a few key master factors. In support of this hypothesis is that of the genes identified as Sertoli cell AR regulated, few result in a lack of sperm production when mutated, although several cause reduced fertility. Loss of Rhox5 impacts fertility, albeit not completely, and germ cells are still able to complete meiosis. In contrast, the androgen-regulated genes *Galgtl*, *PCI*, and *Eppin* are required for fertility but may not directly affect spermatogenesis [112–115]. However, loss of *Lrp8*, another androgen-regulated gene in the testis, leads to sperm defects that develop in the epididymis [120,121]. Despite these limitations, gene expression analysis has identified several common androgen-regulated genes, including *Tsx*, and three serine endopeptidase inhibitors, *Eppin*, *Serpina 3N*, and *PCI*. In addition, *alcohol dehydrogenase I* is essential for local retinoic acid production, which plays a key role in spermatogonial development, meiotic entry, and establishment of the cycle of the seminiferous epithelium [122–124]. *Claudin 3*, an important factor in the formation of the BTB, is also downregulated in more than one SCARKO model (reviewed in [51]).

Together, these data paint a picture of Sertoli cell AR regulation of spermatogenesis occurring at multiple points and at different stages of the spermatogenic cycle. This regulation combines both genomic and nongenomic mechanisms and involves perturbation of more than one downstream factor that is necessary to induce infertility. As such, Sertoli cell AR signaling shows a level of redundancy and plasticity that will require significant further investigation to unravel.

VII. Sertoli cell estrogen signaling (from androgens via aromatase)

The high local concentrations of testosterone in the testis can act as a substrate for the aromatase enzyme and thus be converted to 17β-estradiol [125]. Aromatase is expressed in Sertoli cells, Leydig cells, and germ cells, and, thus, the adult testis is amply equipped to produce estrogen [126–128]. In fact, the concentration of estradiol in rete testis fluid and testicular venous blood is far greater than that in male serum, highlighting the local production of estradiol in the testis [127,129]. Sertoli cells contain both aromatase and estrogen receptors (ERs) and, thus, are capable of producing and responding to estrogen [126,128,130,131]. Sertoli cells are likely to be a significant source of estrogen in the early postnatal period; however, in mature rodents, Leydig cells and germ cells are likely be the predominant source (reviewed in [126,128,130,131]). Testicular estradiol can act on testis cells via a genomic pathway involving the classic ERs ERα and ERβ, which are abundantly expressed in Sertoli cells, Leydig cells, and germ cells [126,128,130]. Estrogen is also able to activate the rapid, nongenomic EGFR/ERK/c-jun signaling cascade in germ cells via ERα and GPR30, a transmembrane, intracellular ER [130].

The role of estradiol in spermatogenesis has been investigated through the use of various genetically modified mouse models. A surprising observation is that estradiol can stimulate spermatogenesis in *hpg* mice, which lack pituitary hormones [132]. However, estradiol treatment in these mice also produced a small but significant rise in FSH [132] that could have had a supportive effect. Importantly, this estradiol-dependent stimulation of spermatogenesis required the presence of a functional AR [133,134] as well as ERα [135]. The latter observation suggested the involvement of an ERα-dependent neuroendocrine effect on FSH secretion. Enhanced Sertoli cell maturation and function also accompanies the estradiol-dependent stimulation of spermatogenesis. Surprisingly, the expression of androgen-dependent genes, such as *Rhox5* and *Eppin*, in Sertoli cells is also increased [135]. One potential explanation for these findings is estradiol stimulation of ERα-AR heterodimers in Sertoli cells, which in turn enhances Sertoli cell function and an ability to support spermatogenesis [135]. However, the existence of such heterodimers in the testis has not been confirmed. An alternative explanation is that the ERα-dependent increase in FSH could stimulate 5α-reductase levels [136], which would promote the conversion of the very low intratesticular testosterone levels in these mice [133] to the more potent androgen dihydrotestosterone (DHT), which could, in turn, theoretically stimulate Sertoli cell AR signaling. However, 5α-reductase and DHT levels have not been measured in this experimental paradigm. Thus, whether estradiol stimulation of spermatogenesis involves increased androgen signaling in Sertoli cells remains to be determined.

Infertility in male mice lacking aromatase and ERα proves that estrogen and ERs are required for spermatogenesis [137,138]. Infertility in mice lacking ERα is caused by efferent duct dysfunction [137] due to disturbances in epithelial cell morphology and ion and water channel function (reviewed in [126]). This dysfunction leads to the inability of the efferent duct to resorb seminiferous tubule fluid, resulting in fluid backup in the testis, which impairs spermatogenesis [137]. However, mice lacking ERβ are fertile [139]. In contrast, aromatase knockout mice are infertile but do not show evidence of disturbed efferent duct function as do ERα knockout mice; instead, spermatogenesis is disrupted with age due to Sertoli cell and/or germ cell dysfunction [138].

The lack of efferent duct dysfunction in aromatase knockout mice appears to be explained by more recent studies indicating that estradiol-independent activation of ERα is essential for efferent duct function [140]. Mice expressing a mutated ERα that is unable to bind estradiol (ENERKI mice) show a similar, late-onset spermatogenic phenotype to aromatase knockout mice, suggesting that estradiol acting on ERα is required for normal spermatogenesis in adulthood [140]. This proposition is supported by the fact that the late-onset spermatogenic phenotype in ENERKI mice is prevented by neonatal, but not postpubertal, treatment with an ERα agonist, an effect that is proposed to be mediated by Sertoli cells [140]. This intriguing finding suggests that neonatal estradiol-dependent signaling through ERα is required to establish normal onset of Sertoli cell maturation and function, which, in turn, is necessary for the maintenance of spermatogenesis in adulthood.

In summary, testosterone can be metabolized within the testis to produce estradiol, which can then act via genomic and nongenomic pathways in testicular cells. Estradiol action via nongenomic pathways in germ cells appear to be important for germ cell function [130], whereas estradiol signaling via ERα in neonatal Sertoli cells appears to be essential for normal adult spermatogenesis.

VIII. Conclusions and future perspectives

Since the last edition of *Sertoli Cell Biology*, significant progress has been made in expanding our understanding of Sertoli cell androgen signaling. The step changes in the past decade have been multifold but are most starkly illustrated by three overarching advances. Arguably, the most significant step forward has been provided by the development of mouse models with Sertoli cell-specific ablation of AR, which has revolutionized our ability to ascribe specific roles to Sertoli cells. Second, the emergence of nonclassical androgen signaling as an important functional paradigm and identification of the roles this plays inside Sertoli cells has provided an entirely new way of understanding the interactions between Sertoli cells and androgen. And third, the widespread use of high-throughput genomics, primarily on new mouse models, has provided a window into the inner workings of androgen signaling inside Sertoli cells. Together, these approaches, along with the other advances discussed above, have allowed us to draw the following, previously impossible, conclusions regarding the specific role of androgen–AR signaling in Sertoli cells:

1. Studies in SCARKO mice have demonstrated that Sertoli cell androgen signaling is dispensable for attainment of final numbers of Sertoli cells, overturning a long-held belief derived from the original Tfm and ARKO mouse studies.
2. Further evidence from SCARKO mice along with the development of mice with premature induction of Sertoli cell AR expression implicate Sertoli cell AR in the promotion of Sertoli cell maturation. However, because maturation eventually occurs in SCARKO mice, Sertoli cell AR signaling is not an essential factor.
3. Sertoli cell androgen signaling is a significant promoter of BTB formation, and both classical and nonclassical androgen signaling can influence this process. This observation has significance for male hormonal contraceptive development and/or male infertility because perturbation to local androgen signaling can lead to disruption of the specialized germ cell environment and exposure of the postmeiotic germ cells to the immune system.
4. Sertoli cell androgen signaling is not required for the initiation of spermatogenesis because the numbers of spermatogonia and the transition to spermatocytes are unaffected following ablation of Sertoli cell AR.
5. Sertoli cell androgen signaling plays a significant role in supporting meiosis. Loss of AR results in increased spermatocyte apoptosis and failure of meiosis I, leading to a significant reduction in the numbers of postmeiotic germ cells—this process essentially blocks meiotic progression. While manipulation of testosterone concentrations has hinted at this role for many years, it is only through the use of SCARKO models that this role in promoting meiosis can be unequivocally attributed to Sertoli cells.
6. Sertoli cell androgen signaling supports spermiation via both classical and nonclassical mechanisms.
7. Sertoli cell androgen signaling controls spermatogenesis via a cascade of myriad downstream effector genes, including miRNAs, with small contributions from each gene acting cumulatively to support spermatogenesis.
8. At the wider level, Sertoli cell androgen signaling influences the development and function of other testicular cell types, such as the Leydig cells, showing that Sertoli cell androgen signaling is an important part of a much wider androgen-dependent paracrine mechanism underpinning testis development and function.
9. Finally, the emergence of a significant role for locally produced estrogens (derived from androgen substrates) in Sertoli cell maturation and support of spermatogenesis has widened the influence and mode of action of androgens.

In conclusion, the first decades of the twenty-first century have seen a significant increase in our understanding of Sertoli cell androgen signaling, arguably the biggest step forward in this field in decades. Yet despite these advances, our goal remains the same as in the last edition of this book—to understand how androgens regulate spermatogenesis. We have undoubtedly traveled significantly further along this road, but if research has revealed anything, it is that androgen control of spermatogenesis is far more complex than we could have imagined. Future research will aim to reconcile these findings, although the complexity is so great that studies in humans, animal models, genetics, molecular biology, and computer modeling will all need to be brought to bear if a complete understanding is ever to be reached. However, encouragingly, the speed of advances in our understanding means this aim no longer seems like an impossible task.

References

[1] Sharpe RM. Regulation of spermatogenesis. In: Knobil E, Neil JD, editors. The physiology of reproduction. New York, NY: Raven Press; 1994. p. 1363–434.
[2] McLachlan RI, O'Donnell L, Meachem SJ, Stanton PG, De Kretser DM, et al. Identification of specific sites of hormonal regulation in spermatogenesis in rats, monkeys, and man. Recent Prog Horm Res 2002;57:149–79.
[3] Heinlein CA, Chang C. Androgen receptor in prostate cancer. Endocr Rev 2004;25:276–308.
[4] Lonergan PE, Tindall DJ. Androgen receptor signaling in prostate cancer development and progression. J Carcinog 2011;10:20.
[5] Shang Y, Myers M, Brown M. Formation of the androgen receptor transcription complex. Mol Cell 2002;9:601–10.
[6] Walker WH. Non-classical actions of testosterone and spermatogenesis. Philos Trans R Soc Lond B Biol Sci 2010;365:1557–69.
[7] Loss ES, Jacobus AP, Wassermann GF. Rapid signaling responses in Sertoli cell membranes induced by follicle stimulating hormone and testosterone: calcium inflow and electrophysiological changes. Life Sci 2011;89:577–83.
[8] Cheng J, Watkins SC, Walker WH. Testosterone activates MAP kinase via Src kinase and the EGF receptor in Sertoli cells. Endocrinology 2007;148:2066–74.
[9] Fix C, Jordan C, Cano P, Walker WH. Testosterone activates mitogen-activated protein kinase and the cAMP response element binding protein transcription factor in Sertoli cells. Proc Natl Acad Sci USA 2004;101:10919–24.
[10] Migliaccio A, Castoria G, Di Domenico M, de Falco A, Barone MV, et al. Steroid-induced androgen receptor–oestradial receptor β-Src complex triggers prostate cancer cell proliferation. EMBO J 2000;20:5406–17.
[11] Gorczynska E, Handelsman DJ. Androgens rapidly increase the cytosolic calcium concentration in Sertoli cells. Endocrinology 1995;136:2052–9.
[12] Lyng FM, Jones GR, Rommerts FFG. Rapid androgen actions on calcium signaling in rat Sertoli cells and two human prostatic cell lines: similar biphasic responses between 1 picomolar and 100 nanomolar concentrations. Biol Reprod 2000;63:736–47.
[13] Von Ledebur EI, Almeida JP, Loss ES, Wassermann GF. Rapid effect of testosterone on rat Sertoli cell membrane potential. Relationship with K^{+}ATP channels. Horm Metab Res 2002;34:550–5.
[14] Loss ES, Jacobsen M, Costa ZS, Jacobus AP, Borelli F, et al. Testosterone modulates K(+)ATP channels in Sertoli cell membrane via the PLC–PIP2 pathway. Horm Metab Res 2004;36:519–25.

[15] Rahman F, Christian HC. Non-classical actions of testosterone: an update. Trends Endocrinol Metab 2007;18:371–8.
[16] Shima Y, Miyabayashi K, Haraguchi S, Arakawa T, Otake H, et al. Contribution of Leydig and Sertoli cells to testosterone production in mouse fetal testes. Mol Endocrinol 2013;27:63–73.
[17] O'Shaughnessy PJ, Baker PJ, Heikkila M, Vainio S, McMahon AP. Localization of 17beta-hydroxysteroid dehydrogenase/17-ketosteroid reductase isoform expression in the developing mouse testis—androstenedione is the major androgen secreted by fetal/neonatal Leydig cells. Endocrinology 2000;141:2631–7.
[18] Tsai MJ, O'Malley BW. Molecular mechanisms of action of steroid/thyroid receptor superfamily members. Ann Rev Biochem 1994;63:451–86.
[19] Comhaire FH, Vermeulen A. Testosterone concentration in the fluids of seminiferous tubules, the interstitium and the rete testis of the rat. J Endocrinol 1976;70:229–35.
[20] Turner TT, Jones CE, Howards SS, Ewing LL, Zegeye B, et al. On the androgen microenvironment of maturing spermatozoa. Endocrinology 1984;115:1925–32.
[21] Awoniyi CA, Santulli R, Sprando RL, Ewing LL, Zirkin BR. Restoration of advanced spermatogenic cells in the experimentally regressed rat testis: quantitative relationship to testosterone concentration within the testis. Endocrinology 1989;124:1217–23.
[22] Maddocks S, Hargreave TB, Reddie K, Fraser HM, Kerr JB, et al. Intratesticular hormone levels and the route of secretion of hormones from the testis of the rat, guinea pig, monkey and human. Int J Androl 1993;16:272–8.
[23] Jarow JP, Chen H, Rosner TW, Trentacoste S, Zirkint BR. Assessment of the androgen environment within the human testis: minimally invasive method to obtain intratesticular fluid. J Androl 2001;22:640–5.
[24] Zirkin BR, Santulli R, Awoniyi CA, Ewing LL. Maintenance of advanced spermatogenic cells in the adult rat testis: quantitative relationship to testosterone concentration within the testis. Endocrinology 1989;124:3043–9.
[25] De Gendt K, Verhoeven G. Tissue- and cell-specific functions of the androgen receptor revealed through conditional knockout models in mice. Mol Cell Endocrinol 2012;352:13–25.
[26] O'Donnell L, McLachlan RI. The role of testosterone in spermatogenesis. In: Nieschlag E, Behre HM, Nieschlag S, editors. Testosterone: action, deficiency, substitution. 4th ed. New York, NY: Cambridge University Press; 2012. p. 123–53.
[27] O'Donnell L, McLachlan RI, Wreford NG, Robertson DM. Testosterone promotes the conversion of round spermatids between stages VII and VIII of the rat spermatogenic cycle. Endocrinology 1994;135:2608–14.
[28] Abbott GW, Sesti F, Splawski I, Buck ME, Lehmann MH, et al. MiRP1 forms IKr potassium channels with HERG and is associated with cardiac arrhythmia. Cell 1999;97:175–87.
[29] Holdcraft RW, Braun RE. Androgen receptor function is required in Sertoli cells for the terminal differentiation of haploid spermatids. Development 2004;131:459–67.
[30] Lim P, Robson M, Spaliviero J, McTavish KJ, Jimenez M, et al. Sertoli cell androgen receptor DNA binding domain is essential for the completion of spermatogenesis. Endocrinology 2009;150:4755–65.
[31] Zhang FP, Poutanen M, Wilbertz J, Huhtaniemi I. Normal prenatal but arrested postnatal sexual development of luteinizing hormone receptor knockout (LuRKO) mice. Mol Endocrinol 2001;15:172–83.
[32] Chang C, Chen YT, Yeh SD, Xu Q, Wang RS, et al. Infertility with defective spermatogenesis and hypotestosteronemia in male mice lacking the androgen receptor in Sertoli cells. Proc Natl Acad Sci USA 2004;101:6876–81.

[33] De Gendt K, Swinnen JV, Saunders PT, Schoonjans L, Dewerchin M, et al. A Sertoli cell-selective knockout of the androgen receptor causes spermatogenic arrest in meiosis. Proc Natl Acad Sci USA 2004;101:1327–32.
[34] Wang RS, Yeh S, Tzeng CR, Chang C. Androgen receptor roles in spermatogenesis and fertility: lessons from testicular cell-specific androgen receptor knockout mice. Endocr Rev 2009;30:119–32.
[35] Lyon MF, Glenister PH, Lamoreux ML. Normal spermatozoa from androgen-resistant germ cells of chimaeric mice and the role of androgen in spermatogenesis. Nature 1975;258:620–2.
[36] Johnston DS, Russell LD, Friel PJ, Griswold MD. Murine germ cells do not require functional androgen receptors to complete spermatogenesis following spermatogonial stem cell transplantation. Endocrinology 2001;142:2405–8.
[37] Tsai MY, Yeh SD, Wang RS, Yeh S, Zhang C, et al. Differential effects of spermatogenesis and fertility in mice lacking androgen receptor in individual testis cells. Proc Natl Acad Sci USA 2006;103:18975–80.
[38] Shapiro E, Huang H, Masch RJ, McFadden DE, Wu XR, et al. Immunolocalization of androgen receptor and estrogen receptors alpha and beta in human fetal testis and epididymis. J Urol 2005;174:1695–8.
[39] Berensztein EB, Baquedano MS, Gonzalez CR, Saraco NI, Rodriguez J, et al. Expression of aromatase, estrogen receptor alpha and beta, androgen receptor, and cytochrome P-450scc in the human early prepubertal testis. Pediatr Res 2006;60:740–4.
[40] Chemes HE, Rey RA, Nistal M, Regadera J, Musse M, et al. Physiological androgen insensitivity of the fetal, neonatal, and early infantile testis is explained by the ontogeny of the androgen receptor expression in Sertoli cells. J Clin Endocrinol Metab 2008;93:4408–12.
[41] Boukari K, Meduri G, Brailly-Tabard S, Guibourdenche J, Ciampi ML, et al. Lack of androgen receptor expression in Sertoli cells accounts for the absence of anti-Mullerian hormone repression during early human testis development. J Clin Endocrinol Metab 2009;94:1818–25.
[42] Buzek SW, Sanborn BM. Increase in testicular androgen receptor during sexual maturation in the rat. Biol Reprod 1988;39:39–49.
[43] Bremner WJ, Millar MR, Sharpe RM, Saunders PTK. Immunohistochemical localization of androgen receptors in the rat testis: evidence for stage-dependent expression and regulation by androgens. Endocrinology 1994;135:1227–34.
[44] Zhou X, Kudo A, Kawakami H, Hirano H. Immunohistochemical localization of androgen receptor in mouse testicular germ cells during fetal and postnatal development. Anat Rec 1996;245:509–18.
[45] You L, Sar M. Androgen receptor expression in the testes and epididymides of prenatal and postnatal Sprague-Dawley rats. Endocrine 1998;9:253–61.
[46] Kato S. Androgen receptor structure and function from knock-out mouse. Clin Paediatr Endocrinol 2002;11:1–7.
[47] Notini AJ, Davey RA, McManus JF, Bate KL, Zajac JD. Genomic actions of the androgen receptor are required for normal male sexual differentiation in a mouse model. J Mol Endocrinol 2005;35:547–55.
[48] Yeh S, Tsai MY, Xu Q, Mu XM, Lardy H, et al. Generation and characterization of androgen receptor knockout (ARKO) mice: an *in vivo* model for the study of androgen functions in selective tissues. Proc Natl Acad Sci USA 2002;99:13498–503.
[49] Walters KA, Simanainen U, Handelsman DJ. Molecular insights into androgen actions in male and female reproductive function from androgen receptor knockout models. Hum Reprod Update 2010;16:543–58.

[50] Matsumoto T, Shiina H, Kawano H, Sato T, Kato S. Androgen receptor functions in male and female physiology. J Steroid Biochem Mol Biol 2008;109:236–41.
[51] Verhoeven G, Willems A, Denolet E, Swinnen JV, De Gendt K. Androgens and spermatogenesis: lessons from transgenic mouse models. Philos Trans R Soc Lond B Biol Sci 2010;365:1537–56.
[52] Zhou X. Roles of androgen receptor in male and female reproduction: lessons from global and cell-specific androgen receptor knockout (ARKO) mice. J Androl 2010;31:235–43.
[53] Lyon MF, Hawkes SG. X-linked gene for testicular feminization in the mouse. Nature 1970;227:1217–9.
[54] Willems A, De Gendt K, Allemeersch J, Smith LB, Welsh M, et al. Early effects of Sertoli cell-selective androgen receptor ablation on testicular gene expression. Int J Androl 2010;33:507–17.
[55] Kaftanovskaya EM, Huang Z, Barbara AM, De Gendt K, Verhoeven G, et al. Cryptorchidism in mice with an androgen receptor ablation in gubernaculum testis. Mol Endocrinol 2012;26:598–607.
[56] Willems A, Batlouni SR, Esnal A, Swinnen JV, Saunders PT, et al. Selective ablation of the androgen receptor in mouse Sertoli cells affects Sertoli cell maturation, barrier formation and cytoskeletal development. PLoS One 2010;5:e14168.
[57] Tan KA, De Gendt K, Atanassova N, Walker M, Sharpe RM, et al. The role of androgens in Sertoli cell proliferation and functional maturation: studies in mice with total or Sertoli cell-selective ablation of the androgen receptor. Endocrinology 2005; 146:2674–83.
[58] Hazra R, Corcoran L, Robson M, McTavish KJ, Upton D, et al. Temporal role of Sertoli cell androgen receptor expression in spermatogenic development. Mol Endocrinol 2013; 27:12–24.
[59] Franca LR, Parreira GG, Gates RJ, Russell LD. Hormonal regulation of spermatogenesis in the hypophysectomized rat: quantitation of germ-cell population and effect of elimination of residual testosterone after long-term hypophysectomy. J Androl 1998;19:335–40.
[60] O'Shaughnessy PJ, Verhoeven G, De Gendt K, Monteiro A, Abel MH. Direct action through the Sertoli cells is essential for androgen stimulation of spermatogenesis. Endocrinology 2010;151:2343–8.
[61] De Gendt K, Atanassova N, Tan KA, de Franca LR, Parreira GG, et al. Development and function of the adult generation of Leydig cells in mice with Sertoli cell-selective or total ablation of the androgen receptor. Endocrinology 2005;146:4117–26.
[62] O'Shaughnessy PJ, Monteiro A, Abel M. Testicular development in mice lacking receptors for follicle stimulating hormone and androgen. PLoS One 2012;7:e35136.
[63] Hazra R, Jimenez M, Desai R, Handelsman DJ, Allan CM. Sertoli cell androgen receptor expression regulates temporal fetal and adult Leydig cell differentiation, function, and population size. Endocrinology 2013;154:3410–22.
[64] Wang RS, Yeh S, Chen LM, Lin HY, Zhang C, et al. Androgen receptor in Sertoli cell is essential for germ cell nursery and junctional complex formation in mouse testes. Endocrinology 2006;147:5624–33.
[65] Meng J, Holdcraft RW, Shima JE, Griswold MD, Braun RE. Androgens regulate the permeability of the blood–testis barrier. Proc Natl Acad Sci USA 2005;102:16696–700.
[66] Meng J, Greenlee AR, Taub CJ, Braun RE. Sertoli cell-specific deletion of the androgen receptor compromises testicular immune privilege in mice. Biol Reprod 2011;85:254–60.

[67] McCabe MJ, Allan CM, Foo CF, Nicholls PK, McTavish KJ, et al. Androgen initiates Sertoli cell tight junction formation in the hypogonadal (hpg) mouse. Biol Reprod 2012;87:38.

[68] Shupe J, Cheng J, Puri P, Kostereva N, Walker WH. Regulation of Sertoli-germ cell adhesion and sperm release by FSH and nonclassical testosterone signaling. Mol Endocrinol 2011;25:238–52.

[69] Li JC, Mruk D, Cheng CY. The inter-Sertoli tight junction permeability barrier is regulated by the interplay of protein phosphatases and kinases: an *in vitro* study. J Androl 2001;22:847–56.

[70] Wong CH, Cheng CY. Mitogen-activated protein kinases, adherens junction dynamics, and spermatogenesis: a review of recent data. Dev Biol 2005;286:1–15.

[71] Siu MK, Wong CH, Lee WM, Cheng CY. Sertoli-germ cell anchoring junction dynamics in the testis are regulated by an interplay of lipid and protein kinases. J Biol Chem 2005;280:25029–47.

[72] Handelsman DJ, Spaliviero JA, Simpson JM, Allan CM, Singh J. Spermatogenesis without gonadotropins: maintenance has a lower testosterone threshold than initiation. Endocrinology 1999;140:3938–46.

[73] Stanton PG, Sluka P, Foo CF, Stephens AN, Smith AI, et al. Proteomic changes in rat spermatogenesis in response to *in vivo* androgen manipulation; impact on meiotic cells. PLoS One 2012;7:e41718.

[74] Kaitu'u-Lino TJ, Sluka P, Foo CF, Stanton PG. Claudin-11 expression and localisation is regulated by androgens in rat Sertoli cells *in vitro*. Reproduction 2007;133:1169–79.

[75] Meng X, Lindahl M, Hyvonen ME, Parvinen M, de Rooij DG, et al. Regulation of cell fate decision of undifferentiated spermatogonia by GDNF. Science 2000;287:1489–93.

[76] Ruwanpura SM, McLachlan RI, Meachem SJ. Hormonal regulation of male germ cell development. J Endocrinol 2010;205:117–31.

[77] Awoniyi CA, Sprando RL, Santulli R, Chandrashekar V, Ewing LL, et al. Restoration of spermatogenesis by exogenously administered testosterone in rats made azoospermic by hypophysectomy or withdrawal of luteinizing hormone alone. Endocrinology 1990;127:177–84.

[78] Huang HF, Marshall GR, Rosenberg R, Nieschlag E. Restoration of spermatogenesis by high levels of testosterone in hypophysectomized rats after long-term regression. Acta Endocrinol (Copenh) 1987;116:433–44.

[79] Sun YT, Wreford NG, Robertson DM, de Kretser DM. Quantitative cytological studies of spermatogenesis in intact and hypophysectomized rats: identification of androgen-dependent stages. Endocrinology 1990;127:1215–23.

[80] Handelsman DJ. Hormonal regulation of spermatogenesis: insights from constructing genetic models. Reprod Fertil Dev 2011;23:507–19.

[81] Pakarainen T, Zhang FP, Makela S, Poutanen M, Huhtaniemi I. Testosterone replacement therapy induces spermatogenesis and partially restores fertility in luteinizing hormone receptor knockout mice. Endocrinology 2005;146:596–606.

[82] O'Donnell L, McLachlan RI, Wreford NG, de Kretser DM, Robertson DM. Testosterone withdrawal promotes stage-specific detachment of round spermatids from the rat seminiferous epithelium. Biol Reprod 1996;55:895–901.

[83] Cameron DF, Muffly KE. Hormonal regulation of spermatid binding. J Cell Sci 1991;100(Pt 3):623–33.

[84] O'Donnell L, Nicholls PK, O'Bryan MK, McLachlan RI, Stanton PG. Spermiation: the process of sperm release. Spermatogenesis 2011;1:14–35.

[85] Beardsley A, O'Donnell L. Characterization of normal spermiation and spermiation failure induced by hormone suppression in adult rats. Biol Reprod 2003;68:1299–307.

[86] Saito K, O'Donnell L, McLachlan RI, Robertson DM. Spermiation failure is a major contributor to early spermatogenic suppression caused by hormone withdrawal in adult rats. Endocrinology 2000;141:2779–85.

[87] O'Donnell L, Pratis K, Wagenfeld A, Gottwald U, Muller J, et al. Transcriptional profiling of the hormone-responsive stages of spermatogenesis reveals cell-, stage-, and hormone-specific events. Endocrinology 2009;150:5074–84.

[88] Chapin RE, Wine RN, Harris MW, Borchers CH, Haseman JK. Structure and control of a cell–cell adhesion complex associated with spermiation in rat seminiferous epithelium. J Androl 2001;22:1030–52.

[89] Zhang J, Wong CH, Xia W, Mruk DD, Lee NP, et al. Regulation of Sertoli-germ cell adherens junction dynamics via changes in protein–protein interactions of the N-cadherin–beta-catenin protein complex which are possibly mediated by c-Src and myotubularin-related protein 2: an *in vivo* study using an androgen suppression model. Endocrinology 2005;146:1268–84.

[90] Wang W, Wine RN, Chapin RE. Rat testicular src: normal distribution and involvement in ethylene glycol monomethyl ether-induced apoptosis. Toxicol Appl Pharmacol 2000;163:125–34.

[91] Roura S, Miravet S, Piedra J, Garcia de Herreros A, Dunach M. Regulation of E-cadherin/catenin association by tyrosine phosphorylation. J Biol Chem 1999;274:36734–40.

[92] Kinch MS, Clark GJ, Der CJ, Burridge K. Tyrosine phosphorylation regulates the adhesions of ras-transformed breast epithelia. J Cell Biol 1995;130:461–71.

[93] Xia W, Cheng CY. TGF-beta3 regulates anchoring junction dynamics in the seminiferous epithelium of the rat testis via the Ras/ERK signaling pathway: an *in vivo* study. Dev Biol 2005;280:321–43.

[94] O'Donnell L, Narula A, Balourdos G, Gu YQ, Wreford NG, et al. Impairment of spermatogonial development and spermiation after testosterone-induced gonadotropin suppression in adult monkeys (*Macaca fascicularis*). J Clin Endocrinol Metab 2001;86:1814–22.

[95] Matthiesson KL, McLachlan RI, O'Donnell L, Frydenberg M, Robertson DM, et al. The relative roles of follicle-stimulating hormone and luteinizing hormone in maintaining spermatogonial maturation and spermiation in normal men. J Clin Endocrinol Metab 2006;91:3962–9.

[96] D'Souza R, Pathak S, Upadhyay R, Gaonkar R, D'Souza S, et al. Disruption of tubulobulbar complex by high intratesticular estrogens leading to failed spermiation. Endocrinology 2009;150:1861–9.

[97] Amann RP. The cycle of the seminiferous epithelium in humans: a need to revisit? J Androl 2008;29:469–87.

[98] Russell LD, Ettlin RA, Sinha Hikim AP, Clegg ED. Histological and histopathological evaluation of the testis. Clearwater, FL: Cache River Press; 1990.

[99] Timmons PM, Rigby PW, Poirier F. The murine seminiferous epithelial cycle is prefigured in the Sertoli cells of the embryonic testis. Development 2002;129:635–47.

[100] Franca LR, Ogawa T, Avarbock MR, Brinster RL, Russell LD. Germ cell genotype controls cell cycle control during spermatogenesis in the rat. Biol Reprod 1998;59:1371–7.

[101] O'Donnell L, Meachem SJ, Stanton PG, McLachlan RI. Endocrine regulation of spermatogenesis. In: Neill JD, editor. Knobil and Neill's physiology of reproduction. 3rd ed. San Diego, CA: Elsevier; 2006. p. 1017–69.
[102] Walker WH, Cheng J. FSH and testosterone signaling in Sertoli cells. Reproduction 2005;130:15–28.
[103] Bremner WJ, Millar MR, Sharpe RM, Saunders PT. Immunohistochemical localization of androgen receptors in the rat testis: evidence for stage-dependent expression and regulation by androgens. Endocrinology 1994;135:1227–34.
[104] McKinnell C, Saunders PT, Fraser HM, Kelnar CJ, Kivlin C, et al. Comparison of androgen receptor and oestrogen receptor beta immunoexpression in the testes of the common marmoset (*Callithrix jacchus*) from birth to adulthood: low androgen receptor immunoexpression in Sertoli cells during the neonatal increase in testosterone concentrations. Reproduction 2001;122:419–29.
[105] Suarez-Quian CA, Martinez-Garcia F, Nistal M, Regadera J. Androgen receptor distribution in adult human testis. J Clin Endocrin Metab 1999;84:350–8.
[106] Lindsey JS, Wilkinson MF. Pem: a testosterone- and LH-regulated homeobox gene expressed in mouse Sertoli cells and epididymus. Dev Biol 1996;179:471–84.
[107] De Gendt K, Denolet E, Willems A, Daniels VW, Clinckemalie L, et al. Expression of Tubb3, a beta-tubulin isotype, is regulated by androgens in mouse and rat Sertoli cells. Biol Reprod 2011;85:934–45.
[108] Johnston DS, Wright WW, Dicandeloro P, Wilson E, Kopf GS, et al. Stage-specific gene expression is a fundamental characteristic of rat spermatogenic cells and Sertoli cells. Proc Natl Acad Sci USA 2008;105:8315–20.
[109] Nicholls PK, Harrison CA, Walton KL, McLachlan RI, O'Donnell L, et al. Hormonal regulation of Sertoli cell micro-RNAs at spermiation. Endocrinology 2011; 152:1670–83.
[110] Panneerdoss S, Chang YF, Buddavarapu KC, Chen HI, Shetty G, et al. Androgen-responsive microRNAs in mouse Sertoli cells. PLoS One 2012;7:e41146.
[111] Denolet E, De Gendt K, Allemeersch J, Engelen K, Marchal K, et al. The effect of a Sertoli cell-selective knockout of the androgen receptor on testicular gene expression in prepubertal mice. Mol Endocrinol 2006;20:321–34.
[112] Takamiya K, Yamamoto A, Furukawa K, Zhao J, Fukumoto S, et al. Complex gangliosides are essential in spermatogenesis of mice: possible roles in the transport of testosterone. Proc Natl Acad Sci USA 1998;95:12147–52.
[113] Sandhoff R, Geyer R, Jennemann R, Paret C, Kiss E, et al. Novel class of glycosphingolipids involved in male fertility. J Biol Chem 2005;280:27310–8.
[114] Uhrin P, Dewerchin M, Hilpert M, Chrenek P, Schofer C, et al. Disruption of the protein C inhibitor gene results in impaired spermatogenesis and male infertility. J Clin Invest 2000;106:1531–9.
[115] O'Rand MG, Widgren EE, Sivashanmugam P, Richardson RT, Hall SH, et al. Reversible immunocontraception in male monkeys immunized with eppin. Science 2004;306:1189–90.
[116] Eacker SM, Shima JE, Connolly CM, Sharma M, Holdcraft RW, et al. Transcriptional profiling of androgen receptor (AR) mutants suggests instructive and permissive roles of AR signaling in germ cell development. Mol Endocrinol 2007;21:895–907.
[117] de Rooij DG, Okabe M, Nishimune Y. Arrest of spermatogonial differentiation in jsd/jsd, Sl17H/Sl17H, and cryptorchid mice. Biol Reprod 1999;61:842–7.
[118] Wang G, Weng CC, Shao SH, Zhou W, de Gendt K, et al. Androgen receptor in Sertoli cells is not required for testosterone-induced suppression of spermatogenesis,

but contributes to Sertoli cell organization in Utp14b jsd mice. J Androl 2009;30:338–48.

[119] Zhou W, Wang G, Small CL, Liu Z, Weng CC, et al. Gene expression alterations by conditional knockout of androgen receptor in adult Sertoli cells of Utp14b jsd/jsd (jsd) mice. Biol Reprod 2011;84:400–8.

[120] Andersen OM, Yeung CH, Vorum H, Wellner M, Andreassen TK, et al. Essential role of the apolipoprotein E receptor-2 in sperm development. J Biol Chem 2003;278:23989–95.

[121] Maclean 2nd JA, Chen MA, Wayne CM, Bruce SR, Rao M, et al. Rhox: a new homeobox gene cluster. Cell 2005;120:369–82.

[122] Sugimoto R, Nabeshima Y, Yoshida S. Retinoic acid metabolism links the periodical differentiation of germ cells with the cycle of Sertoli cells in mouse seminiferous epithelium. Mech Dev 2012;128:610–24.

[123] Morales C, Griswold MD. Retinol-induced stage synchronization in seminiferous tubules of the rat. Endocrinology 1987;121:432–4.

[124] Dann CT, Alvarado AL, Molyneux LA, Denard BS, Garbers DL, et al. Spermatogonial stem cell self-renewal requires OCT4, a factor downregulated during retinoic acid-induced differentiation. Stem Cells 2008;26:2928–37.

[125] Simpson ER, Mahendroo MS, Means GD, Kilgore MW, Hinshelwood MM, et al. Aromatase cytochrome P450, the enzyme responsible for estrogen biosynthesis. Endocr Rev 1994;15:342–55.

[126] Carreau S, Hess RA. Oestrogens and spermatogenesis. Philos Trans R Soc Lond B Biol Sci 2010;365:1517–35.

[127] Hess RA. Estrogen in the adult male reproductive tract: a review. Reprod Biol Endocrinol 2003;1:52.

[128] O'Donnell L, Robertson KM, Jones ME, Simpson ER. Estrogen and spermatogenesis. Endocr Rev 2001;22:289–318.

[129] Hess RA. Oestrogen in fluid transport in efferent ducts of the male reproductive tract. Rev Reprod 2000;5:84–92.

[130] Carreau S, Bois C, Zanatta L, Silva FR, Bouraima-Lelong H, et al. Estrogen signaling in testicular cells. Life Sci 2011;89:584–7.

[131] Carreau S, Lambard S, Delalande C, Denis-Galeraud I, Bilinska B, et al. Aromatase expression and role of estrogens in male gonad: a review. Reprod Biol Endocrinol 2003;1:35.

[132] Ebling FJ, Brooks AN, Cronin AS, Ford H, Kerr JB. Estrogenic induction of spermatogenesis in the hypogonadal mouse. Endocrinology 2000;141:2861–9.

[133] Baines H, Nwagwu MO, Furneaux EC, Stewart J, Kerr JB, et al. Estrogenic induction of spermatogenesis in the hypogonadal (hpg) mouse: role of androgens. Reproduction 2005;130:643–54.

[134] Lim P, Allan CM, Notini AJ, Axell AM, Spaliviero J, et al. Oestradiol-induced spermatogenesis requires a functional androgen receptor. Reprod Fertil Dev 2008;20:861–70.

[135] Allan CM, Couse JF, Simanainen U, Spaliviero J, Jimenez M, et al. Estradiol induction of spermatogenesis is mediated via an estrogen receptor-alpha mechanism involving neuroendocrine activation of follicle-stimulating hormone secretion. Endocrinology 2010;151:2800–10.

[136] Pratis K, O'Donnell L, Ooi GT, Stanton PG, McLachlan RI, et al. Differential regulation of rat testicular 5alpha-reductase type 1 and 2 isoforms by testosterone and FSH. J Endocrinol 2003;176:393–403.

[137] Hess RA, Bunick D, Lee KH, Bahr J, Taylor JA, et al. A role for oestrogens in the male reproductive system. Nature 1997;390:509–12.

[138] Robertson KM, O'Donnell L, Jones ME, Meachem SJ, Boon WC, et al. Impairment of spermatogenesis in mice lacking a functional aromatase (cyp 19) gene. Proc Natl Acad Sci USA 1999;96:7986–91.

[139] Dupont S, Krust A, Gansmuller A, Dierich A, Chambon P, et al. Effect of single and compound knockouts of estrogen receptors alpha (ERalpha) and beta (ERbeta) on mouse reproductive phenotypes. Development 2000;127:4277–91.

[140] Sinkevicius KW, Laine M, Lotan TL, Woloszyn K, Richburg JH, et al. Estrogen-dependent and -independent estrogen receptor-alpha signaling separately regulate male fertility. Endocrinology 2009;150:2898–905.

Activins and inhibins in Sertoli cell biology: implications for testis development and function

7

Kate L. Loveland[a] *and Mark P. Hedger*[b]
[a]Departments of Biochemistry and Molecular Biology & Anatomy and Developmental Biology, Monash University, Clayton, Victoria, Australia, [b]MIMR-PHI Institute of Medical Research, Clayton, Victoria, Australia

I. Introduction: activin and inhibin link multiple cell types to determine male reproductive health

Activins and inhibins are highly conserved members of the transforming growth factor (TGF) β superfamily of pleiotropic signaling molecules. Their actions are effected by membrane receptor subunits and intracellular signaling molecules that are shared with many other TGFβ superfamily growth factors, and their functions are controlled by common soluble and membrane bound signaling antagonists [2,3]. To explore how activin and inhibin influence male reproductive health, this chapter includes an introduction to many of the signaling components involved in TGFβ superfamily signaling in the testis. The structures of activins and inhibins are described, followed by a review of what is known about their influence on testis development and function. The potential for measuring and manipulating activin activity in the testis to achieve improved clinical outcomes is considered, as is the contribution of immune cells to this aspect of testicular functions.

Inhibin was first identified to exert *in vivo* and *in vitro* follicle-stimulating hormone (FSH)-suppressing activity in pituitary cells [4,5] and was subsequently shown to be produced by Sertoli cells and, to a lesser extent, Leydig cells [6,7]. Inhibin regulates FSH secretion in the pituitary gland by a feedback mechanism in conjunction with testosterone and has species-specific features (reviewed in [8–10]). Following the original purification of activin based on its ability to stimulate FSH secretion from rat pituitary cells [11,12], a vast range of crucial processes relating to cell and tissue differentiation has been found to be influenced by inhibin's actions. Genes regulated by activin activity levels control reproductive and immune system function, thereby affecting bone, wound healing, immune cell development, and adult fertility (reviewed in [13–17]).

The central importance of the signaling overlap between activins and other members of the TGFβ superfamily is evident in several biological contexts, spanning organ growth through disease states. In some stages of testis development, there is

Sertoli Cell Biology. DOI: http://dx.doi.org/10.1016/B978-0-12-417047-6.00007-7

information about more than one of these proteins; however, few studies have directly addressed the functional overlap of these proteins. In addition, the broad-ranging actions of activins and inhibins are becoming increasingly evident because these molecules are essentially ubiquitous, and activin functions as both a hormone and as a locally produced factor to influence cellular behavior. This activity is evident in immunobiology, where a new understanding of the powerful impact of activin in determining immune cell fate and function is emerging. In this context, control of bone homeostasis is under the influence of activin, and because testicular function is a main determinant of circulating inhibin levels, interest in monitoring and controlling the levels of these reproductive hormones is increasing. Strategies for manipulating the capacity of activin to signal are on the horizon and may offer new approaches to fertility management; the development of activin signaling modulators to alleviate inflammatory conditions and cancer cachexia also demands that their impact on future fertility and reproductive organ function be understood.

II. General structure and signaling pathways

Activins and inhibins are members of the TGFβ superfamily of over 40 members, many of which are of known importance to reproductive biology [14,17,18]. The classification of activins and inhibins as members of the TGFβ superfamily is based on the presence of six to nine (typically seven) conserved cysteines. In activin, six of these form three intrachain disulfide bonds, and the seventh cysteine is involved in an intermolecular bond that binds the dimer in a double cysteine knot structure. The N-terminal preregion of the nascent polypeptide enables passage through the endoplasmic reticulum membrane into the secretory pathway, which is cleaved to produce a peptide with a proregion that is noncovalently linked to the mature protein subunit (reviewed in [19]).

Activin subunits are encoded by one of four different genes in mammals, with *Inhba* encoding the activin βA subunit (also known as the inhbin βA [inhba] subunit), which dimerizes to form the signaling ligand activin A (Figure 7.1). *Inhbb*, *Inhbc*, and *Inhbe* encode βB, βC, and βE subunits, respectively. In addition to the homodimers that are formed (e.g., two βB subunits form activin B), heterodimers of some of these subunits have been identified (βA:βB forms activin AB) that have distinct signaling properties. The βC subunit is not established to form a directly signaling homodimer, but it can, instead, serve as an antagonist of activin by forming a heterodimer with a βA or βB subunit. The functions of the βE subunit have not been identified in the testis.

The inhibin proteins are composed of one β subunit linked to another TGFβ superfamily protein, the α subunit, produced from the inhibin α (*inha*) gene. Inhibin α subunits are not known to form homodimers or to signal directly. In addition to the activins and inhibins, members of the TGFβ superfamily include TGF-β1, -2, and -3; over 15 distinct BMPs; Müllerian inhibitory substance (also known as anti-Müllerian hormone); growth and differentiation factors (e.g., GDF9, also known as BMP15); and glial-derived neurotropic factor (reviewed in [2,19]).

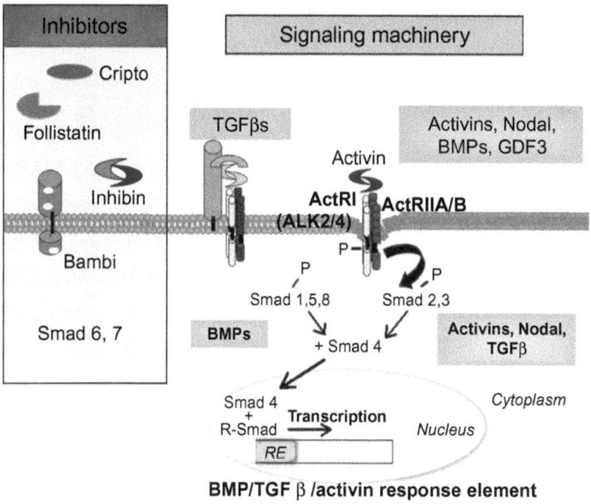

Figure 7.1 Schematic of activin signaling and interactions with other TGFβ ligands and regulators. Activins are dimers of β subunits (blue crescents). Activin signals by binding a constitutively active serine/threonine kinase type II receptor (ActRIIA/IIB) that recruits and activates a type I receptor with serine/threonine kinase activity. These receptors typically phosphorylate Smad 2 and/or Smad 3, which enter the nucleus to selectively activate target genes in association with cell-specific factors. Shared usage of receptors and SMADs among several TGFβ superfamily ligands is indicated in blue boxes. The divergent, preferential utilization of Smads 1, 5, and 8 by bone morphogenetic proteins (BMPs) contrasts with the canonical activation of Smads 2 and 3 by activins, TGFβs, and Nodal. The red box shows a subset of inhibitory proteins, including inhibin, which is formed through dimerization of a β subunit with a distinct α subunit. Cripto is the Nodal coreceptor that also inhibits activin signaling. Betaglycan is a TGFβ coreceptor that also reduces activin bioactivity. (For interpretation of the references to color in this figure legend, the reader is referred to the web version of this book.)

A unique characteristic of the TGFβ signaling pathway consists of the receptor subunits with intracellular serine/threonine kinase activity (for reviews, see [2,20]). Binding of activins as either homo- or heterodimers occurs first through a type 2 receptor subunit with a constitutively active kinase domain. Ligand binding drives recruitment of a type 1 receptor subunit containing a serine threonine kinase domain that becomes activated by the type 2 receptor kinase. Heterotetrameric receptor complexes formed in response to activin-binding drive signaling by the phosphorylating intracellular signaling factors (known as receptor Smads [R-Smads]) Smad2 or Smad3 (Figure 7.1). Through interactions with the coreceptor Smad 4 in the canonical pathway, or with a distinct cofactor, R-Smads form a complex that can enter the nucleus [21]. Inhibitory Smads 6 and 7 can compete for Smad 4 binding to R-Smads in the cytoplasm. Nuclear Smads interact with DNA in conjunction with other transcription factors, including FAST-1, c-Jun, and c-Fos, to activate transcription. Activation of a specific combination of Smads leads to distinct transcriptional outcomes ([22]; reviewed in [23]). The inhibin dimer functions as a potent activin antagonist because the inhibin α subunit does not bind to a type 1 receptor; the β subunit in complex with an α subunit binds to a type 2 receptor without subsequently recruiting the receptor moiety required to activate intracellular signaling.

Activin may signal through different combinations of types I and II receptors to generate an intracellular signal [2]. Its functional crosstalk with other TGFβ superfamily ligands is determined by the use of shared receptor subunits and downstream signaling molecules as well as via common and distinct inhibitory moieties. Activin shares receptor subunits with the BMP proteins, while TGFβ proteins employ entirely distinct type 1 and type 2 receptor proteins as well as a type 3 receptor moiety, betaglycan, for which a soluble isoform has been identified. Inside the cell, Smads 2 and 3 are shared between activin and TGFβ pathways, while BMP signaling predominantly employs the distinct Smad subset of Smads 1, 5, and 8. This potential for competition for receptor occupancy and modulation of signaling outcomes through the actions of other TGFβ superfamily ligands highlights the need to delineate which ones are present in any particular physiological circumstances. A wide variety of signaling antagonists that are soluble and membrane bound and have extracellular or intracellular actions has been identified ([24]; reviewed in [25]).

These dimeric signaling ligands are commonly found in the form of secreted proteins in which the biologically active mature isoform, comprised of the protein carboxy terminus, has been proteolytically separated from the proregion (for the general structure of these proteins, see [19]).

III. Regulation of inhibin and activin production

In adult males, inhibin is almost exclusively produced by the Sertoli cells. Production of inhibin by Sertoli cells is stimulated by FSH acting through the cAMP-dependent protein kinase A signaling pathway [7,26]. This regulation is exerted primarily at the level of the inhibin α subunit gene (encoded by *Inha*), in which the promoter contains a cAMP-response element [26–28]. Androgens augment inhibin production by Sertoli cells but do not appear to be primary regulators of its production [29,30]. The inhibin α subunit is able to heterodimerize with the available β subunits (βA or βB), which are also produced by Sertoli cells, to form either inhibin A or inhibin B [11,31]. Stimulation of Sertoli cells with FSH or cAMP analogs leads to excess production of free α subunit precursor proteins that are not biologically active [32,33], but that cross react in inhibin immunoassays [34,35].

The main form of inhibin produced by Sertoli cells in humans, primates, and commonly used rodent experimental models is inhibin B, but inhibin A is the main form in rams and possibly in other species [36,37]. The reason(s) for these species differences is unknown, but, presumably, these outputs are determined by differences in the coordinated production and availability of each β subunit within Sertoli cells and by differences in the competitive affinity between the subunits [38,39]. Studies in Sertoli cells have identified negative regulators of α subunit gene expression and inhibin production, including interleukin 1 (IL1), tumor necrosis factor (TNF), androgens, and epidermal growth factor [29,40–42]. Available evidence suggests that inhibin is predominantly secreted from the Sertoli cell apical surface and possibly enters the circulation via the seminiferous tubular lumen through the rete testis or efferent ducts [30,43]. Although Leydig cells appear to express the inhibin α subunit [192,193], these cells make only a minor contribution, at best, to inhibin levels in the adult testis [6,44,45], and, in general, levels of inhibin in the blood are directly related to Sertoli cell number and function [44–46].

Unlike the production of inhibin, the production of activin is not induced by FSH action. The βA subunit gene (*Inhba*) promoter appears to possess cAMP-response elements [40,47], but βA subunit expression and activin A production are negatively regulated by FSH/cAMP action in cultured rat Sertoli cells [6,41]. However, there is evidence for stimulation of the βB subunit gene (*Inhbb*) by cAMP, although FSH itself appears to be less effective [27,48,49]. Studies in rat Sertoli cells as well as other cell types that produce activin A in the absence of the α subunit, such as macrophages, have established that activin A production is strongly induced by proinflammatory factors that stimulate transcription of the *Inhba* transcript. These elements act through the Toll-like receptor (TLR)/IL1 signaling pathway via the adapter protein MyD88 and the tumor necrosis factor receptor-associated factors, by TNF itself, and through activation of protein kinase C by phorbol esters [40,42,50–56]. The promoter of the βA subunit gene contains both AP-1 and AP-2 sites, which can account for these responses [28,47,57,58]. Consequently, activin A production by Sertoli cells is stimulated by TLR ligands, such as lipopolysaccharide, IL1, and TNF, and by hormonal and cytokine regulators of protein kinase C [17]. Stimulation of βA subunit mRNA and activin A levels by other immunoregulators, such as TGFβ, interferon-γ, IL13, CD40L, colony-stimulating factor 2, and reactive oxygen species, further highlights the importance of activin A production during inflammatory or immune reactions [50,59–62].

Regulation of the βB subunit and activin B is less well characterized, mostly owing to the previous absence of a specific assay for activin B [63]. Cultured rat Sertoli cells produce activin B at levels similar to those of activin A under basal culture conditions, but, unlike activin A, activin B expression by Sertoli cells does not appear to be directly stimulated by inflammatory cytokines, such as IL1 and TNF (unpublished data). Conversely, pachytene spermatocytes suppress Sertoli cell inhibin B secretion via suppression of βB subunit mRNA production *in vitro* [64]. Although rat Sertoli cells do not increase activin B production when they are stimulated by inflammatory cytokines, circulating activin B is elevated by inflammation *in vivo*, presumably through production by cell types outside the testis [65] (unpublished data).

Inhibin B and activin A display an inverse regulation pattern in Sertoli cells, reflecting the fact that α subunit (*Inha*) expression is stimulated by FSH/cAMP and inhibited by IL1/TNF signaling pathways, while the converse is true for the βA subunit transcript (*Inhba*) [42,66] (Figure 7.2). This reciprocal regulation appears to

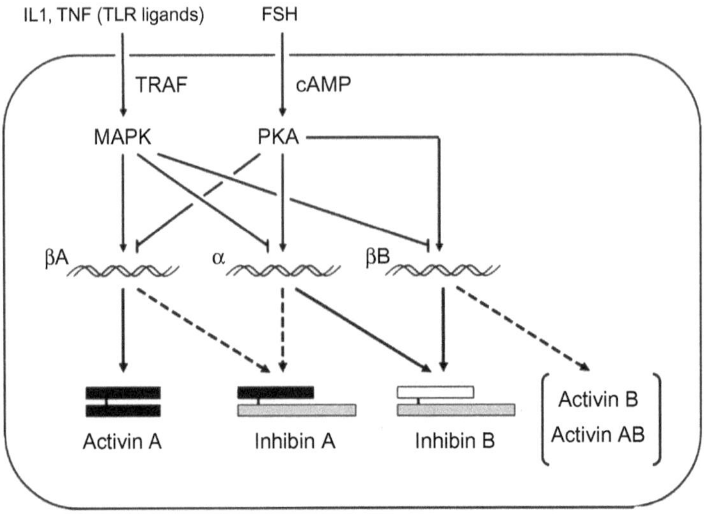

Figure 7.2 Regulation of activin and inhibin protein formation in adult rat Sertoli cells. Synthesis of the βA subunit is stimulated by interleukin 1, TNF, or TLR ligands, which act via the TNF-associated factor signaling pathway, primarily through activation of MAP kinases. Synthesis of the α subunit and βB subunit is stimulated by FSH acting through the cAMP-activated protein kinase A signaling pathway. Reciprocal inhibition of the signaling pathways occurs at the level of each subunit. Normally, inhibin B is the main product of this regulatory process, but significant levels of activin A are also produced. Inhibin A is a relatively minor product of adult rat Sertoli cells. Production of βB homodimers (forming activin B) and βAβB heterodimers (activin AB) also occurs. The relative amounts of these dimers are determined by the relative levels of individual subunits produced as well as the affinities between the subunits. Many of the details regarding the regulatory interactions involved remain to be fully elucidated.

mediate the highly distinctive pattern of stage-specific expression of inhibin B and activin A observed throughout the cycle of the seminiferous epithelium [17,66–71]. While inhibin B is produced by Sertoli cells throughout the cycle of the seminiferous epithelium in rats, its production declines at around the time of spermiation [66,70], when a surge of activin A production occurs that spans stages VIII–XII [66,67]. This rise in activin A production is accompanied by an increase in inflammatory gene expression in the seminiferous epithelium, including increases in IL1 and IL6 expression in Sertoli cells and TNF and NO in germ cells, which is induced by the phagocytosis of the residual cytoplasm at this time [17,72]. The mechanisms linking this process remain to be explained but may involve endogenous ligands from the germ cells activating proinflammatory pattern-recognition receptors in Sertoli cells, such as TLR2 and TLR4 [54,73]. This activation of inflammatory signaling pathways in the seminiferous epithelium has a number of important physiological implications that are discussed in the following section.

In contrast to inhibin, which is produced almost exclusively by Sertoli cells, activin A is also produced by some germ cells, in particular by the pachytene spermatocytes, but at levels that are much lower than in Sertoli cells [74–78] (unpublished data). Activin A is also produced by the peritubular cells, particularly in the developing testis, and by Leydig cells, macrophages, and mast cells in the adult testis [66,77,79]. In humans, at least, there is evidence that pachytene spermatocyte and early spermatids express the βB subunit and may also produce activin B [76], and activin C has been detected in pachytene spermatocytes [78]. Knowledge of how activin subunit production is regulated in germ cells is an area for future research.

In summary, regulation of the inhibin/activin subunits in the adult seminiferous epithelium indicates that inhibin is stimulated primarily by FSH acting upon the Sertoli cells, but this activity is modulated by local factors throughout the cycle of the seminiferous epithelium, with the result that activin A is produced in a very discrete window at around the time of spermiation. As a result, inhibin is the main TGFβ superfamily member produced in the testis. Sertoli cells were first identified as the likely source of this hormone because castration removes inhibin almost completely from the circulation [80], and this finding has been verified in many studies, including by immunohistochemistry and *in situ* hybridization in mice [78]. However, although activin A levels in the adult rat testis appear to be several fold higher than in the circulation, circulating activin A levels are unaffected by castration [81] and presumably arise principally from outside the testis. Production of the βA subunit by Sertoli cells is evident in adult rodents [66,78]. Critically, because of its response to inflammatory stimuli, activin A is upregulated several fold in a broad range of inflammatory conditions.

IV. Activin and inhibin function in the adult testis

Inhibin primarily acts as a circulating hormone that regulates FSH production and secretion—that is, as a feedback regulator of Sertoli cell function in response to FSH and, to a much lesser extent, testosterone [82]. Although several experimental approaches have been employed to find a specific receptor, it is now widely accepted that inhibin exerts its effects solely by blocking the ability of locally produced activins to bind to activin receptors within the pituitary gland [83,84]. This blockage involves a TGFβ type 3 receptor, betaglycan, which facilitates the binding of inhibin αβ heterodimers to the activin receptor [85,86]. Hence, the inhibitory effect of circulating inhibin on FSH is due to the fact that activin B, in particular, is produced within the pituitary gland by the gonadotropes and regulates FSH synthesis and secretion by an autocrine/paracrine mechanism [85,87,88]. Likewise, locally produced inhibin is expected to oppose the actions of activin within the testis. Relevant to this point is that the expression of betaglycan mRNA in the rat seminiferous epithelium is detected in primary spermatocytes and round spermatids during stages I–VI, but spermatogonia, elongated spermatids, and Sertoli cells do not express betaglycan mRNA [89]. Leydig cell clusters also exhibit a high-intensity betaglycan mRNA signal [89,90]. This finding suggests that inhibin exerts stage-specific and cell-specific regulatory effects within the testis itself; however, direct intratesticular actions of inhibin have yet to be formally confirmed.

Activin receptors and/or activin binding have been described in multiple cell types in the adult testis, specifically in the Sertoli cells, spermatogonia, spermatocytes, and round spermatids [66,91–94], with differences in reported outcomes that may reflect differences in the species studied or in the technical approaches employed. Synthesis of the activin type 2 receptors in a stage-dependent manner has been reported within the rodent seminiferous epithelium. The transcript encoding the activin type 2B receptor ACVR2B was readily detected using radioactive *in situ* hybridization in type A spermatogonia and Sertoli cells at stages IX–XI, when activin A levels are highest [66,67,92]. However, the ACVR2A signal encoding the activin type 2A receptor was reported to be most abundant in late pachytene spermatocytes and round spermatids during stages XIII–IV in the adult rat testis using radioactive *in situ* hybridization [67,95], and this receptor was also detected in peritubular myoid cells isolated from immature rat testes [79].

These data implicate activin regulation of specific cell types at select stages of the cycle of the seminiferous epithelium. Data from the adult human testis have not yet been interpreted as providing evidence that the activin receptors have stage-specific expression. Transcripts encoding both type 2 activin receptor subunits have been detected using the highly sensitive RNAse protection assay in samples of normal testis and of testicular germ cell tumors [96]. However, the authors of a study based on histological approaches reported the absence of a signal for either the type 2A mRNA (*ACVR2*) or protein in the normal adult human testis, although proteins corresponding to the type 2B receptor and the ALK4 type 1 receptor subunit were detected in Sertoli cells and spermatogonia [93]. In addition, the inhibitory ALK2 type 1 moiety was detected in these cells in the adult human testis. Significantly, the *ACVR2* transcript was identified in carcinoma *in situ* cells, the precursors to testicular germ cell tumors, while both the transcript and the protein were detected in both seminoma and nonseminoma germ cell tumor subtypes [93]. These findings suggest that synthesis of this receptor subunit may be regulated during the transition to pathological conditions.

Significantly, several critical modulators of activin action also show a cyclical pattern of expression during the seminiferous epithelium cycle. In addition to the previously mentioned increase in betaglycan expression at stages I–VI, the endogenous activin-binding protein follistatin and activin receptor inhibitory subunit BMP and activin membrane bound inhibitor (BAMBI) are most highly expressed in leptotene and/or late pachytene spermatocytes during stages IX–XIV of the rat seminiferous epithelium [67,96,97]. Follistatin is also produced by Sertoli cells, but the protein signal observed with immunohistochemistry is most intense in type B spermatogonia in rat testes [97].

While the full range of activin actions during spermatogenesis remains to be determined, the stage-specific expression pattern of activin A, its receptors, and its regulators—particularly within the seminiferous epithelium—provides important clues regarding its likely roles. Significantly, activin A is minimally expressed at most stages of the cycle, but its peak production level at stages VIII–XII corresponds to the highest radiographic *in situ* hybridization signal intensity for *ACVR2B* in spermatogonia and Sertoli cells [66,67]. These stages feature both DNA synthesis associated with type A spermatogonia and leptotene spermatocytes and the opening and reformation of tight junctions required for spermatocyte passage from the basal to adluminal compartment through the blood–testis barrier [98–100]. Two studies have provided *in vivo* evidence that activin contributes to maintenance of the inter-Sertoli cell tight junctions that create the blood–testis barrier, including by directly regulating the transcription of genes required for male fertility, such as *Gja1* and *Serpina5* [22], and for the production of proteins that determine tight junction integrity [101]. Additional processes that may be influenced by activin are male germ cell mitosis and meiosis [98,102,103] and several functions of postproliferative phase Sertoli cells, including androgen receptor expression, transferrin production, and aromatase activity [79,104]. The central importance of the Smad3 transcription factor for directing activin signaling outcomes in postmitotic Sertoli cells has been established using *in vitro* and *in vivo* approaches [22,105]. Activin may also coordinate the regulation of other locally produced cytokines that influence Sertoli cell and germ cell activity, such as IL1 and IL6 [42,53,54,66,106].

Overall, the available data indicate that there is a specific window within the seminiferous epithelium cycle, coincident with the time of spermiation, when activin A exerts highly localized effects on Sertoli cell and germ cell activity (Figure 7.3). In this model, Sertoli cell activin A production is stimulated by phagocytosis of germ cell cytoplasm, reaching a peak at the time of sperm release, possibly through activation of inflammatory signaling pathways in Sertoli cells. Subsequently, the newly synthesized activin A provokes a wave of spermatogonial and spermatocyte development that includes modulation of Sertoli cell tight junctions to facilitate spermatocyte movement into the adluminal compartment. Inhibitors of activin action, such as inhibin, follistatin and BAMBI, modulate this activity at various stages of the cycle. In this way, activin A plays a fundamental role in regulating the activity of the Sertoli cells and developing germ cells in the mature testis, by coordinating the cycle of the seminiferous epithelium as well as acting as an intermediate between the different generations of developing germ cells.

It remains to be examined whether a similar role for activin B should be proposed. It is also crucial to consider how the regulated actions of activin impact the function of other molecules, particularly with regard to other members of the TGFβ superfamily with which it shares signaling machinery and is demonstrated to have functional overlap in the adult testis. The best established example of this role relates to TGFβ proteins, which are established as critical modulators of Sertoli cell tight junction integrity [107]. Discerning the nature of crosstalk between these signaling molecules and establishing whether this crosstalk contributes to species-specific differences in signaling machinery utilization will be important for gaining a comprehensive picture of how activin functions govern adult male fertility.

In addition to their effects within the seminiferous epithelium, activins produced by the Sertoli cells may influence the function of cells in the interstitial tissue,

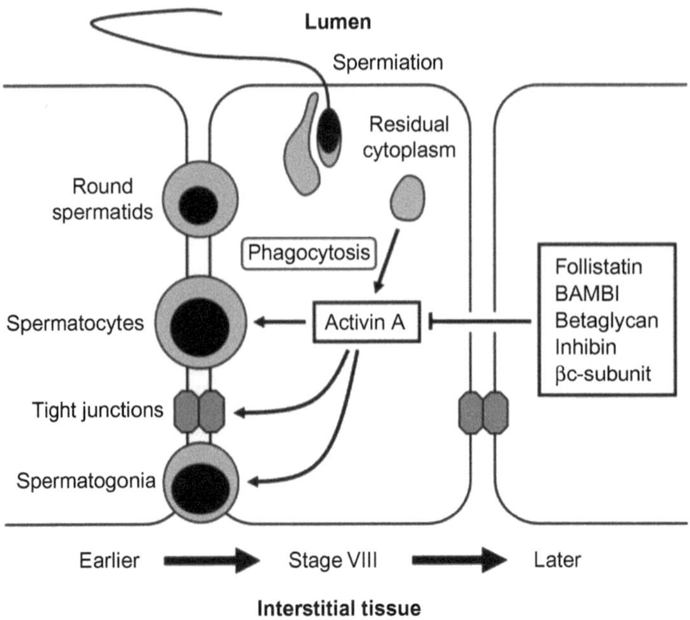

Figure 7.3 Central roles of activin A during the cycle of the adult seminiferous epithelium. A surge of activin A production by Sertoli cells coincides with spermiation at stage VIII of the seminiferous epithelium cycle and appears to be largely in response to phagocytosis of the spermatid residual cytoplasm. During this critical interval, activin A can influence the proliferation and development of both spermatogonia and spermatocytes as well as control reorganization of the inter-Sertoli cell tight junctions. At other times during the cycle, activin A activity may be modulated by the production of molecules that compete with the βA subunit for dimer formation (α subunit, βC subunit), bind to activin directly (follistatin), or inhibit the ability of activin to interact with the activin receptor (inhibin/betaglycan, bone morphogenic protein and activin membrane bound inhibitor, activin C).

particularly the Leydig cells and resident immune cells. Leydig cells express the activin receptor subunits [91,108,109]. Although the affects of activin and inhibin on Leydig cell steroidogenesis have been reported, their significance in the adult testis remains uncertain [110–112]. The well-established inflammatory and immunoregulatory functions of activin A also suggest an important role for Sertoli cell-produced activins in determining the immune environment of the testis [113]. Monocytes/macrophages and dendritic cells throughout the body respond to activin A, which regulates the inflammatory status of these cells both by activating resting monocytes/macrophages and by subsequently modulating their inflammatory actions [113]. It seems highly feasible that activin A may play a role in maintaining the alternatively activated phenotype of the majority of testicular macrophages; this role is believed to contribute to the immune-privileged status of the testis and protection of the developing germ cells from immunological damage [114,115]. It is equally likely that locally produced activins directly influence the activity of circulating

T lymphocytes passing through the testis, which would further contribute to this protection. However, activin A is a potent inducer of myofibroblast differentiation and fibrosis [73], indicating that local production of activin may induce or exacerbate fibrotic damage in testicular disease. The precise roles of the activins in control of testicular inflammation, immunity, and fibrosis have yet to be investigated directly.

The importance of activin for regulating normal testis function has been demonstrated by manipulations of activin activity in several different *in vivo* studies. Inhibiting activin action in the testis by transgenic overexpression of the α subunit caused a reduction in testis size and sperm production in mice [116,117]. In contrast, systemic deletion of the α subunit gene (inha$^{-/-}$ mice) allowed spermatogenesis to begin more or less normally but eventually led to the development of testicular stromal tumors and approximately 100-fold higher levels of circulating activin A that caused these mice to succumb to cancer cachexia [118,119]. Development of these Sertoli cell-derived tumors and the cachexic phenotype was reversed by deletion of Smad3 or by administration of an antibody that selectively blocked ACTR2 function [117,120], reinforcing the selective importance of activin signaling through Smad3 in the events required for Sertoli cell maturation. Mice that overexpress the "nonfunctional" activin βC subunit, which has been localized to Leydig and Sertoli cells and spermatocytes in the normal adult rat and mouse testis [78,121], displayed progressive spermatogenic failure with increased apoptosis at stages V–VIII and IX–XI; reduced numbers of spermatocytes was linked with reduced sperm production, motility and fertility [122]. Similarly, overexpression of follistatin caused spermatogenic disruption, infertility, and Leydig cell hyperplasia [123]. These effects were attributable to antagonism of activin A action at the testis level rather than indirect effects on FSH secretion [122,123].

While each of these studies provides intriguing clues, they are limited by the fact that they represent models in which activin signaling is chronically impaired; we know that the development of the somatic cell niche is altered starting at fetal ages, when activin action is reduced [124,125]. Furthermore, the pace of germ cell and somatic cell development is not directly related to activin signaling capacity but is, instead, advanced in mice that have one Smad3 allele and is reduced in mice lacking both Smad3 alleles [105]. However, normal adult mice in which activin A levels were elevated systemically through femoral muscle transduction with an adeno-associated virus expressing activin A also developed hypospermatogenesis, with a disrupted blood–testis barrier function and a phenotype consistent with the failure of Sertoli cells to support spermatogenesis [101].

Collectively, these studies provide direct experimental support for the physiological importance of endogenous activin in the fine control of spermatogenic development, blood–testis barrier function, and development of intratesticular tumors.

Finally, activin and inhibin produced by Sertoli cell and secreted into the tubule lumen may travel to the epididymis and reach the distal male reproductive tract. Notably, activin A is present in human seminal plasma and is reduced to undetectable levels following vasectomy [126]. There is increasing evidence that the function of the epididymis, and possibly other male reproductive tissues, may be regulated by activin both during development and in adults [127,128]. Consequently, activin and inhibin produced by Sertoli cells may play a role in communication between the testis and the remainder of the male reproductive tract.

V. Activin and inhibin in the developing testis

A. Fetal testis expression and function

Developmental events in the fetal testis are of critical importance for adult male reproductive health. The extent of Sertoli cell proliferation in fetal and juvenile testes is regarded as a primary determinant of adult spermatogenesis due to the role of Sertoli cells in providing the essential niche for male germline development and the capacity for each Sertoli cell to support only a finite number of germ cells [129,130]. Thus, factors that increase or decrease Sertoli cell proliferation can impact fertility by determining the potential for sperm output in adulthood. Studies of rodent models, predominantly rats and mice, have generated most of our knowledge of TGFβ superfamily member roles in governing testis development (reviewed in [1,12,17,18]).

The murine fetal testis is formed from the paired indifferent genital ridges that are adjacent to the mesonephros upon expression of SRY in the Sertoli cell precursors. Within 24 h of this gender-defining event, germ cells are specified irreversibly into the male lineage and become encased by Sertoli cells in the nascent seminiferous cords [131]. This intimate relationship forms the niche that sustains the male germline throughout life. The cords are eventually surrounded by peritubular myoid cells and are increasingly embedded in a loose connective tissue matrix containing Leydig cells, macrophages, vasculature, lymphatic vessels and fibroblasts. Sertoli cells and the other somatic cell populations continue to divide during fetal life.

Sexually indifferent primordial germ cells are specified in early embryogenesis (embryonic day 6.5 [E6.5]) in mice and week 4 of gestation in humans) and form the precursors for gametogenesis [132]. From around E8.5, the primordial germ cells undergo a unique and extensive phase of DNA demethylation that includes erasure of genomic imprints and demethylation of transposable elements, with the exception of intracisternal A particle elements. Demethylation of imprints and transposable elements occurs from E10.5 and is complete by E13.5 [133,134]. This process is unique to germ cells and represents the most extensive epigenetic reprogramming of any *in vivo* cell type. Immediately after this period of demethylation, murine male germline cells cease proliferation (by E14.5 [135]) and remain quiescent until after birth.

Exposure to activin increases the rate of apoptosis by isolated mouse primordial germ cells in culture [136], although *in vivo*, the testes of mice lacking the activin A subunit (called βAKO here) appear to form normally, with an equivalent number of germ and Sertoli cells present at E13.5 in the testes of wild-type, heterozygote (het) and homozygous (knockout [KO]) mutant littermates [124]. However, the level of the *Cdkn1a* transcript encoding the cell cycle inhibitor $p21^{cip1}$ identified as present in male germ cells at this age [135] is significantly lower in het and KO testes at E13.5, while the number of germ cells present at E15.5 is significantly higher in KO testes at E15.5 and at birth (the KO animals die at birth). In contrast, levels of the Sertoli cell *Ccnd2* transcript, which encodes the cell cycle mediator CyclinD2, are significantly lower in KO testes than in het and wild-type littermates at E15.5 and at birth, resulting in a lower number of Sertoli cells at birth in βAKO mice. This outcome of reduced Sertoli cell numbers and an enlarged germ cell population in mice lacking activin A provides the important *in vivo* demonstration that activin A bioactivity sets a balance between the germ cell population and the size of its niche in fetal life. Notably, the apparent interval of impact was different for the two cell populations, and whether other somatic cell types are affected was not elucidated.

The source of activin A of relevance to fetal Sertoli cell proliferation was defined as Leydig cells through selective deletion of the activin A subunit using the anti-Müllerian hormone type 2 receptor-cre [128]. Furthermore, the outcome of cord growth in a complete Inhba knockout model appeared to be similar to the outcome in a model in which *Inhba* was deleted from somatic cells using the Sf1-cre promoter [137]. The latter result leaves unresolved the functional significance of gonocyte-produced activin A protein, which has been detected in rat, pig, and mouse gonocytes [124,138,139]. There is a need to further consider the potential for crosstalk between members of the TGFβ superfamily to orchestrate the complex interactions required to form the fundamental structures of the fetal testis [131]. Transient sex-specific production of the TGFβ ligand Nodal and its signaling regulators Lefty1 and Cripto in germ cells immediately after sex determination has been implicated in regulating germ cell pluripotency and preventing inappropriate entry of male germ line cells into meiosis [137,140]. Both activin A and TGFβ appear to repress proliferation of fetal gonocytes *in vivo* and *in vitro* [124,141], and activin B contributes to vascularization of the gonad [142]. Thus, the interplay between signals from these ligands may contribute to the overall growth of the testis, while each no doubt also serves in highly cell-specific roles.

B. Postnatal testis: activin at the onset of spermatogenesis

The onset of spermatogenesis is poorly defined in humans for ethical reasons, so studies of laboratory rodent species provide knowledge of the key events required for transformation of gonocytes into the spermatogenic cells that form sperm. At the time of birth in rats and in mice, Sertoli cells continue to proliferate, while the gonocytes remain quiescent in the center of the seminiferous cord [130,132]. Within 24 h after birth in mice and 3 days in rats, gonocytes reenter the cell cycle and migrate to the cord perimeter [143]. The germ cells begin their differentiation on the background of proliferating Sertoli cells within a rapidly growing seminiferous cord. Within 12 days after birth in mice and 15–16 days after birth in rats, Sertoli cells cease mitosis [144,145], establish basal tight junctions, and promote seminiferous tubule lumen formation by initiation of apical fluid secretion. The duration and extent of Sertoli cell proliferation during this prepubertal period is a crucial determinant of fertility in the adult.

Detailed analyses of fetal and newborn rats revealed that the quiescent gonocytes synthesize activin βA subunit mRNA and protein in the fetal testis [139]. The quiescent gonocytes appear to store the protein until the time of their transformation into spermatogonia and then begin to synthesize the activin antagonists follistatin and Bambi [97,146]. These early findings were proposed to indicate that the commencement of spermatogonial differentiation requires a downregulation of signaling by activin or other TGFβ superfamily members. In day 3 rat testis fragment cultures, exposure to activin enhances the number of gonocytes present after 1 or 3 days [139], although whether this reflects increased germ cell survival or proliferation has not been determined. Activin exposure enhanced proliferation of rat spermatogonia in cocultures with day 20 Sertoli cells [102], while inhibin enhanced DNA synthesis in Chinese hamster spermatogonia [147]. Addition of activin or BMP4 to spermatogonia isolated from juvenile mice impaired their capacity to function as stem cells in a transplantation assay [148], while a different culture study provided the clue that BMP4 treatment of day 4 spermatogonia would promote their differentiation by enhancing Kit receptor production [149]. These early data demonstrated that activin and inhibin levels could influence the pace of germ cell differentiation and indicated the need for additional research to address what appears to be redundant or overlapping points of action for TGFβ signaling moieties in the postnatal testis (see [1] for a comprehensive historical review).

Analyses of transgenic and KO mice, many of which were created by Matzuk and colleagues, have reinforced the understanding that activin and inhibin are key determinants of postnatal testicular growth and function (reviewed in [14,18,150–152]). While deletion of the inhibin α subunit does not grossly affect postnatal germ cell function, its precise effect on the pace and nature of germ cell maturation has not been vigorously investigated due to the profound effect on Sertoli and adrenal cells that results in tumor formation and cachexia [153]. In addition, the absence of the βB subunit gene (βBKO) does not grossly affect male fertility [154]. When the βA subunit is absent, βAKO mice die immediately after birth due to respiratory complications [155], so these animals cannot be used to investigate activin A function in the postnatal testis. To circumvent this, Brown and colleagues [156] replaced the activin βA subunit coding sequence with that of the activin βB subunit ($Inhba^{BK/BK}$ strain; referred to as "BK" here). Relying on the estimate that activin B has an approximately 10-fold lower affinity for its receptor in cultured cells than activin A [157], this strategy enabled derivation of a mouse strain with reduced activin bioactivity at sites where activin A is normally produced on a background with normal activin βB subunit levels. In contrast to βAKO mice, most BK animals survive to adulthood but exhibit a delay in completion of the first wave of spermatogenesis that defers fertility by 9 days. This early result supported the concept that activin A is important in the first wave of spermatogenesis and prompted more detailed investigations (below, this section). It is important to note that the activin βA subunit mRNA and activin A protein levels are high at birth, highly regulated during the first postnatal week, and significantly downregulated thereafter [77,78,156].

In immature rodent testes, FSH stimulates Sertoli cell proliferation and promotes the onset of Sertoli cell maturation, and this is further promoted by activin in a developmentally regulated fashion. The elegant work of Boitani and colleagues [158,159] first identified synergistic enhancement of rat Sertoli cell proliferation by FSH and activin using *in vitro* cultures of testis fragments from day 3, day 9 and day 18 postpartum rats. Sertoli cell proliferation at day 3 was stimulated by FSH and unaffected by exogenous activin, while proliferation on day 9 required both 100 μg/ml activin A and FSH, and no Sertoli cell proliferation was measured at day 18 in response to either factor as observed *in vivo* [160,161]. An increase in levels of the ACVR2 between days 7 and 9 postpartum in the rat [162] indicates an important role for activin A in driving events at the time when Sertoli cells normally cease proliferation. Additional data indicate that peritubular cells may be the source of activin A at this time in rats [79,161], while *in situ* hybridization and immunohistochemical analyses point to germ and Leydig cells in addition to Sertoli cells as sites of activin βA production in the developing mouse testis [78].

Analysis of the BK mouse strain provided *in vivo* evidence that activin bioactivity determines the extent of Sertoli cell proliferation [125]. The impact of lower levels of activin on the relative proportion of somatic and germ cell types in postnatal testes was shown through quantitation transcripts that were selectively expressed in germ, Leydig and Sertoli cells. These findings indicate that higher numbers of germ and Leydig cells are present in day 0 βAKO and BK testes and in day 7 BK testes compared to wild-type littermates, while the day 7 BK testes contain a lower proportion of Sertoli cells. This finding was reinforced through stereological analyses of day 7 BK testes; similar to observations in βAKO fetal testes, the ratio of Sertoli cells to germ cells in heterozygous and homozygous mutant BK testes was lower than in wild-type littermates, although total spermatogonial numbers were not different. Elevation of germ cell differentiation and meiotic markers in day 7 BK testes, including Kit, Sycp3, and Ccnd3 transcripts measured by quantitative polymerase chain reaction in whole testes and Kit measured by flow cytometry on spermatogonia, highlighted the disruption to normal spermatogenic progression arising from reduced activin signaling. This disruption was further reinforced by the indication of a significant difference in the transitions between germ cell subtypes of wild type and BK samples at day 14, with a relatively lower proportion of type B spermatogonia and higher proportion of spermatocytes present in BK testes (type A:type B:spermatocyte ratios are 1:2.24:1.95 in wild type and 1:1:4.3 in BK littermate testes). The mechanisms underpinning these differences remain to be revealed; however, analyses of mice with one or two mutant Smad3 alleles ([105]; discussed later) indicate that activin signaling through Smad3 can profoundly influence the pace of germ cell maturation at this time.

In the normal postnatal testis, a balance between FSH and inhibin signals is important for cessation of Sertoli cell proliferation. Expression of the cell cycle inhibitor $p27^{Kip1}$ commences when Sertoli cells undergo terminal differentiation [163]. Inhibin normally acts as a "tumor suppressor" by promoting the association of $p27^{Kip1}$ with cylinD2/E1 and Ck4/2, which keeps these cyclin complexes inactive. Mice lacking the inhibin α subunit (inhibin α *KO*) develop Sertoli cell and granulosa cell tumors starting at around 4 weeks postpartum [164]. In the absence of inhibin, the lack of negative feedback to the pituitary gland results in elevated serum FSH and testicular cyclin D2 levels. Double KO mice lacking both $p27^{Kip1}$ and inhibin α have elevated cyclin D2 and Cdk4 mRNA levels. These animals exhibit sustained proliferation and more rapid tumor growth, a finding interpreted as reflecting the increased formation of active cyclin D2/Cdk4 complexes that, in turn, promote active cyclin E1/Cdk2 complex formation and sustained cell cycle progression [165].

In addition to stimulating Sertoli cell proliferation, activin A regulates androgen receptor expression, androgen target gene expression and FSH-induced aromatase activity [22,104,125], and the potential for this regulation to be mediated through Smad3 has been explored ([105]; discussed later). Researchers explored the interactions between activin and androgens *in vivo* by crossing inhibin α KO mice with tfm mice, which have a nonfunctional androgen receptor [166]. These compound mutant mice share characteristics with inhibin α KO mice, including elevated activin secretion from gonadal tumors and cachexia-like wasting syndrome. However, the lifespan of these mice is prolonged (at least 5 weeks) in the absence of a functional androgen receptor, and the hemorrhagic testicular tumor phenotype is reduced. These data indicate that androgens play a modest role in regulating the development of tumors caused by the absence of inhibin, providing a potentially important link between the signaling actions of activin and androgens.

VI. The contribution of Smads to regulation of testis development and growth

Expression profiling of Smads during testis development in the mouse provided ample demonstration that expression of these signaling molecules is tightly regulated during progressive stages of Sertoli and germ cell development (reviewed in [158]). Smad2 and Smad3, R-Smads that mediate signaling by activins, TGFβ proteins, Nodal, myostatin, and GDF 3/9, are highly similar to each other (92% homology). Recent studies of Smad2 and Smad3 in the testis have unequivocally shown that they are functionally distinct and exhibit different expression profiles, which has profound importance for Sertoli cell function and testis maturation. An antibody to detect phosphorylated (i.e., activated) isoforms of Smad2 and Smad3 in fetal mouse testes showed their presence in the nuclei of virtually all somatic and germ cells from the time immediately after sex determination (E13.5) to birth [124]. This interval corresponds to the time when Sertoli and other somatic cells are rapidly dividing and when gonocytes are quiescent (starting at E14.5; [135]). An indistinguishable pattern of nuclear-localized pSmad2/3 was evident in testes from βAKO mice [124], reinforcing the concept that multiple TGFβ ligands contribute to the transcriptional activities of these R-Smads in fetal life. Their actions require Smad4 because selective deletion of Smad4 in mouse Sertoli cells reduces their proliferation, and these animal have short testis cords and small testes [128]. The inhibitory

Smad7 has been identified in fetal testis cords [159], further indicating the likelihood that regulation of TGFβ pathway activities is complex and tightly regulated during fetal development. The cell-specific strategies for coordinating inputs from multiple superfamily ligands are yet to be delineated.

While the authors of an early study reported the presence of Smad2 and Smad3 mRNAs in rat Sertoli cells during postnatal development [167], precise localization by immunocytochemistry in the mouse testis yielded the remarkable finding that only Smad3 is nuclear during the period when Sertoli cells are proliferating [22]. Both Smad2 and Smad3 are phosphorylated in proliferating, juvenile Sertoli cells of day 6 mice; however, Smad2 is excluded from nuclear entry under physiological conditions. This exclusion could be overridden with the addition of excess activin, which led to premature activation of the postmitotic Sertoli cell activin-induced transcriptome. The importance of Smad3 in mediating this distinct responsiveness of day 6 (mitotic) and day 16 (postmitotic) mouse Sertoli cells is revealed by analysis of mice lacking the inhibin α subunit. These mice develop frank gonadal stromal cell tumors by 8 weeks due to the unopposed production of activin, which drives cell proliferation (rather than maturation) and causes the cachexia to which these mice succumb [153]. Li and colleagues made the important discovery that Smad3, but not Smad2, is essential for the sustained activin-induced Sertoli cell proliferation in these animals because mice lacking both alleles of each inhibin α and Smad3 are viable [117]; the absence of Smad2 in inhibin α KO female mice did not protect these mice from gonadal tumors [168].

Several additional observations highlight the central importance of the relative amounts of Smad2 and Smad3 in activin target cells. Evidence relating their relative amounts to developmental changes in their individual functions was obtained by measuring levels in total testicular lysates from juvenile mice; a changing stoichiometry was identified, and the level of Smad3 declined while that of Smad2 was relatively unchanged in the transition through puberty [105]. Smad2 becomes nuclear localized in Sertoli cells isolated from *Smad3+/−* and *Smad3−/−* mice when it is cultured with physiological levels of activin, which leads to a change in activin-induced transcriptional activity [22]. Intriguingly, FSH exposure led to a higher proliferation rate, measured as thymidine uptake, in *Smad3−/−* Sertoli cells compared to cells from wild-type and heterozygous littermates.

Thus, there is an important potential for subtle changes in activin signaling to create different outcomes during testis development. This potential was demonstrated *in vivo* through analyses of Smad3 mutant mice that revealed unexpected complexities. The testes of mice lacking either one or two *Smad3* alleles were smaller than those of their wild-type littermates prior to puberty, but in adulthood, the sizes of these are alleles were indistinguishable. The pace of Sertoli cell maturation was significantly different for each genotype: *Smad3−/+* Sertoli cells matured earlier and supported advanced maturation of germ cells during the first spermatogenic wave, while *Smad3−/−* Sertoli cells matured later than their wild-type counterparts [105]. This finding suggests that the level of activin signaling that occurs during the prepubertal window specifies the timing of pubertal onset, a concept that has yet to be interrogated in humans.

VII. Clinical relevance of activin and inhibin for male reproduction

In general, circulating inhibin levels are believed to reflect Sertoli cell number and activity in adults [44,169,170], whereas activin is a marker of inflammation [78]. It has been suggested that inhibin B levels in serum can be used clinically to monitor Sertoli cell health and spermatogenesis in humans [45,76,171]. Given that it is very widely produced, activin from the testes probably does not contribute significantly to circulating levels, at least under normal conditions or during systemic inflammation. However, we understand very little about how elevated levels of circulating activin impact testicular function [101]. The fact that activin A disappears from human seminal plasma following vasectomy suggests that seminal plasma activin A might be a marker of testis function [81]. However, a recent study in mice suggests that the epididymis is the major source of activin A in seminal plasma [127]. Activin B can also be measured in normal human seminal plasma [63], but its source is unknown. Activin A is increased in the seminal plasma of men with elevated leukocytospermia [172], but it still remains to be seen whether activin levels in either blood or seminal plasma show any relationship to inflammatory disease, especially in the testis.

It has been recognized for many years that serum FSH levels are elevated in men with infertility, and this finding has been attributed to the effects of dysfunction of the spermatogenic process. Based on the presumed biological role of inhibin, it was anticipated that there would be a concomitant fall in serum inhibin levels associated with these elevated serum FSH levels. Early studies [173] found no correlation between serum total inhibin forms and FSH in fertile and infertile men. Serum inhibin assays that were specific for mature inhibin A, mature inhibin B, and the inhibin precursor isoform containing the C-terminal domain pro-αC [174,175] revealed a complex relationship. Inhibin A is undetectable in adult male serum. Serum inhibin B and pro-αC levels have been detected throughout prenatal and postnatal life. Inhibin B is primarily, if not totally, produced by human Sertoli cells, although pachytene spermatocytes may be a contributing source because they produce the inhibin βB subunit [64]. Pro-αC is produced by Leydig and Sertoli cells. In prepubertal boys, there is a positive correlation between serum FSH and inhibin B levels similar to that seen in *in vitro* studies, which may reflect the increase in Sertoli cell number during this period. In adult males, serum inhibin B is inversely correlated with FSH levels in both fertile and infertile men [36,171]. However, serum inhibin levels, like those of FSH, also correlate with testicular size and sperm count.

Additional studies have shown that serum inhibin levels are elevated with FSH, but not LH, treatment and that attempts to reduce gonadotropin levels by steroid treatment in healthy men have led to a partial drop (by 40%) in inhibin levels only [9]. However, severe disruption of the testis by chemotherapy and x-irradiation led to a very rapid and marked depression of serum inhibin levels [176]. These findings suggest that the extent of spermatogenic suppression is an important determinant of circulating inhibin levels. With steroid treatment, many of the steps of spermatogenesis were only partially suppressed, and azoospermia resulted initially from the failure of spermiogenesis [177]. Spermatagonial numbers, for example, fell by only 50%. Chemotherapy and x-irradiation induced marked and immediate lesions at the spermatogonial level, although direct effects on Sertoli cells could not be excluded. These results suggest that germ cells, particularly at the earliest stages, contribute to the regulation of inhibin by mechanisms that remain to be elucidated because existing data have yielded conflicting outcomes [178,179].

Suppression of gonadotropins in monkeys with a gonadotropin-releasing hormone antagonist led to a limited reduction in serum inhibin B levels (by 15–30%), similar to that observed with steroidal treatment [179]. However, the addition of testosterone implants or luteinizing hormone (LH) treatment resulted in a further decline of approximately 50% in inhibin B levels. These findings suggest that LH and testosterone have an inhibitory influence on inhibin B levels, further illustrating the complexity of inhibin B regulation.

While many details of activin and inhibin function relating to development of the testis remain to be established, there are important additional applications for existing and new knowledge to the problems facing men with infertility and testicular cancer. The etiology of testicular cancer and the associated spectrum of phenotypes, collectively termed the "testicular dysgenesis syndrome" [180,181], are believed to be associated with disrupted androgen actions, including through exposure to endocrine disrupting compounds. However, the mechanisms by which these compounds act and the precise pathway by which these male reproductive disorders, such as cryptorchism and hypospadias, occur in humans remain under investigation. The fundamental role of activin in determining testis size is now established to span fetal life through the end of puberty, and it is of great concern that exposure to endocrine disruptors may impact this pathway. Mice fed di-n-butyl phthalate daily in the pre- and peripubertal interval (spanning day 4–14) exhibited reduced Sertoli cell proliferation in addition to smaller anogenital distance, a key indicator of impaired androgen activity [182]. Of additional interest was the observation that total testis inhibin levels were clearly upregulated at the highest doses tested (500 mg/kg/day), reinforcing the important functional link between activin and androgen signaling activity. The potential for other endocrine disruptor compounds to interfere with activin-mediated developmental events is an important topic of ongoing research.

Investigation of the potential link between disrupted activin signaling activity and testicular cancer development [183] has led to the identification of altered activin receptor and activin signaling modulators in testicular germ cell tumors relative to normal adult human spermatogenesis [93,184]. In the latter study, upregulation of the inhibin α subunit and betaglycan, two activin antagonists normally synthesized only in Sertoli cells within the testis, was identified in seminomas from a subset of patients. The capacity for inappropriate Nodal signaling to sustain pluripotent germ cells as adult tumors was highlighted by the detection of Nodal and its coregulator, Cripto (also an activin antagonist), in these cells [185,186]. Such circumstantial evidence suggests that further investigations into conditions that limit activin signaling and enable Nodal actions will provide missing pieces of this very complicated puzzle involving events in the human fetal testis.

Impaired spermatogonial development is a feature of idiopathic infertility in men, yet for many, the cause remains undiagnosed. There is strong clinical interest in the potential to stimulate spermatogonial maturation in men with arrested spermatogenesis as well as in expanding autologous stem cells for replacement in individuals with fertility arrest arising from chemotherapy or x-irradiation. Understanding how activin signaling is controlled to determine male germline fate decisions will offer clues to new approaches for managing these conditions. Development of contraceptives that sustain the stem cell population but prevent completion of spermatogenesis could also be supported by this knowledge. Consideration must also be given to the power of using animal models, such as *Drosophila melanogaster*, to enhance the pace of discovery in which roles for activin and related TGFβ signaling molecules have been established in the functional control of spermatogonial stem cells and their niche [187]. Spermatogenesis is fundamentally conserved across species, and the powerful genetics enabled by work with this model system means that events involving complex molecular interactions can be more readily investigated than by the use of mammalian models.

VIII. Concluding remarks: the need to understand signaling crosstalk in the testis

The importance of activin and its related superfamily members has been unequivocally demonstrated in all aspects of mammalian germline biology, and we are starting to understand the mechanisms by which they govern developmental switches. However, the highly integrated nature of components in this pathway, the multiplicity of regulatory moieties, and the growing literature concerning its capacity to interact with other signaling pathways highlight how great a challenge this will be.

Studies conducted in mice have demonstrated that the relative amounts and functions of Smad2 and Smad3 proteins are tightly coordinated and direct developmentally regulated outcomes that affect male fertility. In studies of embryonic stem cells, selective activation of Smad2 target gene cohorts is dictated by the relative levels of Nodal and activin ligands [188]. Based on data obtained from exposing mitotic and postmitotic mouse Sertoli cells to activin A [22] and additional studies discussed above, we predict that mapping transcriptional outcomes from TGFβ superfamily ligand signaling during periods when somatic and germ cells are responsive to fluctuations in activin levels will reveal crucial steps in testis development that are required for human fertility. Regulated actions by activin, TGFβ superfamily members, and their antagonists drive key events during the development of Sertoli and other somatic cell populations as well as in germ cell differentiation. Some of these factors interact functionally with FSH [139,160,161] and others with androgens, and understanding the stage-specific and species-specific subtleties of this crosstalk may be essential to driving and controlling human fertility.

It is in the adult testis that the local effects of TGFβ superfamily members become even more challenging to investigate due to the extreme complexity of the testicular architecture layered on top of the extravagant number of interacting signaling networks. Advances have been made through the production of KO and transgenic animals in which the expression pattern of one or more genes affects Sertoli cell function or other aspects of testicular physiology. Further careful delineation of the sites of TGFβ superfamily and inhibitor synthesis [24] and the use of selective or broad spectrum signaling pathway inhibitors [189] should continue to yield functional insights, although much fundamental knowledge regarding what controls ligand secretion, signaling outcomes, and signaling moiety turnover is needed. There are few studies of testis biology that compared the impact of different ligand actions on the same functional end point, although it is quite apparent from the survey presented here that there are many potential points of overlap. Competition and enhancement may be the result of integrated signaling by TGFβ superfamily ligands, and future studies should enhance our understanding of this. As a final challenge, developing an understanding of how the TGFβ superfamily interacts with components of other signaling processes, such as the Hedgehog and Wnt pathways (reviewed in [190,191]), will be required to fully understand the mechanisms that underpin testis development and function.

Acknowledgments

Supported by the National Health and Medical Research Council of Australia (fellowships to KL: ID545916 and MH: 1020269; project grant to MH, KL: ID1063843) and by the Victorian government's Operational Infrastructure Support Program.

References

[1] Loveland KL, Robertson DM. The TGFβ superfamily in Sertoli cell biology. In: Griswold M, Skinner M, editors. Sertoli cell biology. USA: Elsevier Science; 2005. p. 227–47.

[2] Mueller TD, Nickel J. Promiscuity and specificity in BMP receptor activation. FEBS Lett 2012;586(14):1846–59.

[3] Oshimori N, Fuchs E. The harmonies played by TGF-β in stem cell biology. Cell Stem Cell 2012;11(6):751–64.

[4] Robertson DM, Foulds LM, Leversha L, Morgan FJ, Hearn MTW, Burger HG, et al. Isolation of inhibin from bovine follicular fluid. Biochem Biophys Res Comm 1985;126 (1):220–6.

[5] Morris PL, Vale WW, Bardin CW. Beta-endorphin regulation of FSH-stimulated inhibin production is a component of a short loop system in testis. Biochem Biophys Res Comm 1987;148(3):1513–9.

[6] Risbridger GP, Clements J, Robertson DM, Drummond AE, Muir J, Burger HG, et al. Immuno- and bioactive inhibin and inhibin α subunit expression in rat Leydig cell cultures. Mol Cell Endocrinol 1989;66(1):119–22.

[7] Bicsak TA, Vale W, Vaughan J, Tucker EM, Cappel S, Hsueh AJ. Hormonal regulation of inhibin production by cultured Sertoli cells. Mol Cell Endocrinol 1987;49(2–3):211–7.
[8] Meachem SJ, Nieschlag E, Simino M. Inhibin B in male reproduction: pathology and clinical relevance. Eur J Endocrinol 2001;145(5):561–71.
[9] Anderson RA. Clinical studies: inhibin in the adult male. Mol Cell Endocrinol 2001;180(1–2):109–16.
[10] Plant TM, Marshall GR. The functional significance of FSH in spermatogenesis and the control of its secretion in male primates. Endocr Rev 2001;22(6):764–86.
[11] Ling N, Ying SY, Ueno N, Shimasaki S, Esch F, Hotta M, et al. Pituitary FSH is released by a heterodimer of the β subunits of the two forms of inhibin. Nature 1986;321(6072):779–82.
[12] Vale W, Rivier J, Vaughan J, McClintock R, Corrigan A, Woo W, et al. Purification and characterization of an FSH releasing protein from porcine follicular fluid. Nature 1986;321(6072):776–9.
[13] Phillips DJ. The activin/inhibin family. In: Thompson A, Lotze MT, editors. The cytokine handbook, vol. 2. 4th ed. London: Academic Press; 2003. p. 1153–77
[14] Barakat B, Itman C, Mendis SH, Loveland KL. Activins and inhibins in mammalian testis development: new models, new insights. Mol Cell Endocrinol 2012;359(1–2):66–77.
[15] Sideras P, Apostolou E, Stavropoulos A, Sountoulidis A, Gavriil A, Apostolidou A, et al. Activin, neutrophils, and inflammation: just coincidence? Semin Immunopathol 2013;35(4):481–99.
[16] Moura J, da Silva L, Cruz MT, Carvalho E. Molecular and cellular mechanisms of bone morphogenetic proteins and activins in the skin: potential benefits for wound healing. Arch Dermatol Res 2013;305(7):557–69.
[17] Hedger MP, Winnall WR. Regulation of activin and inhibin in the adult testis and the evidence for functional roles in spermatogenesis and immunoregulation. Mol Cell Endocrinol 2012;359(1–2):30–42.
[18] Chang H, Brown CW, Matzuk MM. Genetic analysis of the mammalian transforming growth factor-β superfamily. Endocr Rev 2002;23(6):787–823.
[19] Walton KL, Makanji Y, Harrison CA. New insights into the mechanisms of activin action and inhibition. Mol Cell Endocrinol 2012;359(1–2):2–12.
[20] Ikushima H, Miyazono K. Biology of transforming growth factor-β signalling. Curr Pharm Biotechnol 2011;12(12):2099–107.
[21] Heldin CH, Moustakas A. Role of Smads in TGFβ signaling. Cell Tissue Res 2012;347(1):21–36.
[22] Itman C, Small C, Griswold M, Nagaraja AK, Matzuk MM, Brown CW, et al. Developmentally regulated SMAD2 and SMAD3 utilization directs activin signaling outcomes. Dev Dyn 2009;238(7):1688–700.
[23] Morikawa M, Koinuma D, Miyazono K, Heldin CH. Genome-wide mechanisms of Smad binding. Oncogene 2013;32(13):1609–15.
[24] Itman C, Wong C, Whiley PAF, Fernando D, Loveland KL. TGFβ superfamily signalling regulators are differentially expressed in the developing and adult mouse testis. Spermatogenesis 2011;(1)63–72.
[25] Xu P, Liu J, Derynck R. Post-translational regulation of TGF-β receptor and Smad signaling. FEBS Lett 2012;586(14):1871–84.
[26] Klaij IA, Toebosch AM, Themmen AP, Shimasaki S, de Jong FH, Grootegoed JA. Regulation of inhibin α- and β_B-subunit mRNA levels in rat Sertoli cells. Mol Cell Endocrinol 1990;68(1):45–52.

[27] Feng ZM, Li YP, Chen CL. Analysis of the 5'-flanking regions of rat inhibin α- and β-B-subunit genes suggests two different regulatory mechanisms. Mol Endocrinol 1989;3(12):1914–25.

[28] Thompson DA, Cronin CN, Martin F. Genomic cloning and sequence analyses of the bovine α-, β_A- and β_B-inhibin/activin genes. Identification of transcription factor AP-2-binding sites in the 5'-flanking regions by DNase I footprinting. Eur J Biochem 1994;226(3):751–64.

[29] Morris PL, Vale WW, Cappel S, Bardin CW. Inhibin production by primary Sertoli cell-enriched cultures: regulation by follicle-stimulating hormone, androgens, and epidermal growth factor. Endocrinology 1988;122(2):717–25.

[30] Handelsman DJ, Spaliviero JA, Kidston E, Robertson DM. Highly polarized secretion of inhibin by Sertoli cells in vitro. Endocrinology 1989;125(2):721–9.

[31] Forage RG, Ring JM, Brown RW, McInerney BV, Cobon GS, Gregson RP, et al. Cloning and sequence analysis of cDNA species coding for the two subunits of inhibin from bovine follicular fluid. Proc Natl Acad Sci USA 1986;83(10):3091–5.

[32] Hancock AD, Robertson DM, de Kretser DM. Inhibin and inhibin α-chain precursors are produced by immature rat Sertoli cells in culture. Biol Reprod 1992;46(1):155–61.

[33] Grootenhuis AJ, Steenbergen J, Timmerman MA, Dorsman AN, Schaaper WM, Meloen RH, et al. Inhibin and activin-like activity in fluids from male and female gonads: different molecular weight forms and bioactivity/immunoactivity ratios. J Endocrinol 1989;122(1):293–301.

[34] McLachlan RI, Robertson DM, Burger HG, de Kretser DM. The radioimmunoassay of bovine and human follicular fluid and serum inhibin. Mol Cell Endocrinol 1986;46 (2):175–85.

[35] Robertson DM, Giacometti M, Foulds LM, Lahnstein J, Goss NH, Hearn MT, et al. Isolation of inhibin α-subunit precursor proteins from bovine follicular fluid. Endocrinology 1989;125(4):2141–9.

[36] Illingworth PJ, Groome NP, Byrd W, Rainey WE, McNeilly AS, Mather JP, et al. Inhibin-B: a likely candidate for the physiologically important form of inhibin in men. J Clin Endocrinol Metab 1996;81(4):1321–5.

[37] McNeilly AS, Souza CJH, Baird DT, Swanston IA, McVerry J, Crawford J, et al. Production of inhibin A not B in rams: changes in plasma inhibin A during testis growth, and expression of inhibin/activin subunit mRNA and protein in adult testis. Reproduction 2002;123(6):827–35.

[38] Klaij IA, Timmerman MA, Blok LJ, Grootegoed JA, de Jong FH. Regulation of inhibin βB-subunit mRNA expression in rat Sertoli cells: consequences for the production of bioactive and immunoreactive inhibin. Mol Cell Endocrinol 1992;85(3):237–46.

[39] Walton KL, Makanji Y, Robertson DM, Harrison CA. The synthesis and secretion of inhibins. Vitam Horm 2011;85:149–84.

[40] Okuma Y, Saito K, O'Connor AE, Phillips DJ, de Kretser DM, Hedger MP. Reciprocal regulation of activin A and inhibin B by interleukin-1 (IL-1) and follicle-stimulating hormone (FSH) in rat Sertoli cells *in vitro*. J Endocrinol 2005;185(1):99–110.

[41] Mauduit C, Jaspar JM, Poncelet E, Charlet C, Revol A, Franchimont P, et al. Tumor necrosis factor-α antagonizes follicle-stimulating hormone action in cultured Sertoli cells. Endocrinology 1993;133(1):69–76.

[42] Kazutaka S, Winnall WR, Muir JA, Hedger MP. Regulation of Sertoli cell activin A and inhibin B by tumour necrosis factor α and interleukin 1α: interaction with follicle-stimulating hormone/adenosine 3',5'-cyclic phosphate signalling. Mol Cell Endocrinol 2011;335(2):195–203.

[43] Handelsman DJ, Spaliviero JA, Phippard AF. Highly vectorial secretion of inhibin by primate Sertoli cells *in vitro*. J Clin Endocrinol Metab 1990;71(5):1235–8.
[44] Ramaswamy S, Marshall GR, McNeilly AS, Plant TM. Evidence that in a physiological setting Sertoli cell number is the major determinant of circulating concentrations of inhibin B in the adult male rhesus monkey (*Macaca mulatta*). J Androl 1999;20(3):430–4.
[45] Pierik FH, Vreeburg JT, Stijnen T, De Jong FH, Weber RF. Serum inhibin B as a marker of spermatogenesis. J Clin Endocrinol Metab 1998;83(9):3110–4.
[46] Tenover JS, McLachlan RI, Dahl KD, Burger HG, de Kretser DM, Bremner WJ. Decreased serum inhibin levels in normal elderly men: evidence for a decline in Sertoli cell function with aging. J Clin Endocrinol Metab 1988;67(3):455–9.
[47] Ardekani AM, Romanelli JC, Mayo KE. Structure of the rat inhibin and activin β_A-subunit gene and regulation in an ovarian granulosa cell line. Endocrinology 1998;139(7):3271–9.
[48] Najmabadi H, Rosenberg LA, Yuan QX, Reyaz G, Bhasin S. Transcriptional regulation of inhibin βB messenger ribonucleic acid levels in TM.4 or primary rat Sertoli cells by 8-bromo-cyclic adenosine monophosphate. Mol Endocrinol 1993;7(4):561–9.
[49] Toebosch AM, Robertson DM, Trapman J, Klaassen P, de Paus RA, de Jong FH, et al. Effects of FSH and IGF-I on immature rat Sertoli cells: inhibin α- and β-subunit mRNA levels and inhibin secretion. Mol Cell Endocrinol 1988;55(1):101–5.
[50] Shao L, Frigon Jr NL, Sehy DW, Yu AL, Lofgren J, Schwall R, et al. Regulation of production of activin A in human marrow stromal cells and monocytes. Exp Hematol 1992;20(10):1235–42.
[51] Wu H, Chen Y, Winnall WR, Phillips DJ, Hedger MP. Regulation of activin A release from murine bone marrow-derived neutrophil precursors by tumour necrosis factor-α and insulin. Cytokine 2013;61(1):199–204.
[52] Jones KL, Mansell A, Patella S, Scott BJ, Hedger MP, de Kretser DM, et al. Activin A is a critical component of the inflammatory response, and its binding protein, follistatin, reduces mortality in endotoxemia. Proc Natl Acad Sci USA 2007;104(41):16239–44.
[53] Okuma Y, O'Connor AE, Muir JA, Stanton PG, de Kretser DM, Hedger MP. Regulation of activin A and inhibin B secretion by inflammatory mediators in adult rat Sertoli cell cultures. J Endocrinol 2005;187(1):125–34.
[54] Winnall WR, Muir JA, Hedger MP. Differential responses of epithelial Sertoli cells of the rat testis to Toll-like receptor 2 and 4 ligands: implications for studies of testicular inflammation using bacterial lipopolysaccharides. Innate Immun 2011;17(2):123–36.
[55] Takahashi S, Uchimaru K, Harigaya K, Asano S, Yamashita T. Tumor necrosis factor and interleukin-1 induce activin A gene expression in a human bone marrow stromal cell line. Biochem Biophys Res Comm 1992;188(1):310–7.
[56] Bilezikjian LM, Corrigan AZ, Blount AL, Vale WW. Pituitary follistatin and inhibin subunit messenger ribonucleic acid levels are differentially regulated by local and hormonal factors. Endocrinology 1996;137(10):4277–84.
[57] Tanimoto K, Yoshida E, Mita S, Nibu Y, Murakami K, Fukamizu A. Human activin βA gene. Identification of novel 5' exon, functional promoter, and enhancers. J Biol Chem 1996;271(51):32760–9.
[58] Yoshida E, Tanimoto K, Murakami K, Fukamizu A. Isolation and characterization of 5'-regulatory region of mouse activin βA subunit gene. Biochem Mol Biol Int 1998;44 (2):325–32.
[59] Le Magueresse-Battistoni B, Morera AM, Benahmed M. *In vitro* regulation of rat Sertoli cell inhibin messenger RNA levels by transforming growth factor-β1 and tumour necrosis factor α. J Endocrinol 1995;146(3):501–8.

[60] Abe M, Shintani Y, Eto Y, Harada K, Kosaka M, Matsumoto T. Potent induction of activin A secretion from monocytes and bone marrow stromal fibroblasts by cognate interaction with activated T cells. J Leuk Biol 2002;72(2):347–52.

[61] Hardy CL, Lemasurier JS, Olsson F, Dang T, Yao J, Yang M, et al. IL-13 regulates secretion of the tumor growth factor-β superfamily cytokine activin A in allergic airway inflammation. Am J Respir Cell Mol Biol 2010;42(6):667–75.

[62] Lai M, Sirimanne E, Williams CE, Gluckman PD. Sequential patterns of inhibin subunit gene expression following hypoxic-ischemic injury in the rat brain. Neuroscience 1996;70(4):1013–24.

[63] Ludlow H, Phillips DJ, Myers M, McLachlan RI, de Kretser DM, Allan CA, et al. A new 'total' activin B enzyme-linked immunosorbent assay (ELISA): development and validation for human samples. Clin Endocrinol (Oxf) 2009;71(6):867–73.

[64] Clifton RJ, O'Donnell L, Robertson DM. Pachytene spermatocytes in co-culture inhibit rat Sertoli cell synthesis of inhibin βB-subunit and inhibin B but not the inhibin α-subunit. J Endocrinol 2002;172(3):565–74.

[65] Bilezikjian LM, Turnbull AV, Corrigan AZ, Blount AL, Rivier CL, Vale WW. Interleukin-1β regulates pituitary follistatin and inhibin/activin β_B mRNA levels and attenuates FSH secretion in response to activin-A. Endocrinology 1998;139(7):3361–4.

[66] Okuma Y, O'Connor AE, Hayashi T, Loveland KL, de Kretser DM, Hedger MP. Regulated production of activin A and inhibin B throughout the cycle of the seminiferous epithelium in the rat. J Endocrinol 2006;190(2):331–40.

[67] Kaipia A, Penttilä TL, Shimasaki S, Ling N, Parvinen M, Toppari J. Expression of inhibin β_A and β_B, follistatin and activin-A receptor messenger ribonucleic acids in the rat seminiferous epithelium. Endocrinology 1992;131(6):2703–10.

[68] Klaij IA, van Pelt AM, Timmerman MA, Blok LJ, de Rooij DG, de Jong FH. Expression of inhibin subunit mRNAs and inhibin levels in the testes of rats with stage-synchronized spermatogenesis. J Endocrinol 1994;141(1):131–41.

[69] Bhasin S, Krummen LA, Swerdloff RS, Morelos BS, Kim WH, diZerega GS, et al. Stage dependent expression of inhibin α and β-B subunits during the cycle of the rat seminiferous epithelium. Endocrinology 1989;124(2):987–91.

[70] Gonzales GF, Risbridger GP, Hodgson YH, Pöllänen P, de Kretser DM. Stage-specific inhibin secretion by rat seminiferous tubules. Reprod Fertil Dev 1989;1(3):275–9.

[71] Kaipia A, Parvinen M, Shimasaki S, Ling N, Toppari J. Stage-specific cellular regulation of inhibin α-subunit mRNA expression in the rat seminiferous epithelium. Mol Cell Endocrinol 1991;82(2–3):165–73.

[72] Gérard N, Syed V, Jégou B. Lipopolysaccharide, latex beads and residual bodies are potent activators of Sertoli cell interleukin-1α production. Biochem Biophys Res Commun 1992;185(1):154–61.

[73] Hedger MP. Toll-like receptors and signalling in spermatogenesis and testicular responses to inflammation—a perspective. J Reprod Immunol 2011;88(2):130–41.

[74] Shaha C, Morris PL, Chen CL, Vale W, Bardin CW. Immunostainable inhibin subunits are in multiple types of testicular cells. Endocrinology 1989;125(4):1941–50.

[75] Marchetti C, Hamdane M, Mitchell V, Mayo K, Devisme L, Rigot JM, et al. Immunolocalization of inhibin and activin α and β_B subunits and expression of corresponding messenger RNAs in the human adult testis. Biol Reprod 2003;68 (1):230–5.

[76] Andersson AM, Müller J, Skakkebaek NE. Different roles of prepubertal and postpubertal germ cells and Sertoli cells in the regulation of serum inhibin B levels. J Clin Endocrinol Metab 1998;83(12):4451–8.

[77] Buzzard JJ, Loveland KL, O'Bryan MK, O'Connor AE, Bakker M, Hayashi T, et al. Changes in circulating and testicular levels of inhibin A and B and activin a during postnatal development in the rat. Endocrinology 2004;145(7):3532–41.

[78] Barakat B, O'Connor AE, Gold E, de Kretser DM, Loveland KL. Inhibin, activin, follistatin and FSH serum levels and testicular production are highly modulated during the first spermatogenic wave in mice. Reproduction 2008;136(3):345–59.

[79] de Winter JP, Vanderstichele HM, Verhoeven G, Timmerman MA, Wesseling JG, de Jong FH. Peritubular myoid cells from immature rat testes secrete activin-A and express activin receptor type II *in vitro*. Endocrinology 1994;135(2):759–67.

[80] Robertson DM, Hayward S, Irby D, Jacobsen J, Clarke L, McLachlan RI, et al. Radioimmunoassay of rat serum inhibin: changes after PMSG stimulation and gonadectomy. Mol Cell Endocrinol 1988;58(1):1–8.

[81] McFarlane JR, Foulds LM, Pisciotta A, Robertson DM, de Kretser DM. Measurement of activin in biological fluids by radioimmunoassay, utilizing dissociating agents to remove the interference of follistatin. Eur J Endocrinol 1996;134(4):481–9.

[82] Makanji Y, Harrison CA, Robertson DM. Feedback regulation by inhibins A and B of the pituitary secretion of follicle-stimulating hormone. Vitam Horm 2011;85:299–321.

[83] Bilezikjian LM, Blount AL, Leal AM, Donaldson CJ, Fischer WH, Vale WW. Autocrine/paracrine regulation of pituitary function by activin, inhibin and follistatin. Mol Cell Endocrinol 2004;225(1–2):29–36.

[84] Makanji Y, Temple-Smith PD, Walton KL, Harrison CA, Robertson DM. Inhibin B is a more potent suppressor of rat follicle-stimulating hormone release than inhibin A *in vitro* and *in vivo*. Endocrinology 2009;150(10):4784–93.

[85] Lewis KA, Gray PC, Blount AL, MacConell LA, Wiater E, Bilezikjian LM, et al. Betaglycan binds inhibin and can mediate functional antagonism of activin signalling. Nature 2000;404(6776):411–4.

[86] Makanji Y, Walton KL, Wilce MC, Chan KL, Robertson DM, Harrison CA. Suppression of inhibin A biological activity by alterations in the binding site for betaglycan. J Biol Chem 2008;283(24):16743–51.

[87] Corrigan AZ, Bilezikjian LM, Carroll RS, Bald LN, Schmelzer CH, Fendly BM, et al. Evidence for an autocrine role of activin B within rat anterior pituitary cultures. Endocrinology 1991;128(3):1682–4.

[88] Bilezikjian LM, Vaughan JM, Vale WW. Characterization and the regulation of inhibin/activin subunit proteins of cultured rat anterior pituitary cells. Endocrinology 1993;133(6):2545–53.

[89] MacConell LA, Leal AM, Vale WW. The distribution of betaglycan protein and mRNA in rat brain, pituitary, and gonads: implications for a role for betaglycan in inhibin-mediated reproductive functions. Endocrinology 2002;143(3):1066–75.

[90] Goddard I, Bouras M, Keramidas M, Hendrick JC, Feige JJ, Benahmed M. Transforming growth factor-β receptor types I and II in cultured porcine Leydig cells: expression and hormonal regulation. Endocrinology 2000;141(6):2068–74.

[91] de Winter JP, Themmen AP, Hoogerbrugge JW, Klaij IA, Grootegoed JA, de Jong FH. Activin receptor mRNA expression in rat testicular cell types. Mol Cell Endocrinol 1992;83(1):R1–8.

[92] Kaipia A, Parvinen M, Toppari J. Localization of activin receptor (ActR-IIB$_2$) mRNA in the rat seminiferous epithelium. Endocrinology 1993;132(1):477–9.

[93] Dias V, Meachem S, Rajpert-De Meyts E, McLachlan R, Manuelpillai U, Loveland KL. Activin receptor subunits in normal and dysfunctional adult human testis. Hum Reprod 2008;23(2):412–20.

[94] Woodruff TK, Borree J, Attie KM, Cox ET, Rice GC, Mather JP. Stage-specific binding of inhibin and activin to subpopulations of rat germ cells. Endocrinology 1992;130(2):871–81.
[95] Cameron VA, Nishimura E, Mathews LS, Lewis KA, Sawchenko PE, Vale WW. Hybridization histochemical localization of activin receptor subtypes in rat brain, pituitary, ovary, and testis. Endocrinology 1994;134(2):799–808.
[96] van Schaik RH, Wierikx CD, Looijenga LH, Oosterhuis JW, de Jong FH. Human testicular germ cell tumours express inhibin subunits, activin receptors and follistatin mRNAs. Br J Cancer 1997;76(9):1191–8.
[97] Meinhardt A, O'Bryan MK, McFarlane JR, Loveland KL, Mallidis C, Foulds LM, et al. Localization of follistatin in the rat testis. J Reprod Fertil 1998;112(2):233–41.
[98] Hakovirta H, Kaipia A, Söder O, Parvinen M. Effects of activin-A, inhibin-A, and transforming growth factor-β1 on stage-specific deoxyribonucleic acid synthesis during rat seminiferous epithelial cycle. Endocrinology 1993;133(4):1664–8.
[99] Russell L. Movement of spermatocytes from the basal to the adluminal compartment of the rat testis. Am J Anat 1977;148(3):313–28.
[100] Parvinen M. Regulation of the seminiferous epithelium. Endocr Rev 1982;3 (4):404–17.
[101] Nicholls PK, Stanton PG, Chen JL, Olcorn JS, Haverfield JT, Qian H, et al. Activin signaling regulates Sertoli cell differentiation and function. Endocrinology 2012;153 (12):6065–77.
[102] Mather JP, Attie KM, Woodruff TK, Rice GC, Phillips DM. Activin stimulates spermatogonial proliferation in germ-Sertoli cell cocultures from immature rat testis. Endocrinology 1990;127(6):3206–14.
[103] Meinhardt A, McFarlane JR, Seitz J, de Kretser DM. Activin maintains the condensed type of mitochondria in germ cells. Mol Cell Endocrinol 2000;168(1–2):111–7.
[104] de Winter JP, Vanderstichele HM, Timmerman MA, Blok LJ, Themmen AP, de Jong FH. Activin is produced by rat Sertoli cells *in vitro* and can act as an autocrine regulator of Sertoli cell function. Endocrinology 1993;132(3):975–82.
[105] Itman C, Wong C, Hunyadi B, Ernst M, Jans DA, Loveland KL. Smad3 dosage determines androgen responsiveness and sets the pace of postnatal testis development. Endocrinology 2011;152(5):2076–89.
[106] Winnall WR, Okuma Y, Saito K, Muir JA, Hedger MP. Regulation of interleukin 1α, activin and inhibin by lipopolysaccharide in Sertoli cells from prepubertal rats. Mol Cell Endocrinol 2009;307(1–2):169–75.
[107] Lie PP, Cheng CY, Mruk DD. Signalling pathways regulating the blood–testis barrier. Int J Biochem Cell Biol 2013;45(3):621–5.
[108] Chen CL, Pignataro OP, Feng ZM. Inhibin/activin subunits and activin receptor are co-expressed in Leydig tumor cells. Mol Cell Endocrinol 1993;94(1):137–43.
[109] Anderson RA, Cambray N, Hartley PS, McNeilly AS. Expression and localization of inhibin α, inhibin/activin βA and βB and the activin type II and inhibin β-glycan receptors in the developing human testis. Reproduction 2002;123(6):779–88.
[110] Lin T, Calkins JK, Morris PL, Vale W, Bardin CW. Regulation of Leydig cell function in primary culture by inhibin and activin. Endocrinology 1989;125(4):2134–40.
[111] Lejeune H, Chuzel F, Sanchez P, Durand P, Mather JP, Saez JM. Stimulating effect of both human recombinant inhibin A and activin A on immature porcine Leydig cell functions in vitro. Endocrinology 1997;138(11):4783–91.
[112] Hsueh AJ, Dahl KD, Vaughan J, Tucker E, Rivier J, Bardin CW, et al. Heterodimers and homodimers of inhibin subunits have different paracrine action in the modulation

of luteinizing hormone-stimulated androgen biosynthesis. Proc Natl Acad Sci USA 1987;84(14):5082–6.

[113] Hedger MP, Winnall WR, Phillips DJ, de Kretser DM. The regulation and functions of activin and follistatin in inflammation and immunity. Vitam Horm 2011;85:255–97.

[114] Winnall WR, Muir JA, Hedger MP. Rat resident testicular macrophages have an alternatively activated phenotype and constitutively produce interleukin-10 *in vitro*. J Leukoc Biol 2011;90(1):133–43.

[115] Winnall WR, Hedger MP. Phenotypic and functional heterogeneity of the testicular macrophage population: a new regulatory model. J Reprod Immunol 2013;97 (2):147–58.

[116] Cho BN, McMullen ML, Pei L, Yates CJ, Mayo KE. Reproductive deficiencies in transgenic mice expressing the rat inhibin α-subunit gene. Endocrinology 2001;142 (11):4994–5004.

[117] Li Q, Graff JM, O'Connor AE, Loveland KL, Matzuk MM. SMAD3 regulates gonadal tumorigenesis. Mol Endocrinol 2007;21(10):2472–86.

[118] Coerver KA, Woodruff TK, Finegold MJ, Mather J, Bradley A, Matzuk MM. Activin signaling through activin receptor type II causes the cachexia-like symptoms in inhibin-deficient mice. Mol Endocrinol 1996;10(5):534–43.

[119] Matzuk MM, Finegold MJ, Su JG, Hsueh AJ, Bradley A. α-inhibin is a tumour-suppressor gene with gonadal specificity in mice. Nature 1992;360(6402):313–9.

[120] Li Q, Kumar R, Underwood K, O'Connor AE, Loveland KL, Seehra JS, et al. Prevention of cachexia-like syndrome development and reduction of tumor progression in inhibin-deficient mice following administration of a chimeric activin receptor type II-murine Fc protein. Mol Hum Reprod 2007;13(9):675–83.

[121] Gold EJ, O'Bryan MK, Mellor SL, Cranfield M, Risbridger GP, Groome NP, et al. Cell-specific expression of βC-activin in the rat reproductive tract, adrenal and liver. Mol Cell Endocrinol 2004;222(1–2):61–9.

[122] Gold E, Jetly N, O'Bryan MK, Meachem S, Srinivasan D, Behuria S, et al. Activin C antagonizes activin A *in vitro* and overexpression leads to pathologies *in vivo*. Am J Pathol 2009;174(1):184–95.

[123] Guo Q, Kumar TR, Woodruff T, Hadsell LA, DeMayo FJ, Matzuk MM. Overexpression of mouse follistatin causes reproductive defects in transgenic mice. Mol Endocrinol 1998;12(1):96–106.

[124] Mendis SH, Meachem SJ, Sarraj MA, Loveland K. Activin A balances Sertoli and germ cell proliferation in the fetal mouse testis. Biol Reprod 2011;84(2):379–91.

[125] Mithraprabhu S, Mendis S, Meachem SJ, Tubino L, Matzuk MM, Brown CW, et al. Activin bioactivity affects germ cell differentiation in the postnatal mouse testis *in vivo*. Biol Reprod 2010;82(5):980–90.

[126] Anderson RA, Evans LW, Irvine DS, McIntyre MA, Groome NP, Riley SC. Follistatin and activin A production by the male reproductive tract. Hum Reprod 1998;13(12):3319–25.

[127] Winnall WR, Wu H, Sarraj MA, Rogers PA, de Kretser DM, Girling JE, et al. Expression patterns of activin, inhibin and follistatin variants in the adult male mouse reproductive tract suggest important roles in the epididymis and vas deferens. Reprod Fertil Dev 2013;25(3):570–80.

[128] Archambeault DR, Yao HH. Activin A, a product of fetal Leydig cells, is a unique paracrine regulator of Sertoli cell proliferation and fetal testis cord expansion. Proc Natl Acad Sci USA 2010;107(23):10526–31.

[129] Orth JM, Gunsalus GL, Lamperti AA. Evidence from Sertoli cell-depleted rats indicates that spermatid number in adults depends on numbers of Sertoli cells produced during perinatal development. Endocrinology 1988;122(3):787–94.

[130] Orth JM. Cell biology of testicular development in the fetus and neonate. In: Desjardins C, Ewing LL, editors. Cell and molecular biology of the testis. Oxford: Oxford University Press; 1993. p. 3–42.

[131] Cool J, DeFalco T, Capel B. Testis formation in the fetal mouse: dynamic and complex de novo tubulogenesis. Wiley Interdiscip Rev Dev Biol 2012;1(6):847–59.

[132] McCarrey JR. Development of the germ cell. In: Desjardins C, Ewing LL, editors. Cell and molecular biology of the testis. Oxford: Oxford University Press; 1993. p. 58–89.

[133] Hackett JA, Sengupta R, Zylicz JJ, Murakami K, Lee C, Down TA, et al. Germline DNA demethylation dynamics and imprint erasure through 5-hydroxymethylcytosine. Science 2013;339(6118):448–52.

[134] Seisenberger S, Andrews S, Krueger F, Arand J, Walter J, Santos F, et al. The dynamics of genome-wide DNA methylation reprogramming in mouse primordial germ cells. Mol Cell 2012;48(6):849–62.

[135] Western PS, Miles DC, van den Bergen JA, Burton M, Sinclair AH. Dynamic regulation of mitotic arrest in fetal male germ cells. Stem Cells 2008;26(2):339–47.

[136] Richards AJ, Enders GC, Resnick JL. Activin and TGFβ limit murine primordial germ cell proliferation. Dev Biol 1999;207(2):470–5.

[137] Archambeault DR, Tomaszewski J, Childs AJ, Anderson RA, Yao HH. Testicular somatic cells, not gonocytes, are the major source of functional activin a during testis morphogenesis. Endocrinology 2011;152(11):4358–67.

[138] Jarred RA, Cancilla B, Richards M, Groome NP, McNatty KP, Risbridger GP. Differential localization of inhibin subunit proteins in the ovine testis during fetal gonadal development. Endocrinology 1999;140(2):979–86.

[139] Meehan T, Schlatt S, O'Bryan MK, de Kretser DM, Loveland KL. Regulation of germ cell and Sertoli cell development by activin, follistatin, and FSH. Dev Biol 2000;220(2):225–37.

[140] Souquet B, Tourpin S, Messiaen S, Moison D, Habert R, Livera G. Nodal signaling regulates the entry into meiosis in fetal germ cells. Endocrinology 2012;153(5):2466–73.

[141] Moreno SG, Attali M, Allemand I, Messiaen S, Fouchet P, Coffigny H, et al. TGFβ signaling in male germ cells regulates gonocyte quiescence and fertility in mice. Dev Biol 2010;342(1):74–84.

[142] Yao HH, Aardema J, Holthusen K. Sexually dimorphic regulation of inhibin βB in establishing gonadal vasculature in mice. Biol Reprod 2006;74(5):978–83.

[143] McGuinness MP, Orth JM. Reinitiation of gonocyte mitosis and movement of gonocytes to the basement membrane in testes of newborn rats *in vivo* and *in vitro*. Anat Rec 1992;233(4):527–37.

[144] Orth JM. Proliferation of Sertoli cells in fetal and postnatal rats: a quantitative autoradiographic study. Anat Rec 1982;203(4):485–92.

[145] Steinberger A, Steinberger E. Replication pattern of Sertoli cells in maturing rat testis *in vivo* and in organ culture. Biol Reprod 1971;4(1):84–7.

[146] Loveland KL, Bakker M, Meehan T, Christy E, von Schönfeldt V, Drummond A, et al. Expression of Bambi is widespread in juvenile and adult rat tissues and is regulated in male germ cells. Endocrinology 2003;144(9):4180–6.

[147] van Dissel-Emiliani FM, Grootenhuis AJ, de Jong FH, de Rooij DG. Inhibin reduces spermatogonial numbers in testes of adult mice and Chinese hamsters. Endocrinology 1989;125(4):1898–903.

[148] Nagano M, Ryu BY, Brinster CJ, Avarbock MR, Brinster RL. Maintenance of mouse male germ line stem cells *in vitro*. Biol Reprod 2003;68(6):2207–14.
[149] Pellegrini M, Grimaldi P, Rossi P, Geremia R, Dolci S. Developmental expression of BMP4/ALK3/SMAD5 signaling pathway in the mouse testis: a potential role of BMP4 in spermatogonia differentiation. J Cell Sci 2003;116(Pt 16):3363–72.
[150] Itman CL, Mendis S, Barakat B, Loveland KL. All in the family: transforming growth factor β action in testis development. Reproduction 2006;132(2):233–46.
[151] Matzuk MM, Lamb DJ. Genetic dissection of mammalian fertility pathways. Nat Cell Biol 2002;(4 Suppl.)s41–9.
[152] Matzuk MM, Kumar TR, Shou W, Coerver KA, Lau AL, Behringer RR, et al. Transgenic models to study the roles of inhibins and activins in reproduction, oncogenesis, and development. Recent Prog Horm Res 1996;51:123–54 [discussion 155–7].
[153] Matzuk MM, Finegold MJ, Mather JP, Krummen L, Lu H, Bradley A. Development of cancer cachexia-like syndrome and adrenal tumors in inhibin-deficient mice. Proc Natl Acad Sci USA 1994;91(19):8817–21.
[154] Vassalli A, Matzuk MM, Gardner HA, Lee KF, Jaenisch R. Activin/inhibin βB subunit gene disruption leads to defects in eyelid development and female reproduction. Genes Dev 1994;8(4):414–27.
[155] Matzuk MM, Kumar TR, Vassalli A, Bickenbach JR, Roop DR, Jaenisch R, et al. Functional analysis of activins in mammalian development. Nature 1995;374 (6520):354–6.
[156] Brown CW, Houston-Hawkins DE, Woodruff TK, Matzuk MM. Insertion of Inhbb into the Inhba locus rescues the Inhba-null phenotype and reveals new activin functions. Nat Genet 2000;25(4):453–7.
[157] Matthews LS, Vale WW. Expression cloning of an activin receptor, a predicted transmembrane serine kinase. Cell 1991;65(6):973–82.
[158] Itman C, Loveland KL. Smads and cell fate: distinct roles in specification, development and tumourigenesis in the testis. IUBMB Life 2013;65(2):85–97.
[159] Luukko K, Ylikorkala A, Mäkelä TP. Developmentally regulated expression of Smad3, Smad4, Smad6, and Smad7 involved in TGF-beta signaling. Mech Dev 2001;101(1–2):209–12.
[160] Boitani C, Stefanini M, Fragale A, Morena AR. Activin stimulates Sertoli cell proliferation in a defined period of rat testis development. Endocrinology 1995;136 (12):5438–44.
[161] Buzzard JJ, Farnworth PG, de Kretser DM, O'Connor AE, Wreford NG, Morrison JR. Proliferative phase Sertoli cells display a developmentally regulated response to activin *in vitro*. Endocrinology 2003;144(2):474–83.
[162] Fragale A, Puglisi R, Morena AR, Stefanini M, Boitani C. Age-dependent activin receptor expression pinpoints activin A as a physiological regulator of rat Sertoli cell proliferation. Mol Hum Reprod 2001;7(12):1107–14.
[163] Beumer TL, Kiyokawa H, Roepers-Gajadien HL, van den Bos LA, Lock TM, Gademan IS, et al. Regulatory role of p27kip1 in the mouse and human testis. Endocrinology 1999;140(4):1834–40.
[164] Matzuk MM, Finegold MJ, Su JJ, Hsueh AJW, Bradley A. α-inhibin is a tumor-suppressor gene with gonadal specificity in mice. Nature 1992;360(6402):313–9.
[165] Cipriano SC, Chen L, Burns KH, Koff A, Matzuk MM. Inhibin and p27 interact to regulate gonadal tumorigenesis. Mol Endocrinol 2001;15(6):985–96.
[166] Shou W, Woodruff TK, Matzuk MM. Role of androgens in testicular tumor development in inhibin-deficient mice. Endocrinology 1997;138(11):5000–5.

[167] Xu J, Beyer AR, Walker WH, McGee EA. Developmental and stage-specific expression of Smad2 and Smad3 in rat testis. J Androl 2003;24(2):192–200.

[168] Rajanahally S, Agno JE, Nalam RL, Weinstein MB, Loveland KL, Matzuk MM, et al. Genetic evidence that SMAD2 is not required for gonadal tumor development in inhibin-deficient mice. Reprod Biol Endocrinol 2010;8:69.

[169] Pfaff T, Rhodes J, Bergmann M, Weinbauer GF. Inhibin B as a marker of Sertoli cell damage and spermatogenic disturbance in the rat. Birth Defects Res B Dev Reprod Toxicol 2013;98(1):91–103.

[170] Sharpe RM, Turner KJ, McKinnell C, Groome NP, Atanassova N, Millar MR, et al. Inhibin B levels in plasma of the male rat from birth to adulthood: effect of experimental manipulation of Sertoli cell number. J Androl 1999;20(1):94–101.

[171] Anawalt BD, Bebb RA, Matsumoto AM, Groome NP, Illingworth PJ, McNeilly AS, et al. Serum inhibin B levels reflect Sertoli cell function in normal men and men with testicular dysfunction. J Clin Endocrinol Metab 1996;81(9):3341–5.

[172] Ochsenkühn R, O'Connor AE, Hirst JJ, Baker HW, de Kretser DM, Hedger MP. The relationship between immunosuppressive activity and immunoregulatory cytokines in seminal plasma: influence of sperm autoimmunity and seminal leukocytes. J Reprod Immunol 2006;71(1):57–74.

[173] de Kretser DM, McLachlan RI, Robertson DM, Burger HG. Serum inhibin levels in normal men and men with testicular disorders. J Endocrinol 1989;120(3):517–23.

[174] Groome NP, Illingworth PJ, O'Brien M, Pai R, Rodger FE, Mather J, et al. Measurement of dimeric inhibin B throughout the human menstrual cycle. J Clin Endocrinol Metab 1996;81(4):1401–5.

[175] Groome NP, Illingworth PJ, O'Brien M, Priddle J, Weaver K, McNeilly AS. Quantitation of inhibin Pro-αC-containing forms in human serum by a new ultrasensitive two-site enzyme-linked immunosorbent assay. J Clin Endocrinol Metab 1995;80 (10):2926–32.

[176] Wallace EM, Groome NP, Riley SC, Parker AC, Wu FC. Effects of chemotherapy-induced testicular damage on inhibin, gonadotropin, and testosterone secretion: a prospective longitudinal study. J Clin Endocrinol Metab 1997;82(9):3111–5.

[177] McLachlan RI, O'Donnell L, Meachem SJ, Stanton PG, de Kretser DM, Pratis K, et al. Identification of specific sites of hormonal regulation in spermatogenesis in rats, monkeys and man. Recent Prog Horm Res 2002;57:149–79.

[178] Pineau C, Sharpe RM, Saunders PTK, Gerard N, Jegou B. Regulation of Sertoli cell inhibin production and of inhibin α-subunit mRNA levels by specific germ cell types. Mol Cell Endocrinol 1990;72(1):13–22.

[179] Ramaswamy S, Marshall GR, Pohl CR, Friedman RL, Plant TM. Inhibitory and stimulatory regulation of testicular inhibin B secretion by luteinizing hormone and follicle stimulating hormone, respectively in the rhesus monkey (*Macaca mulatta*). Endocrinology 2003;144(4):1175–85.

[180] Skakkebaek NE, Rajpert-De Meyts E, Jørgensen N, Main KM, Leffers H, Andersson AM, et al. Testicular cancer trends as 'whistle blowers' of testicular developmental problems in populations. Int J Androl 2007;30(4):198–204 [discussion 204–5].

[181] Sharpe RM, Skakkebaek NE. Testicular dysgenesis syndrome: mechanistic insights and potential new downstream effects. Fertil Steril 2008;89(2 Suppl.):e33–8.

[182] Moody S, Goh H, Bielanowicz A, Rippon P, Loveland KL, Itman C. Prepubertal mouse testis growth and maturation and androgen production are acutely sensitive to di-n-butyl phthalate. Endocrinology 2013;154(9):3460–75.

[183] Loveland KL, Dias V, Meachem S, Rajpert-De Meyts E. The transforming growth factor-beta superfamily in early spermatogenesis: potential relevance to testicular dysgenesis. Int J Androl 2007;30(4):377–84 [discussion 384].

[184] Dias VL, Rajpert-De Meyts E, McLachlan R, Loveland KL. Analysis of activin/TGFβ signaling modulators within the normal and dysfunctional adult human testis reveal evidence of altered signaling capacity in a subset of seminomas. Reproduction 2009;138(5):801–11.

[185] Spiller CM, Feng CW, Jackson A, Gillis AJ, Rolland AD, Looijenga LH, et al. Endogenous Nodal signaling regulates germ cell potency during mammalian testis development. Development 2012;139(22):4123–32.

[186] Spiller CM, Bowles J, Koopman P. Nodal/Cripto signaling in fetal male germ cell development: implications for testicular germ cell tumors. Int J Dev Biol 2013;57 (2–4):211–9.

[187] Hime GR, Loveland KL, Abud HE. Drosophila spermatogenesis: insights into testicular cancer. Int J Androl 2007;30(4):265–74 [discussion 274].

[188] Lee KL, Lim SK, Orlov YL, Yit le Y, Yang H, Ang LT, et al. Graded Nodal/Activin signaling titrates conversion of quantitative phospho-Smad2 levels into qualitative embryonic stem cell fate decisions. PLoS Genet 2011;7(6):e1002130.

[189] Miles DC, Wakeling SI, Stringer JM, van den Bergen JA, Wilhelm D, Sinclair AH, et al. Signaling through the TGF beta-activin receptors ALK4/5/7 regulates testis formation and male germ cell development. PLoS One 2013;8(1):e54606.

[190] Perrot CY, Javelaud D, Mauviel A. Overlapping activities of TGF-β and Hedgehog signaling in cancer: therapeutic targets for cancer treatment. Pharmacol Ther 2013;137 (2):183–99.

[191] Dalton S. Signaling networks in human pluripotent stem cells. Curr Opin Cell Biol 2013;25(2):241–6.

[192] Tilbrook AJ, de Kretser DM, Clarke IJ. Studies on the testicular source of inhibin and its route of secretion in rams: failure of the Leydig cell to secrete inhibin in response to a human chorionic gonadotrophin/LH stimulus. J Endocrinol 1991;130(1):107–14.

[193] de Winter JP, Timmerman MA, Vanderstichele HM, Klaij IA, Grootenhuis AJ, Rommerts FF, et al. Testicular Leydig cells *in vitro* secrete only inhibin α-subunits, whereas Leydig cell tumors can secrete bioactive inhibin. Mol Cell Endocrinol 1992;83(2–3):105–15.

The initiation of spermatogenesis and the cycle of the seminiferous epithelium

8

Michael D. Griswold
School of Molecular Biosciences, Center for Reproductive Biology, Washington State University, Pullman, WA

I. Introduction and highlights since the last volume

Spermatogenesis is a well-organized and tightly regulated process composed of three distinct biological activities: (1) continuous stem cell renewal and the production and expansion of progenitor cells (mitosis), (2) the production of haploid cells from diploid progenitor cells (meiosis), and (3) the unique differentiation of haploid cells into spermatozoa (spermiogenesis). This chapter focuses on the processes and timing that lead to the initiation of meiosis in mice and how this initiation leads to the organization of the seminiferous epithelium that results in continuous sperm production.

In the first edition of *Sertoli Cell Biology* [1], there was little mention of the role of retinoic acid (RA) in germ cell development and the onset of meiosis. A flurry of articles were published on vitamin A and the testis in the late 1980s and early 1990s, including the first demonstration that the cycle of the seminiferous epithelium could be synchronized by vitamin A deficiency followed by retinol or RA supplementation [1a]. The action of RA in the differentiation of spermatogonia was demonstrated in the early 1980s [2,3]. There was very little new information in the field until the demonstration that RA and the induction of stimulated by RA 8 (STRA8) protein were essential for the entry of germ cells into meiosis in both males and females [4]. More recent work has focused on the role of RA in the initiation of spermatogenesis and the differentiation of A spermatogonia into A1 spermatogonia.

II. Differentiation of spermatogonia

Stem cells and early progenitor cells, which are known as A spermatogonia in mice, are described as undifferentiated. It has been pointed out that the term "undifferentiated" is somewhat misleading because both the spermatogonial stem cells and progenitor cells are on a differentiation pathway that leads away from pluripotency [5]. Undifferentiated single spermatogonia divide mitotically but incompletely to form chains of 2, 4, 8, or 16 cells (A aligned (Aal) spermatogonia) connected by

Sertoli Cell Biology. DOI: http://dx.doi.org/10.1016/B978-0-12-417047-6.00008-9

intracellular bridges [5]. These mitotic divisions are random with respect to the cycle of the seminiferous epithelium [6].

At some point, the Aal cells become A1 spermatogonia and enter a "differentiation" pathway that leads to meiosis (Figure 8.1). The progression into A1 cells requires action on the germ cells by at least one extrinsic factor (RA) and results in the progression of A1 cells through a series of irreversible differentiation steps that lead to meiosis. Once spermatogonia enter the differentiation pathway to become A1 spermatogonia in adult mice, they normally undergo six successive mitotic divisions to form A2, A3, A4, intermediate, and B spermatogonia, and, ultimately, preleptotene spermatocytes. These mitotic divisions are synchronized with the cycle of the seminiferous epithelium, and the formation of preleptotene spermatocytes is generally considered to be the initiation of meiosis. The cells enter this irreversible differentiation pathway that occurs at stages VII–VIII of the cycle of the seminiferous epithelium, and the subsequent mitotic divisions, and ultimate formation of preleptotene spermatocytes, again, at stages VII–VIII is strictly timed.

If the formation of preleptotene spermatocytes is considered to be the initiation of meiosis, then the transition of A undifferentiated spermatogonia to A1 differentiating spermatogonia could be considered to be the first commitment to meiosis. Discussed in more detail below, this irreversible progression, starting from A spermatogonia and resulting in the formation of preleptotene spermatocytes, requires the action of RA and the resulting induction of STRA8 protein that is uniquely expressed in premeiotic germ cells [7–9]. In addition, the variable and periodic action of RA along the seminiferous tubule results in the establishment of the cycle

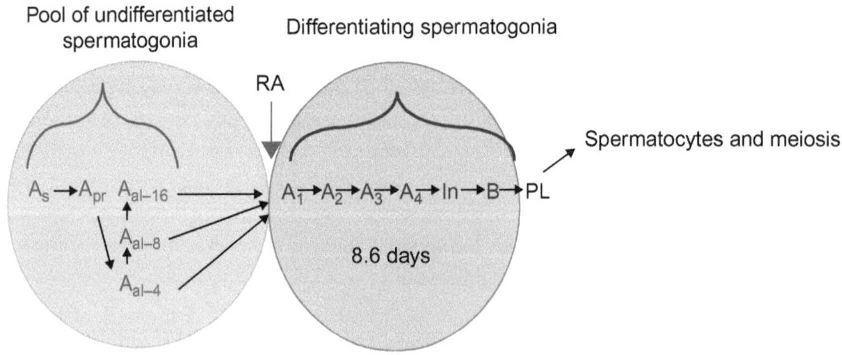

Figure 8.1 Overview of spermatogonial differentiation in murine testes. The spermatogonial stem cells (A_s) divide to self-renew and produce progenitor cells that comprise the population of undifferentiated spermatogonia (red). A_{pr}, A paired; A_{al}, A aligned. When acted on by RA, the A undifferentiated pool can form the A1 differentiating spermatogonia (blue) that undergo a series of cell divisions and differentiation steps during one cycle of the seminiferous epithelium to ultimately form the first spermatocytes. In, intermediate; B, B spermatogonia. (For interpretation of the references to color in this figure legend, the reader is referred to the web version of this book.)

of the seminiferous epithelium. Thus, unless acted upon by RA, undifferentiated spermatogonia cannot progress into meiosis.

III. Evidence that RA is required for the initiation of meiosis

What is the evidence that RA is an extrinsic factor during the initiation of meiosis, and what do we know about when and how this initiation occurs? The incontrovertible evidence for the involvement of RA in meiosis in both males and females has been reviewed [10]. The requirement for RA in male germ cell meiosis is supported by many observations and experiments in a number of species. When examined thoroughly, the results from these studies lead to several conclusions.

First, the initiation of meiosis in spermatogenesis is blocked by a deficiency of dietary retinoids and is triggered by exogenous and endogenous retinoids. It has been known for decades that dietary vitamin A deprivation results in spermatogenic arrest; this arrest occurs at the spermatogonial A-to-A1 transition (undifferentiated to differentiated spermatogonia) in mice [11–16]. In rats, the initial arrest occurs at both the A-to-A1 transition and at the preleptotene spermatocytes. When retinol (ROL) is provided to vitamin-A-deficient (VAD) rodents, meiosis is reinitiated in a synchronous manner [11,13,15]. Large doses of RA can also reinitiate meiosis in this system [16]. ROL injection into VAD mice dramatically induced the expression of *Stra8* over a 24-h period. The expression of a number of meiotic markers showed that the exogenous RA can stimulate germ cells in the embryonic testis to enter meiosis [4,17–19].

Second, gene knockout studies have clearly shown that the expression of *Stra8*, a premeiotic gene, is required for the switch from mitosis to meiosis in the adult testis [7,20–22]. The gene designation *Stra8* is derived from "stimulated by retinoic acid gene 8". The expression of *Stra8 in vivo* is confined to germ cells in the testis undergoing the A-to-A1 transition and to preleptotene spermatocytes. Treatment with RA of mouse fetal gonadal tissue, adult testis tissue, or VAD mice also leads to upregulation of *Stra8* [4,8,17,19,23,24]. While the function of STRA8 is unknown, the expression of this gene is an excellent marker for the onset of meiosis and the action of RA. *Stra8* encodes an acidic, glutamic acid-rich, highly phosphorylated protein with no recognizable homologies [22].

Third, reagents that limit the breakdown of RA stimulate spermatogonial differentiation and the onset of meiosis, and reagents that reduce the synthesis of RA inhibit spermatogonial differentiation. RA is normally metabolized to inactive products by P450 enzymes from the CYP26 family. These enzymes can be inhibited nonspecifically by ketoconazole or, more specifically, by the drug R115866. In the prepubertal testis, inhibition of CYP26 with R115866 resulted in an increase in differentiating spermatogonia and a decrease in undifferentiated spermatogonia [25]. Bis-(dichloroacetyl)-diamines (BDADs) are compounds that inhibit retinaldehyde dehydrogenases and block the conversion of retinaldehyde to RA. Administration of a specific BDAD, WIN 18,446, to neonatal mice, adult mice, or isolated germ cells blocked the expression of *Stra8* and germ cell differentiation [26]. Expression of *Stra8* can be rescued by exogenous RA but not by ROL.

Fourth, agents or procedures that interfere with the function of RA receptors (RARs) block spermatogenesis. Both Sertoli cells and germ cells express RARs; RARα is primarily expressed in Sertoli cells and RARγ is primarily expressed in germ cells [27–29]. Both *Stra8* induction and the entry into meiosis can be inhibited in the embryonic ovary by the action of RAR antagonists or induced with RAR agonists [4,17].

IV. The initiation of asynchronous spermatogenesis by RA

A few studies have addressed the role of RA signaling during the initiation of the first round of spermatogenesis. In mammals, spermatogenesis (as defined by the developmental progression of undifferentiated spermatogonia) is initiated during the juvenile period, starting with puberty in men and at 2–4 days after birth in mice [30]. In mice, while it is clear that neonatal (juvenile) spermatogenesis is derived from slightly different cell populations (gonocytes in neonates and spermatogonial stem cells in adults), there is evidence showing that similar mechanisms drive neonatal and adult spermatogenesis regardless of the originating cell population or timing of initiation. As described above, the evidence that RA acts as a regulator of spermatogenesis is overwhelming, leading to the question of how, when, and where RA acts within the testis.

Snyder and colleagues addressed these questions using a transgenic *RARE-hsplacZ* mouse model that expressed β-galactosidase under the control of a RA response element [25,31]. In these mice, if the RA signaling pathway is present in the form of active receptors and ligand, the synthesis of β-galactosidase is initiated and can be detected by routine histochemical methods. Note that this procedure denotes the place where RA acts but not where synthesis or storage occurs. Snyder and colleagues found that β-galactosidase staining could be detected in murine prospermatogonia shortly after birth. Within 1–2 days after birth, the presence of β-galactosidase-positive germ cells revealed an intact and active RA signaling system. These β-galactosidase-positive cells were also positive for the presence of STRA8, the definitive marker for differentiating spermatogonia (Figure 8.2). Prospermatogonia or spermatogonia that were β-galactosidase negative were positive for markers of undifferentiated spermatogonia. Interestingly, in these studies, the β-galactosidase-positive cells within the tubule were almost exclusively spermatogonia with very little to no staining detected in the Sertoli cells. In addition, the β-galactosidase-positive spermatogonia were distributed in a nonuniform or patchy manner along the length of the tubule. In some regions along the tubule, there were large clusters of differentiating spermatogonia, while adjacent regions were devoid of these β-galactosidase-positive cells. Snyder and colleagues interpreted these results as representing the asynchronous initiation of spermatogenesis that ultimately leads to the stages of

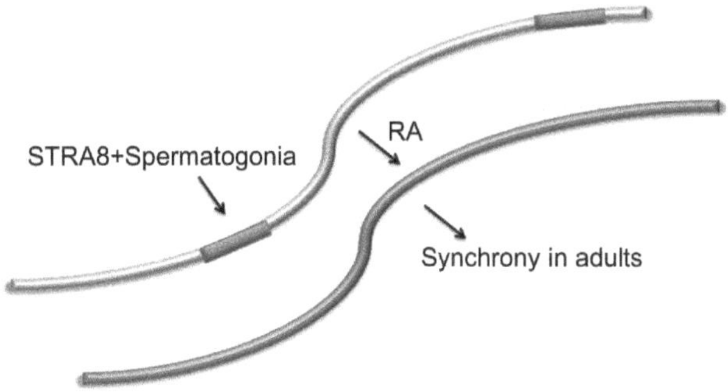

Figure 8.2 Illustration of the results from Snyder and colleagues [25]. RARE-*hsplacZ* transgenic mice were used to examine RA signaling along the tubule in 2-day-old mice. Patches of β-galactosidase-positive and stimulated by RA 8 (STRA8)-positive spermatogonia were coincident, indicating that active RA signaling and transition from A to A1 spermatogonia was occurring. These patches were interpreted to represent the onset of asynchronous spermatogenesis and the beginning of the cycle of the seminiferous epithelium. If the mice were treated 24 h earlier with RA, the entire tubule was positive for β-galactosidase and STRA8. When the mice treated at 2 dpp with RA became adults, the spermatogenic wave was absent and spermatogenesis was synchronous.

the cycle of the seminiferous epithelium and the continuous sperm production in the adult testis. Therefore, the spermatogenic cycle that is characteristic of the adult testis was initiated by the spatial distribution of the action of RA on germ cells along the neonatal seminiferous tubules.

If the ligand is limiting in the incidence of β-galactosidase-positive cells, then the patches of A1 cells also denote a relatively higher concentration of RA in that portion of the tubule. It follows that the pulse of a higher concentration of RA moves along the tubule and is regulated by controlled synthesis and degradation of active ligand.

Snyder and colleagues treated neonatal animals from the same strain of RARE-hsplacZ mice with either RA or inhibitors of the CYP26 family of enzymes that degrade the RA signal. Both treatments resulted in the expression of β-galactosidase and STRA8 in nearly all of the germ cells along the tubule and eliminated the periodicity of A1 spermatogonia. These results support the idea that the concentration of RA was limiting. In addition, mice that were treated with exogenous RA as neonates (2 dpp) demonstrated highly synchronized spermatogenesis as adults with a completely abnormal cycle of the seminiferous epithelium. Once the conversion of A spermatogonia to A1 spermatogonia occurs as a result of the action of RA, the progressive development of germ cells through meiosis appears to be initiated. The treatment with exogenous RA results in the entire first wave of A spermatogonia progressing through meiosis at the same time without regard to the spatial distribution along the tubule and in synchronous spermatogenesis.

In an extension of these studies, Davis and colleagues showed that treatment with exogenous RA results in synchronous spermatogenesis in prepubertal mice until the appearance of preleptotene spermatocytes [32]. Once the first preleptotene spermatocytes are formed, at 6–8 days after birth, exogenous RA does not

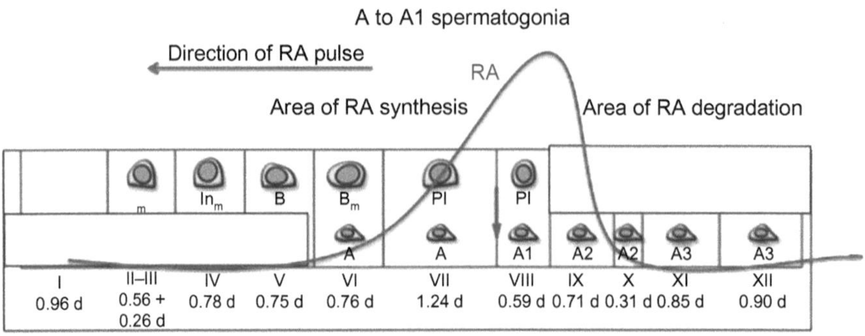

Figure 8.3 RA pulse during the initiation of spermatogenesis. The results from Snyder and colleagues [25] can be interpreted to be the result of a pulse of RA moving along the seminiferous tubules in a periodic manner. The pulse stimulates the transition from A to A1 spermatogonia and reflects the beginning of stage VII–VIII of the cycle of the seminiferous epithelium. The generation of the pulse requires active RA synthesis and degradation.

stimulate additional A spermatogonia to induce STRA8 and enter the differentiation pathway. Thus, the appearance of preleptotene spermatocytes exactly one cycle after the transition of A spermatogonia to A1 spermatogonia somehow alters the environment of the tubule so as to maintain the asynchronous status.

In summary, the studies described above suggest that the first wave of spermatogenesis is initiated within a few days after birth in mice, when patches or groups of adjacent prospermatogonia and/or undifferentiated A spermatogonia are stimulated by RA to make the transition into A1 differentiating spermatogonia and, ultimately, to enter meiosis. It is the spatial distribution of the RA signal that leads to asynchronous spermatogenesis. For the spermatogenic wave to be initiated, the RA signal must progress along the tubule in a carefully controlled and timed manner. That signal could be a pulse of RA that is generated at the pulse front and degraded as the pulse passes (Figure 8.3). The source of that RA signal and how its synthesis and degradation are controlled are key research gaps. Both Sertoli cells and spermatogonia have RARs and can respond to RA, and both appear to contain enzymes that are necessary for the oxidation of ROL to RA.

V. Regulation of RA synthesis and degradation in the developing testis

The action of RA is intimately tied to its synthesis and degradation because the availability of the ligand appears to be rate limiting. Because of the absolute requirement for this vitamin, its metabolism is subject to many biological controls and genetic redundancies. Important steps in the metabolism of RA and potential control points for its actions include the absorption of precursors, storage of retinyl esters, oxidation of these esters to the primary active metabolite RA, and degradation of RA to inactive metabolites.

Briefly, the oxidation of ROL to RA requires two sequential steps catalyzed by ROL dehydrogenases (RDHs) or alcohol dehydrogenases and retinaldehyde dehydrogenases. RA can then bind and activate cells through two families of nuclear receptors, the RARs and the retinoid X receptors (RXRs). RA in tissue undergoes oxidative degradation by the cytochrome P450 enzyme CYP26 family [33]. In the variety of systems studied thus far, RA appears to function in a paracrine manner. One cell type appears to control the storage and oxidation of ROL, while the target cell is nearby but of a different cell type [33]. The half-life of RA in tissue has been estimated to be about 30 min.

In the developing murine testis, the components for RA biosynthesis, action, and degradation have been found in many different cell types. Using several approaches, such as immunohistochemistry and *in situ* hybridization coupled with cell-specific gene knockouts, investigators have reported the localization and physiological importance of many of the components. More recently, the use of RiboTag mice has produced additional evidence supporting the localization of specific components [34,35].

When RiboTag mice are crossed with the appropriate Cre-expressing mice, the ribosomes in the target cells are tagged and can be isolated with antibodies to the tag. Expression arrays then provide evidence regarding which RNA sequences are present on polysomes in the target cells.

In all of these approaches from immunohistochemistry to RiboTag mice, the data are limited by the sensitivity of the assay and by the large redundancy in the RA metabolic and activation pathway. Therefore, the interpretation of the retinoid signaling pathway, shown in Figure 8.4, must be considered in light of these limitations. Even though a component, such as retinaldehyde dehydrogenase (ALDH) 1a1, can be localized by these techniques to the Sertoli cells, there could be low levels of enzyme in germ cells, and enzymes that are unknown to be part of the RA pathway may even play a role. In addition, the expression and localization of metabolic components may change throughout testis development.

ROL circulates in the form of a complex of ROL binding protein 4-transthyretin (ROL-RBP4-TTR). This protein is generally stored in the liver, and transthyretin makes the complex large enough to avoid being filtered out by the kidneys [36]. The only putative receptor for the ROL-RBP4-TTR complex that is expressed in the testis is STRA6, which is expressed primarily in the Sertoli cells [34,37–39]. It has been shown that STRA6 interacts with lecithin ROL transferase (LRAT) and cellular retinol binding protein 1 and promotes bidirectional transport of ROL [40]. However, knockout of the *Stra6* gene does not affect levels of ROL to a measurable extent in the testis, and the function of STRA6 may involve intracellular signaling [41].

Once inside the Sertoli cells, ROL can be stored in the form of retinyl esters in a reaction catalyzed by LRAT [42]. During the initiation of spermatogenesis, the first step in the synthesis of active RA apparently occurs via RDH10, which is localized primarily in the Sertoli cells [43]. If RDH10 is knocked out in Sertoli cells but not in germ cells, there is a depletion of germ cells. A deficiency of RDH10 in both Sertoli cells and germ cells resulted in a phenotype similar to in VAD animals—a complete blockage of germ cell development. As these deficient mice achieved maturity, their spermatogenesis recovered, suggesting a developmental change in either the source of RA or in the gene responsible for the initial oxidation of ROL

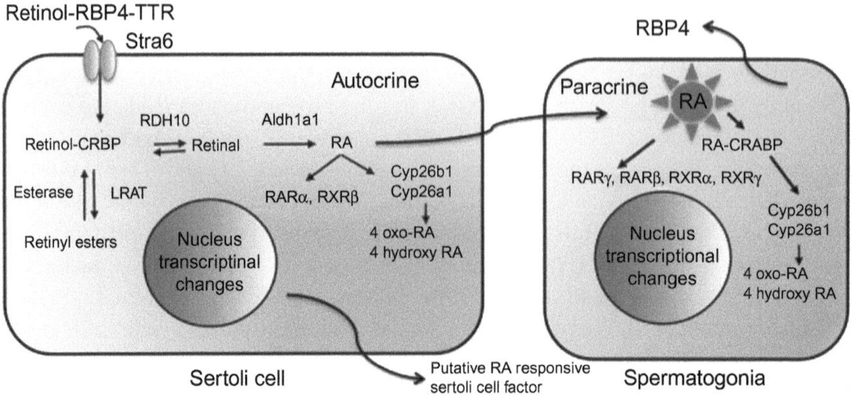

Figure 8.4 Possible pathway for the flow of retinoids during the initiation of spermatogenesis. The cellular localizations of the enzymes and binding proteins are based on immunocytochemistry and expression arrays of RiboTag mice. Gene knockouts of components in red result in severe a phenotype. Abbreviations: RBP4-TTR, retinol binding protein 4, transthyretin complex; STRA6, stimulated by RA gene 6 cell membrane receptor; LRAT, lecithin retinol transferase; CRBP, cellular retinol binding protein; RDH10, retinol dehydrogenase 10; ALDH1a1, aldehyde dehydrogenase 1a1; RA, retinoic acid; CYP26, cytochrome P450 enzymes from the cyp26 family; RAR, retinoic acid receptor; RXR, rexinoid receptor; CRABP, cellular retinoic acid receptor; RBP4, retinol binding protein 4. (For interpretation of the references to color in this figure legend, the reader is referred to the web version of this book.)

to retinal (RAL). The primary enzyme responsible for the conversion of RAL to RA that is present in the Sertoli cells is ALDH1a1 [29,34].

When all three known ALDHs are deficient in Sertoli cells, the conversion of spermatogonia A to spermatogonia A1 is blocked [44]. Once RA is produced, it can interact with RARα and RXRβ, which are the primary receptors in the Sertoli cells, or it can be shuttled to spermatogonia to initiate the transition of A spermatogonia to A1 spermatogonia [29,45–47]. When a dominant-negative RAR receptor that dimerized with all RAR and RXR monomers but could not bind to DNA was expressed specifically in Sertoli cells, spermatogenesis was initiated normally but could not be sustained [48]. This result supports the concept that RA action in Sertoli cells is not necessary during the initiation of spermatogenesis but is necessary for its maintenance. It may be that components of the structural support of Sertoli cells, such as tight junctions, require RA action in Sertoli cells. If the RA is shuttled to spermatogonia, it can interact with CRABP or with RARγ and several RXRs [29,34,49].

It appears that CRABPI, which is found exclusively in germinal cells, binds RA and delivers it to the Cyp26 enzymes for degradation. As mentioned previously, the half-life of RA is very short (around 30 min), and the CYP26 family of P450 enzymes appears to be responsible for the degradation of active RA to 4-oxo-RA and 4-hydroxy-RA [50]. For a review on the degradation of RA, see Ref. [51].

Finally, there is evidence that the spermatogonia in the developing testis also make RBP4, which may be useful in the transport of RA between cells [34].

In a comprehensive review, Duester [33] discussed evidence showing that in most cases, RA acts via a paracrine mechanism through which the RA is synthesized by one cell type and acts on an adjacent cell type. In the testis, one hypothesis is that Sertoli cells could be the source of RA and the target cells are undifferentiated germ cells.

A few studies have tested various aspects of this hypothesis, and these authors have come to varying conclusions. Using a combination of cell-specific gene knockouts and pharmacological approaches, Raverdeau and colleagues concluded that while the initiation of spermatogenesis is dependent on the synthesis of RA in Sertoli cells, it is a cell-autonomous event [44]. These investigators deleted genes for ALDH1A1, ALDH1A2, and ALDH1A3 specifically in Sertoli cells and found that spermatogonia never underwent the A-to-A1 transition that is characteristic of spermatogonial differentiation. They then used agonists of RARα and RARγ in these mice to show that the differentiation of spermatogonia only occurred when the RARα agonist was used. Because these investigators had reported previously that RARα was exclusive to Sertoli cells and RARγ was exclusive to germ cells, they concluded that autocrine activation of the RARα in Sertoli cells was all that was necessary to drive spermatogonial differentiation during the first wave [45,49].

In contrast, when a similar experiment was done on VAD mice, the RARα-specific agonist had no effect, suggesting the need for paracrine action by RA synthesized by Sertoli cells. The investigators reconciled these findings by suggesting that in the first wave of spermatogenesis, the signal committing prospermatogonia to differentiation depends on a Sertoli-cell-autonomous activation of RARα, but undifferentiated spermatogonia also can rely on either ROL or RA from other sources to enter spermatogenesis. Isolated undifferentiated spermatogonia respond in culture to either added RA or ROL with increased STRA8 expression as well as other markers of spermatogonial differentiation [9]. The conclusion that RA acting on but not coming from Sertoli cells (an autocrine process) was necessary to initiate spermatogonial differentiation was dependent on the pharmacological specificity of the agonists and will require further investigation.

Tong and colleagues knocked out RDH10 in Sertoli cells and germ cells of the murine testis [43]. They chose RDH10 from many possible RDHs based on its expression profile in the testis. These investigators found that the knockout of RDH10 in germ cells had little effect, but that if the knockout was done in Sertoli cells, there was a complete blockage of spermatogonial differentiation. If the RDH10 was knocked out in both germ cells and Sertoli cells, there was a block in germ cell differentiation similar to that seen in vitamin A deficiency. On the basis of this study, the investigators concluded that in juvenile mice, Sertoli cells are the primary source of the RA that acts in a paracrine manner on germ cells.

In the studies by Raverdeau and colleagues [44] and Tong and colleagues [43] in which the ability of Sertoli cells to synthesize RA was compromised, germ cell development was blocked at the transition of A spermatogonia to A1 spermatogonia. In both studies, a single injection of RA resulted in the completion and maintenance of spermatogenesis. These results suggest that once initiated, the continual cyclic process of spermatogenesis can be maintained in the absence of RA from Sertoli cells.

Advanced germ cells, especially spermatocytes, have the metabolic enzymes necessary to synthesize RA and are the likely source of RA for the maintenance of spermatogenesis [52]. It therefore appears that the initiation of spermatogenesis and the maintenance of the cycle of the seminiferous epithelium both require RA but possibly from different sources. In mice, it is clear that RA is necessary to move the A spermatogonia into the A1 differentiation pathway at stage VIII of the cycle. The maintenance of the cycle could then depend on RA from advanced germ cells that also appear in stage VIII. Once established, the cycle would continue. But if advanced germ cells were lost, the cycle would require reinitiation by RA from Sertoli cells or an exogenous source.

VI. Extrinsic versus intrinsic factors

The induction of STRA8 by RA in A spermatogonia and preleptotene spermatocytes is an excellent marker for the action of RA and the commitment of cells to meiosis. The action of RA on germ cells is not confined to the induction of STRA8 and is not sufficient to trigger the commitment to meiosis in cells other than germ cells. Thus, the importance of intrinsic cellular factors, such as the RNA binding protein DAZL, must be considered [23,53]. Germ cells that lack the *DAZL* gene cannot respond to RA by inducing STRA8 and cannot initiate meiosis. DAZL may function as a competence factor because germ cells in *Dazl* knockout mice maintain the primordial-germ-cell-like morphology [54]. *Doublesex and maleabnormal-3 related transcription factor* (*DMRT1*) also appears to regulate meiosis. In the postnatal murine testis, where DMRT1 expression is eliminated in germ cells, there is an increase in STRA8 expression, suggesting that the role of Dmrt1 is to oppose the action of RA [55–57]. Another proposed inhibitor of RA action is fibroblast growth factor 9 (FGF9) [58]. The addition of exogenous FGF9 to cultured germ cells from ovaries or testes suppresses the expression of STRA8 [54].

References

[1] Skinner M, Griswold M, editors. Sertoli Cell Biology. London: Elsevier Academic Press; 2005. p. 494.

[1a] Morales CR, Griswold MD. Retinol induces stage synchronization in seminiferous tubules of vitamin A deficient rats. Ann N Y Acad Sci 1987;513:292–3.

[2] Haneji T, Maekawa M, Nishimune Y. *In vitro* differentiation of Type A spermatogonia from mouse cryptorchid testes in serum-free media. Biol Reprod 1983;28:1217–23.

[3] Haneji T, Maekawa M, Nishimune Y. Retinoids induce differentiation of type A spermatogonia *in vitro*: organ culture of mouse cryptorchid testes. J Nutr 1983;113:1119–23.

[4] Koubova J, Menke DB, Zhou Q, Capel B, Griswold MD, Page DC. Retinoic acid regulates sex-specific timing of meiotic initiation in mice. Proc Natl Acad Sci USA 2006;103:2474–9.

[5] de Rooij DG, Griswold MD. Questions about spermatogonia posed and answered since 2000. J Androl 2012;33:1085–95.
[6] de Rooij DG, van Dissel Emiliani FM, van Pelt AM. Regulation of spermatogonial proliferation. Ann N Y Acad Sci 1989;564:140–53.
[7] Anderson EL, Baltus AE, Roepers-Gajadien HL, Hassold TJ, de Rooij DG, van Pelt AM, et al. Stra8 and its inducer, retinoic acid, regulate meiotic initiation in both spermatogenesis and oogenesis in mice. Proc Natl Acad Sci USA 2008;105:14976–80.
[8] Zhou Q, Li Y, Nie R, Friel P, Mitchell D, Evanoff RM, et al. Expression of stimulated by retinoic acid gene 8 (*Stra8*) and maturation of murine gonocytes and spermatogonia induced by retinoic acid *in vitro*. Biol Reprod 2008;78:537–45.
[9] Zhou Q, Nie R, Li Y, Friel P, Mitchell D, Hess RA, et al. Expression of stimulated by retinoic acid gene 8 (*Stra8*) in spermatogenic cells induced by retinoic acid: an *in vivo* study in vitamin A-sufficient postnatal murine testes. Biol Reprod 2008;79:35–42.
[10] Griswold MD, Hogarth CA, Bowles J, Koopman P. Initiating meiosis: the case for retinoic acid. Biol Reprod 2012;86:35.
[11] Griswold MD, Bishop PD, Kim KH, Ping R, Siiteri JE, Morales C. Function of vitamin A in normal and synchronized seminiferous tubules. Ann N Y Acad Sci 1989;564:154–72.
[12] Huang HF, Hembree WC. Spermatogenic response to vitamin A in vitamin A deficient rats. Biol Reprod 1979;21:891–904.
[13] Morales C, Griswold MD. Retinol-induced stage synchronization in seminiferous tubules of the rat. Endocrinology 1987;121:432–4.
[14] Thomson JN, Howell JM, Pitt GAJ. Vitamin A and reproduction in rats. Proc R Soc Lond (Biol) 1964;159:510–35.
[15] van Pelt AM, de Rooij DG. Synchronization of the seminiferous epithelium after vitamin A replacement in vitamin A-deficient mice. Biol Reprod 1990;43:363–7.
[16] van Pelt AM, de Rooij DG. Retinoic acid is able to reinitiate spermatogenesis in vitamin A-deficient rats and high replicate doses support the full development of spermatogenic cells. Endocrinology 1991;128:697–704.
[17] Bowles J, Knight D, Smith C, Wilhelm D, Richman J, Mamiya S, et al. Retinoid signaling determines germ cell fate in mice. Science 2006;312:596–600.
[18] MacLean G, Li H, Metzger D, Chambon P, Petkovich M. Apoptotic extinction of germ cells in testes of Cyp26b1 knockout mice. Endocrinology 2007;148:4560–7.
[19] Trautmann E, Guerquin MJ, Duquenne C, Lahaye JB, Habert R, Livera G. Retinoic acid prevents germ cell mitotic arrest in mouse fetal testes. Cell Cycle 2008;7:656–64.
[20] Mark M, Jacobs H, Oulad-Abdelghani M, Dennefeld C, Feret B, Vernet N, et al. STRA8-deficient spermatocytes initiate, but fail to complete, meiosis and undergo premature chromosome condensation. J Cell Sci 2008;121:3233–42.
[21] Bouillet P, Oulad-Abdelghani M, Vicaire S, Garnier JM, Schuhbaur B, Dolle P, et al. Efficient cloning of cDNAs of retinoic acid-responsive genes in P19 embryonal carcinoma cells and characterization of a novel mouse gene, Stra1 (mouse LERK-2/Eplg2). Dev Biol 1995;170:420–33.
[22] Oulad-Abdelghani M, Bouillet P, Decimo D, Gansmuller A, Heyberger S, Dolle P, et al. Characterization of a premeiotic germ cell-specific cytoplasmic protein encoded by *Stra8*, a novel retinoic acid-responsive gene. J Cell Biol 1996;135:469–77.
[23] Lin Y, Gill ME, Koubova J, Page DC. Germ cell-intrinsic and -extrinsic factors govern meiotic initiation in mouse embryos. Science 2008;322:1685–7.
[24] Guerquin MJ, Duquenne C, Lahaye JB, Tourpin S, Habert R, Livera G. New testicular mechanisms involved in the prevention of fetal meiotic initiation in mice. Dev Biol 2010;346:320–30.

[25] Snyder EM, Small C, Griswold MD. Retinoic acid availability drives the asynchronous initiation of spermatogonial differentiation in the mouse. Biol Reprod 2010;83:783–90.

[26] Hogarth CA, Evanoff R, Snyder E, Kent T, Mitchell D, Small C, et al. Suppression of Stra8 expression in the mouse gonad by WIN 18,446. Biol Reprod 2011;84:957–65.

[27] Boulogne B, Levacher C, Durand P, Habert R. Retinoic acid receptors and retinoid X receptors in the rat testis during fetal and postnatal development: immunolocalization and implication in the control of the number of gonocytes. Biol Reprod 1999;61:1548–57.

[28] Dufour JM, Kim KH. Cellular and subcellular localization of six retinoid receptors in rat testis during postnatal development: identification of potential heterodimeric receptors. Biol Reprod 1999;61:1300–8.

[29] Vernet N, Dennefeld C, Rochette-Egly C, Oulad-Abdelghani M, Chambon P, Ghyselinck NB, et al. Retinoic acid metabolism and signaling pathways in the adult and developing mouse testis. Endocrinology 2006;147:96–110.

[30] Bellvé AR, Cavicchia JC, Millette CF, O'Brien DA, Bhatnagal YM, Dym M. Spermatogenic cells of the prepuberal mouse: isolation and morphological characterization. J Cell Biol 1977;74:68–85.

[31] Rossant J, Zirngibl R, Cado D, Shago M, Giguere V. Expression of a retinoic acid response element-*hsplacZ* transgene defines specific domains of transcriptional activity during mouse embryogenesis. Genes Dev 1991;5:1333–44.

[32] Davis JC, Snyder EM, Hogarth CA, Small C, Griswold MD. Induction of spermatogenic synchrony by retinoic acid in neonatal mice. Spermatogenesis 2013;3:e23180.

[33] Duester G. Retinoic acid synthesis and signaling during early organogenesis. Cell 2008;134:921–31.

[34] Evans E, Hogarth C, Mitchell D, Griswold M. Riding the spermatogenic wave: profiling gene expression within neonatal germ and Sertoli cells during a synchronized initial wave of spermatogenesis. Biol Reprod 2014;90 [Epub 2014 Apr 9].

[35] Sanz E, Evanoff R, Quintana A, Evans E, Miller JA, Ko C, et al. RiboTag analysis of actively translated mRNAs in Sertoli and Leydig cells *in vivo*. PloS One 2013;8: e66179.

[36] Bellovino D, Morimoto T, Tosetti F, Gaetani S. Retinol binding protein and transthyretin are secreted as a complex formed in the endoplasmic reticulum in HepG2 human hepatocarcinoma cells. Exp Cell Res 1996;222:77–83.

[37] Bouillet P, Sapin V, Chazaud C, Messaddeq N, Decimo D, Dolle P, et al. Developmental expression pattern of Stra6, a retinoic acid-responsive gene encoding a new type of membrane protein. Mech Dev 1997;63:173–86.

[38] Kawaguchi R, Zhong M, Kassai M, Ter-Stepanian M, Sun H. STRA6-catalyzed vitamin A influx, efflux, and exchange. J Membr Biol 2012;245:731–45.

[39] Pasutto F, Sticht H, Hammersen G, Gillessen-Kaesbach G, Fitzpatrick DR, Nurnberg G, et al. Mutations in STRA6 cause a broad spectrum of malformations including anophthalmia, congenital heart defects, diaphragmatic hernia, alveolar capillary dysplasia, lung hypoplasia, and mental retardation. Am J Hum Genet 2007;80:550–60.

[40] Isken A, Golczak M, Oberhauser V, Hunzelmann S, Driever W, Imanishi Y, et al. RBP4 disrupts vitamin A uptake homeostasis in a STRA6-deficient animal model for Matthew-Wood syndrome. Cell Metab 2008;7:258–68.

[41] Berry DC, Jacobs H, Marwarha G, Gely-Pernot A, O'Byrne SM, DeSantis D, et al. The STRA6 receptor is essential for retinol-binding protein-induced insulin resistance but not for maintaining vitamin A homeostasis in tissues other than the eye. J Biol Chem 2013;288:24528–39.

[42] Bishop PD, Griswold MD. Uptake and metabolism of retinol in cultured Sertoli cells: evidence for a kinetic model. Biochemistry 1987;26:7511–8.
[43] Tong MH, Yang QE, Davis JC, Griswold MD. Retinol dehydrogenase 10 is indispensible for spermatogenesis in juvenile males. Proc Natl Acad Sci USA 2013;110:543–8.
[44] Raverdeau M, Gely-Pernot A, Feret B, Dennefeld C, Benoit G, Davidson I, et al. Retinoic acid induces Sertoli cell paracrine signals for spermatogonia differentiation but cell autonomously drives spermatocyte meiosis. Proc Natl Acad Sci USA 2012;109:16582–7.
[45] Vernet N, Dennefeld C, Guillou F, Chambon P, Ghyselinck NB, Mark M. Prepubertal testis development relies on retinoic acid but not rexinoid receptors in Sertoli cells. EMBO J 2006;25:5816–25.
[46] Vernet N, Dennefeld C, Klopfenstein M, Ruiz A, Bok D, Ghyselinck NB, et al. Retinoid X receptor beta (RXRB) expression in Sertoli cells controls cholesterol homeostasis and spermiation. Reproduction 2008;136:619–26.
[47] Kim KH, Griswold MD. The regulation of retinoic acid receptor mRNA levels during spermatogenesis. Mol Endocrinol 1990;4:1679–88.
[48] Tong M, Hogarth C, Griswold M. RA signaling cell autonomously drives spermatogonial differentiation 2014, unpublished observations.
[49] Gely-Pernot A, Raverdeau M, Celebi C, Dennefeld C, Feret B, Klopfenstein M, et al. Spermatogonia differentiation requires retinoic acid receptor gamma. Endocrinology 2012;153:438–49.
[50] Fiorella PD, Napoli JL. Microsomal retinoic acid metabolism. Effects of cellular retinoic acid-binding protein (type I) and C18-hydroxylation as an initial step. J Biol Chem 1994;269:10538–44.
[51] Kedishvili NY. Enzymology of retinoic acid biosynthesis and degradation. J Lipid Res 2013;54:1744–60.
[52] Sugimoto R, Nabeshima Y, Yoshida S. Retinoic acid metabolism links the periodical differentiation of germ cells with the cycle of Sertoli cells in mouse seminiferous epithelium. Mech Dev 2012;128:610–24.
[53] Feng CW, Bowles J, Koopman P. Control of mammalian germ cell entry into meiosis. Mol Cell Endocrinol 2014;382:488–97.
[54] Gill ME, Hu YC, Lin Y, Page DC. Licensing of gametogenesis, dependent on RNA binding protein DAZL, as a gateway to sexual differentiation of fetal germ cells. Proc Natl Acad Sci USA 2011;108:7443–8.
[55] Krentz AD, Murphy MW, Zhang T, Sarver AL, Jain S, Griswold MD, et al. Interaction between DMRT1 function and genetic background modulates signaling and pluripotency to control tumor susceptibility in the fetal germ line. Dev Biol 2013;377:67–78.
[56] Matson CK, Murphy MW, Sarver AL, Griswold MD, Bardwell VJ, Zarkower D. DMRT1 prevents female reprogramming in the postnatal mammalian testis. Nature 2011;476:101–4.
[57] Matson CK, Murphy MW, Griswold MD, Yoshida S, Bardwell VJ, Zarkower D. The mammalian doublesex homolog DMRT1 is a transcriptional gatekeeper that controls the mitosis versus meiosis decision in male germ cells. Dev Cell 2010;19:612–24.
[58] Bowles J, Feng CW, Spiller C, Davidson TL, Jackson A, Koopman P. FGF9 suppresses meiosis and promotes male germ cell fate in mice. Dev Cell 2010;19:440–9.

Retinoic acid metabolism, signaling, and function in the adult testis

9

Cathryn Hogarth
School of Molecular Biosciences and the Center for Reproductive Biology, Washington State University, Pullman, WA

I. Introduction

The active metabolite of vitamin A, retinoic acid (RA), is absolutely required for the production of fertile sperm throughout the reproductive life of mammalian males. The importance of vitamin A to male fertility was first described in 1925, when male rats were found to be infertile following the removal of vitamin A from their diets [1]. Studies performed in the 1980s and 1990s determined that only Sertoli cells and undifferentiated spermatogonia remain within the seminiferous epithelium of vitamin-A-deficient (VAD) rodent testes [2–5]. Since then, our understanding of the regulation of spermatogenesis by vitamin A has grown substantially. We now know that vitamin A must be converted to RA within the testis to exert its effects, which include regulating several key developmental events during spermatogenesis and both initiating and maintaining the cyclic nature of this process. The enzymes and receptors responsible for RA synthesis, metabolism, and signaling are well characterized, and extensive localization studies have shed some light on which cell types have the ability to generate, degrade, and/or respond to RA. However, we still understand very little about which of these proteins, and which testicular cell types, are responsible for normal RA function and what the direct outcomes of RA signaling are within the adult testis. Chapter 8 focused on how RA production and signaling within Sertoli and germ cells drive the initiation of spermatogenesis. This chapter extends this discussion to the adult testis and the maintenance of the cycle of the seminiferous epithelium and reviews what is currently known about how vitamin A affects Sertoli cell development and function directly.

II. RA synthesis, signaling, and degradation

Vitamin A was first discovered in the early twentieth century and is now known to be essential for mammalian development and the health of many different organs and biological systems. Since the early 2000s, major steps forward have been taken toward understanding how dietary retinoids are taken up via the gut, carried within

Sertoli Cell Biology. DOI: http://dx.doi.org/10.1016/B978-0-12-417047-6.00009-0

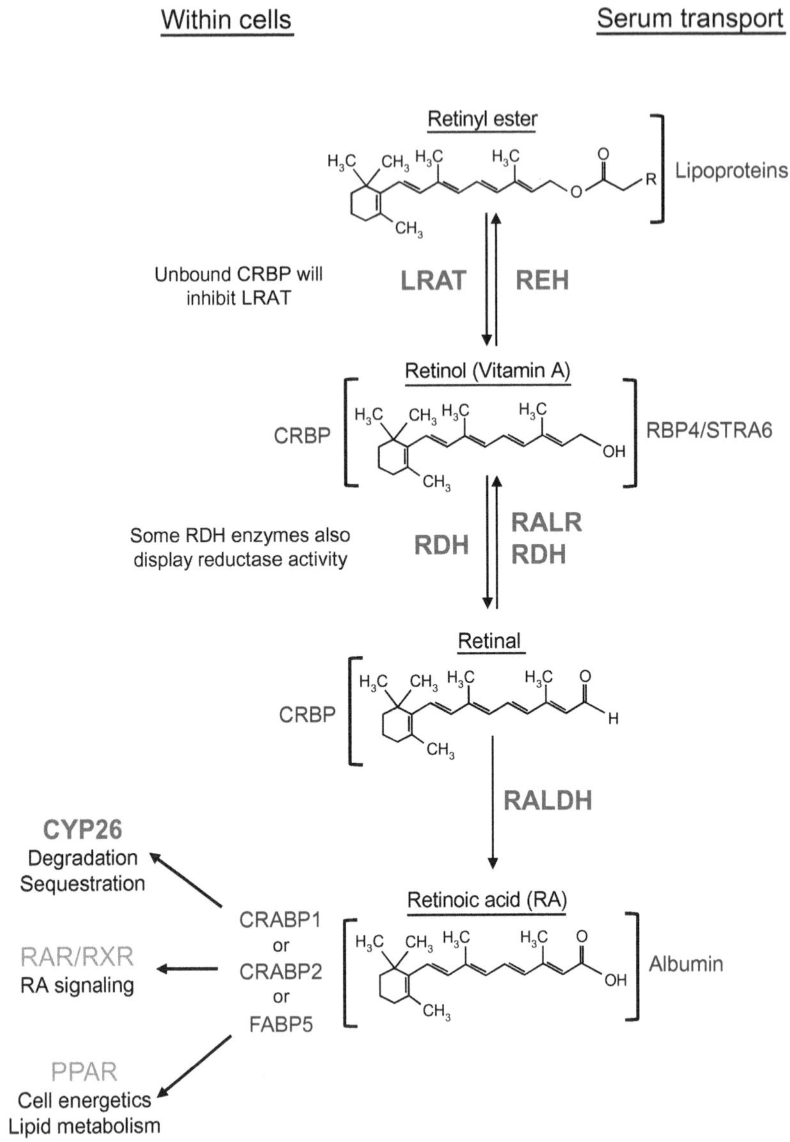

Figure 9.1 Mammalian vitamin A metabolism and signaling. Schematic of the enzymes (red) and transport (blue) and signaling proteins (green) involved in metabolizing vitamin A in mammalian cells. Dietary vitamin A is transported through the circulation as retinol bound to retinol-binding protein 4 (RBP4), and its uptake into cells is facilitated by an interaction between RBP4 and the transmembrane receptor stimulated by retinoic acid (STRA) 6. Cells can also gain their required retinoids via spontaneous transfer of free retinol across the membrane, uptake of retinyl esters transported in complex with lipoproteins, and/or uptake of retinoic acid (RA) bound to albumin. Once inside cells, retinol is bound by cellular RBPs (CRBPs) and follows one of two fates: (1) it is converted to retinyl esters by lecithin:retinol

the circulation, and delivered to different organs within the body (reviewed in Ref. [6]). However, very little is known about the molecular events that regulate retinoid storage, metabolism, and signaling within specific cell types.

Figure 9.1 offers an overview of our current understanding of vitamin A metabolism, signaling, binding, and storage in mammalian cells. The large majority of dietary vitamin A (more than 95%) is transported through the circulation as retinol bound to retinol-binding protein 4 (RBP4), and its uptake into cells is facilitated by an interaction between RBP4 and the transmembrane receptor stimulated by RA (STRA) 6. There is evidence to indicate that cells within the testis can also gain their required retinoids via spontaneous transfer of free retinol across the membrane and uptake of retinyl esters (transported in complexes with lipoproteins) and/or RA (bound to albumin) [7]. These alternative modes are sufficient to deliver enough retinoid for normal male germ cell development because male mice lacking either RBP4 [8] or STRA6 [9] are fertile. Once inside cells, retinol is bound by cellular retinol-binding proteins (CRBPs) and follows one of two fates: (1) it is converted to retinyl esters by lecithin:retinol acyltransferase (LRAT), or (2) it is oxidized to RA via a two-step enzymatic process. The storage versus oxidation decision is chaperoned by CRBP because unbound CRBP can inhibit LRAT activity, thereby signaling to the cell that retinoid levels are low and promoting RA synthesis [10].

The conversion of retinol to retinal, the first of two dehydrogenation steps required to synthesize RA, is rate limiting, takes place mostly within microsomes and is catalyzed by the retinol dehydrogenases (RDHs). The second step is catalyzed by the retinaldehyde dehydrogenases (RALDHs), occurs rapidly (relative to the dehydrogenation of retinol), and is irreversible. In addition to retinol, the CRBPs bind retinal, and *in vitro* studies have demonstrated that retinal bound to CRBP is recognized more readily by the RALDHs than free retinal alone [11].

◀ acyltransferase (LRAT) or (2) it is oxidized to RA via a two-step enzymatic process. The storage-versus-oxidation decision is chaperoned by CRBP because unbound CRBP can inhibit LRAT activity, and the mobilization of cellular retinyl ester stores is driven by the activity of retinyl ester hydrolase (REH). In addition to retinol, the CRBPs also bind retinal within cells. The conversion of retinol to retinal is reversible, with the dehydrogenation reaction catalyzed by the retinol dehydrogenases (RDHs) and the reduction reaction catalyzed by the retinal reductases. Interestingly, some RDH enzymes can also display reductase activity and can convert retinal back to retinol. The retinal-to-RA step is irreversible and regulated by the retinaldehyde dehydrogenases (RALDHs). Once generated, the function of RA within cells is directed by the fatty-acid-binding protein (FABP) family member that it is bound by. RA bound to CRABP2 is directed toward the RA receptors (RARs) and can therefore drive classical RA signaling. CRABP1 possibly targets RA for degradation by directing it toward the cytochrome p450, family 26 (CYP26) family of enzymes or sequesters RA to prevent its association with RARs. FABP5 transfers RA to the peroxisome proliferator activated receptor (PPAR) β/δ as a means of promoting cell survival and regulating cell energetics and lipid metabolism. (For interpretation of the references to color in this figure legend, the reader is referred to the web version of this book.)

Once generated, the function of RA within cells is directed by the fatty-acid-binding protein (FABP) family member that it binds to, which is either the cellular RA binding proteins (CRABPs) or FABP5 [12]. RA bound to CRABP2 is directed toward the RA receptors (RARs) and can therefore drive classical RA signaling [13]. This signaling is characterized by the binding of RA-bound RAR to a retinoid X receptor (RXR) and the interaction of the resulting heterodimer with DNA via RA response elements (RAREs) within gene promoters or association with chromatin-bound proteins to exert effects on gene expression (for a recent review, see Ref. [14]).

The second of the two CRABP family members, CRABP1, does not appear to facilitate the delivery of RA to RAR but, instead, possibly targets RA for degradation or sequesters RA and prevents its association with RARs [15,16]. In contrast, FABP5 transfers RA to the peroxisome proliferator activated receptor (PPAR) β/δ as a means of promoting cell survival and regulating cell energetics and lipid metabolism [17,18]. There is some evidence to imply that RA can signal within the cell that is responsible for its synthesis [19], yet most research indicates that RA acts in a paracrine manner [20].

Finally, RA signaling is stopped or, in some cases, prevented [21,22] by the action of the cytochrome p450, family 26 (CYP26) degradation enzymes. There are three family members—CYP26A1, CYP26B1, and CYP26C1—and all three display widespread tissue distribution [23]. Global deletions of either CYP26A1 or CYP26B1 result in embryonic or early postnatal lethality [24,25], indicating the importance of these enzymes to RA homeostasis.

III. Components of the RA pathway within the adult testis

Microarray, *in situ* hybridization, and immunohistochemistry studies indicate that multiple cell types within the adult testis are capable of performing all aspects of retinoid metabolism and signaling (summarized in Table 9.1). In the following sections, we review our current understanding of which retinoid metabolizing and signaling proteins are likely to play a key role in adult testis function.

A. RA synthesis and degradation

Very little is currently known about the expression and function of the RDH enzymes within the testis. Of the numerous alcohol dehydrogenase enzymes expressed in mammalian cells, only three—RDH1, RDH10, and dehydrogenase/reductase (SDR family) member 9 (DHRS9)—are thought to oxidize the biologically active form of retinol, all-*trans* retinol, that is believed to be important for testis function [7]. However, it is highly likely that other forms that are important for testis function remain to be discovered. mRNAs encoding two of the three forms, *Rdh1* and *Rdh10*, have been detected within Sertoli cells of rodent testis [26,27],

Table 9.1 Localization of vitamin A metabolizing, signaling, binding and storage proteins in the mammalian testis

Gene	Experiment	Cell type	References
Synthesis			
Rdh1	ISH	Sertoli cells	[26]
Rdh10	ISH	Sertoli cells Spermatogonia of the juvenile testis	[26,27,36]
Aldh1a1	ISH	Leydig cells, spermatogonia, spermatocytes, spermatids	[28–30,36]
Aldh1a2	ISH Western blotting	Spermatogonia, preleptotene spermatocytes, pachytene spermatocytes, spermatids, Sertoli cells of the 5 dpp testis Purified preleptotene spermatocytes	[19,29,30,34]
Aldh1a3	ISH	Leydig cells	[29]
Degradation			
Cyp26a1	ISH	Peritubular myoid cells, Sertoli cells	[29,34,36]
Cyp26b1	ISH and IHC	Peritubular myoid cells, spermatogonia	[29,39]
Signaling			
Rarα	ISH and IHC	Sertoli cells, spermatocytes, spermatids, purified spermatogonia	[29,40,45]
Rarβ	ISH and IHC	Round spermatids, Sertoli cells	[29,53]
Rarγ	IHC	Undifferentiated spermatogonia	[48]
Rxrα	ISH and IHC	Round spermatids, most germ cells	[29,53]
Rxrβ	ISH and IHC	Sertoli cells	[29,40]
Rxrγ	ISH and IHC	Round spermatids, most germ cells	[29,53]
Pparβ/δ	ISH and IHC	Sertoli cells, Leydig cells	[54,55]
Fabp5	ISH	Sertoli cells, Leydig cells	[56]
Binding and storage			
Crbp1	IHC	Sertoli cells, spermatogonia, spermatocytes	[26,29,60,61]
Crabp1	ISH	Spermatogonia	[62,70]
Crabp2	IHC	Sertoli cells	[62]
Lrat	ISH, primary Sertoli cells cultures	Sertoli cells	[34,69]
Stra6	ISH	Sertoli cells	[34]

with *Rdh10* also present in spermatogonia of the juvenile testis [27]. The Sertoli-cell-specific deletion of *Rdh10* induces errors in the initiation of spermatogenesis in neonatal mice [27] (see Chapter 8 for further discussion). However, the testes of these mice recover so that spermatogenesis proceeds normally by adulthood, indicating that while RDH10 is important for spermatogenesis in the juvenile testis, subsequent waves of spermatogenesis can be initiated and completed in its absence.

Significantly more information is available regarding the expression and function of the RALDHs within the testis. The mRNA localization pattern of three of the four ALDH enzymes capable of oxidizing retinaldehyde—*Aldh1a1*, *Aldh1a2*, and *Aldh1a3*—has been described for the developing and adult testis. Conflicting localization patterns have been published for *Aldh1a1* because it was originally reported to be Leydig cell specific [28], yet more recent studies have also detected transcript in Sertoli cells [29], spermatogonia, spermatocytes, and spermatids [30]. Interestingly, global deletion of *Aldh1a1* in mice results in no phenotypic abnormalities, indicating that this enzyme is dispensable for male fertility. Yet, when all three ALDH1 isoforms are deleted from Sertoli cells, the initiation of spermatogenesis is impaired (as discussed further in Chapter 8), and errors associated with spermiogenesis occur [19] (discussed further in Section V.B). These observations indicate that compensation among the different ALDH1 enzymes takes place and that ALDH1 activity within Sertoli cells is essential for the normal progression of spermatogenesis.

ALDH1A2 is of interest as a target for male contraceptive development. The authors of three different studies have reported that *Aldh1a2* mRNA is highly enriched in the testis compared to other organs in both mice and men [31–33]. *Aldh1a2* transcript localization in the mouse testis has been investigated in three separate studies. In these studies, mRNA was present in spermatogonia and spermatocytes [30], in preleptotene and pachytene spermatocytes and spermatids in the adult testis [29,34], but only in Sertoli cells at 5 days postpartum (dpp) in one study [29]. *Aldh1a3* has only been detected in Leydig cells [29], and the fourth enzyme, *Aldh8a1*, is known to be present in the testis [31]; however, its cellular localization pattern has yet to be determined.

Within cells, the degradation of RA is controlled by CYP26 enzymes. All three isoforms, CYP26A1, CYP26B1, and CYP26C1, are present in the testis [35,36]. The majority of the information available demonstrating the importance of these enzymes to testicular function stems from studies of the embryonic gonad. Multiple published reports involving a combination of localization studies, gene knockouts, and chemical inhibitors indicate that CYP26B1 activity in Sertoli cells of the embryonic testis is essential to inhibit male germ cells from prematurely entering meiosis, whereas the absence of this enzyme in the embryonic ovary allows for oogonia to begin the meiotic prophase [21,22,25,37]. Analyses of testes isolated from mice lacking CYP26B1 activity only in Sertoli cells revealed that while some sperm were present, spermatogenesis did not proceed normally, and some testis tubules were devoid of advanced germ cells [38]. This observation suggests that compensation occurs between CYP26 isoforms within Sertoli cells to guarantee that RA is degraded, which allows normal spermatogenesis to take place. Interestingly, transcripts for either *Cyp26a1* or *Cyp26b1* were not detectable in Sertoli cells in *in situ* hybridization studies but were present within the peritubular myoid cells [29], and CYP26B1 has been localized to spermatogonia in the neonatal testis [39]. Further protein-based localization studies of the developing and adult mouse testis as well as elimination of these enzymes from the different cell types of the testis will be necessary to advance our understanding of how RA levels are regulated within the seminiferous epithelium.

B. RA signaling

There are multiple receptors for RA within the adult testis. RARα and RXRβ mRNA and protein, which are members of the classical RA signaling pathway, have been detected in Sertoli cells in both the developing and adult rodent testis [29,40]. Null mutations in either of these genes (both global and Sertoli cell specific) lead to male infertility (discussed further in Section V) [41–44].

Whether RARα is present in germ cells has been a topic of extensive debate. Studies in the rat testis have localized *Rarα* mRNA and protein to round and elongated spermatids in the adult testis and to meiotic prophase germ cells and Sertoli cells in juvenile animals [40]. *Rarα* mRNAs have been detected in cultures of purified mouse spermatogonia [45]. However, the authors of a more recent study argued that any signal detected in germ cells using immunohistochemistry was the result of nonspecific antibody binding [29]. Germ cell transplantation studies have shown that *Rarα*-deficient germ cells rarely colonize a recipient testis; however, once the testis is colonized, spermatogenesis proceeds normally. In the reciprocal experiment, wild-type germ cells transplanted into a *Rarα*-deficient testis colonized and proliferated normally; however, their development was mostly halted during the meiotic prophase [46]. A second transplantation study also identified improper cellular associations and abnormal sperm production when *Rarα*-deficient germ cells were introduced into a germ-cell-depleted wild-type testis [47]. Clearly, RARα is a key signaling molecule required within Sertoli cells for normal spermatogenesis, and additional studies are required to definitively show whether it is present and functions within male germ cells.

Through the use of immunohistochemistry and *LacZ* reporter gene studies in mice, RARγ was found in A spermatogonia [29,48]. While no fertility issues were originally reported for *Rarγ*-deficient male mice [49], a more recent study suggests that this receptor is crucial for spermatogonial differentiation following the first wave of spermatogenesis [48]. When Gely-Pernot and colleagues examined *Rarγ*-deficient male mice and mice lacking *Rarγ* only in germ cells, they observed an impairment in spermatogonial differentiation, resulting in testis tubules containing only Sertoli cells and undifferentiated spermatogonia in older animals [48]. Interestingly, spermatogenesis appeared to progress normally in testes isolated from 6-week-old *Rarγ*-deficient animals, indicating that other receptors can compensate for the loss of RARγ during the first few waves of spermatogenesis. Deficiencies in any of the other three receptors (*Rarβ*, *Rxrα*, *Rxrγ*) have not yet been reported to affect male fertility [50–52], and there are differing reports regarding the localization of these receptors in mammalian testes. The authors of one study reported that each of these receptors is exclusively expressed in round spermatids of the adult mouse testis [29]. However, a second study in the rat showed staining in Sertoli cells and germ cells for RARβ and in the majority of germ cell populations for both RXRα and RXRγ [53].

RA also appears to be able to signal through PPARs within Sertoli cells. However, this mechanism is not critical for male fertility. PPARβ/δ transcript and protein is made by Sertoli cells and Leydig cells [54,55], as is the mRNA that encodes FABP5 [56], the binding protein responsible for delivering RA to PPARs. *In vitro* studies have demonstrated that PPARβ/δ can control Sertoli cell energetics by regulating fatty acid metabolism and lactate production [57]. However, null mutations in the genes encoding either this receptor or *Fabp5* had no effect on fertility [58,59], indicating that compensation mechanisms are in place to ensure that Sertoli cell energetics are maintained via signaling through other ligands and/or receptors.

C. Retinoid binding and storage

Our current understanding of the function of the retinoid binding and storage enzymes in the adult testis has been mostly derived from mRNA and protein localization studies and primary Sertoli cell cultures. Rodent Sertoli cells express CRBP1 [29,60,61] and CRABP2 [62] and CRBP is also present in bovine Sertoli cells [63]. CRBP1 was also detected within spermatogonia and primary spermatocytes of the adult rat testis [26]. In addition, cultured rat Sertoli cells have been found to secrete CRBP [64]; however, the administration of hormones (follicle-stimulating hormone, insulin, or testosterone) and/or retinol changed the levels of secreted CRBP in culture [65]. Genetic studies have demonstrated that neither CRBP1 and CRBP2 nor RBP4 are essential for spermatogenesis [66–68], but the testes of RBP4-deficient animals are more vulnerable to dietary-induced vitamin A deficiency than their wild-type siblings [8]. This is also the case for mice that are deficient in the retinoid storage enzyme LRAT, and Sertoli cells have been shown to synthesize the active form of this enzyme [69].

An interesting relationship appears when the localization of the CRABP mRNAs and proteins are examined. *Crabp1*, which is likely responsible for targeting RA for degradation, has only been detected within spermatogonia [62,70] and CRABP2, which promotes RA signaling via the RARs, appears to be confined to Sertoli cells [62]. These observations imply that the cell-specific expression of these two proteins may play a role in how RA signaling is regulated in the different cell types of the seminiferous epithelium. However, neither CRABP1 nor CRABP2 is essential for male fertility [71], indicating that additional mechanisms must be in place within germ and Sertoli cells to ensure that RA is directed toward either signaling or degradation in an appropriate manner.

What has yet to be examined is whether germ cell development is arranged normally in the testes of adult mice that are deficient in either the retinoid binding or storage enzymes. It is possible that without the proper delivery or storage of retinoids within the testis, the timing of RA availability may not support the cyclic nature of spermatogenesis (discussed further in Section IV). As a result, spermatozoa may still be produced, but the arrangement of the stages of the seminiferous epithelium may be distorted. This distorted arrangement of seminiferous epithelial stages occurs in adult mouse testes when RA production is restored following a brief period of RA deficiency shortly after birth [72]. Further histological examination of the testes of retinoid-binding-enzyme-deficient or retinoid-storage-enzyme-deficient mice will be required to determine the importance of these enzymes for the normal cyclic nature of spermatogenesis.

IV. Maintenance of the spermatogenic cycle by RA

The generation of spermatozoa within the adult mammalian testis occurs in a highly organized manner. Germ cells at specific steps in their differentiation pathway always appear together at any given point along a testis tubule; these well-defined germ cell associations are known as "stages" [73,74]. Studies often utilize a staging chart (Figure 9.2B) to identify stage-specific patterns in the localization of an mRNA or protein or whether there are stage-specific alterations in the progression of spermatogenesis (Figure 9.2A). Such studies include those that characterize the cellular localization of mRNAs and proteins in the adult testis or testis histology in mice when spermatogenesis is affected by genetic mutations or drug treatments. A staging chart is a pictorial representation of each of the 12 stages of the spermatogenic cycle in the mouse testis and displays the male germ cell types that should always be found together within the seminiferous epithelium [74].

When a cross section of a testis is viewed (Figure 9.2A), these stages can be visually distinguished from one another based on a set of well-characterized morphological differences, including germ cell nuclear morphology and positioning within the seminiferous epithelium as well as stage of spermatid differentiation.

If a testis tubule is analyzed longitudinally, these stages are arranged sequentially, indicating that the germ cells in each stage differentiate naturally as a group to the next phase of their development, the subsequent stage, with the progression of germ cells through each stage along a tubule known as the "spermatogenic wave" (Figure 9.2C). It is hypothesized that this wave is generated by the temporally controlled entry of germ cells into their differentiation pathway along the length of the testis tubules. In addition, if we could observe germ cell differentiation over time at a single position within a tubule, we would see a specific stage, followed by several others, and then the reappearance of the original stage, constituting the cycle of the seminiferous epithelium. In mice, this cycle takes 8.6 days, and there is now extensive evidence to indicate that RA both generates and maintains this cycle (see Chapter 8 for a discussion of RA regulation of the initiation of the cycle).

Current evidence to support a role for RA in the maintenance of the mammalian cycle of the seminiferous epithelium is derived from analysis of both VAD and vitamin-A-sufficient rodents. VAD rodents display a block in spermatogenesis at spermatogonial differentiation. When these animals are rescued with exogenous retinoids, spermatogenesis is reinitiated by the stimulation of spermatogonial differentiation in a synchronized manner throughout the testis [4,5]. The testis does not recover in such a way that all 12 stages can be identified in tubules among a

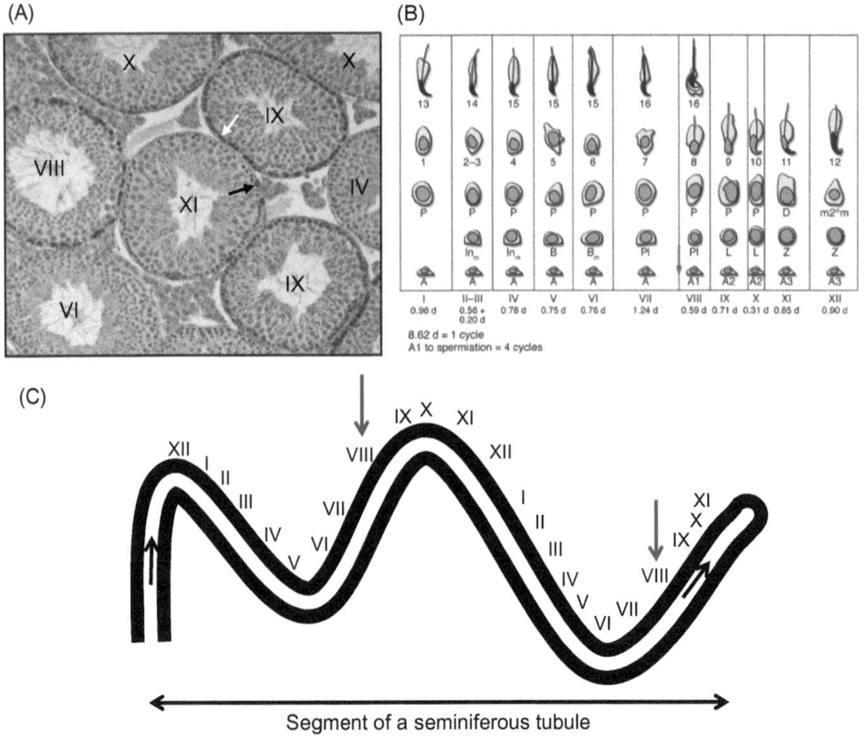

Figure 9.2 The stages of the cycle of the seminiferous epithelium. (A) Cross section of an adult mouse testis stained to detect stimulated by retinoic acid (STRA) 8, the classic marker of a response to retinoic acid (RA) within the testis. The black arrow shows a STRA8-positive spermatogonium and the yellow arrow shows a STRA8-positive preleptotene spermatocyte. Each tubule is numbered according to the stage of the cycle of the seminiferous epithelium that it depicts. All stages are usually represented in a cross section of a wild-type mouse testis. (B) A schematic of the stages of the cycle of the seminiferous epithelium, including the duration of each stage. The red arrow indicates where RA is required to induce spermatogonial differentiation and, therefore, the maintenance of the cycle. (C) Schematic of a testis tubule illustrating the spermatogenic wave. The stages of the cycle of the seminiferous epithelium are arranged sequentially along the tubules of the mouse testis, indicating that the germ cells in each stage naturally differentiate as a group to the next phase of their development, the subsequent stage. It is hypothesized that the spermatogenic wave is generated by the temporally controlled entry of germ cells into their differentiation pathway, a process driven by pulses of RA at stage VIII (red arrows), along the length of the testis tubules, and this process results in continuous sperm production. (For interpretation of the references to color in this figure legend, the reader is referred to the web version of this book.)
(A) and (B) Adapted from Refs. [74,108].

histological cross section. Instead, only three or four sequential stages are present, and this synchronization is maintained for many months. In addition, this process results in the release of spermatozoa only every 8.6 days, and, therefore, vitamin-A-driven synchronization in the mouse testis eliminates continual sperm production.

Synchronized spermatogenesis is also seen following treatment with an inhibitor, WIN 18,446, of the ALDH enzymes to create a RA-deficient testis and recovery induced via an injection of RA [72]. Data now indicate that the synchronized response of the testis following exposure to exogenous retinoids occurs only in the absence of meiotic germ cells. When juvenile mice aged 2, 4, or 6 dpp, prior to the appearance of meiotic germ cells within the testis, were treated with a single injection of RA, either spermatogenesis was synchronized or only 4 or 5 of the 12 stages were present within cross sections of their testes following recovery until 90 dpp [75,76]. However, no synchronization was observed when mice aged 8 dpp, coinciding with the appearance of the first meiotic cells (the preleptotene spermatocytes), were treated with RA [75]. This result implies that the cycle of the seminiferous epithelium can no longer be altered by exogenous RA when advanced germ cells are present.

The hypothesis that the advanced germ cells contribute to the way in which RA regulates the maintenance of the spermatogenic cycle is also supported by studies that have reported stage-specific and testis-cell-specific localization of mRNAs and proteins encoding RA synthesizing, degrading, and storage enzymes [19,29,34] (Figure 9.3). *In situ* hybridization studies, first by Vernet and colleagues and more recently by Sugimoto and colleagues, imply that there are very low or undetectable levels of transcripts coding for RA-synthesizing enzymes present in stages I–VI, just before spermatogonial differentiation takes place [29,34]. Interestingly, transcripts coding for the only known RA uptake protein, STRA6, are present in Sertoli cells, with the highest levels detected in stages VII–IX, which coincides with when RA would be required for spermatogonial differentiation [34]. *Aldh1a2* transcripts were also observed in preleptotene and pachytene spermatocytes in stages VII–XII, suggesting that RA is produced by the differentiating germ cells during these stages [19,29,34]. RA degradation appears to be promoted after spermatogonial differentiation has taken place because the highest levels of *Cyp26a1* mRNA can be detected in stages VIII–XI [34].

Taken together, these expression studies imply that there is a sharp increase in endogenous RA production followed by a rapid decrease in local RA levels at stages VII–IX of the seminiferous epithelium. Coincidently, all of the spermatogenic processes that require RA (discussed in Section V) take place during stage VIII. Therefore, it is highly likely that RA activity at stage VIII is responsible for maintaining the cycle of the seminiferous epithelium, driven perhaps by the synthesis of RA by the preleptotene and pachytene spermatocytes and signaling via RARγ in the spermatogonia. The task for researchers in the coming years will be to determine which enzymes and confirm which testicular cell types are responsible for generating and regulating these pulses of RA required at stage VIII.

V. Sertoli cell contributions to RA function within the adult testis

Extensive data now exist to support the hypothesis that RA is a key regulator of three important steps during spermatogenesis: (1) spermatogonial differentiation

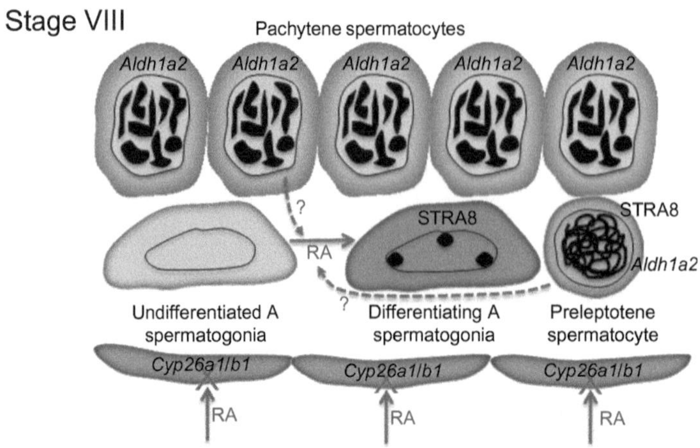

Figure 9.3 Current model of the maintenance of the cycle of the seminiferous epithelium in the adult testis. Schematic combining data from multiple publications that models a way in which advanced germ cells may be responsible for maintaining the cycle of the seminiferous epithelium. The retinoic acid (RA) required to drive spermatogonial differentiation in the adult testis, and therefore the cycle, appears to be generated by the activity of aldehyde dehydrogenase (ALDH) 1A2 in either pachytene or preleptotene spermatocytes or possibly by the activity of both of these cell types. RA receptor (RAR) γ appears to be the predominant RAR driving spermatogonial differentiation after the first spermatogenic wave [48]. In addition, cytochrome p450, family 26 (CYP26) activity within the peritubular myoid cells is hypothesized to create a barrier to the influx of RA from the serum, thereby allowing the seminiferous epithelium to precisely regulate a pulse of RA. This model has been generated via gene expression and protein localization studies [19,29,34] and by studies of treatments of wild-type mice with exogenous RA [75,76]. Studies of animals that are deficient in different retinoid-metabolizing enzymes specifically in Sertoli cells suggest that Sertoli-cell-derived RA may not be required after the first wave of spermatogonial differentiation [19,27].

and meiotic initiation (to both initiate and maintain spermatogenesis), (2) reorganization of the blood–testis barrier (BTB), and (3) spermiation. All three events occur during stages VII–IX of the cycle of the seminiferous epithelium. Based on the localization studies described above and the *ex vivo* cultures and knockout mouse models summarized below, Sertoli cells are clearly a source of RA within the testis, but the requirement for Sertoli-cell-derived RA differs between the juvenile and adult testis. Sertoli-cell-derived RA is essential for the initial differentiation of spermatogonia and the onset of meiosis within the neonatal testis but not for the ongoing production of sperm [19,27]. How the Sertoli cells and RA regulate the initiation of spermatogenesis is discussed in Chapter 8. This section summarizes our current understanding of how Sertoli cells contribute to RA-regulated events in the adult testis: BTB formation and maintenance, and spermiation.

A. BTB formation and maintenance

The BTB is essential for normal male germ cell development. In mice, the BTB is formed between 15 and 18 dpp and is established almost exclusively from specialized junctions close to the basement membrane between adjacent Sertoli cells (the BTB is discussed in more detail in Chapter 13; for a recent review see Ref. [77]). Germ cells are arranged within the seminiferous epithelium such that spermatogonia reside outside of the BTB (the basal compartment) and most spermatocytes (zygotene, pachytene, and diplotene), spermatids, and spermatozoa are located inside the BTB (the adluminal compartment), placing these meiotic and haploid cells in an immune-privileged environment. Germ cells transit through the barrier, moving from the basal to the adluminal side, as preleptotene spermatocytes during stages VII and VIII of the cycle of the seminiferous epithelium. This movement occurs via new tight junctions that form between Sertoli cells on the basal side of preleptotene spermatocytes before the older junctions are dissolved, releasing the germ cells into the adluminal compartment [78]. As discussed above, RA levels are thought to be highest at stages VII and VIII of the cycle, and there is growing evidence to support an important role for RA in BTB formation and integrity.

The majority of published data to implicate RA signaling in regulating BTB function has been derived from studies of RARα-null mice and mice that overexpress a dominant-negative form of this receptor specifically in Sertoli cells. Chung and colleagues utilized a hypertonic fixation technique and a dye-coupling assay to investigate the structure and permeability of the BTB in male mice lacking RARα [79]. Hypertonic fixation revealed defects in the integrity of the BTB. The detection of rhodamine-dextran, a molecule that is normally too big to diffuse across the BTB, in the adluminal compartment in mutant mice indicated that tight junctions are improperly formed in the absence of RARα signaling. In addition, the RARα–mutant mice displayed a delay in the appearance and adluminal localization of ZO-1, a marker for the establishment of Sertoli-cell tight junctions, and perturbed stage-specific localization of vimentin, an intermediate filament protein that ensures the correct localization of the elongated spermatid within the seminiferous epithelium. An *in silico* search for RAREs within the promoters of genes encoding gap and tight-junction proteins revealed the potential for RA to regulate the expression of these proteins directly [79].

In fact, retinoid signaling has been found to initiate tight-junction formation in primary Sertoli-cell cultures established from 10 and 20 dpp rat testes [80]. In addition, a RA-inducible gene product, RAI14, has been localized to the BTB within Sertoli cells but only at stages VII and VIII of the cycle, and blocking *Rai14* translation via siRNAs in cultures of Sertoli cells disrupts the tight junctions between cells [81]. An examination of testes isolated from adult male mice injected with a lentivirus that disrupted RA signaling in a Sertoli-cell-specific manner revealed that Sertoli cell nuclei were detached from the tubule basement membrane [82], a phenotype often observed when the BTB is abnormal. Furthermore, biotin tracer assays revealed increased BTB permeability in mice with Sertoli-cell-specific, disrupted RA signaling, and the expression of BTB markers was reduced in a stage-specific manner [82]. While it is possible that the BTB defects seen in the absence of RA signaling in Sertoli cells could be side effects associated with disruptions of the germ cell population that also occur in these animals, the observation that RA can induce Sertoli cells to begin to form tight junctions in culture and the identification of RAREs in the promoters of gap- and tight-junction genes indicates that further research on whether and/or how RA can directly regulate the BTB, especially during stages VII and VIII, is required.

B. Spermiogenesis

The end result of the meiotic division of a spermatocyte is the generation of four round, haploid spermatids that must then undergo extensive morphological changes to become spermatozoa. This process, known as spermiogenesis, begins at stage I and occurs throughout a cycle before culminating in the generation and release of mature spermatozoa from the seminiferous epithelium at stage VIII of the following cycle [74]. During spermiogenesis, drastic changes to nuclear architecture, cell structure, and spermatid alignment within the seminiferous epithelium take place, and RA synthesis and activity within Sertoli cells is critical for the normal progression of these events.

Studies in mice lacking components of the RA metabolism and signaling pathway only in Sertoli cells indicate that the Sertoli cells are the major source of the RA required for spermiogenesis and spermiation and that the complex that is most likely to transmit the signals required for these events is a heterodimer of RARα and RXRβ. When all three ALDH isoforms are ablated specifically in Sertoli cells, thereby eliminating the ability of Sertoli cells to synthesize RA, the elongating spermatids fail to align at the luminal edge and are retained within the seminiferous epithelium beyond their normal time of spermiation [19]. In addition, the epididymides of these animals contain only abnormal spermatozoa displaying signs of necrosis (e.g., separation of the acrosome from the nucleus), indicating that RA produced by the Sertoli cell is critical for normal spermiogenesis [19].

Mice carrying Sertoli-cell-specific deletions of either *Rara* or *Rxrb* demonstrate that these two receptors are critical for RA regulation of haploid male germ cells. In testes from both mouse lines, and consistent with the ALDH knockout phenotype, elongated spermatids failed to align at the luminal edge of the tubules, and spermatozoa were not effectively released from the seminiferous epithelium [43,44]. In addition, increased levels of spermatid apoptosis were observed in the testes of the Sertoli-cell-specific knockout animals of either receptor [43,44].

A comparison of the two mouse lines indicated that the effect of RARα loss appears to be more severe. In addition to failed spermatid alignment and release, the testes of Sertoli-cell-specific *Rara*-null animals contained very few elongated spermatids. The elongated spermatids that were present were mostly blocked at step 8–9 of the differentiation process, exhibited large vacuoles and sloughing of the round spermatid cell layer from the epithelium, and lacked the normal germ cell associations seen in control testes [43]. In addition, the very few mature spermatozoa that were present displayed ultrastructural abnormalities indicative of necrosis [43].

Interestingly, no defects were observed in mice carrying one mutant and one wild-type allele for both *Rara* and *Rxrb*, indicating that one gene copy of either receptor within the Sertoli cells is sufficient for normal spermatogenesis. No compounding phenotypic effects were seen when both receptors were eliminated from Sertoli cells simultaneously.

An additional study of the testicular phenotype of mice expressing RXRβ with a mutation in its transcriptional activation domain showed no spermiation defects in the absence of transcriptionally active RXRβ [83]. This observation indicates that spermiogenesis is likely to be controlled by the ligand-receptor-transcriptional activation interaction between RA and RARα, with RXRβ providing stability to the complex. This hypothesis is also supported by the observation that the *Rara*-null phenotype is more severe.

VI. The effects of retinoids on Sertoli cell function

While RA synthesis and signaling within Sertoli cells is critical for male germ-cell development, there is also evidence to suggest that RA is important for the normal function of Sertoli cells. RA appears to drive cyclic Sertoli cell gene expression, ensuring that the correct environment is provided for the differentiating germ cells as they transition through the cycle of the seminiferous epithelium. There is now a growing body of evidence to suggest that retinoids control the proliferative activity and terminal differentiation of Sertoli cells, and cultures of primary Sertoli cells serve as a model system for the investigation of a potential role for retinoids in cellular and systemic oxidative balance.

A. The adult Sertoli cell cycle

Multiple cyclic gene expression events occur within Sertoli cells during the cycle of the seminiferous epithelium [34,84,85], ensuring that differentiating germ cells are provided with a microenvironment that is suitable for their development. Numerous studies have generated data to indicate that RA is critical for the normal cycling of Sertoli cells, and this cycling appears to be regulated via RARα.

Examination of testes from VAD mice treated with RA for recovery revealed that the Sertoli cell cycle can be modified by RA [34]. Upon treatment with exogenous RA, spermatogenesis resumed within the testis of a VAD rodent, but in a synchronous manner in that all germ cells followed their normal differentiation pathway simultaneously. *In situ* hybridization experiments performed by Sugimoto and colleagues have demonstrated that the Sertoli cell cycle is also rendered synchronous following RA treatment of a VAD mouse [34]. Closer examination revealed that exogenous RA reset Sertoli cell cyclic gene expression such that 24 h after treatment, Sertoli cell gene expression displayed a pattern reminiscent of a stage VII–VIII testis tubule cross section and continued with its normal cycle, albeit synchronously, with the differentiating germ cells. This observation indicates that RA, in the absence of advanced germ cells, can reset Sertoli cell gene expression to stage VII from any stage. Further investigations utilizing mouse models with vitamin-A-sufficient testes lacking advanced germ cells (cryptorchids, *Nanos3* mutants, and W/W^{v} mutants) demonstrated that exogenous RA did reset the Sertoli cell cycle independently of the endogenous levels of RA when there were no advanced germ cells. However, in the presence of advanced germ cells (i.e., in wild-type mice), exogenous RA had no effect on cyclic Sertoli cell gene expression. Therefore, differentiating germ cells are likely part of the regulatory mechanism that maintains the Sertoli cell cycle [34,75,76]. Interestingly, cyclic Sertoli cell gene expression remained intact in the testes of VAD mice [34], probably because the Sertoli cell cycle was established embryonically [84] before the mice became VAD.

Elimination of RARα signaling in Sertoli cells, either via a conditional deletion of the gene [43] or lentiviral inhibition of gene expression [82], impairs cyclic gene expression within Sertoli cells,. In Sertoli-cell-specific, *Rara*-null mice, the cyclical nature is lost as early as 5 dpp. Interestingly, no effect on gene expression within Sertoli cells was observed in animals lacking all three RXR isoforms, indicating that not only are the RXRs dispensable for the cyclic activity of Sertoli cells, but also that RARα must regulate these events via a mechanism that is independent of the RAR/RXR heterodimer.

B. Sertoli cell proliferation

Precise regulation of Sertoli cell proliferation is critical for normal spermatogenesis. Each Sertoli cell has a finite number of germ cells that it can support [86]. Therefore, sperm output and testis size are dependent on the total number of Sertoli cells within the testis. In rodents, Sertoli cells are highly mitotic throughout embryonic and early postnatal testis development, but they then begin to terminally differentiate (i.e., their proliferation rate declines), resulting in mitotic quiescence approximately 2–3 weeks after birth [87]. This cessation of cell division immediately precedes the formation of tight junctions between Sertoli cells and the establishment of the BTB.

While it is well known that Sertoli cells undergo terminal differentiation, the regulatory mechanisms driving this process are not well understood. However, there is emerging evidence to indicate that RA is one of the factors controlling this process. Most of the data to implicate RA in regulating Sertoli cell terminal differentiation has been generated from cultures of primary rodent Sertoli cells. When both Buzzard and colleagues and Zanotto-Filho and colleagues treated primary Sertoli cell cultures with RA, they observed a suppressive effect on proliferation, rate of division, and DNA synthesis; an accumulation of cell cycle inhibitors (*p27Kip1* and *p21Cip1*), and increased expression of markers of Sertoli cell differentiation (*Gata1* and *Gata4*) [88,89]. In contrast, treatment of cultured Sertoli cells with retinol stimulated proliferation, as measured by increased DNA synthesis and decreased p21 levels. These data also indicated that induced proliferation was mediated through nonclassical retinoid signaling and redox-dependent mechanisms [89,90]. A more recent report from Nicholls and colleagues has extended these studies and linked RA and activin to the regulation of Sertoli cell terminal differentiation [80].

In both the embryonic and juvenile rodent testis, activin stimulates Sertoli cell proliferation [91–93] and can induce terminally differentiated Sertoli cells to reenter the cell cycle [94]. When primary rat Sertoli cells were cultured with both RA and activin, a dose-dependent inhibition of cell proliferation was observed, regardless of whether the cells were collected prior to (10 dpp) or following (20 dpp) Sertoli cell terminal differentiation. Gene expression analyses indicated that while activin could induce the expression of multiple cyclins, indicative of its ability to promote the cell cycle, RA blocked the expression of cyclin E1, suggesting that RA stops Sertoli cells from entering S phase [80]. Further examination of the gene expression changes in cultured primary Sertoli cells following activin treatment revealed increases in RA-responsive genes, *Stra6* and *Rbp1*, suggestive of a role for activin in promoting RA uptake and storage within Sertoli cells. This finding, combined with the observation that the normal timing of activin signaling in both the prepubertal and adult testis either immediately precedes [95,96], or occurs during, RA-regulated events (e.g., activin secretion peaks at stage VIII, when RA-regulated processes take place) [97], suggests that temporal changes take place in activin and retinoid signaling control when Sertoli cells enter quiescence and form tight junctions in the juvenile testis and when BTB reorganization occurs in the adult testis. Further analysis of Sertoli cell proliferation rates and numbers in animal models lacking RA within the juvenile testis, either induced via the diet in genetically modified mice [8,98] or via chemical inhibition of RA synthesis [72], will be useful for determining whether Sertoli cells continue to proliferate and never terminally differentiate in the absence of RA.

C. Sertoli cells as a model for investigating retinoid-regulated oxidative balance

The role of antioxidants in the maintenance of general human health and protection against multiple different diseases has been the focus of extensive research for decades. Numerous vitamins, including vitamin A and its derivatives, are known to have antioxidant properties, but there have been recent reports of prooxidant effects resulting from retinol treatment at levels only slightly above concentrations that would normally be present endogenously. Because the postnatal testis is one of the main sites of retinoid action, cultured primary Sertoli cells are used as a tool for investigating the oxidant effects of retinoids and the mechanisms affected by vitamin A supplementation (e.g., cosmetic products, dietary supplements, acne treatments, and cancer therapies) at a cellular level. Several studies have shown increased free-radical production by Sertoli cells in culture following retinol treatment. The activity of catalase, one of the main enzymes involved in protecting cells from oxidative damage, was significantly enhanced by retinol within Sertoli cells [99,100]. Increases in the expression of RAGE (receptor for advanced glycation endproducts) proteins, which are responsible for binding toxic end products resulting from glycation and are associated with unbalanced production of reactive oxygen species, were also detected in these cultures [101]. Retinol treatment was shown to alter the structure of actin filaments within Sertoli cells [102]. In fact, de Oliveria and colleagues concluded that structural changes within actin filaments in treated Sertoli cells enabled electrons produced via retinol metabolism to be transmitted to the mitochondria. This process, in turn, led to overloading this organelle with a supply of electrons and induced superoxide radical production, which initiated oxidative stress [102] and disturbed mitochondrial metabolism, detected via a reversible drop in mitochondrial activity and cellular adenosine triphosphate levels [103,104].

The end result of retinol-induced oxidative damage in Sertoli cells appears to be apoptosis. The classical, intrinsic mitochondrial apoptotic pathway (i.e., the release of death factors, such as cytochrome c, from the inner mitochondrial membrane to the cytosol) was initiated in Sertoli cells treated with retinol [90,103], as were single- and double-stranded DNA breaks, both of which are hallmarks of a normal apoptotic response. Taken together, these data suggest that excess levels of retinol can induce intracellular reactive oxygen species production, changes in cytoskeleton structure, mitochondrial damage, and apoptosis. While these observations may only be partially relevant to how Sertoli cells react to elevated levels of retinol within the testis, because Sertoli cells behave differently in culture than *in vivo*, these studies do show how useful primary Sertoli cell cultures have been to furthering our understanding of how retinol affects cellular oxidative balance.

While retinol has been shown to induce the production of free radicals in cultured Sertoli cells, the same is not true for RA. No increases in reactive oxygen species were observed following treatment of Sertoli cells with RA. However, a dose-dependent decrease in cell viability was observed [105], and this pro-apoptotic response appeared to be mediated via a nonclassical, nongenomic mechanism. Phosphorylated ERK, one of the major signaling complexes in the apoptosis pathway, was detected as soon as 15 min following RA treatment of cultured Sertoli cells, and the observation that cycloheximide did not block this phosphorylation event indicated that *de novo* protein synthesis was not required. Treatment of Sertoli cells with an inhibitor of the ERK pathway blocked RA-induced apoptosis, and mobility shift assays determined that nongenomic activation of the ERK pathway, induced by RA, increased the DNA binding activity of the transcription factor AP-1, a known key player in apoptosis [105]. Because the retinoids are lipophilic and can dissolve within the lipid bilayer of cell or mitochondrial membranes, it is possible that these molecules interact directly with transmembrane receptors to rapidly induce signal transduction or they affect cell energetics within the mitochondria. There is evidence to suggest that retinol can have rapid effects on the membrane potential of Sertoli cells (e.g., stimulating calcium and amino acid uptake (reviewed in Ref. [106])); however, the underlying nongenomic actions of the retinoids are highly speculative and remain to be further investigated.

VII. Conclusions and remaining questions

Significant steps toward understanding how vitamin A regulates sperm development have been made since the early 2000s. Based on studies of both neonatal and adult rodents, the ways in which Sertoli cells participate in this regulation appear to differ between the first and subsequent waves of spermatogenesis. While knockout mouse models suggest that Sertoli-cell-derived RA drives the initiation of spermatogenesis and the cycle of the seminiferous epithelium, this RA does not seem to be required for spermatogonial differentiation in subsequent waves or for maintenance of the cycle. Instead, the advanced germ cells are the likely candidate source of RA for these processes, and further investigation of whether this source consists of preleptotene, pachytene, or a combination of both types of spermatocytes is required. In the adult testis, the Sertoli cells contain all the machinery required to synthesize and respond to RA, and these components play a critical role in the formation and maintenance of the BTB and in spermiation, both of which take place during stage VIII of the seminiferous cycle. Although not regulated by Sertoli-cell-derived RA in the adult testis, spermatogonial differentiation also takes place during stage VIII. Altogether, these observations lead to the hypothesis that RA synthesis is required in both Sertoli and germ cells in the adult testis to ensure that levels are at their peak when germ cells need to progress through stage VIII. Further studies of the enzymes responsible for RA synthesis and degradation will help determine the ways in which RA levels are so tightly controlled across the cycle.

RA also appears to influence the proliferation and differentiation of Sertoli cells directly. Primary Sertoli-cell cultures have been useful in demonstrating that RA can inhibit Sertoli cell proliferation and induce formation of the types of cell junctions present in the BTB *in vivo*. This has led to the hypothesis that RA may play a role in the terminal differentiation of Sertoli cells in the juvenile testis. However, studies of juvenile animal models with modulated testicular RA levels will be crucial in determining whether these *ex vivo* studies are relevant to the regulation of the proliferative activity of Sertoli cells and establishment of the BTB *in vivo*.

Cultures of primary Sertoli cells also serve as a model system for investigating how retinoids affect cellular oxidative balance and the nongenomic actions of RA. Therapies designed to induce higher levels of retinoids *in vivo* are being studied in clinical trials for various cancers and skin diseases. While the antioxidant properties of retinoids are well established, there are now published data to imply that the different retinoid derivatives may have variable effects on cellular redox balance. It is vital that we further our understanding of the direct effects of retinoids on cells if retinoids are to become viable therapeutic options. Given that the testis is one of the main sites of retinoid action in adults, cells derived from this organ could be an excellent tool for these investigations.

In addition, the nongenomic actions of RA are a new avenue of research that could help explain specific phenotypes observed in animals with deficiencies in retinoid machinery. For example, not all of the defects (e.g., maintenance of the Sertoli cell cycle) seen in *Rarα*-null mice are the result of a lack of RARα/RXR heterodimers. Recent studies have indicated that interactions of RARα with proteins within cell membranes and the phosphorylation of this receptor can induce protein kinase signaling cascades that are independent of interactions with DNA [107]. Whether nongenomic RA signaling mechanisms are involved in the regulation of spermatogenesis remains to be determined, but this involvement may be the reason why one molecule can have such varied effects on the different cell types of the testis.

References

[1] Wolbach SB, Howe PR. Tissue changes following deprivation of fat-soluble A vitamin. J Exp Med 1925;42:753–77.

[2] Mitranond V, Sobhon P, Tosukhowong P, Chindaduangrat W. Cytological changes in the testes of vitamin-A-deficient rats. I. Quantitation of germinal cells in the seminiferous tubules. Acta Anat (Basel) 1979;103:159–68.

[3] Unni E, Rao MR, Ganguly J. Histological & ultrastructural studies on the effect of vitamin A depletion & subsequent repletion with vitamin A on germ cells & Sertoli cells in rat testis. Indian J Exp Biol 1983;21:180–92.

[4] Griswold MD, Bishop PD, Kim KH, Ping R, Siiteri JE, Morales C. Function of vitamin A in normal and synchronized seminiferous tubules. Ann N Y Acad Sci 1989; 564:154–72.

[5] van Pelt AM, de Rooij DG. Synchronization of the seminiferous epithelium after vitamin A replacement in vitamin A-deficient mice. Biol Reprod 1990;43:363–7.

[6] D'Ambrosio DN, Clugston RD, Blaner WS. Vitamin A metabolism: an update. Nutrients 2011;3:63–103.

[7] Napoli JL. Physiological insights into all-trans-retinoic acid biosynthesis. Biochim Biophys Acta 2012;1821:152–67.

[8] Ghyselinck NB, Vernet N, Dennefeld C, Giese N, Nau H, Chambon P, et al. Retinoids and spermatogenesis: lessons from mutant mice lacking the plasma retinol binding protein. Dev Dyn 2006;235:1608–22.

[9] Berry DC, Jacobs H, Marwarha G, Gely-Pernot A, O'Byrne SM, DeSantis D, et al. The STRA6 receptor is essential for retinol-binding protein-induced insulin resistance but not for maintaining vitamin A homeostasis in tissues other than the eye. J Biol Chem 2013;288:24528–39.

[10] Herr FM, Ong DE. Differential interaction of lecithin-retinol acyltransferase with cellular retinol binding proteins. Biochemistry (Mosc) 1992;31:6748–55.

[11] Posch KC, Burns RD, Napoli JL. Biosynthesis of all-trans-retinoic acid from retinal. Recognition of retinal bound to cellular retinol binding protein (type I) as substrate by a purified cytosolic dehydrogenase. J Biol Chem 1992;267:19676–82.

[12] Noy N. Retinoid-binding proteins: mediators of retinoid action. Biochem J 2000;348(Pt 3): 481–95.

[13] Dong D, Ruuska SE, Levinthal DJ, Noy N. Distinct roles for cellular retinoic acid-binding proteins I and II in regulating signaling by retinoic acid. J Biol Chem 1999;274: 23695–8.

[14] Mendoza-Parra MA, Gronemeyer H. Genome-wide studies of nuclear receptors in cell fate decisions. Semin Cell Dev Biol 2013;24:706–15.

[15] Boylan JF, Gudas LJ. The level of CRABP-I expression influences the amounts and types of all-trans-retinoic acid metabolites in F9 teratocarcinoma stem cells. J Biol Chem 1992;267:21486–91.

[16] Boylan JF, Gudas LJ. Overexpression of the cellular retinoic acid binding protein-I (CRABP-I) results in a reduction in differentiation-specific gene expression in F9 teratocarcinoma cells. J Cell Biol 1991;112:965–79.

[17] Schug TT, Berry DC, Shaw NS, Travis SN, Noy N. Opposing effects of retinoic acid on cell growth result from alternate activation of two different nuclear receptors. Cell 2007; 129:723–33.

[18] Noy N. The one-two punch: retinoic acid suppresses obesity both by promoting energy expenditure and by inhibiting adipogenesis. Adipocyte 2013;2:184–7.

[19] Raverdeau M, Gely-Pernot A, Feret B, Dennefeld C, Benoit G, Davidson I, et al. Retinoic acid induces Sertoli cell paracrine signals for spermatogonia differentiation but cell autonomously drives spermatocyte meiosis. Proc Natl Acad Sci USA 2012; 109:16582–7.

[20] Duester G. Retinoid signaling in control of progenitor cell differentiation during mouse development. Semin Cell Dev Biol 2013;24:694–700.

[21] Bowles J, Knight D, Smith C, Wilhelm D, Richman J, Mamiya S, et al. Retinoid signaling determines germ cell fate in mice. Science 2006;312:596–600.

[22] Koubova J, Menke DB, Zhou Q, Capel B, Griswold MD, Page DC. Retinoic acid regulates sex-specific timing of meiotic initiation in mice. Proc Natl Acad Sci USA 2006; 103:2474–9.

[23] Topletz AR, Thatcher JE, Zelter A, Lutz JD, Tay S, Nelson WL, et al. Comparison of the function and expression of CYP26A1 and CYP26B1, the two retinoic acid hydroxylases. Biochem Pharmacol 2012;83:149–63.

[24] Abu-Abed S, Dolle P, Metzger D, Beckett B, Chambon P, Petkovich M. The retinoic acid-metabolizing enzyme, CYP26A1, is essential for normal hindbrain patterning, vertebral identity, and development of posterior structures. Genes Dev 2001;15: 226–40.

[25] MacLean G, Li H, Metzger D, Chambon P, Petkovich M. Apoptotic extinction of germ cells in testes of Cyp26b1 knockout mice. Endocrinology 2007;148:4560–7.

[26] Zhai Y, Higgins D, Napoli JL. Coexpression of the mRNAs encoding retinol dehydrogenase isozymes and cellular retinol-binding protein. J Cell Physiol 1997;173:36–43.

[27] Tong MH, Yang QE, Davis JC, Griswold MD. Retinol dehydrogenase 10 is indispensible for spermatogenesis in juvenile males. Proc Natl Acad Sci USA 2013;110:543–8.

[28] Lopez-Fernandez LA, del Mazo J. The cytosolic aldehyde dehydrogenase gene (Aldh1) is developmentally expressed in Leydig cells. FEBS Lett 1997;407:225–9.

[29] Vernet N, Dennefeld C, Rochette-Egly C, Oulad-Abdelghani M, Chambon P, Ghyselinck NB, et al. Retinoic acid metabolism and signaling pathways in the adult and developing mouse testis. Endocrinology 2006;147:96–110.

[30] Zhai Y, Sperkova Z, Napoli JL. Cellular expression of retinal dehydrogenase types 1 and 2: effects of vitamin A status on testis mRNA. J Cell Physiol 2001;186:220–32.

[31] Nishimura M, Naito S. Tissue-specific mRNA expression profiles of human phase I metabolizing enzymes except for cytochrome P450 and phase II metabolizing enzymes. Drug Metab Pharmacokinet 2006;21:357–74.

[32] Alnouti Y, Klaassen CD. Tissue distribution, ontogeny, and regulation of aldehyde dehydrogenase (Aldh) enzymes mRNA by prototypical microsomal enzyme inducers in mice. Toxicol Sci 2008;101:51–64.

[33] Maly IP, Crotet V, Toranelli M. The so-called "testis-specific aldehyde dehydrogenase" corresponds to type 2 retinaldehyde dehydrogenase in the mouse. Histochem Cell Biol 2003;119:169–74.

[34] Sugimoto R, Nabeshima Y, Yoshida S. Retinoic acid metabolism links the periodical differentiation of germ cells with the cycle of Sertoli cells in mouse seminiferous epithelium. Mech Dev 2012;128:610–24.

[35] Shima JE, McLean DJ, McCarrey JR, Griswold MD. The murine testicular transcriptome: characterizing gene expression in the testis during the progression of spermatogenesis. Biol Reprod 2004;71:319–30.

[36] Evans E, Hogarth C, Mitchell D, Griswold M. Riding the spermatogenic wave: profiling gene expression within neonatal germ and sertoli cells during a synchronized initial wave of spermatogenesis in mice. Biol Reprod 2014;90:108.

[37] Kumar S, Chatzi C, Brade T, Cunningham TJ, Zhao X, Duester G. Sex-specific timing of meiotic initiation is regulated by Cyp26b1 independent of retinoic acid signalling. Nat Commun 2011;2:151.
[38] Li H, MacLean G, Cameron D, Clagett-Dame M, Petkovich M. Cyp26b1 expression in murine Sertoli cells is required to maintain male germ cells in an undifferentiated state during embryogenesis. PLoS One 2009;4:e7501.
[39] Snyder EM, Small C, Griswold MD. Retinoic acid availability drives the asynchronous initiation of spermatogonial differentiation in the mouse. Biol Reprod 2010;83:783–90.
[40] Akmal KM, Dufour JM, Kim KH. Retinoic acid receptor alpha gene expression in the rat testis: potential role during the prophase of meiosis and in the transition from round to elongating spermatids. Biol Reprod 1997;56:549–56.
[41] Kastner P, Mark M, Leid M, Gansmuller A, Chin W, Grondona JM, et al. Abnormal spermatogenesis in RXR beta mutant mice. Genes Dev 1996;10:80–92.
[42] Lufkin T, Lohnes D, Mark M, Dierich A, Gorry P, Gaub MP, et al. High postnatal lethality and testis degeneration in retinoic acid receptor alpha mutant mice. Proc Natl Acad Sci USA 1993;90:7225–9.
[43] Vernet N, Dennefeld C, Guillou F, Chambon P, Ghyselinck NB, Mark M. Prepubertal testis development relies on retinoic acid but not rexinoid receptors in Sertoli cells. EMBO J 2006;25:5816–25.
[44] Vernet N, Dennefeld C, Klopfenstein M, Ruiz A, Bok D, Ghyselinck NB, et al. Retinoid X receptor beta (RXRB) expression in Sertoli cells controls cholesterol homeostasis and spermiation. Reproduction 2008;136:619–26.
[45] Baleato RM, Aitken RJ, Roman SD. Vitamin A regulation of BMP4 expression in the male germ line. Dev Biol 2005;286:78–90.
[46] Doyle TJ, Braun KW, McLean DJ, Wright RW, Griswold MD, Kim KH. Potential functions of retinoic acid receptor A in Sertoli cells and germ cells during spermatogenesis. Ann N Y Acad Sci 2007;1120:114–30.
[47] Chung SS, Wang X, Wolgemuth DJ. Expression of retinoic acid receptor alpha in the germline is essential for proper cellular association and spermiogenesis during spermatogenesis. Development 2009;136:2091–100.
[48] Gely-Pernot A, Raverdeau M, Celebi C, Dennefeld C, Feret B, Klopfenstein M, et al. Spermatogonia differentiation requires retinoic acid receptor gamma. Endocrinology 2012;153:438–49.
[49] Purton LE, Dworkin S, Olsen GH, Walkley CR, Fabb SA, Collins SJ, et al. RARgamma is critical for maintaining a balance between hematopoietic stem cell self-renewal and differentiation. J Exp Med 2006;203:1283–93.
[50] Wendling O, Chambon P, Mark M. Retinoid X receptors are essential for early mouse development and placentogenesis. Proc Natl Acad Sci USA 1999;96:547–51.
[51] Snyder JM, Jenkins-Moore M, Jackson SK, Goss KL, Dai HH, Bangsund PJ, et al. Alveolarization in retinoic acid receptor-beta-deficient mice. Pediatr Res 2005;57:384–91.
[52] Krezel W, Dupe V, Mark M, Dierich A, Kastner P, Chambon P, et al. RXR gamma null mice are apparently normal and compound RXR alpha +/−/RXR beta −/−/RXR gamma −/− mutant mice are viable. Proc Natl Acad Sci USA 1996;93:9010–4.
[53] Dufour JM, Kim KH. Cellular and subcellular localization of six retinoid receptors in rat testis during postnatal development: identification of potential heterodimeric receptors. Biol Reprod 1999;61:1300–8.
[54] Braissant O, Foufelle F, Scotto C, Dauca M, Wahli W. Differential expression of peroxisome proliferator-activated receptors (PPARs): tissue distribution of PPAR-alpha, -beta, and -gamma in the adult rat. Endocrinology 1996;137:354–66.

[55] Schultz R, Yan W, Toppari J, Volkl A, Gustafsson JA, Pelto-Huikko M. Expression of peroxisome proliferator-activated receptor alpha messenger ribonucleic acid and protein in human and rat testis. Endocrinology 1999;140:2968–75.

[56] Oresti GM, Garcia-Lopez J, Aveldano MI, Del Mazo J. Cell-type-specific regulation of genes involved in testicular lipid metabolism: fatty acid-binding proteins, diacylglycerol acyltransferases, and perilipin 2. Reproduction 2013;146:471–80.

[57] Regueira M, Riera MF, Galardo MN, Pellizzari EH, Cigorraga SB, Meroni SB. Activation of PPAR alpha and PPAR beta/delta regulates Sertoli cell metabolism. Mol Cell Endocrinol 2014;382:271–81.

[58] Peters JM, Lee SS, Li W, Ward JM, Gavrilova O, Everett C, et al. Growth, adipose, brain, and skin alterations resulting from targeted disruption of the mouse peroxisome proliferator-activated receptor beta(delta). Mol Cell Biol 2000;20:5119–28.

[59] Maeda K, Uysal KT, Makowski L, Gorgun CZ, Atsumi G, Parker RA, et al. Role of the fatty acid binding protein mal1 in obesity and insulin resistance. Diabetes 2003; 52:300–7.

[60] Kato M, Sung WK, Kato K, Goodman DS. Immunohistochemical studies on the localization of cellular retinol-binding protein in rat testis and epididymis. Biol Reprod 1985;32:173–89.

[61] Eskild W, Ree AH, Levy FO, Jahnsen T, Hansson V. Cellular localization of mRNAs for retinoic acid receptor-alpha, cellular retinol-binding protein, and cellular retinoic acid-binding protein in rat testis: evidence for germ cell-specific mRNAs. Biol Reprod 1991;44:53–61.

[62] Zheng WL, Bucco RA, Schmitt MC, Wardlaw SA, Ong DE. Localization of cellular retinoic acid-binding protein (CRABP) II and CRABP in developing rat testis. Endocrinology 1996;137:5028–35.

[63] Moura AA, Dore Jr JJ, Eberhardt DM, Erickson BH, Godkin JD. Immunolocalization of retinol-binding protein and profiles of its mRNA as related to testicular development in the beef bull. Domest Anim Endocrinol 1998;15:35–44.

[64] Davis JT, Ong DE. Synthesis and secretion of retinol-binding protein by cultured rat Sertoli cells. Biol Reprod 1992;47:528–33.

[65] Galdieri M, Piantedosi R, Blaner WS. Levels of binding proteins for retinoids in cultured Sertoli cells: effect of medium composition. Biochim Biophys Acta 1989;1011:168–70.

[66] E X, Zhang L, Lu J, Tso P, Blaner WS, Levin MS, et al. Increased neonatal mortality in mice lacking cellular retinol-binding protein II. J Biol Chem 2002;277:36617–23.

[67] Ghyselinck NB, Bavik C, Sapin V, Mark M, Bonnier D, Hindelang C, et al. Cellular retinol-binding protein I is essential for vitamin A homeostasis. EMBO J 1999; 18:4903–14.

[68] Quadro L, Blaner WS, Salchow DJ, Vogel S, Piantedosi R, Gouras P, et al. Impaired retinal function and vitamin A availability in mice lacking retinol-binding protein. EMBO J 1999;18:4633–44.

[69] Shingleton JL, Skinner MK, Ong DE. Retinol esterification in Sertoli cells by lecithin-retinol acyltransferase. Biochemistry (Mosc) 1989;28:9647–53.

[70] Rajan N, Kidd GL, Talmage DA, Blaner WS, Suhara A, Goodman DS. Cellular retinoic acid-binding protein messenger RNA: levels in rat tissues and localization in rat testis. J Lipid Res 1991;32:1195–204.

[71] Lampron C, Rochette-Egly C, Gorry P, Dolle P, Mark M, Lufkin T, et al. Mice deficient in cellular retinoic acid binding protein II (CRABPII) or in both CRABPI and CRABPII are essentially normal. Development 1995;121:539–48.

[72] Hogarth CA, Evanoff R, Mitchell D, Kent T, Small C, Amory JK, et al. Turning a spermatogenic wave into a tsunami: synchronizing murine spermatogenesis using WIN 18,446. Biol Reprod 2013;88:40.
[73] Clermont Y. Kinetics of spermatogenesis in mammals: seminiferous epithelium cycle and spermatogonial renewal. Physiol Rev 1972;52:198–236.
[74] Russell LD, Ettlin RA, Sinha Hikim AD, Clegg EP. Histological and histopathological evaluation of the testis. 1st ed. St. Louis, MO, USA: Cache River Press; 1990.
[75] Davis JC, Snyder EM, Hogarth CA, Small C, Griswold MD. Induction of spermatogenic synchrony by retinoic acid in neonatal mice. Spermatogenesis 2013;3:e23180.
[76] Snyder EM, Davis JC, Zhou Q, Evanoff R, Griswold MD. Exposure to retinoic acid in the neonatal but not adult mouse results in synchronous spermatogenesis. Biol Reprod 2011;84:886–93.
[77] Cheng CY, Mruk DD. The blood-testis barrier and its implications for male contraception. Pharmacol Rev 2012;64:16–64.
[78] Smith BE, Braun RE. Germ cell migration across Sertoli cell tight junctions. Science 2012;338:798–802.
[79] Chung SS, Choi C, Wang X, Hallock L, Wolgemuth DJ. Aberrant distribution of junctional complex components in retinoic acid receptor alpha-deficient mice. Microsc Res Tech 2010;73:583–96.
[80] Nicholls PK, Harrison CA, Rainczuk KE, Wayne Vogl A, Stanton PG. Retinoic acid promotes Sertoli cell differentiation and antagonises activin-induced proliferation. Mol Cell Endocrinol 2013;377:33–43.
[81] Qian X, Mruk DD, Cheng CY. Rai14 (retinoic acid induced protein 14) is involved in regulating f-actin dynamics at the ectoplasmic specialization in the rat testis*. PLoS One 2013;8:e60656.
[82] Hasegawa K, Saga Y. Retinoic acid signaling in Sertoli cells regulates organization of the blood-testis barrier through cyclical changes in gene expression. Development 2012;139:4347–55.
[83] Mascrez B, Ghyselinck NB, Watanabe M, Annicotte JS, Chambon P, Auwerx J, et al. Ligand-dependent contribution of RXRbeta to cholesterol homeostasis in Sertoli cells. EMBO Rep 2004;5:285–90.
[84] Timmons PM, Rigby PW, Poirier F. The murine seminiferous epithelial cycle is pre-figured in the Sertoli cells of the embryonic testis. Development 2002;129: 635–47.
[85] Johnston DS, Wright WW, Dicandeloro P, Wilson E, Kopf GS, Jelinsky SA, et al. Stage-specific gene expression is a fundamental characteristic of rat spermatogenic cells and Sertoli cells. Proc Natl Acad Sci USA 2008;105:8315–20.
[86] Orth JM, Gunsalus GL, Lamperti AA. Evidence from Sertoli cell-depleted rats indicates that spermatid number in adults depends on numbers of Sertoli cells produced during perinatal development. Endocrinology 1988;122:787–94.
[87] Vergouwen RP, Jacobs SG, Huiskamp R, Davids JA, de Rooij DG. Proliferative activity of gonocytes, Sertoli cells and interstitial cells during testicular development in mice. J Reprod Fertil 1991;93:233–43.
[88] Buzzard JJ, Wreford NG, Morrison JR. Thyroid hormone, retinoic acid, and testosterone suppress proliferation and induce markers of differentiation in cultured rat sertoli cells. Endocrinology 2003;144:3722–31.
[89] Zanotto-Filho A, Schroder R, Moreira JC. Differential effects of retinol and retinoic acid on cell proliferation: a role for reactive species and redox-dependent mechanisms in retinol supplementation. Free Radic Res 2008;42:778–88.

[90] Klamt F, Dal-Pizzol F, Roehrs R, de Oliveira RB, Dalmolin R, Henriques JA, et al. Genotoxicity, recombinogenicity and cellular preneoplasic transformation induced by vitamin A supplementation. Mutat Res 2003;539:117–25.

[91] Mendis SH, Meachem SJ, Sarraj MA, Loveland KL. Activin A balances Sertoli and germ cell proliferation in the fetal mouse testis. Biol Reprod 2011;84:379–91.

[92] Mithraprabhu S, Mendis S, Meachem SJ, Tubino L, Matzuk MM, Brown CW, et al. Activin bioactivity affects germ cell differentiation in the postnatal mouse testis *in vivo*. Biol Reprod 2010;82:980–90.

[93] Archambeault DR, Yao HH, Activin A. A product of fetal Leydig cells, is a unique paracrine regulator of Sertoli cell proliferation and fetal testis cord expansion. Proc Natl Acad Sci USA 2010;107:10526–31.

[94] Nicholls PK, Stanton PG, Chen JL, Olcorn JS, Haverfield JT, Qian H, et al. Activin signaling regulates Sertoli cell differentiation and function. Endocrinology 2012;153: 6065–77.

[95] Barakat B, O'Connor AE, Gold E, de Kretser DM, Loveland KL. Inhibin, activin, follistatin and FSH serum levels and testicular production are highly modulated during the first spermatogenic wave in mice. Reproduction 2008;136:345–59.

[96] Buzzard JJ, Loveland KL, O'Bryan MK, O'Connor AE, Bakker M, Hayashi T, et al. Changes in circulating and testicular levels of inhibin A and B and activin A during postnatal development in the rat. Endocrinology 2004;145:3532–41.

[97] Okuma Y, O'Connor AE, Hayashi T, Loveland KL, de Kretser DM, Hedger MP. Regulated production of activin A and inhibin B throughout the cycle of the seminiferous epithelium in the rat. J Endocrinol 2006;190:331–40.

[98] Li H, Palczewski K, Baehr W, Clagett-Dame M. Vitamin A deficiency results in meiotic failure and accumulation of undifferentiated spermatogonia in prepubertal mouse testis. Biol Reprod 2011;84:336–41.

[99] Pasquali MA, Gelain DP, Zanotto-Filho A, de Souza LF, de Oliveira RB, Klamt F, et al. Retinol and retinoic acid modulate catalase activity in Sertoli cells by distinct and gene expression-independent mechanisms. Toxicol In Vitro 2008;22: 1177–83.

[100] Gelain DP, de Bittencourt Pasquali MA, Zanotto Filho A, de Souza LF, de Oliveira RB, Klamt F, et al. Retinol increases catalase activity and protein content by a reactive species-dependent mechanism in Sertoli cells. Chem Biol Interact 2008;174:38–43.

[101] Gelain DP, de Bittencourt Pasquali MA, Caregnato FF, Zanotto-Filho A, Moreira JC. Retinol up regulates the receptor for advanced glycation endproducts (RAGE) by increasing intracellular reactive species. Toxicol In Vitro 2008;22:1123–7.

[102] de Oliveira RB, de Bittencourt Pasquali MA, Filho AZ, Dalmolin RJ, Gelain DP, Gottfried C, et al. Can electrons travel through actin microfilaments and generate oxidative stress in retinol treated Sertoli cell? Mol Cell Biochem 2007;301:33–45.

[103] Klamt F, Dal-Pizzol F, Gelain DP, Dalmolin RS, Birnfeld de Oliveira R, Bastiani M, et al. Vitamin A treatment induces apoptosis through an oxidant-dependent activation of the mitochondrial pathway. Cell Biol Int 2008;32:100–6.

[104] Zanotto-Filho A, Gelain DP, Schroder R, Souza LF, Pasquali MA, Klamt F, et al. The NF kappa B-mediated control of RS and JNK signaling in vitamin A-treated cells: duration of JNK-AP-1 pathway activation may determine cell death or proliferation. Biochem Pharmacol 2009;77:1291–301.

[105] Zanotto-Filho A, Cammarota M, Gelain DP, Oliveira RB, Delgado-Canedo A, Dalmolin RJ, et al. Retinoic acid induces apoptosis by a non-classical mechanism of ERK1/2 activation. Toxicol In Vitro 2008;22:1205–12.

[106] Silva FR, Leite LD, Wassermann GF. Rapid signal transduction in Sertoli cells. Eur J Endocrinol 2002;147:425–33.
[107] Piskunov A, Rochette-Egly C. A retinoic acid receptor RARalpha pool present in membrane lipid rafts forms complexes with G protein alphaQ to activate p38MAPK. Oncogene 2012;31:3333–45.
[108] Hogarth CA, Griswold MD. The key role of vitamin A in spermatogenesis. J Clin Invest 2010;120:956–62.

Stage-specific gene expression by Sertoli cells

10

William W. Wright
Division of Reproductive Biology, Department of Biochemistry and Molecular Biology, Johns Hopkins Bloomberg School of Public Health, Baltimore, MD

I. Introduction

A. Male fertility requirement for the production of millions of sperm per day

A man's fertility is, in part, a numbers game because the probability that any one sperm will fertilize an ovulated egg is extremely low. Only 1 in 25,000 ejaculated sperm enters the oviduct, the site of fertilization [1]. This fact is one reason why, when the concentration of sperm in the ejaculate drops below 20 million/ml, a man is generally considered to be infertile [2]. To achieve fertility, a healthy man aged 21–50 years must produce approximately 160 million sperm per day or 1800 sperm per heartbeat. This prodigious production occurs in a paired organ with a combined volume of 30–50 ml [3]. The ability of testes to continuously produce such large numbers of gametes is due, in part, to the exquisite organization of the seminiferous epithelium, the tissue in which spermatogenesis occurs.

B. Morphological basis for the production of large numbers of sperm required for male fertility

The general organization of the seminiferous epithelium is the same for all male mammals. This epithelium is organized into tubules whose general structures are dictated by two somatic cells, the stromal peritubular myoid cells that reside on the exterior of the tubule and the Sertoli cells, which are highly branched cells that extend from the basement membrane to the lumen of the tubule and spatially organize the spermatogenic cells (Figure 10.1). Spermatogonia reside in the basal compartment of the tubule between the Sertoli cells and the basement membrane. Spermatocytes reside along the basal–lateral plasma membrane of the Sertoli cells, while one or two generations of spermatids reside along their more apical plasma membrane. A remarkable feature of this epithelium is that in all mammals, spermatogonia, spermatocytes, and spermatids at specific phases of development are always found together, and these cells develop synchronously. In developing synchronously, these cells progress through a series of cellular associations called the "stages of the cycle of the seminiferous epithelium" [4,5]. Each of the specific associations of

Sertoli Cell Biology. DOI: http://dx.doi.org/10.1016/B978-0-12-417047-6.00010-7

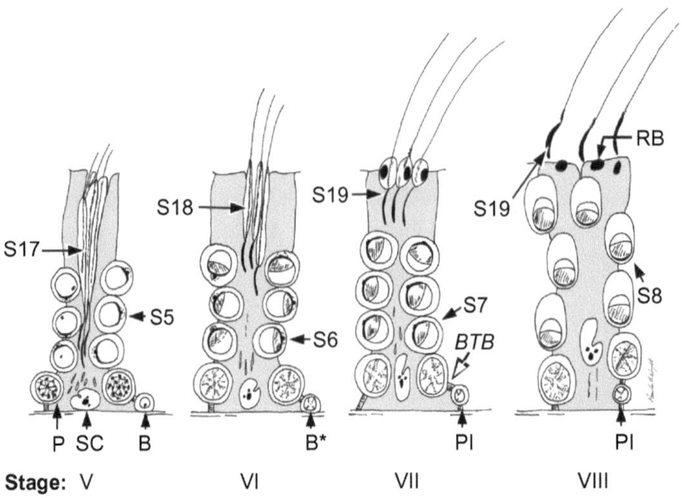

Figure 10.1 Illustration of 4 of the 14 stages of the cycle of the seminiferous epithelium of the rat. SC, a Sertoli cell whose cytoplasm is shaded gray; B, B spermatogonia; B*, B spermatogonia undergoing mitosis; P, pachytene spermatocyte; S5, step 5 spermatid; S17, step 17 spermatid; S6, step 6 spermatid; S18, step 18 spermatid; Pl, preleptotene spermatocyte; BTB, blood–testis barrier (pointed to by a white arrow); S7 = step 7 spermatid; S19, step 19 spermatid; S8, step 8 spermatid; RB, residual body.

spermatogonia, spermatocytes, and spermatids define a stage, and, with the exception of men and a few other higher primates, all germ cells in a cross section of a tubule are at the same stage [6,7]. In most mammals, one stage follows another along the length of a tubule, thereby creating a wave of the seminiferous epithelium [8].

While the coordinated development of the germ cells surrounding a given Sertoli cell is a continuous and not a stochastic process, the seminiferous epithelium of rats can be conveniently described as exhibiting 14 distinct stages of the cycle of the seminiferous epithelium [4]. Figure 10.1 depicts four of these stages and illustrates the coordinated development of the spermatogonia, spermatocytes, and spermatids around a single Sertoli cell. At stage V, step 17 spermatids are present in crypts formed by deep invaginations of the Sertoli cell's plasma membrane. By stage VI, these invaginations begin to disappear as the spermatids move centripetally toward the apical surface of the Sertoli cells, which they reach by stage VII. At the end of stage VIII, the step 19 spermatids are released into the tubule lumen in a process called "spermiation." Also during stages V–VII, steps 5–7 spermatids complete their formation of their acrosome cap. At stage VIII, the step 8 spermatids rotate so that their nuclei and acrosome face the basement membrane of the seminiferous tubule. Finally, the last mitotic division of spermatogonia occurs at stage VI, thereby forming preleptotene spermatocytes. At stage VII, the preleptotene spermatocytes begin to cross the blood–testis barrier, a complex of intercellular junctions between adjacent Sertoli cells that physiologically isolate the adluminal compartment of the seminiferous tubule.

As the spermatogonia, spermatocytes, and spermatids develop synchronously, the Sertoli cells with which they are associated exhibit significant morphological changes. There are changes in the shape and volume of the Sertoli cells as well as changes in the volumes of the rough endoplasmic reticulum, smooth endoplasmic reticulum, Golgi apparatus, mitochondria, lysosomes, lipid droplets, and cytoskeleton [9–12]. These marked stage-specific changes in Sertoli cell structure are suggestive of significant changes in Sertoli cell function, and they are consistent with the hypothesis that Sertoli cells recognize and respond to the developmental status of their surrounding spermatogenic cells.

C. Scope of this review

This review explores data supporting the general hypothesis that stage-specific changes in gene expression are a fundamental characteristic of Sertoli cells and that the resulting changes in Sertoli cell function affect the development of the adjacent spermatogonia, spermatocytes, and/or spermatids. This review is conducted in three steps. First, we briefly describe the initial evidence showing that Sertoli cell functions change as the surrounding spermatogenic cells progress through the stages of the cycle of the seminiferous epithelium. Second, we review what we have learned about the regulation of stage-specific gene expressions from studies of the cathepsin L (*Ctsl*) gene. Third, we discuss new insights into stage-specific Sertoli cell gene expression that come from a reanalysis of a published dataset that was generated by a genome-wide analysis of stage-specific gene expression by Sertoli cells, spermatogonia, early spermatocytes, pachytene spermatocytes, and round spermatids [13].

II. Evidence that spermatogenic cells regulate biologically important, stage-specific functions of Sertoli cells

A. The pioneering work of Parvinen and colleagues

While stage-specific changes in Sertoli cell morphology have suggested significant changes in the function of these cells, direct analysis of function required the development by Martti Parvinen (University of Turku, Finland) of a method for transillumination-assisted microdissection of seminiferous tubules. Because the position of compacted spermatids changes with the progression of the stages of the cycle and because their position affects light adsorption by the tubules, this method allowed for the isolation of living tubules at defined stages of the cycle [14]. With his many collaborators, Parvinen documented many large stage-specific changes in Sertoli cell function, including the secretion of androgen-binding protein, plasminogen activator, and 15 other proteins there were identified by two-dimensional gel electrophoresis [15–17]. By using irradiation-induced maturation depletion and repletion of the seminiferous epithelium, Parvinen generated the first direct evidence that stage-specific changes in Sertoli cell function were driven by signals from specific spermatogenic cells (reviewed in Ref. [18]). Furthermore, Parvinen and his collaborators demonstrated that when growth factors known to be secreted by Sertoli cells were added *in vitro* to tubules at defined stages, DNA synthesis by spermatogonia or spermatocytes was affected [19–22].

B. Consequences of genetically altering the expression of two stage-specific genes

The significance of stage-specific Sertoli gene expression was further emphasized by the fact that some genes that are expressed in a stage-specific manner appear to be essential for male fertility. For example, androgen receptor expression by Sertoli cells is maximal at stages VII–VIII, and loss of androgen stimulation to the Sertoli cells results in detachment of round spermatids at stages VI–VIII and the failure of spermiation at stage VIII [23–25].

In summary, the pioneering work by Parvinen and colleagues and subsequent studies of knockout mice provide strong support for the hypothesis that spermatogenic cells regulate stage-specific gene expression by Sertoli cells. Furthermore, some of the Sertoli cell products that are expressed in a stage-specific manner appear to be essential for spermatogenesis. This fact raises the question of which mechanisms are used by spermatogenic cells to regulate Sertoli cell function.

III. CTSL, a model for the analysis of the function and regulation of stage-specific gene expression

A. Identification of CTSL as a stage-specific secretory product of Sertoli cells

One of the stage-specific secretory proteins that were identified by two-dimensional gel electrophoresis of medium from cultured tubules was initially called "cyclic protein 2" (CP-2) [17]. CP-2 was the predominant protein secreted by stages VI and VII tubules, but it was not secreted by stages XIII–XIV tubules. To develop CP-2 as a model in which to study stage-specific germ-cell/Sertoli-cell interactions, the protein was isolated and used to generate a monospecific antibody [26,27]. Through the use of this antibody, it was demonstrated that in the rodent testis, this protein was only synthesized by Sertoli cells and, once synthesized by these cells, it was quantitatively secreted [27,28]. The antibody was also used for expression cloning of cDNAs encoding this protein. DNA sequencing showed that CP-2 was the proenzyme form of cathepsin L (CTSL), a lysosomal protein that is secreted when it is expressed at high levels [29].

Fortuitously, soon afterward these results were obtained, the spontaneous *furless* mutation was identified as rendering CTSL catalytically inactive [30]. Morphologic examination of testes of homozygous adult *furless* mice revealed a substantial increase in death rates of germ cells and development of Sertoli-cell-only tubules (Figure 10.2) [31]. In addition, it was shown that apparently normal, nonatrophic tubules contained 16% fewer preleptotene spermatocytes and 25% fewer pachytene spermatocytes than wild-type control tubules.

To gain further insight into the function of CTSL in the seminiferous epithelium, a detailed analysis of the distribution of this protein was performed by immunocytochemistry and confocal microscopy. Results showed that CTSL was loaded into residual bodies, plasma membrane-enclosed structures that contain the organelles and excess cytoplasm shed by step 19 spermatids at spermiation (see Figure 10.1) [32]. The significance of this finding included that residual bodies exhibit many of the biochemical characteristics of apoptotic cells, CTSL is also loaded into apoptotic germ cells, and failure of Sertoli cells to engulf apoptotic germ cells leads to infertility [33–36]. Thus, we conclude that the function of stage-specific expression of CTSL is to insure that the Sertoli cells degrade membrane-bound structures with the characteristics of apoptotic bodies efficiently and thoroughly, thereby preventing damage to the epithelium that could lead to infertility.

If efficient spermatogenesis requires CTSL, do germ cells regulate the transcription of this gene and thereby promote their own development? Metabolic labeling studies proved that stage-specific changes in *Ctsl* mRNA expression reflected the rate of gene transcription [37]. Furthermore, analysis of the effect of maturation depletion and repletion of spermatogenic cells proved that at stages I–V and IX–XIV, late pachytene spermatocytes and/or round spermatids suppressed *Ctsl* gene expression by Sertoli cell (Figure 10.3) [37]. This finding raised the question of the mechanisms by which this suppression occurred.

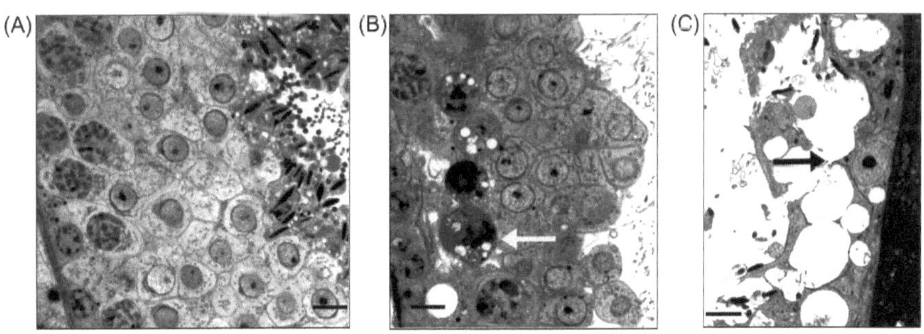

Figure 10.2 The morphology of the seminiferous epithelium of the *furless* mouse. (A) A cross section of normal stage VII seminiferous tubule in the testis of a *furless* mouse. (B) A cross section of an atrophying stage VII seminiferous tubule in the testis of the same *furless* mouse. The white arrow points to an apoptotic pachytene spermatocyte. Note the complete lack of step 19 spermatids. (C) A seminiferous tubule from the same *furless* mouse as in (A) and (B). The only cells within this tubule are Sertoli cells (see black arrow). The black bar on each part of this figure is equal to 10 μm.
Data are from Ref. 31.

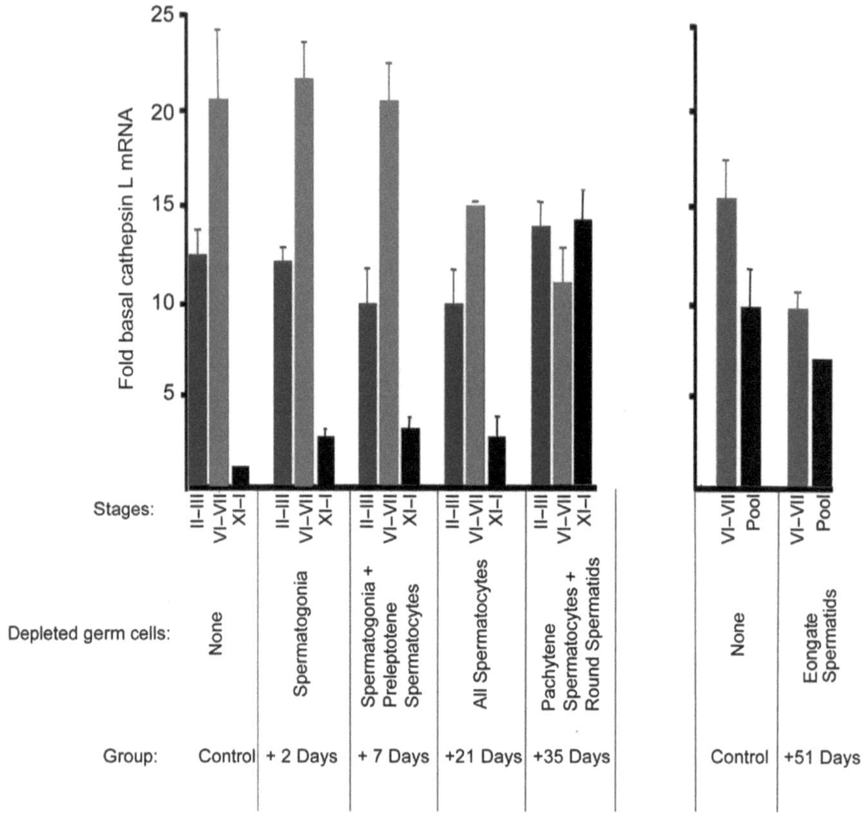

Figure 10.3 Stage-specific expression of Cathepsin L mRNA by Sertoli cells regulated by signals from spermatogenic cells. The testes of adult male rats were subjected to 3 Gy of γ-irradiation or were sham manipulated (control). At 2, 7, 21, 35, or 51 days thereafter, one testis was collected for microscopy and the cell types absent from the testis subsequent to irradiation-induced maturation depletion and repletion were identified. Seminiferous tubules at defined stages of the cycle were collected from the other testis, and the cathepsin L mRNA and clusterin mRNA content of the samples was measured. Clusterin mRNA levels were used to normalize the amounts of cathepsin L mRNA for equal numbers of Sertoli cells.
Data are from Ref. 37.

B. *Identification of domains within the* CTSL *gene that regulate stage-specific gene expression*

Because the full range of interactions between Sertoli cells and spermatogenic cells cannot be reconstituted *in vitro*, transgenic analysis was required to identify the cis-acting elements that regulate stage-specific *Ctsl* gene expression. The first step in this identification was to isolate a fragment of the *Ctsl* gene that drove accurate stage-specific reporter gene expression *in vivo* in Sertoli cells. The second step was to identify domains within the gene that responded to signals from spermatogenic cells. The first transgene tested was *Ctsl* (−2065/+977)-LacZ (Figure 10.4). This genomic fragment included 2065 bp of sequence upstream from the transcription start site, the first exon, the first intron, and the 12 bp of the second exon that immediately preceded the translation start site [38]. In two independent lines of transgenic mice, Sertoli cells expressed LacZ in an accurate, stage-specific manner; no other cells in these transgenic mice expressed the transgene. This result was a surprise because CTSL is expressed in many other organs, especially the brain and kidney [28]. Therefore, *Ctsl* (−2065/+977)-LacZ contains all of the regulatory cis-acting elements required for stage-specific gene expression by Sertoli cells but lacks the elements required for expression of the *Ctsl* gene by other cell types.

To begin to identify the location of the cis-acting transcriptional repressors by which germ cells exert their stage-specific effects on *Ctsl* gene transcription, two other transgenes were tested: *Ctsl* (−935/+977)-Lac Z and *Ctsl* (−451/+977)-Lac Z. The first of these two transgenes was expressed by Sertoli cells at all stages of the cycle and by some spermatogenic cells; the second transgene was not expressed by Sertoli cells but was expressed by spermatogenic cells [39]. Furthermore, both of these transgenes were expressed in many other organs (Figure 10.4). Taken together, the analyses of expression of these three transgenes led to two important conclusions. First, as predicted by our studies of the effects of spermatogenic cell maturation depletion and repletion on stage-specific expression of *Ctsl* mRNA, stage-specific expression of the *Ctsl* gene results primarily from transcriptional repression at stages I–IV and IX–XIV of the cycle. Second, stage-specific transcription of the *Ctsl* gene requires the interaction of cis-acting transcriptional repressor elements that are located between −2065 and −935 and cis-acting enhancer elements that are located between −935 and −451. We propose that

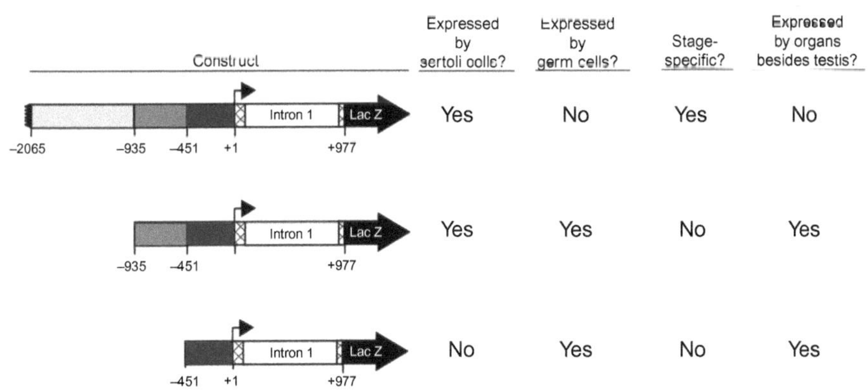

Figure 10.4 Summary of the expression in adult mice of three different *Cathepsin L-LacZ* transgenes. Shown are the general structures of the three transgenes, including the fragments of the *cathepsin L* gene that they contain and whether the transgene is expressed by Sertoli cells, by germ cells, by Sertoli cells in a stage-specific manner, or by organs other than the testis.

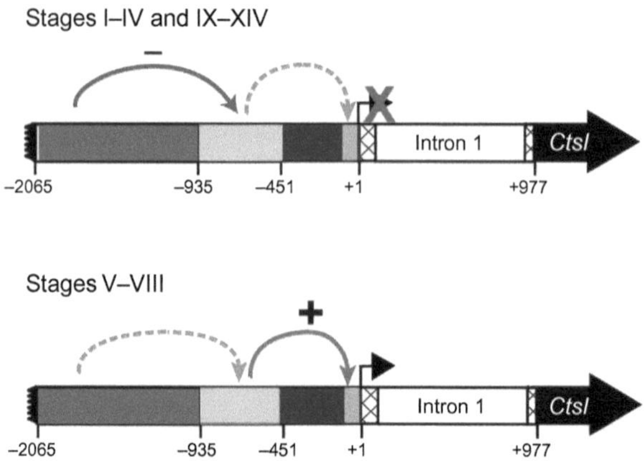

Figure 10.5 Model of the regulation of the stage-specific expression of the *Cathepsin L* gene. The diagram depicts four functional regions upstream from the transcription start site (TSS, bent arrow). Region 1 spans −451 to +1 (relative to the TSS) and contains the proximal promoter, which is denoted by a light gray box overlapping the TSS. Region 2 spans −935 to −451. Region 3 spans −2065 to −935. The model depicts the repression of transcription at stages I–IV and stages IX–XIV and activation of transcription at stages V through VIII. We propose that at stages I–IV and stages IX–XIV, transacting factors bind to cis-acting transcriptional repressors in region 3. We further propose that this event prevents the binding of other factors to cis-acting transcriptional activators in region 2. We also propose that at stages V–VIII, there is no expression of the factors that bind to the cis-acting repressor elements in region 3, which allows for the binding of other factors to the cis-acting transcriptional activators in region 2. These factors then interact with the core promoter and transcription ensues.
This figure summarizes data presented in Refs. 38 and 39.

when the cis-acting factors are bound to the transcriptional repressor elements, binding of other factors to the downstream enhancers is blocked. It follows that at stages VI–VIII, the cis-acting transcriptional repressors must be unoccupied, allowing binding of transacting factors to the enhancer elements and, consequently, expression of the *Ctsl* gene (Figure 10.5).

IV. Stage-specific gene expression as a fundamental characteristic of Sertoli cells

A. Genome-wide analysis of stage-specific gene expression by Sertoli cells

While the preceding discussion proves that Sertoli cells express genes in a stage-specific manner, it sheds no light on the numbers of genes with this characteristic, whether groups of genes within a particular pathway are similarly regulated, or whether groups of genes with similar functions share a common, stage-specific pattern of expression. These issues are important because they address the extent to which Sertoli cell functions are regulated by the coordinated development of spermatogonia, spermatocytes, and spermatids and the potential consequences of these responses to the germ cells.

To address these issues and to explore the extent to which spermatogenic cells also express genes in a stage-specific manner, Affymetrix array analysis was used to obtain a genome-wide view of gene expression by rat spermatogonia and early spermatocytes, pachytene spermatocytes, round spermatids, and Sertoli cells and by seminiferous tubules at defined stages of the cycle. Probe sets whose expression varied at least fourfold between two cell types and between two stages of the cycle were defined as stage specific for a specific cell type. (For the purpose of this analysis, a probe set was defined as a transcript that binds to a set of oligonucleotides on the array that are complementary to a specific nucleotide sequence.) A total of 1286 probe sets were identified that met these criteria [13].

Since the publication of that dataset, there have been considerable advances in the annotation of the rat genome and the identification of the genes recognized by the probe sets on the Affymetrix RAE230 2.0 array. This fact and the necessarily brief discussion of the data on Sertoli cells in the published analysis [13] prompted a reanalysis of the data on Sertoli cells. This reanalysis was conducted as follows. The transcript identified by each probe set was verified through the use of the NetAffx™ Analysis Center at Affymetrix (http://www.affymetrix.com/estore/) and one of the Gene Ontology (GO) biological process, molecular function, and cellular component terms listed for each probe set was selected to gain insight into the function of each of the genes in Sertoli cells. The relevance of the GO terms to each gene was further evaluated by comparisons of these genes to the Entrez Gene Database summary, the GeneCards® summary, and the Tocris Bioscience summary provided in GeneCards (http://www.genecards.org). In some cases, further evaluation was conducted by consultation of the primary literature. Finally, pathway and functional annotation clustering analysis was performed using the Database for Annotation, Visualization and Integrated Discovery (DAVID), version 6.7 Functional Bioinformatics Resources (http://david.abcc.ncifcrf.gov). When annotations in DAVID were inconsistent with the annotations from Affymetrix, the probe set was removed from further bioinformatic analyses.

The array analysis described above identified 3113 probe sets that were expressed at higher levels in Sertoli cells than in spermatogenic cells. The expression of 193 of these 3113 probe sets varied at least fourfold between two stages of the cycle (Figure 10.6 and Table 10.1). Based on the assumption that these 3113 probe sets are representative of the entire Sertoli cell transcriptome, we concluded that approximately 6% of a Sertoli cell's transcriptome is expressed in a stage-specific manner. According to the current annotation of the rat genome, 157 genes are represented in the 193 probe sets (Table 10.1). The data also show that distinct probe sets and, thus, the genes that they recognize, are maximally expressed at each of the stages of the cycle of the seminiferous epithelium. In addition, based on the assumption that the expressions of most of these genes by Sertoli cells are regulated by signals from spermatogenic cells, the data suggest that many different

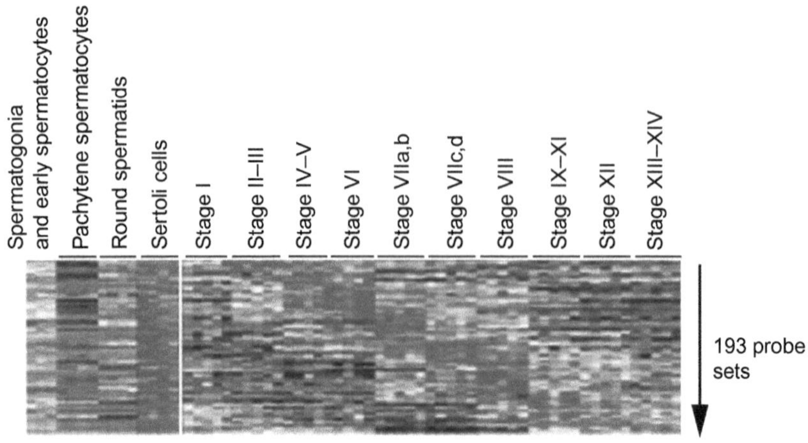

Figure 10.6 Heat map showing the relative levels of 193 probe sets expressed in a stage-specific manner by Sertoli cells. The data show the pattern of expression of each probe set, with progression of the stages of the cycle. The heat map also compares the expression of the probe set by Sertoli cells with the expression by spermatogonia, early spermatocytes, pachytene spermatocytes, and round spermatids.
The data are from Ref. 13.

Table 10.1 Number of probe sets and currently annotated genes maximally expressed by Sertoli cells per stage

Stage	Number of probe sets	Number of currently annotated genes
I	3	3
II–III	13	13
IV–V	13	10
VI	32	24
VIIa,b	29	20
VIIc,d	25	22
VIII	36	27
IX–XI	26	23
XII	8	7
XIII–XIV	8	8
Total	193	157

paracrine and juxtracrine signals from spermatogenic cells affect gene expression by Sertoli cells.

The next fact that is evident from the data is that some of these stage-specific changes in gene expression are very large. If the changes in signal intensity on the array actually reflect the rates of gene transcription, as is the case for the *Ctsl* gene, there must be large stage-specific variations in the extrinsic and intrinsic signals regulating transcription by Sertoli cells. Table 10.2 identifies 14 probe sets and their

Table 10.2 Genes whose expressions by Sertoli cells vary at least 15-fold as the surrounding spermatogenic cells progress through the stages of the cycle of the seminiferous epithelium

Probe set	Gene description	Gene symbol	Max stage	Fold change	GO biological process	GO molecular function	GO cellular component
1384924_at	Prion protein dublet	Prnd	VIIa.b	717	Cellular copper ion homeostasis	Copper ion binding	Plasma membrane
1391252_at	Ring finger protein 41	Rnf41	VIIa.b	65	Protein polyubiquitination	Ubiquitin-protein ligase activity	
1367939_at	Retinol binding protein 1, cellular	Rbp1	IX–XI	59.47	Transport	Transporter activity	Cytoplasm
1374966_at	Doublecortin	Dcx	VI	57.79	Neuron migration	Microtubule binding	Microtubule
1371754_at	Solute carrier family 25 (mitochondrial carrier, phosphate carrier), member 25	Slc25a25	IV–V	56.118	Transmembrane transport	Calcium ion binding	Mitochondrial inner membrane
1390525_a_at	Stimulated by retinoic acid 6	Stra6	VIII	24.73	Developmental growth	Receptor activity; vitamin transporter activity	Plasma membrane
1368079_at	Pyruvate dehydrogenase kinase 1	Pdk1	VIII	23.75	Glucose metabolic process	Pyruvate dehydrogenase (acetyl-transferring) kinase activity	Mitochondrial matrix
1373886_at	Receptor accessory protein 1	Reep1	XIII–XIV	22.93	Protein insertion into membrane	Microtubule binding	Endoplasmic reticulum

(Continued)

Table 10.2 (Continued)

Probe set	Gene description	Gene symbol	Max stage	Fold change	GO biological process	GO molecular function	GO cellular component
1387061_at	Junction plakoglobin	Jup	VIII	22.7	Desmosome assembly; cell−cell adhesion	Alpha-catenin binding; cadherin binding	Cell−cell junction
1380229_at	V-maf musculoaponeurotic fibrosarcoma oncogene homolog F (avian)	Maff	VIII	20.41	Regulation of transcription, DNA-dependent	Sequence-specific DNA binding	Nucleus
1388122_at	Glutathione S-transferase, pi 2	Gstp1, Gstp2	XII	17.55	Response to reactive oxygen species	Glutathione transferase activity	Cytoplasm; mitochondrion; nucleus
1367729_at	Ornithine aminotransferase	Oat	VIIa.b	17.218	Ornithine metabolic process	*N*2-acetyl-L-ornithine:2-oxoglutarate 5-aminotransferase activity	Mitochrondrial matrix
1372592_at	Histone deacetylase 6	Hdac6	VIIc.d	16.02	Histone deacetylation	Histone deacetylase activity	Nucleus
1373175_at	Phospholipase B domain containing 1	Plbd1	VIII	15.12	Lipid metabolic process	Hydrolase activity	Extracellular region

annotated genes whose expressions by Sertoli cells vary more than 15-fold with progression of the stages of the cycle. This table also identifies the gene symbol, the stage of maximal expression, the fold change in gene expression with the progression of the stages, and relevant GO descriptors. The gene that exhibits the greatest stage-specific expression is Prion Protein 2 Doublet (*Prnd*). Expression of this gene varies approximately 700-fold as the stages of the cycle progress. An interesting fact is that the *Prnd* gene was generated by a duplication of the Prion Protein (*Prnp*) gene, and the two genes are separated by only 15 kb on the chromosome. Thus, it is noteworthy that these two genes exhibit very similar patterns of expression (Figure 10.7). The peak of *Prnd* expression occurs at stage VI and the peak of *Prnp* expression at stage VIIa,b, suggesting that the two genes are regulated in a similar manner. It is also noteworthy that *Prnd*-null mice are infertile, suggesting that the stage-specific expression of *Prnd* by Sertoli cells is fundamentally important to male fertility [40,41].

The rest of the genes identified in Table 10.2 encode proteins with a wide variety of biological functions, including protein degradation, metabolism, gene expression regulation, and regulation of cell structure. These gene products are targeted to many different domains in the cell—cytoplasm, microtubules, mitochondria, and nucleus. Thus, signals from germ cells potentially have significant effects on many different functions and in many different compartments of Sertoli cells.

B. Stage-specific regulation of the lysosome pathway

Two approaches initially were used for bioinformatics analysis of the stage-specific transcriptome of Sertoli cells, pathway analysis and functional annotation clustering. Because all of the genes in a pathway function together toward one biological end point, pathway analysis was conducted first. The results show that genes in one

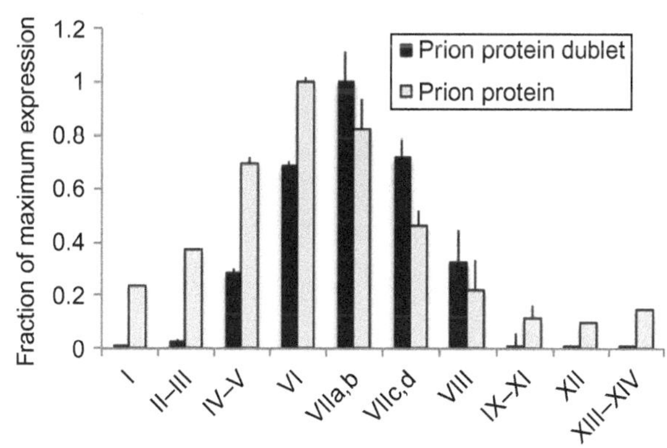

Figure 10.7 A comparison of the stage-specific expression of transcripts encoding prion protein 2 doublet (PRND) and prion protein (PRNP). Data on the expression at each stage (mean + SEM) are normalized to the level of maximal expression.

KEGG pathway, lysosome, were enriched in the dataset. Three genes, *Ctsl*, cathepsin K (*Ctsk*), and lysosomal-associated membrane protein 2 (*Lamp2*), are maximally expressed at stages VI or VIIIa,b. One gene, sphingomyelin phosphodiesterase I, acid lysosomal (*Smpd1*), is maximally expressed at stage VIII. Two genes, mannosidase, beta A (*Manba*) and GM2 ganglioside activator (*Gm2a*), are maximally expressed at stage XII (Figure 10.8). These data are consistent with morphometric data showing that there are two stage-dependent changes in lysosome volume [42]. From stage VI to stage VIII, there is an increase in the volume density of primary lysosomes in the basal portion and midportion of Sertoli cells; at the end of the cycle, there is a second peak in primary lysosome volume. In addition, there is a marked but transient increase at stage IX in the volume density of secondary lysosomes (i.e., lysosomes fused with the residual bodies that have been engulfed by Sertoli cells).

The particular functions in Sertoli cells of the proteins encoded by these genes can be deduced from their biochemistry and their subcellular localization. CTSL

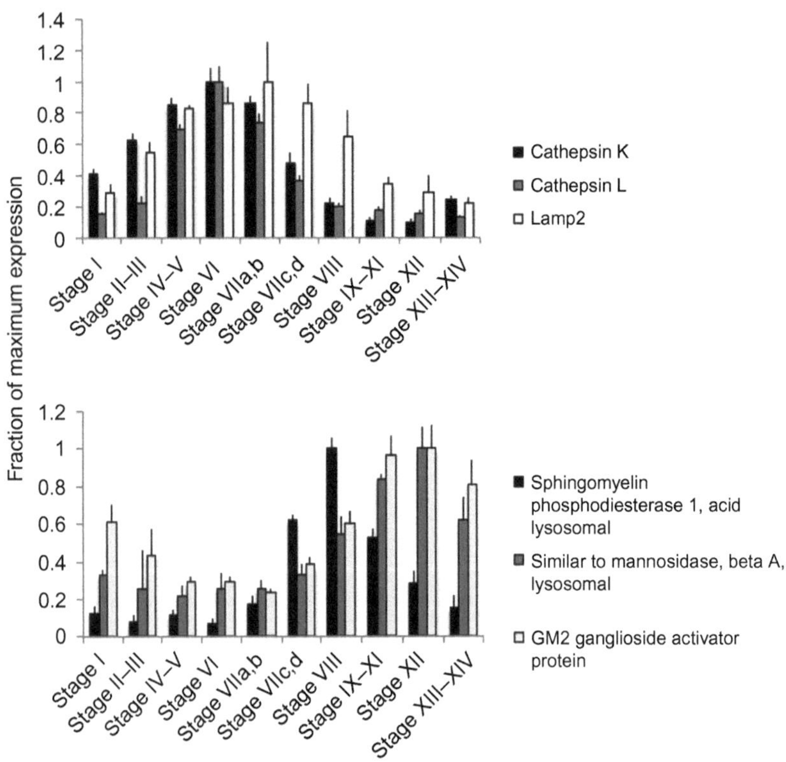

Figure 10.8 Stage-specific expresion by Sertoli cells of genes in the lysosome pathway. Data are shown for cathepsin (*Ctsl*), cathepsin K (*Ctsk*), lysosomal-associated membrane protein 2 (*Lamp2*) sphingomyelin phosphodiesterase 1, acid lysosomal (*Smpd1*), mannosidase, beta A (*Manba*), and GM2 ganglioside activator protein (*Gm2a*).
Data for expression at each stage (mean + SEM) are normalized to the level of maximal expression.

and CTSK are loaded into residual bodies once they are released from step 19 spermatids and engulfed by Sertoli cells, thereby priming the residual bodies for degradation [32]. LAMP2 is one of the two major lysosomal membrane proteins, and the increase in its expression partially explains the increase in the volume of primary lysosomes at stages VI–VIII [43]. SMPD1 converts sphingomelin into ceramide and, thus, participates in the breakdown of residual bodies. The increased expressions of *Manba* and *Gm2a* occur coincidently with the formation of secondary lysosomes. MANBA cleaves the nonreducing terminal mannose from carbohydrates [44]. GM2A is a glycolipid transport protein that acts as an essential cofactor for β-hexosaminidase [45]. Taken together, these data partially explain the cycle of lysosome biogenesis that occurs in Sertoli cells and suggests that in response to signals from spermatogenic cells, Sertoli cells mount a coordinated response that ensures that residual bodies are fully degraded and, therefore, that the seminiferous epithelium is protected from apoptotic bodies that, if they are not degraded, could threaten this tissue's integrity.

C. Stage-specific expression of genes with related functions

Functional annotation clustering allows one to identify genes with related functions that are enriched in a list of genes. To identify genes with similar functions that are expressed by Sertoli cells in a stage-specific manner, we used the DAVID Functional Bioinformatics Resource with the highest classification stringency. Table 10.3 shows the 24 clusters that were identified, the enrichment score (a measure of how enriched the probe sets in the cluster were in the entire dataset), the numbers of GO terms identified in that cluster, the GO term with the highest *P*-value, the number of genes in the cluster (in all cases, the same genes were identified for each GO term in a cluster), and the statistical significance of the enrichment as defined by a *P*-value.

The clusters shown in Table 10.3 can be grouped into four general categories: cell structure, cell development, regulation, and homeostasis. Not surprisingly, genes with the GO term "lysosome" were the most enriched in this set of data. While some of the GO terms do not appear to be immediately relevant to Sertoli cells (e.g., Z disc), this and other clusters contain genes that encode cytoskeletal or signaling proteins.

Given the stage-specific changes in Sertoli cell structure (Figure 10.1) and that the diverse stage-specific patterns of gene expression imply a complex signaling network, more focused analyses of genes involved in cytoskeleton and cell signaling were warranted. As described below, these analyses suggested that Sertoli cells coordinate many different functions with development of the adjacent spermatogenic cells.

1. Cytoskeleton (Table 10.4)

The cytoskeleton plays a crucial role in maintaining the polarity of Sertoli cells, in their positioning of spermatogenic cells within the epithelium, and in anchoring

Table 10.3 Functional annotation clustering of genes expressed by Sertoli in a stage-specific manner

Annotation cluster	Enrichment score	Number of GO terms in cluster	GO term with highest *P*-value	Number of genes/ cluster	*P*-value
1	1.62	3	Lysosome	6	1.9×10^{-2}
2	1.48	6	Z disc	4	8.2×10^{-3}
3	1.46	4	Negative regulator of proteolysis	4	2.4×10^{-3}
4	1.02	5	Cellular ion homeostasis	8	6.1×10^{-2}
5	0.89	3	Striated muscle cell development	3	6.5×10^{-2}
6	0.78	3	Tissue homeostasis	3	1×10^{-1}
7	0.77	4	Cellular di-, tri-valent inorganic cation homeostasis	5	1.3×10^{-1}
8	0.67	3	Negative regulation of apoptosis	6	2.1×10^{-1}
9	0.49	3	DNA catabolic process	3	7.8×10^{-2}
10	0.42	3	Chromatin modification	4	2.1×10^{-1}
11	0.37	3	Striated muscle tissue development	3	3.5×10^{-1}
12	0.35	0[a]	EGF-like, type 3	3	3.8×10^{-1}
13	0.30	3	Regulation of cell migration	3	4.6×10^{-1}
14	0.24	4	Regulation of neuron differentiation	3	4.5×10^{-1}
15	0.23	4	Cell migration	4	4.4×10^{-1}
16	0.22	2	Peptidase activity, action on L-amino acids	5	6.6×10^{-1}
17	0.21	3	Blood vessel morphogenesis	3	5.3×10^{-1}
18	0.21	3	Positive regulation of apoptosis	4	6.1×10^{-1}
19	0.21	4	Axonogenesis	3	5.3×10^{-1}
20	0.15	3	Purine ribonucleotide binding	11	6.9×10^{-1}

(*Continued*)

Table 10.3 (Continued)

Annotation cluster	Enrichment score	Number of GO terms in cluster	GO term with highest *P*-value	Number of genes/ cluster	*P*-value
21	0.14	4	Adenyl ribonucleotide binding	11	6.5×10^{-1}
22	0.14	0[b]	Serine/threonine-protein kinase, active site	3	7.0×10^{-1}
23	0.13	4	Regulation of phosphorylation	4	7.3×10^{-1}
24	0.08	5	Positive regulation of transcription from RNA polymerase II promoter	3	6.6×10^{-1}

[a]SP_PIR_KEYWORDS: EGF-like domain.
[b]SP_PIR_KEYWORDS: serine/threonine-protein kinase.

Sertoli cells to germ cells and to one another. The ectoplasmic specialization (ES), an actin-rich structure that mediates the adhesion of Sertoli cells to each other (basally) and to spermatids (apically), is important to germ-cell/Sertoli-cell interactions. As the stages of the cycle progress, there are dramatic changes in both the shape of Sertoli cells and their morphological relationships with other cell types (see Figure 10.1; [46]). We, therefore, asked whether there were significant changes with the progression of the stages in the expression of genes that encode important cytoskeletal components. In this analysis, we identified all probe sets that encoded genes associated with GO descriptors—cytoskeleton, microtubule, actin, or intermediate filament. Table 10.4 lists the probe sets, gene descriptions, gene symbols, stages of maximum and minimum expression, and fold change in transcript expression with the progression of the stages of the cycle. In addition, a relevant GO biology process, molecular function, and cellular component are listed for each gene.

a. The actin cytoskeleton

Much of the actin in Sertoli cells is located in the ES, subjacent to the plasma membrane that is associated with the blood–testis barrier or with sites of attachment to spermatids. Morphological analysis indicates that in both locations, the structure of the ES and the organization of its actin filaments are dynamic. For example, actin is distributed along the entire length of the crypts of the Sertoli cells that envelop the elongating spermatids at stage II, while at stage VII, actin is concentrated in the apex of the Sertoli cells around step 19 spermatid heads [12]. This finding suggests that the actin filaments within Sertoli cells' apical ESs become more tightly

Table 10.4 **Cytoskeletal elements encoded by genes whose expressions by Sertoli cells vary in a stage-specific manner**

Probe set	Gene description	Gene symbol	Max stage	Min stage	Fold change	GO biological process	GO molecular function	GO cellular component
1381575_at	Nebulin	NEB	I	VIIa,b	4.18		Protein binding	Contractile fiber
1389107_at	Cingulin-like 1	Cgnl1	II–III	VIII	4		Motor activity	Actin cytoskeleton
1374966_at	Doublecortin	Dcx	VI	XIII–XIV	33.4	Protein kinase binding	Protein kinase binding	Microtubule
1367970_at	Profilin 2	Pfn2	VIIa,b	XIII–XIV	7.4	Actin cytoskeleton organization	Actin binding	Actin cytoskeleton
1373363_at	Microtubule-associated protein 1B	Map1b	VIIa,b	XIII–XIV	3.8	Microtubule cytoskeleton organization	Microtubule binding	Cytoskeleton
1371360_at	N-myc downstream regulated 1	Ndrg1	VIIc,d	I	5	Positive regulation of spindle checkpoint	Microtubule binding	Microtubule cytoskeleton
1386647_at	Microtubule-associated protein-7	Map7	VIII	I	4.06	Microtubule bundle formation	Protein binding	Microtubule cytoskeleton
1387061_at	Junction plakoglobin	Jup	VIII	I	22.7	Cell–cell adhesion	Cadherin binding	Cell–cell junction
1367574_at	Vimentin	Vim	IX–XI	VIIa,b	6.1	Intermediate filament organization	Structural constituent of cytoskeleton	Intermediate filament
1388433_at	Keratin 19	Krt19	XII	VIIc,d	5.3	Sarcomere organization	Structural constituent of muscle	Intermediate filament
1373886_at	Receptor accessory protein 1	Reep1	XIII–XIV	VIIc,d	22.93	Protein insertion into membrane	Microtubule binding	Endoplasmic reticulum

bundled as their associated spermatids move apically. The ES associated with the blood–testis barrier must also be restructured for every 12.5 days that preleptotene and leptotene spermatocytes cross this barrier [47].

The genomic analysis described here identified four genes that encode actin-binding proteins. These genes are nebulin (*Neb*), cingulin-like 1 (*Cgnl1*), A kinase anchor protein 2 (*Akap2*), and profilin 2). The first two of these genes are maximally expressed at stages I and II–III, respectively. NEB stabilizes actin filaments and prevents depolymerization of actin bundles once they are formed [48]. CGNL1's function is to link cell junctions to cytoskeleton [49].

We therefore hypothesize that at stages I and II, as the elongate spermatids penetrate the crypts formed by Sertoli cells, CGNL1 increases the strength of the adhesive forces binding these two cells to each other, while NEB reshapes ES as the compacted spermatids are drawn into crypts formed by invaginations of the Sertoli cell plasma membrane. The gene encoding the actin bundling protein PFN2 is maximally expressed at stages VIIa and VIIb. Thus, PFN2 may facilitate the final compaction of the ES around the spermatids once they have reached the apex of Sertoli cells.

b. The microtubule cytoskeleton

Sertoli cells exhibit stage-dependent changes in both amounts of tubulin and the location of microtubules. Microtubules form a parallel array from the base to the apex of the Sertoli cells and are present in processes that surround spermatids. The microtubules in these processes are dynamic structures, and the amounts of tubulin they contain increase as their associated spermatids elongate [50]. Furthermore, as the elongate spermatids move toward the tubule lumen (stages VI–VIII), a ring of tubulin is formed and then lost from the apex of the Sertoli cells [51]. Thus, microtubules in these apical processes of Sertoli cells are highly dynamic [50]. Because microtubules are stabilized by microtubule-binding proteins, changes in their amount or post-translational modification must play important roles in the shaping of Sertoli cells and in determining these cells' association with the adjacent spermatogenic cells.

The array data identified four genes encoding microtubule-binding proteins that are expressed in a stage-specific manner by Sertoli cells: doublecortin (*Dcx*), microtubule-associated protein 1B (*Map1b*), n-myc downstream regulated 1 (*Ndrg1*), and microtubule-associated protein 7 (*Map7*) (Table 10.4). The overlapping, sequential pattern of expression of these four genes is noteworthy (Figure 10.9.). *Dcx* expression rises sharply at stages IV–V, peaks at stage VI, and then falls dramatically to a nadir at stage VIII. Stage-specific changes in the expression of *Map1b* are more gradual, with *Map1b* increasing from stage I to stage VIIa, b, and decreasing more sharply thereafter. *Ndrg1* expression does not begin to increase until stage VIIa,b, and it then peaks at stages VIIb,c and VIII. *Map7* expression is also increased at stages VIIa,b but does not peak until stage VIII. This sequential expression illustrates the precision of stage-specific gene expression by Sertoli cells. The patterns of expression of these genes and the known functions of the proteins they encode suggest that the genes have distinct functions in Sertoli cells. As a microtubule-binding protein, DCX nucleates the formation of

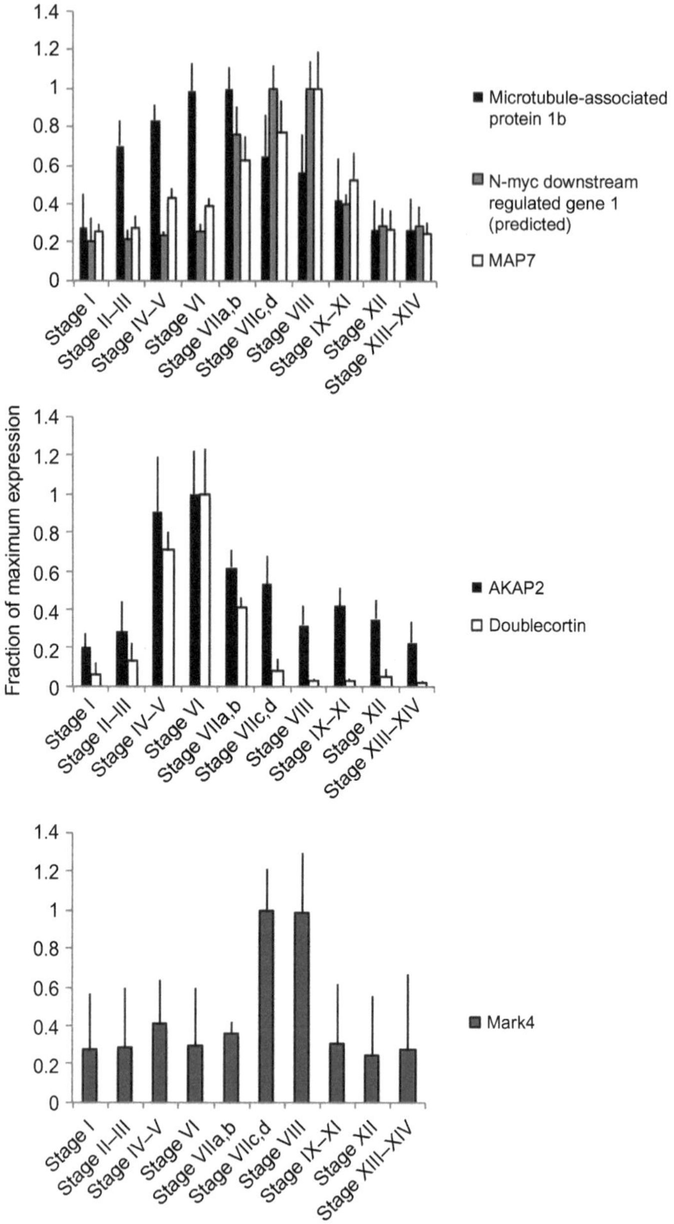

Figure 10.9 Stage-specific expression by Sertoli cells of genes that encode important components of the Sertoli cell ctoskeleton or signaling molecules. Data are shown for Doublecortin (*Dcx*), microtubule-associated protein 1B (*Map1b*), n-myc downstream regulated 1 (*Ndrg1*), and microtubule-associated protein-7 (*Map7*), A kinase anchor protein-2 (*Akap2*), doublecortin (*Dcx*), and MAP/microtubule affinity-regulating kinase 4 (*Mark4*). Data on the expression at each stage (mean + SEM) are normalized to the level of maximal expression.

microtubules and then stabilizes them [52] (Figure 10.9). As with all microtubule-binding proteins, phosphorylation of DCX decreases its affinity for microtubules, which, as a consequence, begin to depolymerize. In the processes of developing neurons, interactions of DCX, microtubules, and the kinase CDK5 result in the rapid turnover of microtubules, thereby promoting the rapid growth and development of axons [52,53]. It is noteworthy that CDK5 is localized to Sertoli cell microtubules [54]. Therefore, we hypothesize that from stages IV to VIIa,b, DCX is involved in generating and then eliminating the deep crypts in which the elongate spermatids are held.

MAP1B promotes interactions between microtubules and actin in developing neurons [55] and, thus, may link microtubules to the ES in Sertoli cells. NDRG1 is a major target for protein kinases C and A and, thus, may mediate their effects on the bundling of microtubules in the Sertoli cells [56]. Because phosphorylation of microtubule-binding proteins decreases their affinity for tubulin and destabilizes microtubules, the binding of NDRG1 may be a prerequisite for the dissolution of the microtubules in the apex of the Sertoli cells [57]. MAP7 is a kinesin-binding protein, and indirect evidence indicates that kinesin along with dynein is expressed by Sertoli cells [58]. Given that the expression of *Map7* mRNA peaks at stage VIII, it is tempting to speculate that microtubule motors may play at role in the rearrangement of the apical cytoskeleton immediately prior to spermiation.

c. Intermediate filaments

Intermediate filaments provide structural support, regulate key signaling pathways, and facilitate the movement of proteins to specific domains of polarized cells, such as Sertoli cells [59]. To date, no keratins have been localized to mature Sertoli cells, although keratin 19 mRNA was detected in the genomic array analysis described above (Table 10.4) [60]. Thus, vimentin is the primary, if not the sole, intermediate filament protein in Sertoli cells, and its stage-specific pattern of expression matches that of its transcript [61] (Table 10.4). Vimentin mRNA is maximally expressed from stages XII to I, and, at those stages, the protein extends from the base to the apex of the Sertoli cells [61]. At the nadir of vimentin expression (stages IX–XI), this cytoskeletal protein is only detected in the base of the cell. At all stages, vimentin filaments extend from the nucleus to the basal–lateral processes that form the blood–testis barrier [61].

Junction plakoblobin (*Jup*) encodes the one intermediate filament-binding protein expressed in a stage-specific manner by Sertoli cells (Table 10.4). Because vimentin is bound by JUP, it must play an important role in the intermediate filament dynamics of Sertoli cells [62]. This suggestion is supported by the fact that in other cell types, JUP links intermediate filaments to desmosomes [63]. Because desmosomes are functional components of the blood–testis barrier [64], we hypothesize that JUP links the blood–testis barrier to the basal vimentin filaments. Because *Jup* and *vimentin* mRNA expressions are maximal at stages VIII and IX–XI, respectively, we hypothesize that the coordinated synthesis of their protein products allows for an efficient assembly of the desmosomal junctions within the blood–testis barrier.

2. Signaling molecules

Stage-specific changes in Sertoli cell shape and function must be triggered by extrinsic or intrinsic signals. Given the Sertoli cells' diverse patterns of stage-specific gene expression (Figure 10.6 and Table 10.1), it seemed likely that some of the stage-specific genes encode components of signaling pathways. Therefore, we identified all probe sets and their associated genes with GO molecular terms that contained the words "kinase," "transcription," "signal," "receptor," "DNA binding," or "channel." Table 10.5 lists the 23 probe sets that were identified through this search. This table also provides the gene descriptions for the probe set, the gene symbols; the stages of maximal and minimal expression; fold changes in expression with progression of the stages of the cycle; and relevant GO biological processes, molecular functions, and cellular components. The 23 genes identified in Table 10.5 can be generally divided into five groups: kinase anchoring proteins, kinases, transcriptional regulators, receptors, and receptor ligands. One plasma-membrane-associated kinase substrate is also identified.

a. Kinase anchoring proteins

Genes encoding three anchoring proteins were identified: breast cancer anti-estrogen resistance 1 (*Bcar1*), Src homology 2 domain containing family, member 4 (*Shc4*), and A kinase anchor protein-2 (*Akap2*). *Bcar1* and *Sch4* are maximally expressed at stages I and II–III, respectively. BCAR1 is known to be concentrated in the ES and to be associated with dynamin and Src proto-oncogene nonreceptor tyrosine kinase (SRC) [65–67]. This association is potentially important to the seminiferous epithelium because in neurons, phosphorylation of dynamin stimulates bulk endocytosis [68]. We therefore hypothesize that BCAR1 regulates turnover in the plasma membrane of the Sertoli cells that overlie the ES because spermatids are drawn into the invaginations of the Sertoli cell plasma membrane at stage II.

Sch4 is maximally expressed at stages II–III, and the docking protein is concentrated at the plasma membrane, where it facilitates ligand-independent phosphorylation of the epidermal growth factor (EGF) receptor, which is also expressed by Sertoli cells [69,70]. This phosphorylation allows for cross talk between different receptor systems, leading us to suggest that the stage-dependent increase in SCH4 allows the efficient integration of multiple signaling pathways in Sertoli cells [71].

Akap2 mRNA is maximally expressed at stage VI and encodes an actin-binding protein that serves as a docking site for the RII regulatory subunit of PKA. Thus, AKAP2 represents a major link between cAMP-dependent signaling pathways and the Sertoli cell cytoskeleton [72,73].

b. Plasma-membrane-associated kinase substrate

Myristoylated alanine-rich protein kinase C substrate (*Marcks*) is maximally expressed at stage VI. MARCKS is targeted to the plasma membrane by myristoylation and serves as a major target for protein kinase C [74]. Gene knockout studies indicate that in the neurons of the developing brain, MARCKS is required for the targeting of N-cadherin and of β-catenin to the proper location within the plasma membrane of these highly polarized cell [75]. It is therefore noteworthy that both β-catenin and N-cadherin are expressed by Sertoli cells and are essential to male fertility [67,76].

Table 10.5 Signaling molecules encoded by genes whose expressions by Sertoli cells vary in a stage-specific manner

Probe set	Gene description	Gene symbol	Max stage	Min stage	Fold change	GO biological process	Go molecular function
1367752_at	Breast cancer anti-estrogen resistance 1	Bcar1	I	IX–XI	8.1	G-protein-coupled receptor signaling pathway	Protein kinase binding
1375205_at	p300/CBP-associated factor	Pcaf	II–III	XII	4.7	Regulation of transcription, DNA-dependent	Transcription cofactor activity
1382613_at	SHC (Src homology 2 domain containing) family, member 4	Shc4/Shcd	II–III	VIIa,b	5	Intracellular signal transduction	Receptor tyrosine kinase binding
1367816_at	HOP homeobox	Hopx	VI	XIII–XIV	7.6	Negative regulation of transcription from RNA polymerase II	Negative regulation of transcription from RNA polymerase II
1370948_a_at	Myristoylated alanine-rich protein kinase C substrate	Marcks	VI	XII	4.8		Protein kinase C binding
1371696_at	G-protein-coupled receptor 56	Gpr56	VI	IX–XI	4.2	Protein kinase C signaling cascade	Signal transducer activity
1383047_at	Growth arrest specific 6	Gas6	VI	XII	3.8	Activation of protein kinase B activity	Receptor tyrosine kinase binding
1386764_at	A kinase (PRKA) anchor protein 2	Akap2	VI	I	5	Transmembrane receptor protein serine/threonine kinase signaling	Receptor tyrosine kinase binding
1367628_at	Lectin, galactoside-binding, soluble, 1	Lgals1/ galectin1	VIIa,b	IX–XI	6.4	Positive regulation of I-kappaB kinase/NF-kappaB cascade	Signal transducer activity

(*Continued*)

Table 10.5 **(Continued)**

Probe set	Gene description	Gene symbol	Max stage	Min stage	Fold change	GO biological process	Go molecular function
1368243_at	Anti-Mullerian hormone receptor, type II	Amhr2	VIIa,b	XII−XIII	6.5	Transmembrane receptor protein serine/threonine	Transmembrane receptor protein serine/threonine kinase activity
1372038_at	MAP kinase-interacting serine/threonine kinase 2	Mknk2	VIIc,d	I	6.66	Intracellular protein kinase cascade	Protein serine/threonine kinase activity
1375399_a_at	MAP/microtubule affinity-regulating kinase 4	Mark4	VIIc,d	I	4.1	Microtubule cytoskeleton organization	Protein serine/threonine kinase activity
1377961_at	Zinc finger homeobox 3	Zfhx3/ Atbf1	VIIc,d	II−III	12.4	Negative regulation of transcription from RNA polymerase II	Sequence-specific DNA binding
1374189_at	Zinc finger protein 219	Zfp219/ Znf219	VIIc,d	XIII−XIV	4.5	Transcription, DNA-dependent	Sequence-specific DNA binding transcription
1367609_at	Macrophage migration inhibitory factor	Mif	VIII	I	7.7	Positive regulation of protein kinase A signaling cascade	Cytokine receptor binding
1373745_at	General transcription factor IIH, polypeptide 3	Gtf2h4	VIII	II−III	6.06	Regulation of transcription, DNA-dependent	Protein kinase activity

1387578_a_at	Purinergic receptor P2X, ligand-gated ion channel, 2	P2rx2	VIII	II–III	14	Chemoreceptor signaling	Extracellular ATP-gated cation channel activity
1389244_x_at	Chemokine (C-X-C motif) receptor 4	Cxcr4	VIII	II–III	4.6	G-protein-coupled receptor signaling pathway	G-protein-coupled receptor activity
1380229_at	V-maf musculoaponeurotic fibrosarcoma oncogene homolog F	Maff	VIII	I	20.3	Regulation of transcription, DNA-dependent	Sequence-specific DNA binding transcription
1386162_at	TXK tyrosine kinase	Txk	IX–XI	VIIa,b	9.8	Regulation of transcription from RNA polymerase II promoter	Protein kinase activity
1388255_x_at	RT1 class I, locus CE5	RT1-CE5	IX–XI	VIIc,d	6.6	Antigen processing and presentation of peptide antigen via MHC	Receptor binding
1388774_at	Methyl-CpG binding domain protein 2	Mbd2	IX–XI	VIIc,d	4.1	Negative regulation of transcription from RNA polymerase II	DNA binding
1389651_at	Apelin	Apln	IX–XI	IV–V	4.8	Signal transduction	Apelin receptor binding
1370216_at	Discoidin domain receptor tyrosine kinase 1	Ddr1	XIII–XIV	IX–XI	5.5	Transmembrane receptor protein tyrosine kinase signaling pathway	Protein tyrosine kinase activity

c. Kinases

The array data reveal two kinases that are maximally expressed by Sertoli cells at stage VIIc,d: Map kinase-interacting serine/threonine kinase 2 (*Mknk2*) and MAP/microtubule affinity-regulating kinase 4 (*Mark4*). MKNK2 phosphorylates the eukaryotic translation initiation factor eIF4E, thereby increasing its affinity for the mRNA cap structure [77,78]. Data indicate that MKNK2 also regulates the splicing and turnover of RNA and modulates tyrosine kinase signaling [78]. MARK4 phosphorylates microtubule-associated proteins, thereby causing them to dissociate and destabilize the microtubules [79]. MARK4 is found in the apical ES of Sertoli cells, and recent data indicate that by regulating microtubule dynamics, Mark4 indirectly controls the integrity of the apical ES that tethers elongate spermatids to Sertoli cells [80].

Finally, Txk tyrosine kinase (*Txk*) is maximally expressed at stages IX–XI. TXK phosphorylates phospholipase C gamma and thereby contributes to signaling from many different receptors. One downstream target of pathways of Txk is the actin cytoskeleton [81].

A comparison of the stage-specific patterns of expression of *Map1*, *Ndrg1*, *Map7*, *Dcx*, and *Mark4* suggests how the microtubule cytoskeleton is regulated in a stage-specific manner by Sertoli cells and, thereby, the morphological relationship of these cells to spermatids (Figure 10.8). The expression of the microtubule-binding protein MAP1b increases approximately threefold from stage I to stages II–III and then by an additional 40% by stage VI. These increases are followed by the expression of *Dcx*, which is maximally expressed at stage VI. We hypothesize that sequential increases in MAP1b and DCX synthesis increases stability. At stages VII–VIIcd, expression of NDRG1 is increased but so too is an important kinase that targets these proteins. The simultaneous expression of NDRG1 and MARK4 may be responsible for a rapid turnover in microtubules that would be expected lead to the reformation of the morphological relationship between Sertoli cells and spermatids prior to spermiation.

d. Transcriptional activators and repressors

There are seven transcriptional regulators whose peak levels of expression occur between stages II–III and IX–XI (Table 10.5), providing distinct targets for different signals from upstream regulatory pathways. p300/CBP-associated factor, which is maximally expressed at stages II–III, encodes a histone acetyltransferase and transcriptional co-activator [82]. Three genes encode transcription factors that can act as transcriptional activators or repressors, depending on the other regulatory molecules bound to the specific targeted gene: zinc finger protein 219 methyl-CpG binding domain protein 2, methyl-CpG binding domain protein 2, and v-maf musculoaponeurotic fibrosarcoma oncogene homolog F (*Maff*) [83–89]. Maximal expression of these genes occurs at stages VIIc,d, VIII, and IX–XI, respectively. *Maff* is noteworthy because its protein product suppresses EGF-driven gene transcription and because its stage-specific pattern of expression is the opposite of that of *Sch4*, which facilitates ligand-independent phosphorylation and activation of the EGF -receptor [69]. Two transcriptional repressors complete this list: HOP homeobox (*Hopx*) and zinc finger homeobox 3 (*Zfhx3*), which are maximally expressed at stages

VI and VIIc,d, respectively. *Zfhx3* encodes a protein involved in neuronal differentiation [90], and Hopx acts with serum response factor (SRF) and histone deacetylase 2 (HDAC2) to suppress the transcription of genes, including *GATA 4* [91]. These characteristics of Hopx are significant to this discussion because Sertoli cells express SRF and GATA binding protein 4 (GATA4) [92–94].

e. Receptors and their ligands

Table 10.5 lists three receptor ligands and six cell receptors that are expressed by Sertoli cells in a stage-specific manner. The receptor ligands are encoded by the genes: growth arrest 6 (*Gas6*), macrophage migration inhibitor factor (*Mif*), and apelin (*Apln*). *Gas6*, which is maximally expressed at stage VI, may encode an autocrine factor for Sertoli cells because they express GAS6 receptors and respond to this cytokine with increased phagocytosis [95,96]. GAS6 also inhibits interleukin (IL) 1 and IL6 secretion by macrophages and may, therefore, have a similar effect on Sertoli cells [97,98]. *Mif* is maximally expressed at stage VIII and has important functions in the testis. *Mif*-null male mice exhibit significantly lower intratesticular testosterone levels than wild-type mice and an abnormally high incidence of abnormal spermatid morphology [99]. Thus, both Leydig cells and spermatids may be targets for MIF. *Apln* is maximally expressed at stages IX–XI, and the protein is expressed at high levels in the testis, brain, skeletal muscle, and a number of other organs [100]. While the Apln receptor has yet to be identified in the testis, it stimulates glucose uptake in skeletal muscle [100]. This raises the possibility that Apln regulates the metabolism of testicular cells.

Sertoli cells also express six receptors in a stage-specific manner: P2X, ligand-gated ion channel 2 (*P2rx2*), chemokine (C-X-C motif) receptor 4 (*Cxcr4*), G-protein-coupled receptor 56 (*Gpr56*), discoidin domain receptor tyrosine kinase 1 (*Ddr1*), RT1 class I, locus Ce5 (*RT1-CE5*), and anti-Mullerian hormone receptor, type II (*Amhr2*). Both *P2rx2* and *Cxcr4* are maximally expressed at stage VIII, and P2RX2 and CXCR4 bind extracellular adenosine triphosphate and chemokine (C-X-C motif) ligand 12 (CXCL12), respectively [101,102]. (The array data document that *Cxcl12* is expressed by spermatogonia and/or early spermatocytes.) Interestingly, when bound, both receptors open calcium channels on the plasma membrane [101,102]. *Gpr56* and *Ddr1* are maximally expressed at stages VI and VIII, respectively. Interestingly, these two receptors mediate the responses of cells to collagen or other basement membrane components [103–105]. In the developing brain, both of these receptors promote neural differentiation, and *Gpr56* also promotes the lamination of the cerebral cortex [104,105]. Such effects undoubtedly involve the regulation of cytoskeletal structure. *Amhr2* is maximally expressed at stage VIIa,b. While it is generally accepted that its ligand, AMH, is not expressed by Sertoli cells in the testes of fertile men, low levels of its transcript have been detected in the testes of adult rats [2,106]. Thus, in rats, AMHII may be involved in a very short feedback loop that regulates Sertoli cell function. *Mdb2* is maximally expressed at stages IX–XI. As in the immune system, MDB2 is involved in antigen presentation, and this receptor may facilitate immune tolerance in the testis [107].

Taken together, the large numbers of kinase-anchoring proteins, kinases, kinase substrates, receptor ligands, receptors, and transcriptional activators and repressors that are expressed in a stage-specific manner by Sertoli cells provide a strong rationale for the hypothesis that Sertoli cells are capable of responding in a stage-specific manner to many different signals from their adjacent spermatogenic cells. We therefore hypothesize that the successful integration of the cells into a functional seminiferous epithelium involves multiple cell–cell interactions and signaling pathways.

V. Future directions

A goal of this review has been to build the case that stage-specific gene expression by Sertoli cells is a fundamental characteristic of these cells. Clearly, many genes are expressed in this manner, and some of these genes are essential to male fertility. However, what has yet to be tested for any of these genes is the hypothesis that the stage-specific pattern of expression of a gene is essential, rather than just the expression at all stages of the cycle. As discussed in this review, many of the genes are expressed at stages when their protein products would seem to be needed. These correlations are certainly consistent with the hypothesis. However, a direct test of this hypothesis requires that the stage-specific pattern of expression of a gene be changed without causing the gene to be generally overexpressed or underexpressed. This will require the identification of upstream regions of genes that contain the regulatory sequences that drive gene expression at specific stages of the cycle. Much work needs to be done before this hypothesis can be tested.

We close this chapter with an acknowledgment that a new and important area of research was not addressed in this review, the potential role of microRNAs in regulating stage-specific translation [108]. Thus, it is quite likely that in the near future, we will realize that the regulation of stage-specific Sertoli cell function is even more complicated and interesting than described in this review.

Acknowledgments

Research in the author's laboratory is funded by the Eunice Kennedy Shriver NICHD/NIH through cooperative agreement U01HD07452-01 as part of the Specialized Cooperative Centers Program in Reproduction and Infertility Research.

References

[1] Williams M, Hill CJ, Scudamore I, Dunphy B, Cooke ID, Barratt CL. Sperm numbers and distribution within the human fallopian tube around ovulation. Hum Reprod 1993;8:2019–26.

[2] Sabanegh E, Agarwal A. Male infertility. In: Wein AJ, Kavoussi LR, Novick AC, Partin AW, Peters CA, editors. Campbell-walsh urology. 10th ed. Washington DC: Elsevier Sanders; 2012. p. 616–47.

[3] Prader A. Testicular size: assessment and clinical importance. Triangle 1966;7: 240–3.
[4] Leblond CP, Clermont Y. Definition of the stages of the cycle of the seminiferous epithelium in the rat. Ann N Y Acad Sci 1952;55:548–73.
[5] Clermont Y. Kinetics of spermatogenesis in mammals: seminiferous epithelium cycle and spermatogonial renewal. Physiol Rev 1972;52:198–263.
[6] Schulze W. Evidence of a wave of spermatogenesis in human testis. Andrologia 1982; 14:200–7.
[7] Fujii-Hanamoto H, Matsubayashi K, Nakano M, Kusunoki H, Enomoto T. A comparative study on testicular microstructure and relative sperm production in gorillas, chimpanzees, and orangutans. Am J Primatol 2011;73:570–7.
[8] Perey B, Clermont Y, Leblond CP. The wave of the seminiferous epithelium in the rat. Am J Anat 1961;108:47–77.
[9] Kerr JB. A light microscopic and morphometric analysis of the Sertoli cell during the spermatogenic cycle of the rat. Anat Embryol (Berl) 1988;177:341–8.
[10] Kerr JB. An ultrastructural and morphometric analysis of the Sertoli cell during the spermatogenic cycle of the rat. Anat Embryol (Berl) 1988;179:191–203.
[11] Chemes H. The phagocytic function of Sertoli cells: a morphological, biochemical, and endocrinological study of lysosomes and acid phosphatase localization in the rat testis. Endocrinology 1986;119:1673–81.
[12] Cavicchia JC, Foscolo M, Ibanez J, Lillig C, Capani F. The actin filament network associated to Sertoli cell ectoplasmic specializations. Biocell 2011;35:81–9.
[13] Johnston DS, Wright WW, Dicandeloro P, Wilson E, Kopf GS, Jelinski SA. Stage-specific gene expression is a fundamental characteristic of rat spermatogenic cells and Sertoli cells. Proc Natl Acad Sci USA 2008;105:8315–20.
[14] Soderstrom KO, Parvinen M. RNA synthesis in different stages of rat seminiferous epithelial cycle. Mol Cell Endocrinol 1976;5:181–99.
[15] Ritzen EM, Boitani C, Parvinen M, French FC, Feldman M. Stage-dependent secretion of ABP by rat seminiferous tubules. Mol Cell Endocrinol 1982;25:25–33.
[16] Vihko KK, Suominen JJ, Parvinen M. Cellular regulation of plasminogen activator secretion during spermatogenesis. Biol Reprod 1984;31:383–9.
[17] Wright WW, Parvinen M, Musto NA, Gunsalus GL, Phillips DM, Mather JP, et al. Identification of stage-specific proteins synthesized by rat seminiferous tubules. Biol Reprod 1983;29:257–70.
[18] Toppari J, Kangasniemi M, Kaipia A, Mali P, Huhtaniemi I, Parvinen M. Stage- and cell-specific gene expression and hormone regulation of the seminiferous epithelium. J Electron Microsc Tech 1991;19:203–14.
[19] Hakovirta H, Syed V, Jegou B, Parvinen M. Function of interleukin-6 as an inhibitor of meiotic DNA synthesis in the rat seminiferous epithelium. Mol Cell Endocrinol 1995;108:193–8.
[20] Hakovirta H, Kaipia A, Soder O, Parvinen M. Effects of activin-A, inhibin-A, and transforming growth factor-beta 1 on stage-specific deoxyribonucleic acid synthesis during rat seminiferous epithelial cycle. Endocrinology 1993;133:1664–8.
[21] Soder O, Bang P, Wahab A, Parvinen M. Insulin-like growth factors selectively stimulate spermatogonial, but not meiotic, deoxyribonucleic acid synthesis during rat spermatogenesis. Endocrinology 1992;131:2344–50.
[22] Parvinen M, Soder O, Mali P, Froysa B, Ritzen EM. In vitro stimulation of stage-specific deoxyribonucleic acid synthesis in rat seminiferous tubule segments by interleukin-1 alpha. Endocrinology 1991;129:1614–20.

[23] Hill CM, Anway MD, Zirkin BR, Brown TR. Intratesticular androgen levels, androgen receptor localization, and androgen receptor expression in adult rat Sertoli cells. Biol Reprod 2004;71:1348–58.
[24] Beardsley A, O'Donnell L. Characterization of normal spermiation and spermiation failure induced by hormone suppression in adult rats. Biol Reprod 2003;68:1299–307.
[25] O'Donnell L, McLachlan RI, Wreford NG, de Kretser DM, Robertson DM. Testosterone withdrawal promotes stage-specific detachment of round spermatids from the rat seminiferous epithelium. Biol Reprod 1996;55:895–901.
[26] Wright WW, Luzarraga ML. Isolation of cyclic protein-2 from rat seminiferous tubule fluid and Sertoli cell culture medium. Biol Reprod 1986;35:761–72.
[27] Wright WW. Germ cell-Sertoli cell interactions: analysis of the biosynthesis and secretion of cyclic protein-2. Dev Biol 1988;130:45–56.
[28] Zabludoff SD, Erickson-Lawrence M, Wright WW. Sertoli cells, proximal convoluted tubules in the kidney, and neurons in the brain contain cyclic protein-2. Biol Reprod 1990;43:15–24.
[29] Erickson-Lawrence M, Zabludoff SD, Wright WW. Cyclic protein-2, a secretory product of rat Sertoli cells, is the proenzyme form of cathepsin L. Mol Endocrinol 1991;5:1789–98.
[30] Roth W, Deussing J, Botchkarev VA, Pauly-Evers M, Saftig P, Hafner A, et al. Cathepsin L deficiency as molecular defect of furless: hyperproliferation of keratinocytes and pertubation of hair follicle cycling. FASEB J 2000;14:2075–86.
[31] Wright WW, Smith L, Kerr C, Charron M. Mice that express enzymatically inactive cathepsin L exhibit abnormal spermatogenesis. Biol Reprod 2003;68:680–7.
[32] Anway MD, Wright WW, Zirkin BR, Korah N, Mort JS, Hermo L. Expression and localization of cathepsin k in adult rat sertoli cells. Biol Reprod 2004;70:562–9.
[33] Kim GH, Wright WW. A comparison of the effects of testicular maturation and aging on the stage-specific expression of CP-2/cathepsin L messenger ribonucleic acid by Sertoli cells of the Brown Norway rat. Biol Reprod 1997;57:1467–77.
[34] Blanco-Rodriguez J, Martinez-Garcia C. Apoptosis is physiologically restricted to a specialized cytoplasmic compartment in rat spermatids. Biol Reprod 1999;61:1541–7.
[35] Elliott MR, Zheng S, Park D, Woodson RI, Reardon MA, Juncadella IJ, et al. Unexpected requirement for ELMO1 in clearance of apoptotic germ cells in vivo. Nature 2010;467:333–7.
[36] Park D, Han CZ, Elliott MR, Kinchen JM, Trampont PC, Das S, et al. Continued clearance of apoptotic cells critically depends on the phagocyte Ucp2 protein. Nature 2011;477:220–4.
[37] Wright WW, Zabludoff SD, Penttila TL, Parvinen M. Germ cell-Sertoli cell interactions: regulation by germ cells of the stage-specific expression of CP-2/cathepsin L mRNA by Sertoli cells. Dev Genet 1995;16:104–13.
[38] Charron M, Folmer JS, Wright WW. A 3-kilobase region derived from the rat cathepsin L gene directs in vivo expression of a reporter gene in sertoli cells in a manner comparable to that of the endogenous gene. Biol Reprod 2003;68:1641–8.
[39] Visone T, Charron M, Wright WW. Activation and repression domains within the promoter of the rat cathepsin L gene orchestrate sertoli cell-specific and stage-specific gene transcription in transgenic mice. Biol Reprod 2009;81:571–9.
[40] Behrens A, Genoud N, Naumann H, Rulicke T, Janett F, Heppner FL, et al. Absence of the prion protein homologue Doppel causes male sterility. Embo J 2002;21:3652–8.

[41] Paisley D, Banks S, Selfridge J, McLennan NF, Ritchie AM, McEwan C, et al. Male infertility and DNA damage in Doppel knockout and prion protein/Doppel double-knockout mice. Am J Pathol 2004;164:2279–88.

[42] Ueno H, Mori H. Morphological Analysis of Sertoli Cell Ultrastructure during the Seminiferous epithelial Cycle in Rats. Biol Reprod 1990;43:769–76.

[43] Eskelinen EL. Roles of LAMP-1 and LAMP-2 in lysosome biogenesis and autophagy. Mol Aspects Med 2006;27:495–502.

[44] Percheron F, Foglietti MJ, Bernard M, Ricard B. Mammalian beta-D-mannosidase and beta-mannosidosis. Biochimie 1992;74:5–11.

[45] Hirabayashi Y, Li YT, Li SC. The protein activator specific for the enzymic hydrolysis of GM2 ganglioside in normal human brain and brains of three types of GM2 gangliosidosis. J Neurochem 1983;40:168–75.

[46] Vogel AW, Vaid KS, Guttman JA. The sertoli cell cytoskeleton. In: Cheng CY, editor. Molecular mechanisms in spermatogenesis. New York: Landis Science and Springer Science + Business Media; 2008. p. 168–75.

[47] Smith BE, Braun RE. Germ cell migration across Sertoli cell tight junctions. Science 2012;338:798–802.

[48] Pappas CT, Bliss KT, Zieseniss A, Gregorio CC. The Nebulin family: an actin support group. Trends Cell Biol 2011;21:29–37.

[49] Paschoud S, Yu D, Pulimeno P, Jond L, Turner JR, Citi S. Cingulin and paracingulin show similar dynamic behaviour, but are recruited independently to junctions. Mol Membr Biol 2011;28:123–35.

[50] O'Donnell L, O'Bryan MK. Microtubules and spermatogenesis. Semin Cell Dev Biol 2014;30:45–54.

[51] Wenz JR, Hess RA. Characterization of stage-specific tyrosinated alpha-tubulin immunoperoxidase staining patterns in Sertoli cells of rat seminiferous tubules by light microscopic image analysis. Tissue Cell 1998;30:492–501.

[52] Moores CA, Perderiset M, Francis F, Chelly J, Houdusse A, Milligan RA. Mechanism of microtubule stabilization by doublecortin. Mol Cell 2004;14:833–9.

[53] Tanaka T, Serneo FF, Tseng HC, Kulkarni AB, Tsai LH, Gleeson JG. Cdk5 phosphorylation of doublecortin ser297 regulates its effect on neuronal migration. Neuron 2004;41:215–27.

[54] Session DR, Fautsch MP, Avula R, Jones WR, Nehra A, Wieben ED. Cyclin-dependent kinase 5 is expressed in both Sertoli cells and metaphase spermatocytes. Fertil Steril 2001;75:669–73.

[55] Montenegro-Venegas C, Tortosa E, Rosso S, Peretti D, Bollati F, Bisbal M, et al. MAP1B regulates axonal development by modulating Rho-GTPase Rac1 activity. Mol Biol Cell 2010;21:3518–28.

[56] Kim KT, Ongusaha PP, Hong YK, Kurdistani SK, Nakamura M, Lu KP, et al. Function of Drg1/Rit42 in p53-dependent mitotic spindle checkpoint. J Biol Chem 2004;279:38597–602.

[57] Drewes G, Ebneth A, Mandelkow EM. MAPs, MARKs and microtubule dynamics. Trends Biochem Sci 1998;23:307–11.

[58] Guttman JA, Kimel GH, Vogl AW. Dynein and plus-end microtubule-dependent motors are associated with specialized Sertoli cell junction plaques (ectoplasmic specializations). J Cell Sci 2000;113(Pt 12):2167–76.

[59] Kim S, Coulombe PA. Intermediate filament scaffolds fulfill mechanical, organizational, and signaling functions in the cytoplasm. Genes Dev 2007;21:1581–97.

[60] Appert A, Fridmacher V, Locquet O, Magre S. Patterns of keratins 8, 18 and 19 during gonadal differentiation in the mouse: sex- and time-dependent expression of keratin 19. Differentiation 1998;63:273–84.

[61] Show MD, Anway MD, Folmer JS, Zirkin BR. Reduced intratesticular testosterone concentration alters the polymerization state of the Sertoli cell intermediate filament cytoskeleton by degradation of vimentin. Endocrinology 2003;144:5530–6.

[62] Wang R, Li Q, Tang DD. Role of vimentin in smooth muscle force development. Am J Physiol Cell Physiol 2006;291:C483–9.

[63] Kowalczyk AP, Green KJ. Structure, function, and regulation of desmosomes. In: van Roy F, editor. Progress in molecular biology and translational medicine. Waltham: Elsevier; 2013. p. 95–118.

[64] Cheng CY, Wong EW, Lie PP, Li MW, Mruk DD, Yan HH, et al. Regulation of blood-testis barrier dynamics by desmosome, gap junction, hemidesmosome and polarity proteins: An unexpected turn of events. Spermatogenesis 2011;1:105–15.

[65] Lie PP, Xia W, Wang CQ, Mruk DD, Yan HH, Wong CH, et al. Dynamin II interacts with the cadherin- and occludin-based protein complexes at the blood-testis barrier in adult rat testes. J Endocrinol 2006;191:571–86.

[66] Lie PP, Mruk DD, Lee WM, Cheng CY. Cytoskeletal dynamics and spermatogenesis. Philos Trans R Soc Lond B Biol Sci 2010;365:1581–92.

[67] Xiao X, Mruk DD, Cheng FL, Cheng CY. C-Src and c-Yes are two unlikely partners of spermatogenesis and their roles in blood-testis barrier dynamics. Adv Exp Med Biol 2012;763:295–317.

[68] Clayton EL, Sue N, Smillie KJ, O'Leary T, Bache N, Cheung G, et al. Dynamin I phosphorylation by GSK3 controls activity-dependent bulk endocytosis of synaptic vesicles. Nat Neurosci 2010;13:845–51.

[69] Wills MK, Tong J, Tremblay SL, Moran MF, Jones N. The ShcD Signaling Adaptor Facilitates Ligand-Independent Phosphorylation of the EGF Receptor. Mol Biol Cell 2014;25:739–52.

[70] Suarez-Quian CA, Niklinski W. Immunocytochemical localization of the epidermal growth factor receptor in mouse testis. Biol Reprod 1990;43:1087–97.

[71] Tsutsumi O, Kurachi H, Oka T. A physiological role of epidermal growth factor in male reproductive function. Science 1986;233:975–7.

[72] Diviani D, Scott JD. AKAP signaling complexes at the cytoskeleton. J Cell Sci 2001;114:1431–7.

[73] Gold MG, Fowler DM, Means CK, Pawson CT, Stephany JJ, Langeberg LK, et al. Engineering A-kinase anchoring protein (AKAP)-selective regulatory subunits of protein kinase A (PKA) through structure-based phage selection. J Biol Chem 2013;288:17111–21.

[74] Blackshear PJ. The MARCKS family of cellular protein kinase C substrates. J Biol Chem 1993;268:1501–4.

[75] Weimer JM, Yokota Y, Stanco A, Stumpo DJ, Blackshear PJ, Anton ES. MARCKS modulates radial progenitor placement, proliferation and organization in the developing cerebral cortex. Development 2009;136:2965–75.

[76] Kerr GE, Young JC, Horvay K, Abud HE, Loveland KL. Regulated wnt/beta-catenin signaling sustains adult spermatogenesis in mice. Biol Reprod 2014;90:3.

[77] Jackson RJ, Wickens M. Translational controls impinging on the 5'-untranslated region and initiation factor proteins. Curr Opin Genet Dev 1997;7:233–41.

[78] Buxade M, Parra-Palau JL, Proud CG. The Mnks: MAP kinase-interacting kinases (MAP kinase signal-integrating kinases). Front Biosci 2008;13:5359–73.

[79] Naz F, Anjum F, Islam A, Ahmad F, Hassan MI. Microtubule affinity-regulating kinase 4: structure, function, and regulation. Cell Biochem Biophys 2013;67:485–99.

[80] Tang EI, Xiao X, Mruk DD, Qian XJ, Mok KW, Jenardhanan P, et al. Microtubule affinity-regulating kinase 4 (MARK4) is a component of the ectoplasmic specialization in the rat testis. Spermatogenesis 2012;2:117–26.

[81] Readinger JA, Mueller KL, Venegas AM, Horai R, Schwartzberg PL. Tec kinases regulate T-lymphocyte development and function: new insights into the roles of Itk and Rlk/Txk. Immunol Rev 2009;228:93–114.

[82] Nagy Z, Tora L. Distinct GCN5/PCAF-containing complexes function as co-activators and are involved in transcription factor and global histone acetylation. Oncogene 2007;26:5341–57.

[83] Sakai T, Hino K, Wada S, Maeda H. Identification of the DNA binding specificity of the human ZNF219 protein and its function as a transcriptional repressor. DNA Res 2003;10:155–65.

[84] Takigawa Y, Hata K, Muramatsu S, Amano K, Ono K, Wakabayashi M, et al. The transcription factor Znf219 regulates chondrocyte differentiation by assembling a transcription factory with Sox9. J Cell Sci 2010;123:3780–8.

[85] Rupon JW, Wang SZ, Gnanapragasam M, Labropoulos S, Ginder GD. MBD2 contributes to developmental silencing of the human epsilon-globin gene. Blood Cells Mol Dis 2011;46:212–9.

[86] Stefanska B, Suderman M, Machnes Z, Bhattacharyya B, Hallett M, Szyf M. Transcription onset of genes critical in liver carcinogenesis is epigenetically regulated by methylated DNA-binding protein MBD2. Carcinogenesis 2013;34:2738–49.

[87] Motohashi H, Katsuoka F, Engel JD, Yamamoto M. Small Maf proteins serve as transcriptional cofactors for keratinocyte differentiation in the Keap1-Nrf2 regulatory pathway. Proc Natl Acad Sci USA 2004;101:6379–84.

[88] Itoh K, Chiba T, Takahashi S, Ishii T, Igarashi K, Katoh Y, et al. An Nrf2/small Maf heterodimer mediates the induction of phase II detoxifying enzyme genes through antioxidant response elements. Biochem Biophys Res Commun 1997;236: 313–22.

[89] Amit I, Citri A, Shay T, Lu Y, Katz M, Zhang F, et al. A module of negative feedback regulators defines growth factor signaling. Nat Genet 2007;39:503–12.

[90] Jung CG, Kim HJ, Kawaguchi M, Khanna KK, Hida H, Asai K, et al. Homeotic factor ATBF1 induces the cell cycle arrest associated with neuronal differentiation. Development 2005;132:5137–45.

[91] Yamashita K, Katoh H, Watanabe M. The homeobox only protein homeobox (HOPX) and colorectal cancer. Int J Mol Sci 2013;14:23231–43.

[92] Imai T, Kawai Y, Tadokoro Y, Yamamoto M, Nishimune Y, Yomogida K. In vivo and in vitro constant expression of GATA-4 in mouse postnatal Sertoli cells. Mol Cell Endocrinol 2004;214:107–15.

[93] Tremblay JJ, Robert NM, Viger RS. Modulation of endogenous GATA-4 activity reveals its dual contribution to Mullerian inhibiting substance gene transcription in Sertoli cells. Mol Endocrinol 2001;15:1636–50.

[94] Chaudhary J, Skinner MK. Basic helix-loop-helix proteins can act at the E-box within the serum response element of the c-fos promoter to influence hormone-induced promoter activation in Sertoli cells. Mol Endocrinol 1999;13:774–86.

[95] Chan MC, Mather JP, McCray G, Lee WM. Identification and regulation of receptor tyrosine kinases Rse and Mer and their ligand Gas6 in testicular somatic cells. J Androl 2000;21:291–302.

[96] Xiong W, Chen Y, Wang H, Wu H, Lu Q, Han D. Gas6 and the Tyro 3 receptor tyrosine kinase subfamily regulate the phagocytic function of Sertoli cells. Reproduction 2008;135:77−87.
[97] Alciato F, Sainaghi PP, Sola D, Castello L, Avanzi GC. TNF-alpha, IL-6, and IL-1 expression is inhibited by GAS6 in monocytes/macrophages. J Leukoc Biol 2010;87:869−75.
[98] Stephan JP, Syed V, Jegou B. Regulation of Sertoli cell IL-1 and IL-6 production in vitro. Mol Cell Endocrinol 1997;134:109−18.
[99] Anahara R, Toyama Y, Koda M, Honma S, Nishihira J, Toshimori K, et al. Deletion of macrophage migration inhibitory factor gene induces down regulation of sex hormones and ultrastructural abnormalities in mouse testes. Reprod Toxicol 2006;21:167−70.
[100] O'Carroll AM, Lolait SJ, Harris LE, Pope GR. The apelin receptor APJ: journey from an orphan to a multifaceted regulator of homeostasis. J Endocrinol 2013;219:R13−35.
[101] Lamont C, Wier WG. Evoked and spontaneous purinergic junctional Ca2+ transients (jCaTs) in rat small arteries. Circ Res 2002;91:454−6.
[102] Drury LJ, Ziarek JJ, Gravel S, Veldkamp CT, Takekoshi T, Hwang ST, et al. Monomeric and dimeric CXCL12 inhibit metastasis through distinct CXCR4 interactions and signaling pathways. Proc Natl Acad Sci USA 2011;108:17655−60.
[103] Thomazy V, Fesus L. Differential expression of tissue transglutaminase in human cells. An immunohistochemical study. Cell Tissue Res 1989;255:215−24.
[104] Luo R, Jeong SJ, Jin Z, Strokes N, Li S, Piao X. G protein-coupled receptor 56 and collagen III, a receptor-ligand pair, regulates cortical development and lamination. Proc Natl Acad Sci USA 2011;108:12925−30.
[105] Wang CZ, Yeh YC, Tang MJ. DDR1/E-cadherin complex regulates the activation of DDR1 and cell spreading. Am J Physiol Cell Physiol 2009;297:C419−29.
[106] Benoit G, Trasler JM. Developmental expression of DNA methyltransferase messenger ribonucleic acid, protein, and enzyme activity in the mouse testis. Biol Reprod 1994;50:1312−9.
[107] Dal Secco V, Riccioli A, Padula F, Ziparo E, Filippini A. Mouse Sertoli cells display phenotypical and functional traits of antigen-presenting cells in response to interferon gamma. Biol Reprod 2008;78:234−42.
[108] Nicholls PK, Harrison CA, Walton KL, McLachlan RI, O'Donnell L, Stanton PG. Hormonal regulation of sertoli cell micro-RNAs at spermiation. Endocrinology 2011;152:1670−83.

MicroRNAs and Sertoli cells

11

Madhuvanthi Ramaiah and Miles F. Wilkinson
Department of Reproductive Medicine, University of California San Diego, La Jolla, CA

I. Noncoding RNAs

Noncoding RNAs (ncRNAs) have stolen much of the limelight from their well-studied, protein-coding counterparts since the early 2000 s. Genome-wide expression analysis has revealed that there is pervasive transcription of almost the entire mammalian genome. Because protein-coding genes account only for ~2% of the genome, by definition, ncRNAs account for the vast majority of transcription products derived from mammalian genomes [1–3]. Intriguingly, ncRNA content (rather than protein-coding gene content) correlates with the increasing developmental complexity of eukaryotic organisms, providing evidence for the functional importance of ncRNAs in biological diversification [4,5].

ncRNAs are derived from both intragenic and intergenic regions of eukaryotic genomes. Some ncRNAs, such as ribosomal RNA, transfer RNA, small nuclear RNA, and small nucleolar RNA, serve as structural and catalytic RNAs that are critical for widely used molecular machines in the cell, such as the spliceosome and the ribosome. Other ncRNAs have been shown to serve in regulatory roles. One such class of "regulatory ncRNAs" are long ncRNAs (lncRNAs), which are defined as RNAs longer than 200 nucleotides (nt) that lack the ability to code for proteins. There has been an explosion of publications on this topic that, together, have shown that lncRNAs function in a wide variety of both nuclear and cytoplasmic events [6–9]. Some nuclear lncRNAs regulate the transcription of specific genomic loci by recruiting chromatin modifiers. Cytoplasmic lncRNAs regulate mRNA stability and translation. Some cytoplasmic lncRNAs also regulate the function of other ncRNAs by, for example, sequestering them and thereby inhibiting their action.

Short ncRNAs include microRNAs (miRNAs), small-interfering RNAs (siRNAs), and Piwi-interacting RNAs (piRNAs) [10]. miRNAs are the subject of this review and are described below. siRNAs are ~21 nt long and are cleaved from long double-stranded precursor RNAs by the RNase III endonuclease DICER. siRNAs bind with perfect complementarity to their target RNAs and mediate their destruction by a widely studied process called "RNA interference." siRNAs were initially implicated in the defense against transgenes and viruses because they are generated from long RNAs transcribed from these exogenous sources. Now we know that siRNAs are also generated from repetitive elements, including transposons, hairpin RNAs, and convergent mRNA transcripts; such siRNAs are generally referred to as

Sertoli Cell Biology. DOI: http://dx.doi.org/10.1016/B978-0-12-417047-6.00011-9

"endo-siRNAs" [11–13]. piRNAs are ~25–33 nt long and are particularly prevalent in male germ cells. piRNAs are derived in a DICER-independent manner by processing of long RNAs transcribed from discrete loci in the genome. After these "primary piRNAs" are processed, they are amplified by a so-called "ping-pong cycle," whereupon they protect germline cells from transposons. piRNAs are likely to have other functions in addition to defense from transposons [14].

II. miRNAs

miRNAs are small RNAs (~21–24 nt long) first identified in *Caenorhabditis elegans* [15] that are now known to be present in eukaryotes spanning the phylogenetic scale [16]. To date, approximately 2600 and 1900 mature miRNAs have been identified in humans and mice, respectively (miRBase, Release 20). The vast majority of miRNAs serve as negative regulators of gene expression. They accomplish this regulation by binding to partially complementary sequences in their target mRNAs and recruiting effector proteins that trigger translational repression and/or an increased rate of mRNA decay. While miRNAs typically repress the expression of their targets only modestly, they potentially have widespread effects in cells, given that a typical single miRNA is predicted to regulate the expression of hundreds of target mRNAs [17].

A large proportion of miRNAs are processed from the introns of protein-coding genes, while other miRNAs are generated from intronic or exonic regions of ncRNA transcription units. The vast majority of miRNAs are transcribed by RNA polymerase II, the same polymerase that transcribes protein-coding mRNAs. In rare cases, RNA polymerase III (which is responsible for the generation of small RNAs, such as tRNAs) drives the expression of miRNAs [16,18,19].

miRNAs are typically generated from long primary transcripts (pri-miRNAs) that are processed to generate ~55–70 nt precursor miRNAs (pre-miRNAs), which then undergo further processing to generate the mature (~21–24 nt) miRNA [18,20,21]. This canonical miRNA biogenesis pathway requires DROSHA and DICER, which cleave the pri-miRNAs and pre-miRNAs, respectively. DROSHA, in combination with the double-stranded RNA-binding protein, DiGeorge Syndrome Critical Region Gene 8, binds to the pri-miRNA in the nucleus, liberating a double-stranded hairpin—the pre-miRNA. This pre-miRNA is transported to the cytoplasm via the action of exportin-5. The loop is cleaved off by DICER, leaving the double-stranded stem, which is composed of two strands called the "miRNA" and "miRNA*" strands. Both of these strands have the potential to become an "active" mature miRNA. To become active, these strands must be released from the duplex and then loaded into a so-called "RNA-induced silencing complex" (RISC). The strand that is incorporated into the RISC and binds its target is called the "guide strand" [22–25].

miRNA guide strands are directed to specific mRNAs by virtue of sequence complementarity. miRNAs typically have base complementarity with the 3' untranslated region (UTR) of their target mRNAs, but they can also bind to the coding region or even the 5' UTR of mRNAs [26]. Plant miRNAs have either perfect or near-perfect complementarity with their targets, making it relatively easy to predict their targets. In contrast, animal miRNAs have limited complementarity with their targets, making it more difficult to predict their targets with high reliability [26].

Several parameters have been identified that are associated with *bona fide* mammalian miRNA targets and have been incorporated into algorithms designed to predict miRNA targets. One powerful approach that greatly aids in the identification of *bona fide* miRNA targets is comparing orthologous mRNAs for the presence of conserved miRNA seed sequences in their 3' UTRs. While this is useful for conserved miRNAs, many miRNAs are rapidly evolving, including a large number of those expressed in the male reproductive tract. This is an interesting property of male-biased genes in general (including protein-coding genes) that may result from strong selection pressure emanating from sexual conflict and postcopulatory sexual selection events [27–29]. However, the rapidly evolving nature of both miRNAs and protein-coding genes expressed in the male reproductive tract greatly compromises the ability of investigators to identify *bona fide* miRNA targets by *in silico* methods. Thus, it is important to complement *in silico* analysis with empirical approaches to identify miRNA targets. Biochemical approaches to identify miRNA targets at a genome-wide level include high-throughput sequencing of RNAs isolated by cross-linking immunoprecipitation of Argonaute (Ago HITS-CLIP), Photoactivatable-ribonucleoside-enhanced crosslinking and immunoprecipitation (PAR-CLIP), Ribonucleoprotein immunoprecipitation microarray (RIP-Chip), and RNA induced silencing complex-trap (RISCtrap) [30–33].

III. The role of miRNAs in spermatogenesis *in vivo*

As described above, DROSHA and DICER are both required for the canonical miRNA biogenesis pathway. Thus, conditional knockout (cKO) of either of these factors in mice provides a means to determine the functional importance of miRNAs in a given cell type or tissue. To define the role of miRNAs in spermatogenesis, a number of groups have analyzed cKO mice lacking DROSHA or DICER in Sertoli cells (SCs) and germ cells (Table 11.1). Below, we describe the biological consequences of the loss of these enzymes in these cell types.

A. Spermatogenic defects resulting from loss of Dicer in SCs

Three studies have examined the consequences of inactivating *Dicer* in fetal SCs (Table 11.2). *Dicer* was knocked out in early fetal SCs using the *steroidogenic factor-1* (*Sf1*) promoter to drive the expression of CRE recombinase. Later-stage fetal SCs were targeted using the *anti-Müllerian hormone* (*Amh*; Müllerian-inhibiting substance) promoter. Below, we discuss these three studies in detail.

Table 11.1 Conditional knockout (cKO) mice lacking miRNA biogenesis enzymes in Sertoli cells and germ cells

Gene targeted	Promoter driving CRE	cKO cell type and stage	Fertility	Reference
Dicer	*Sf1*	SCs, Leydig cells (~E10.5)	Not tested (lethality)	[34]
Dicer	*Amh*	SCs (~E14.5)	Infertile	[35]
Dicer	*Amh*	SCs (~E14.5)	Infertile	[36]
Dicer	*Tnap*	PGCs (~E9.5)	Subfertile	[37]
Dicer	*Tnap*	PGCs (~E9.5)	Subfertile	[38]
Dicer	*Vasa/Ddx4*	Pro-spermatogonia (~E15.5)	Infertile	[39]
Dicer	*Vasa/Ddx4*	Pro-spermatogonia (~E15.5)	Infertile	[40]
Dicer	*Pgk2*	Spermatocytes/spermatids	Fertile	[39]
Dicer	*Stra8*	A-spermatogonia (~P3)	Subfertile	[39]
Dicer	*Stra8*	A-spermatogonia (~P3)	Infertile	[41]
Drosha	*Stra8*	A-spermatogonia (~P3)	Infertile	[42]
Dicer	*Stra8*	A-spermatogonia (~P3)	Infertile	[42]
Dicer	*Ngn3*	A-spermatogonia (~P5)	Infertile	[43]

The *Dicer* and *Drosha* genes encode endonucleases essential for the biogenesis of most miRNAs. These genes were genetically inactivated using *Cre*-driver mice expressing CRE recombinase in the cell types indicated. Embryonic (E) and postnatal (P) days indicate approximately when the promoters are first significantly expressed.

Huang and Yao examined *Dicer*$^{fl/fl}$;*Sf1*-Cre mice, which lose *Dicer* in both fetal SCs and Leydig cells during midgestation because the *Sf1* promoter is first expressed at approximately embryonic day 10.5 (E10.5) in these cell types [34]. Surprisingly, these mutant mice showed no testis defects during the fetal stage. Indeed, even at birth, there were no obvious morphological defects in the testes or epididymis of these mutant mice, although some SCs were apoptotic (based on cleaved caspase-3 staining), suggesting that miRNAs promote SC survival. Apoptosis was also evident in the fetal period (at E18.5), although the type of cells undergoing apoptosis was not determined. On postnatal day 2 (P2), mutant testes had clear morphological defects, including cord degeneration, and they were smaller than control testes. These defects are likely to be secondary to SC dysfunction because mutant testes had apoptotic SCs and fewer proliferating SCs than control testes (the latter based on fewer SOX9/Ki-67 double-stained cells).

In addition, many germ cells were apoptotic at P2, which was also likely to be secondary to a SC defect because germ cells are in direct contact with SCs and depend on them for support. P2 mutant testes also had reduced numbers of Leydig cells, perhaps reflecting a direct effect of reduced miRNA expression in Leydig cells, as the *Sf1* promoter is expressed in Leydig cells. By P5, testicular defects were more severe, with few testis cords remaining and no Leydig cells observable. The mutant mice died at ~P5 as a result of adrenal insufficiency, precluding analysis of the effect of loss of DICER at later time points of spermatogenesis.

Table 11.2 Testis defects in cKO mice lacking miRNA biogenesis in SCs

Study	Huang and Yao [34]	Papaionnou et al. [35]	Kim et al. [36]
Gene targeted	*Dicer*	*Dicer*	*Dicer*
Promoter driving *Cre*	*Sf1*	*Amh*	*Amh*
Initial *Cre* expression	SCs, Leydig cells (~E10.5)	SCs (~E14.5)	SCs (~E14.5)
Basic characterization	Postnatal (~P5) death	Viable, normal genitalia, normal testosterone levels, infertile	Viable, normal genitalia, infertile
Fetal defects	*E18.5*: Cell apoptosis (measured by cleaved CASP3 staining), but no obvious defects in SC differentiation or testis cord formation		
Perinatal defects	*P0*: SC apoptosis, but no obvious morphological defects in testis or the seminiferous epithelium (SE)	*P0*: No obvious morphological defects in the testis or SE despite twofold reduction in SC-specific miRNAs. Increased SC proliferation (~1.3-fold, as measured by Ki-67 staining). 77 genes upregulated and 68 genes downregulated (≥ twofold regulated)	*P0*: No obvious morphological defects in the testis or SE despite reduced expression of several SC-expressed genes. Some apoptotic (CASP3 positive) cells. No detectable cell proliferation defect
Very early postnatal defects	*P2*: Testes fail to increase in size after P0. Testis cords show first signs of degeneration. More apoptotic and less proliferating SCs (the latter measured by Ki-67 staining). More apoptotic germ cells. Fewer Leydig cells		*P3*: More interstitial space than controls. Some SC nuclei abnormally localized to the center of the SE tubule. Many apoptotic SCs. Less mitotic cells (measured using phosph-H3)

(*Continued*)

Table 11.2 **(Continued)**

Study	Huang and Yao [34]	Papaionnou et al. [35]	Kim et al. [36]
Early postnatal	*P5*: Testes size of mutant similar to P0 control testes. Few testes cords; those present are abnormal. More testicular apoptotic cells. No detectable Leydig cells	*P5*: KO completely manifests at this point, as SC-specific miRNAs are virtually absent. First morphological defects in the testis detected. Nuclei of SCs mislocalized to the center of SE tubules, suggestive of SC maturation defect. Increased SC proliferation (1.5-fold). Massive increase in SC apoptosis. Many pycnotic cells. Abnormal Leydig cell clustering (persists into adulthood). 787 upregulated and 796 downregulated genes ($\geq$twofold regulated)	*P6*: Small testes. Disorganized and fibrotic SE tubule regions. Most SC nuclei abnormally localized at the center of SE tubules. More apoptotic cells (both SCs and germ cells)
Mid postnatal		*P15*: 80% reduction in testis size. Abnormal germ cell layering in SE. Lack of SE lumen. Delayed meiotic progression (pachytene spermatocytes not visible). Many pycnotic cells. SC cells appear to be immature based on their increased proliferation (2.6-fold) and the mislocalization of their nuclei to the center of the SE tubule. More SC and germ cell apoptosis	*P12*: Most SE tubules lack lumen. Most SC nuclei abnormally localized at the center of SE tubules. Few PLZF+ cells. Some PLZF+ cells abnormally located outside of SE tubules. Asynchronous meiotic progression based on several meiotic markers
Late postnatal defects		*P21*: More evidence of immature SCs but SC apoptosis no longer detectable. More evidence for delayed meiotic progression. Most SE regions looked abnormal. Many pycnotic cells. Increased germ cell apoptosis	*P30*: Testes much smaller than controls with many SE abnormalities, including regions with only SCs and no lumen. No elongated spermatids observed.

			SC nuclei are now normally localized at the SE tubule periphery. ~1/2 of SE tubule regions lack PLZF + cells. Some SE tubule regions lack postmeiotic germ cells, others lack elongated spermatids, and few lack all germ cells
Young adult defects		*P42*: Severe SE disorganization and vacuolization. Few round spermatids; no detectable elongated spermatids. Elevated numbers of apoptotic germ cells. High % of SE regions with SCO phenotype	
Adult defects		*P60*: 90% reduction in testis size. SE exhibits same defects as at earlier stages, including vacuolization and many SE regions with SCO phenotype or meiotic arrest. No elongated spermatids. No sperm in the epididymis. Normal androgen-dependent behavior	*P60*: Severely fibrotic and sparsely packed SE tubules. Most SE tubules lack a lumen and few elongated spermatids are present. No sperm in the epididymis. While external genitalia are normal, the mice are sterile
Adult defects		*P90*: More severe testis degeneration than at D60. Most regions of SE have SCO phenotype	
Adult defects		*P180*: Severe testis degeneration (95% reduced mass). Mostly composed of interstitial cells with few germ cells	

The *Dicer* gene was genetically inactivated using the *Cre*-driver mice indicated (see also Table 11.1).

The phenotypes of these *Dicer*$^{fl/fl}$;*Sf1*-Cre mice support the notion that miRNAs expressed in SCs are directly responsible for both the survival and proliferation of SCs and that these miRNAs indirectly (through SCs) drive the development of normal testis cord architecture and germ cell survival. However, it is also possible that some or all of the testis cord and germ cell defects in these mutant mice could be secondary defects resulting from miRNA loss in Leydig cells rather than in SCs. As noted above, it was surprising that these mutant mice did not suffer from any detectable morphological abnormalities in the fetal testes. This lack of abnormalities could indicate that miRNAs have no role in fetal SCs. However, it is possible that miRNAs have roles in fetal SCs that were not revealed in these cKO mice for technical reasons, as discussed in detail below.

Papaioannou et al. examined the effect of loss of DICER using *Amh*-Cre mice, which first express CRE in fetal SCs at E14.5 [35] These *Dicer*$^{fl/fl}$;*Amh*-Cre mice had some of the same defects as the *Dicer*$^{fl/fl}$;*Sf1*-Cre mice, which express CRE at E11.5 (Table 11.2). For example, both mutant mice exhibited SC apoptosis during early postnatal period and neither exhibited testicular defects during the fetal period despite the loss of DICER during the fetal stage. Indeed, in both mutant mouse models, there was an 8- to 10-day delay between the onset of CRE transgene expression and observable defects.

These findings suggest that miRNAs function specifically in postnatal SCs, not fetal SCs. By analogy, neural progenitor cells are selectively affected by the loss of DICER only at later stages of development [44]. However, it is also possible that miRNAs *do* function in fetal SCs, but overt effects of their loss, such as apoptosis, are not observable until after birth. Given that CRE-mediated deletion is a relatively inefficient process *in vivo*, another possibility is that a sufficient number of SCs escaped CRE-mediated mutation of *Dicer* during the fetal period to allow normal SC function. Another possibility is that DICER and/or fetal miRNAs are very stable in SCs, and, thus, they are not sufficiently depleted to generate defects until after the fetal period. In support of the latter two possibilities, SC-specific miRNA levels were found to be only ~ two-fold decreased at birth, with complete loss not observable until P5 [35].

Unlike *Dicer*$^{fl/fl}$;*Sf1*-Cre mice, which die at approximately P5, *Dicer*$^{fl/fl}$;*Amh*-Cre mice survive into adulthood, which allowed Papaioannou et al. to examine the effect of loss of DICER during a variety of postnatal and adult ages [35]. Interestingly, this analysis revealed that SC apoptosis was no longer observable by P21, a developmental time point just after SCs normally stop proliferating and undergo terminal maturation [45]. This finding suggested that miRNAs are specifically required for the survival of SCs only during their proliferative phase. While the SC apoptosis defect was reversed by the mid-postnatal period, most other testicular defects worsened with age (Table 11.2). For example, while testes in the mutant mice were already dramatically smaller than in controls at P15 (an approximately 80% reduction), this size defect became even more severe by adulthood. Seminiferous epithelial (SE) morphology abnormalities, which were first detected at P5, worsened progressively with age such that, by day 180, the testes of mutant mice were largely devoid of SE tubules, and the few tubules that were present typically lacked germ cells. No sperm were evident at any age, and the mutant mice were infertile.

Several lines of evidence suggest that these progressively worsening aberrations were, at least in part, the result of a SC maturation defect. First, the mutant mice had *increased* numbers of proliferating SCs (as measured by double-immunofluorescence analysis with the SC marker GATA4 and the proliferation marker Ki-67) relative to control mice during the early postnatal period (P0 to P15). This finding was consistent with a maturation defect because SCs normally cease proliferating when they undergo terminal maturation [45]. Second, the nuclei of mutant SCs were mislocalized at the center of the SE tubules, indicating a failure to migrate to the periphery during postnatal development. Third, SCs in the mutant mouse testes exhibited abnormal staining with the basolateral marker TJP1, such that the normal ring pattern was replaced by a discontinuous staining pattern. Fourth, mutant SCs appeared to have dysfunctional secretory activity, given that there was no lumen in mutant SE tubules. Finally, the inability of mutant SCs to support germ cell progression (Table 11.2) was consistent with a SC maturation defect.

Like Papaioannou et al. [35], Kim et al. [36] examined the effect of the loss of DICER using *Amh*-Cre mice and observed most of the same testes defects (Table 11.2). These investigators studied the nature of germ cell defects in these mutant mice in detail and observed abnormalities at various stages of germ cell development, including at both the meiotic and postmeiotic stages. In addition, they discovered that the mutant mice had significantly decreased numbers of cells marked by the spermatogonial stem cell (SSC)/spermatogonial progenitor markers PLZF and OCT4 at P12. This finding suggested that SCs express one or more miRNAs that are required to maintain SSCs and/or promote the survival or proliferation of spermatogonial progenitors, which normally greatly outnumber SSCs. The authors also observed that the mutant postnatal mouse testes had increased testicular cell apoptosis and decreased proliferation, consistent with the notion that miRNA expression in SCs is required for these somatic cells to promote both the survival and proliferation of the adjacent germ cells. In addition, they observed that mutant mice had many PLZF + cells outside of SE tubules, which suggests that one or more miRNAs expressed in SCs generates signals that prevent the adjacent SSCs and/or spermatogonial progenitors from abnormally migrating outside of the SE tubules.

One result obtained by Kim et al. that differed from that obtained by Papaioannou et al. was that the mice recovered from the SC nuclei mislocalization defect at P30. This difference might be due to the fact that the two studies used floxed-*Dicer* mice that differed with regard to the position of the loxP sites in the *Dicer* gene. The genetic background also differed because the mice used by Kim et al. were in a pure C57BL/6 background, while the mice used by Papaioannou et al. were in a mixed C57BL/6 and SV129 background.

B. mRNA and protein dysregulation as a result of loss of DICER in SCs

To determine the effect of loss of DICER at the molecular level, Papaioannou et al., and Kim et al., used microarray and real-time polymerase chain reaction analyses, respectively, to compare the expression of mRNAs in SC *Dicer*-cKO versus control testes [35,36]. Papaioannou et al. found that several categories of genes, including those implicated in cell signaling, cell death, organ development, cell movement, and tissue development, were dysregulated in the mutant testes. In addition, a number of genes known to function in SCs or serve as SC markers were shown to be downregulated in response to the loss of DICER at P0 in both studies. Among these downregulated genes were *Wt1*, which is critical for SC development and function [46]; *Dhh*, which is essential for normal spermatogenesis [47]; and the sex determination gene *Sox9*, which is highly expressed in adult SCs [48]. Two particularly intriguing downregulated genes were *Claudin 11* and *Connexin 43*, which are involved in the formation of the blood–testis barrier (BTB) and gap junctions, respectively, and, thus, their dysregulation may be responsible for some of the specific defects observed in SC *Dicer*-cKO mice. Other downregulated genes identified were *Gdnf* and *Kitl*, which drive SSC self-renewal and differentiation, respectively [49].

Loss of miRNAs in SC *Dicer*-cKO mice would be predicted to upregulate the targets of SC-expressed miRNAs. However, Papaioannou et al. did not find an enrichment of miRNA-binding sites in the 3' UTRs of the upregulated mRNAs that they identified in the mutant mice. This finding could be the result of any number of nonmutually-exclusive reasons, including that (1) the accuracy of miRNA-prediction programs is imperfect (see above); (2) the indirect effects of the loss of miRNAs obscure the direct effects; and (3) SC-expressed miRNAs may primarily repress the translation of their target mRNAs, rather than degrading them, thereby making them "invisible" to microarray analysis.

To begin to address the latter possibility, Papaioannou et al. examined the effect of SC-specific loss of DICER on the proteome in the testis [50]. Using an isotope-coded protein labeling approach, they identified 50 proteins that were significantly upregulated (at least 1.3-fold) and three proteins that were downregulated in *Dicer*$^{fl/fl}$;*Amh*-Cre mice relative to control mice. Intriguingly, the only functional category significantly overrepresented was cytoskeletal proteins, which is of interest given that an elaborate cytoskeletal network—comprised of intermediate filaments and microtubules—is found in terminally differentiated SCs. The cytoskeletal proteins identified as being upregulated in SC *Dicer*-cKO mice were the microtubule-binding protein, tubulin-β5 (TUBB5); the microfilament proteins, actin-g1 (ACTG1) and actin-γ2 (ACTG2); the actin-binding protein, profilin-2; the intermediate filament protein, Vimentin; and the cytoskeleton-linker protein, Annexin a2 (ANXA2). This finding is important, given the evidence that these and other proteins in the cytoskeletal network drive changes in SC morphology that facilitate spermatogenesis. Such cytoskeletal changes have been implicated in the maturation of SCs [51], which, in turn, is consistent with the many lines of evidence showing that SCs lacking miRNAs have a profound maturation defect (see above). Other proteins found to be dysregulated in *Dicer* SC-cKO mice include the oxidative stress protector Cu/Zn superoxide dismutase and the RNA metabolism regulators splicing factor arginine/serine-rich-1 and poly rC-binding protein-1.

C. Spermatogenic defects resulting from loss of DICER or DROSHA in germ cells

Many studies have examined the effect of loss of DICER in germ cells (Table 11.1). The earliest developmental time point at which the *Dicer* gene has been inactivated is ~E9.5 using *Tnap*-Cre mice [37,38]. Hayashi et al. found that such *Dicer*$^{fl/fl}$;*Tnap*-Cre mice had decreased numbers of primordial germ cells relative to control mice as early as E12.5. These mutant mice also had half as many spermatogonia as control mice (as a result of reduced proliferation) at neonatal stages, they lacked germ cells in some tubule sections during late postnatal development (P28), and mice older than 8 months appeared to completely lack germ cells [38]. Using the same Cre-driver mice, another study by Maatoouk et al. found abnormal SE tubules, including those with only elongating spermatids and others lacking germ cells. Testes of these *Dicer*$^{fl/fl}$;*Tnap*-Cre mice were smaller than those of control mice at all developmental time points tested. Both studies found that these mutant mice were subfertile.

Romero et al. and Liu et al. examined the effect of loss of DICER at a later point of embryonic development (~E15.5) using *Vasa*-Cre mice [39,40]. Both groups found that these *Dicer*$^{fl/fl}$;*Vasa*-Cre mice had a more severe phenotype, including complete infertility, than the *Dicer*$^{fl/fl}$;*Tnap*-Cre mice. A likely explanation is the more efficient deletion of *Dicer* in the former (more than 90%) than the latter (approximately 60%) mutant mice [37,38,40].

Liu et al. reported that the *Dicer*$^{fl/fl}$;*Vasa*-Cre mice displayed a variety of defects, including a likely reduction in undifferentiated spermatogonial survival (based on reduced POU5F1+ cells and increased basal SE cell apoptosis) and an unknown defect resulting in decreased numbers of elongated spermatids. Romero et al. also observed that these mutant mice had few elongated spermatids and greatly elevated apoptosis of spermatocytes. Both groups found that the mice lacked proper germ cells associations at particular SE stages and that there was a large number of vacuoles in SE tubules. Of note, these mutant mice did not display morphological defects until P15 (germ cell sloughing from the SE), a time point when germ cells undergo the zygotene–pachytene transition during the first wave of spermatogenesis [40]. Thus, there was a long delay (approximately 3 weeks) between the conditional mutation of *Dicer* in germ cells and observable defects in these cells. The reason for this delay might have been that miRNAs do not function in germ cells until mid-postnatal development, but this possibility is inconsistent with the fetal and perinatal defects observed in mice in which *Dicer* was knocked out earlier in germ cell development [37,38]. It would appear more likely that technical limitations of the *Vasa*-Cre mice prevent earlier functions of miRNAs from being uncovered (by analogy with what was discussed, above, for SC *Dicer*-cKO mice).

Wu et al., Liu et al., and Greenlee et al. conditionally ablated the *Dicer* gene in germ cells during early postnatal development (P3) using *Stra8*-Cre mice [41,42]. As in the studies that ablated *Dicer* in germ cells during late embryonic development, loss of DICER postnatally caused dramatically reduced testes size and infertility. These *Dicer*$^{fl/fl}$;*Stra8*-Cre mice had defects in spermatocytes. The authors of one study reported elevated numbers of apoptotic spermatocytes [42], and those of another study reported an accumulation of early spermatocytes and reduced numbers of later-stage spermatocytes, suggesting inhibited progression of germ cells through meiosis [41]. Wu et al. observed that spermatids were sloughing off from the SE, which explained the almost complete lack of elongated spermatids in the SE. Like the other *Dicer*-cKO mice, these mice had a delay in the appearance of phenotypic defects, with no observable defects occurring until P14 (this delay might have been due, in part, to the inefficient CRE-mediated cleavage of *Dicer* that the authors observed at earlier postnatal time points).

Korhonen et al. ablated *Dicer* using *Ngn3*-Cre mice, which turn on CRE slightly later (at ~P5) than *Stra8*-Cre mice (~P3) [43]. These *Dicer*$^{fl/fl}$; *Ngn3*-Cre mice were infertile but did not have an obvious meiotic progression defect (unlike *Dicer*$^{fl/fl}$;*Stra8*-Cre mice) but did have increased spermatocyte apoptosis at some stages of the SE cycle. They also had defects in spermatids, including an accumulation of abnormal elongated spermatids and inappropriate stages of haploid cells in some stages of the SE cycle.

Because DICER is required not only for miRNA biogenesis, but also for the generation of siRNAs, it is possible that some of the defects observed in *Dicer*-cKO mice resulted from an inability to generate siRNAs. siRNAs in mammals were originally thought to be generated only in oocytes and early embryos [12,52]. But it has come to light that siRNAs are also made in germ cells in mouse testis [53].

In an attempt to specifically address the role of miRNAs in male germ cells, Wu and colleagues conditionally knocked out *Drosha*, which is required for the biogenesis of miRNAs but not siRNAs [42]. Using *Stra8*-iCre mice to conditionally knock out *Drosha* specifically in postnatal germ cells, they observed that the loss of *Drosha* caused severe disruptions of the SE, with very few elongating spermatids. The abnormalities in the SE included a loss of spermatocytes and early spermatids. Surprisingly, all of these defects were more severe in *Drosha*$^{fl/fl}$;*Stra8*-iCre mice than in *Dicer*$^{fl/fl}$;*Stra8-iCre* mice. This finding was unexpected because the latter mice conditionally ablate *Dicer* rather than *Drosha* and thus have defects in both miRNA and siRNA biogenesis. An explanation derives from the fact that DROSHA has roles in addition to miRNA biogenesis. For example, DROSHA cleaves and thereby inactivates mRNAs harboring miRNA precursor-like stem loops [54]. DROSHA also binds to a subset of promoters and modulates their transcription rate [55]. Thus, loss of either or both of these activities in *Drosha*$^{fl/fl}$;*Stra8*-iCre mice could be responsible for the more severe spermatogenic defects.

Hayashi et al. used another approach to specifically examine the role of miRNAs in germ cells. In their study, they conditionally knocked out the *Ago2* gene, which encodes the canonical protein thought to be essential for the action of most miRNAs [38]. Unexpectedly, they found that the targeted loss of *Ago2* in germ cells did not cause any detectable defects in spermatogenesis. This finding suggests that AGO2 has a role that can be compensated for by other AGO proteins (there are four known AGO proteins), or, alternatively, an AGO other than AGO2 is mainly responsible for miRNA action in germ cells. Consistent with either possibility, *Ago2* mRNA is only modestly expressed in the germ cell subsets that have been examined [56]. Instead, germ cells express high levels of *Ago4* mRNA and testis has higher levels of *Ago4* mRNA than other tissues that have been examined [56]. Primary SCs have the inverse expression pattern: high levels of *Ago2* mRNA and relatively low levels of *Ago4* mRNA. To examine the functional relevance of this, it will be intriguing to conditionally knock out *Ago2* in SCs *in vivo*.

IV. SC-expressed miRNAs and their functions

Genome-wide studies conducted using microarray analysis and sequencing of small RNA libraries have identified testicular miRNAs in a wide variety of mammals including mice, rats, pigs, goats, sheep, cattle, primates, and humans [57–63]. The X chromosome is particularly enriched for such testis-expressed miRNA genes, many of which are present in large clusters [58,62,64]. Interestingly, male-biased, protein-coding genes are also enriched on the X chromosome [65–67]. This enrichment of both testis-expressed miRNA and protein-coding genes on the X chromosome may stem from the fact that males have a single X chromosome, thereby allowing male beneficial genes to become more rapidly fixed on the X chromosome than autosomal chromosomes during evolutionary time [68,69].

Given that the testis has a number of different somatic and germ cell types, understanding the cell-type-specific expression and regulation of miRNA targets in these cell types will take us closer to understanding the roles of miRNAs in spermatogenesis. In the following sections, we discuss what is known about testis-expressed miRNAs, with an emphasis on SC-expressed miRNAs.

A. *Targets and potential functions of SC-expressed miRNAs*

A number of studies have cloned and sequenced small RNA libraries from enriched SCs from both early postnatal (P6 and P7) testes [35,58,70] and adult testes [71]. Table 11.3 provides a list of SC-expressed miRNAs, including those that have been shown to be more highly expressed in SCs than whole testes and those that are androgen responsive. Of note, only a small number of the miRNAs listed in Table 11.3 have been identified by more than one study, indicating that saturation has not been reached. Thus, it is likely that many more SC-enriched miRNAs will be identified in the future.

To begin to assess the potential role for these SC-expressed miRNAs, we identified high-confidence direct targets using well-established miRNA-target prediction programs. We then conducted a literature search to narrow down the predicted mRNA targets known to be expressed in the testis and/or have known functions in the testis. Table 11.3 lists some of these predicted mRNA targets of SC-expressed miRNAs.

To date, no SC-expressed miRNA has been demonstrated to have a function in SCs. However, several miRNAs expressed in SCs have been shown to exert functions in other cell types. This finding may provide clues to what these miRNAs do in SCs. For example, the SC-expressed miRNA miR-34c has been shown to have several roles. In germ cells, the blockade of miR-34c function with a specific inhibitor decreases germ cell apoptosis *in vivo*, indicating that miR-34c can serve as a prosurvival factor [72]. The SC-expressed miR-449 cluster targets mRNAs encoding proteins in the E2F transcription factor-retinoblastoma protein pathway,

Table 11.3 SC-expressed miRNAs and some of their predicted target mRNAs

miRNA	Predicted targets	miRNA	Predicted targets
miR-15b[a,b]	*Atp7a, Nonagouti*	miR-431[c]	*Lhx8, Opn3*
miR-19a[a,b]	*Spirel, Zmynd 11, Sin3b*	miR-449a[a]	*Notchl, Fgf9*
miR-19b[a,b]	*Taokl*	miR-463[a]	*Ubfdl, Aebp2*
miR-25[a,b]	*Map2K4, Clndn11, Dcaf6*	miR-465a-3p[d]	*Actr3, Pdelc*
miR-29c[a]	*Nasp, Mkrn2*	miR-465c[d,e]	*Kdm2a, Cyp26al*
miR-34a[a,e]	*Notchl, Fam126b, Rras*	miR-465c-3p[a]	*Actr3, Arhgap36*
miR-34b[a,e]	*Nampt, Zdhhc16, Numbl*	miR-466e-3p[a]	*Uty, Cxcl5, Ilf2*
miR-34c[a,e]	*Notchl, Nampt*	miR-468[a]	*Ube2d2, Hnrnpa2b1*
miR-34c*[a]	*Ppmld, Nampt*	miR-470[a,d,e]	*Emx2, Tgifl, Notchl*
miR-92a[a]	*Dcaf6, Cpeb3, Clndn11*	miR-470*[a,d]	*Hifla, Abil, Ube2q2*
miR-129-3p[a]	*Ift122, Memol, Purpl, Rhobtbl*	miR-471[a,d,e]	*Sfpq,Pafahlb2*
miR-135b[a]	*Itm2b, Cpsf6*	miR-471-3p[d]	*Ccngl, Mtap2*
miR-181d[a]	*Eif4a2, Armcx3, Ddx3x*	miR-487b[c]	*Ovoll,Pitx2*
miR-181b[a]	*Eif4a2, Armcx3, Ddx3x*	miR-547[a]	*Armcx2, Inpp4a, Igfl*
miR-196b[a]	*Epha7, Hdac9*	miR-674*[c]	*Pckl, Bcl6*
miR-201[a,e]	*Ubxn4*	miR-679[c]	*Snapc3,Fmrl*
miR-202[a,d]	*Cnot2, Pdhal*	miR-741[a,d,e]	*Vegfa, Foxil, Fmrl*
miR-299[c]	*Cuxl,Surf2, Paxipl*	miR-742[d,e]	*Scin, Ube2d2*
miR-328[a]	*Myo19, Tesk2*	miR-743a[a,e]	*Fmrl, Atad2, Top2b*
miR-335-5p[c]	*Hoxd12, Eif4a2*	miR-743b-3p[a,e]	*Fmrl, Elavl2, Wtl*
miR-341[a,c]	*Arpc2, Ryr3*	miR-871[a,e]	*Slitrk5, Lamcl*
miR-375[a,e]	*Ahr, Qk*	miR-878[a,e]	*Bazla, Wdr19*
miR-376a[c]	*Stcl,Bmp2*	miR-880[a,e]	*Atrn, Mtap2*
miR-381[c]	*Zfpm2, Ocln*	miR-883a-3p[a,d]	*Gpr64, Tmfl*
miR-409-5p[c]	*Cpsf6, Lsm14a*	miR-883a[a,d,e]	*Ubqlnl, Rbm41, Arpc5*
		miR-883b-3p[a]	*Potla, Tmsb4x*

[a]Androgen-responsive miRNAs expressed in SCs purified from adult mouse testes [71].
[b]miRNAs expressed in enriched SCs from P6 mice [70].
[c]miRNAs more highly expressed in enriched P6 SCs than total P6 testis [35].
[d]miRNAs expressed in enriched SCs from P6 mice [58].
[e]Developmentally regulated miRNAs expressed at higher level in P14 testes compared to P7 testes [62].
High-confidence direct targets were determined using TargetScan.org and microRNA.org.

suggesting that this miRNA cluster controls the proliferation of testicular cells, although whether this is the case in SCs is not known [73]. The targeted deletion of the miR-449 cluster was not found to disrupt spermatogenesis, perhaps because of functional redundancy with miR-34b/c, which is upregulated by loss of the miR-449 cluster and targets many of the same mRNAs as the miR-449 cluster. The SC-expressed miRNA miR-883a targets the mRNA encoding actin-related protein 5, a ribonucleoprotein-associated translational repressor that has been shown to inhibit 80S formation [74]. By regulating the translation of specific subsets of mRNAs in SCs, miR-883a could potentially have profound effects on spermatogenesis. The SC-expressed miRNA miR-202 is implicated in roles downstream of the sex-determining factor SOX9 by virtue of being a likely direct target of SOX9 [75]. SOX9 is expressed at high levels in adult SCs, and, thus, it may operate in a circuit with miR-202 to control the expression of miR-202 target genes. Of note is that a high-confidence predicted target of miR-202 is the mRNA encoding the mRNA deadenylase *CNOT2* (Table 11.3). Because mRNA deadenylation is a rate-limiting step for mRNA decay, it is possible that a SOX9-miR-202 circuit dictates the rate of mRNA decay (and, thus, the steady-state levels) of large subsets of mRNAs in SCs.

Many of the predicted targets of SC-expressed miRNAs encode proteins with roles in testis development and spermatogenesis (Table 11.3). Among these are mRNAs that have already been demonstrated to be expressed in SCs, including *Fmr1*, *Vegfa*, *Wt1*, *Igf1*, *Cldn11*, *Ubqln1*, *Notch1*, and *Nampt*. At least two SC-expressed miRNAs, miR-25 and miR-92a, are predicted to target *Cldn11*, which encodes a SC-expressed protein required for the establishment of SC tight junctions and whose absence prevents normal spermatogenesis [76]. miR-547 targets insulin growth factor-1 (*IGF1*), raising the possibility that this miRNA represses insulin/IGF signaling in SCs, a signaling event that is essential for follicle-stimulating hormone (FSH)-driven SC proliferation. Indeed, it is known that repressed IGF signaling is necessary to generate a testis of normal size as well as for normal sperm counts [77]. Alternatively, miR-547 may serve to fine-tune IGF1 levels in response to environmental and genetic perturbation. This notion is consistent with the tendency of most miRNAs to only modestly repress their targets, a property that allows miRNAs to operate in feedback networks that confer robustness [78].

B. Protein classes predicted to be regulated by SC-expressed miRNAs

Are there functional classes of proteins that SC-expressed miRNAs tend to regulate? While this question has not yet been definitively addressed, emerging evidence suggests that the answer is affirmative. Using Ingenuity® Pathway Analysis (IPA), Papaioannou et al. identified significantly enriched classes of proteins encoded by mRNAs that are differentially expressed in SC-cKO *Dicer* postnatal testes versus control postnatal testes [35] (Table 11.4). Ortogero et al. identified enriched functional classes of proteins encoded by predicted targets of miRNAs expressed in postnatal SCs [70] (using the microcosm miRNA-target prediction program and Gene Ontology analysis; Table 11.4). The same approach was adopted by Paneerdoss et al. to identify functional classes of proteins predicted to be targeted by androgen-regulated miRNAs [71] (using the miRNA-target prediction programs PicTar, MirTarget2, TargetScan, miRanda, and miTarget, coupled with IPA; Table 11.4).

Several themes are suggested by these analyses. First, signaling molecules appear to be common targets of miRNAs in both postnatal and adult SCs. This finding raises the possibility that miRNAs are critical for signaling events in SCs in general. Second, SC-expressed miRNAs expressed postnatally appear to commonly regulate molecules that dictate life-or-death decisions. Third, SC-expressed miRNAs that are expressed postnatally commonly regulate genes encoding transcriptional and posttranscriptional regulatory factors. This finding suggests that SC-expressed miRNAs are interwoven in circuits with gene regulatory factors devoted to controlling SC development and/or proliferation. The possibility that SC-expressed miRNAs have roles in proliferation is reinforced by the finding that such miRNAs commonly target cell cycle genes. Finally, adult SC-expressed miRNAs regulate molecules that are important in mature SCs, such as proteins involved in the formation of SC–SC junctions, germ cell–SC junctions, and the cytoskeleton.

C. *Functions of germ-cell-expressed miRNAs*

Table 11.5 lists germ-cell-expressed miRNAs whose role in the testis have been characterized. Among these miRNAs is the let-7 family, a well-studied family of miRNAs that are important for promoting differentiation in many cell types in organisms across the phylogenetic scale. Many members of this family are expressed in the testis, particularly in spermatogonia and spermatocytes in mice.

Table 11.4 Functional classes of proteins encoded by predicted SC-expressed miRNA targets

Postnatal (P6) SC-expressed miRNAs [70]	Postnatal (P0/P5) SC-expressed miRNAs [35]	Adult SC-expressed miRNAs [71]
Cell cycle processes	Cell signaling	PTEN signaling
Phosphorylation	Cell death	Wnt/β-catenin signaling
RNA processing	Organ development	Germ cell–SC junction
Negative regulation of gene expression	Cellular movement	Actin cytoskeleton signaling
Phosphate metabolic process	Tissue development	SC–SC junction
		Gap junction

The text explains how the functional classes of miRNA targets were defined.

These miRNAs are known to be induced by retinoic acid treatment and are predicted to target mRNAs encoding several proteins that are important for spermatogonial development, including MYCN, CCND1, and COLLα2 [79]. It will therefore be interesting to determine, in the future, whether Let-7 family members have roles in retinoic-acid-induced spermatogonial differentiation.

V. Regulation of SC-expressed miRNAs

Testicular development and function are influenced by a number of regulatory factors. In this section, we summarize what is known about the regulation of SC-expressed miRNAs. This may ultimately provide insights into how these miRNAs are integrated into molecular circuits important for spermatogenesis.

A. Developmental regulation

The notion that miRNAs regulate SC development has been bolstered by the emerging evidence that some SC-expressed miRNAs are developmentally regulated. Paneerdoss et al. identified 9 SC-expressed miRNAs that differ in their levels during postnatal development. Many of them peak in level during mid-postnatal (P13 and P21) development [71]. Using genome-profiling methods, Buchold et al. also identified a number of miRNAs that are highly expressed during mid-postnatal development (P14) [62], many of which were shown by Paneerdoss et al. to be SC expressed [71]. The timing of this peak expression raises the possibility that such miRNAs have roles in promoting the transition of proliferative SCs to terminally differentiated, postmitotic SCs. A nonmutually-exclusive possibility is that such SC-expressed miRNAs indirectly influence meiotic and differentiation events in the neighboring germ cells.

B. Hormonal regulation

A hallmark of spermatogenesis is its control by hormones. For example, androgens act directly on SCs, Leydig, and peritubular myoid cells to dictate the development and survival of germ cells. In SCs, androgens bind to the androgen receptor to promote meiotic progression and spermatid differentiation events in the directly adjacent germ cells. Collaborating with androgens are luteinizing hormone (LH) and Follicle stimulating hormone (FSH). LH drives the production of androgen by Leydig cells, while FSH promotes the development, proliferation, and function of SCs.

The molecular mechanism of action of FSH and androgens in SCs remains largely obscure. Most research has focused on the ability of these hormones to regulate signaling and transcriptional events. The discovery of miRNAs has raised the

Table 11.5 Germ-cell-expressed miRNAs and their functions

miRNA	Implicated role	Reference
Let-7 family	Retinoic acid (RA) induced spermatogonial differentiation.	[79]
miR-17-92	Expression suppressed during RA-induced spermatogonial differentiation. KO results in small testes and low sperm count.	[80]
miR-21	Essential for SSC maintenance. Regulated by the transcription factor ETV5.	[81]
miR-22	Downregulates the translational repressor ARPC5.	[74]
miR-34c	Enhances the expression of germ cell markers in differentiating ES cells. Sperm borne-miR-34c is required for the first cleavage division of zygotes. Promotes germ cell survival.	[72,82,83]
miR-34b/c	Targets the E2F transcription factor-retinoblastoma protein (E2F-pRb) pathway. Expression is upregulated during meiosis.	[73]
miR-106b-25	Suppressed during RA-induced spermatogonial differentiation.	[80]
miR-140-5p/3p	KO has increased Leydig cell numbers. Implicated in testis differentiation.	[84]
miR-146	Negatively regulates spermatogonial differentiation by downregulating Med1—a co regulator for retinoid receptors.	[85]
miR-202-5p/3p	Positively regulated by SOX9 during male sex determination in primordial germ cells.	[75]
miR-221/222	Promotes SSC self-renewal.	[86]
miR-383	Targets *cyclinD1* and *interferon regulatory factor-1* mRNA. Physically associates with FMRP in the testis. Negatively regulated by FMRP.	[87]
miR-449 cluster	Expression is upregulated during the meiotic phase of the first wave of spermatogenesis. Targets many members of the E2F transcription factor-retinoblastoma protein (E2F-pRb) pathway. Loss of miR-449 cluster upregulates miR-34b/c in the testis *in vivo*.	[73]
miR-469	Directly targets *transition protein-2* and *protamine-2*.	[88]
miR-883a	Directly targets *Arpc5*, which encodes a translational repressor.	[74]

While all these miRNAs are expressed in male germ cells, many are also expressed in other cell types. The functions and targets of these miRNAs listed are necessarily incomplete, with a focus on male reproductive functions.

possibility that androgens and FSH also regulate functions in SCs through these small RNAs. To begin to examine this issue, Paneerdoss and colleagues used miRNA-microarray analysis to screen for androgen-regulated SC-expressed miRNAs [71]. They identified 28 miRNAs that exhibited significantly altered expression in the testis in response to combination treatment with the androgen antagonist flutamide and the gonadotropin-releasing hormone antagonist acyline. The regulation of these miRNAs by androgen was validated by several approaches, including analysis of LHβ KO mice and rescue of their normal expression after injection of exogenous androgen. Interestingly, 18 of these 28 androgen-regulated miRNAs are encoded on the X chromosome, suggesting that this sex chromosome is highly enriched for androgen-regulated miRNA genes. Twenty of these miRNAs were found to be negatively regulated by androgen. Given that miRNAs typically repress their mRNA targets, this finding implies the existence of a circuit in which androgens *elevate* the level of the proteins encoded by the targets of these miRNAs. Using IPA analysis, Paneerdoss et al. found that these androgen-regulated miRNAs tend to target mRNAs encoding proteins involved in signaling events between SCs as well as between SCs and germ cells (Table 11.4). This role may be functionally relevant because inter-SC tight junctions form the so-called BTB in the SE during postnatal development. This barrier is a dynamic entity because it is locally broken down and reestablished as the preleptotene/leptotene spermatocytes pass from the basal to the adluminal compartment of the SE [89]. By regulating the levels of signaling and structural components in the BTB, miRNAs could act either as switches to control the BTB-dependent steps of spermatogenesis or confer robustness from environmental, developmental, and/or genetic insults.

In another study, Nicholls et al. screened rat testes for hormonally regulated miRNAs [90]. These investigators used a well-established hormone suppression regime involving daily injections of antibodies against FSH along with high doses of testosterone and estrogen to trigger feedback inhibition of the hypothalamic-pituitary-testis axis, followed by injection of the androgen antagonist flutamide to block the action of residual androgen. Using microarray analysis, they identified 56 upregulated miRNAs and 5 downregulated miRNAs (regulated by more than two-fold) in response to this treatment. This finding indicated that these hormones tend to repress the expression of far more miRNAs than they upregulate, which is in agreement with the findings of Paneerdoss et al. [71].

In a parallel experiment, Nicholls et al. screened primary SC cultures for miRNAs regulated by testosterone and recombinant FSH. Microarray analysis revealed that these hormones downregulated 122 miRNAs and upregulated 41 miRNAs. A core group of 24 miRNAs was found to be hormone regulated in both *in vivo* and *in vitro* systems. This core group of miRNAs will be good subjects for further study. Indeed, these investigators chose to focus on five of these miRNAs—miR-23b, −30c, 30d, 125b-3p, and −690—to examine their response to FSH alone and androgen alone versus FSH and androgen in combination. They found that each of these miRNAs responded differently to hormone treatment. For example, while miR-23b was only repressed by the combination of FSH and androgen, miR-30d was strongly repressed by FSH alone (more strongly than either androgen alone or FSH in combination with androgen), and miR-690 levels were repressed nearly equally by androgen alone and FSH alone. It is possible that these miRNAs are expressed from promoters that are directly responsive to these hormones, given that shifts in miRNA levels were relatively rapid (6−12 h after hormone treatment or withdrawal).

To assess the potential downstream effects of the large number of "hormone-suppressed" miRNAs that they identified, Nicholls et al. used DIANA-miRPath analysis to identify classes of proteins encoded by mRNA that were predicted to be targeted by these miRNAs [90]. These functional classes included actin cytoskeleton, transforming growth factor β (TGF-β) signaling, focal adhesions, and mitogen-activated protein kinase signaling. An intriguing possibility is that some of these hormone-responsive miRNAs have a role in spermiation, given that spermiation failure is a classic defect resulting from hormone suppression [91−93]. A candidate for having a role is miR-23b, which directly targets the mRNA encoding phosphatase and tensin homolog, a component of the focal adhesions that are integral for the spermiation machinery [92].

C. TGF-β signaling regulation

Olcorn et al. reported that the TGF-β signaling activator, activin A, downregulated 60 miRNAs and upregulated 1 miRNA (more than 1.5-fold), among the 411 miRNAs they detected as being expressed in P20 rat SCs *in vitro* [94]. This result has the potential to be physiologically important for spermatogenesis because activin A is crucial for promoting the proliferation of fetal SCs [95]; indeed, activin A can convert differentiated postmitotic SCs back into the proliferative mode [94,96]. Furthermore, evidence suggests that SC tight junctions and the BTB are established postnatally by reduced activin signaling that normally occurs during this phase of development [96]. It will be interesting in the future to determine whether any of the activin-regulated miRNAs identified by Olcorn et al. mediate such fetal and postnatal actions of activin in SCs.

VI. Perspective

In this chapter, we have summarized studies exploring the role of miRNAs in SCs and spermatogenesis. Considerable information on this topic has been generated but large gaps in our knowledge remain. While studies that genetically ablated miRNA biogenesis specifically in SCs *in vivo* have identified several steps in spermatogenesis that are likely to depend on SC-expressed miRNAs, it is not known which specific miRNAs function in these steps or what they do. Indeed, to our knowledge, the field has not yet assigned a definitive SC function to any of the miRNAs so far shown to be expressed in SCs. Furthermore, while many predicted target mRNAs of SC-expressed miRNAs have been identified, few have been empirically proven.

We suggest that while mouse models should continue to be used to define miRNA functions in SCs, they need to be refined. The *Dicer*-cKO models that have so far been examined block not only miRNA biogenesis but also other RNA pathways. Therefore, future studies need to be directed toward generating and analyzing other mouse models, such as *Drosha*- and *Ago*-deficient mice. An additional future goal is to understand why there are no observable fetal SC defects in mice that genetically ablate *Dicer* in fetal SCs. Is it because SC-expressed miRNAs only function in postnatal and adult SCs? Or are fetal functions of SC-expressed miRNAs not exposed with the cKO approaches so far employed?

Once miRNAs have been identified that have specific functions in SCs, more in-depth questions will arise. For example, do such miRNAs primarily act in a subtle manner to maintain homeostasis in the testes, or do they, instead, serve as biological switches to control specific developmental steps and/or functions in the testis?

Another future goal is to identify clinical applications for testis-expressed miRNAs. A number of miRNAs have been found in human sperm [63,97], raising the possibility that miRNA profiling of sperm could be useful as a diagnostic tool for infertility. For example, miRNA signatures may be identified that correlate with specific types of male subfertility or infertility. Indeed, studies have already identified specific miRNAs that are dysregulated in the seminal plasma and testes of patients with azoospermia and/or asthenozoospermia [98,99]. miRNAs may also prove useful for clinical treatment of infertility and other testicular conditions, including testicular cancer. With regard to the latter, Liu et al. reported that overexpression of miR-302 in a testicular germ-cell tumor-derived cell line, rendered these cells highly susceptible to the chemotherapeutic agent cisplatin [100]. This is an exciting finding because it raises the possibility that miR-302 (and possibly other testis-expressed miRNAs) can be used along with chemotherapeutic drugs to reduce the dosage and, thereby, the harmful side effects of such toxic drugs. Clearly, the field is only just beginning to realize the potential clinical applications of testis-expressed miRNAs

References

[1] Amaral PP, Dinger ME, Mercer TR, Mattick JS. The eukaryotic genome as an RNA machine. Science 2008;319:1787–9.

[2] Dinger ME, Amaral PP, Mercer TR, Mattick JS. Pervasive transcription of the eukaryotic genome: functional indices and conceptual implications. Brief Funct Genomic Proteom 2009;8:407–23.

[3] Jacquier A. The complex eukaryotic transcriptome: unexpected pervasive transcription and novel small RNAs. Nat Rev Genet 2009;10:833–44.

[4] Mattick JS, Makunin IV. Non-coding RNA. Hum Mol Genet 2006;15 Spec No: R17–29.

[5] Mattick JS. The genetic signatures of noncoding RNAs. PLoS Genet 2009;5:e1000459.

[6] Rinn JL, Chang HY. Genome regulation by long noncoding RNAs. Ann Rev Biochem 2012;81:145–66.

[7] Guttman M, Rinn JL. Modular regulatory principles of large non-coding RNAs. Nature 2012;482:339–46.

[8] Ulitsky I, Bartel DP. lincRNAs: genomics, evolution, and mechanisms. Cell 2013;154:26–46.

[9] Fatica A, Bozzoni I. Long non-coding RNAs: new players in cell differentiation and development. Nat Rev Genet 2014;15:7–21.

[10] Carthew RW, Sontheimer EJ. Origins and mechanisms of miRNAs and siRNAs. Cell 2009;136:642–55.

[11] Golden DE, Gerbasi VR, Sontheimer EJ. An inside job for siRNAs. Mol Cell 2008;31:309–12.

[12] Watanabe T, Totoki Y, Toyoda A, Kaneda M, Kuramochi-Miyagawa S, Obata Y, et al. Endogenous siRNAs from naturally formed dsRNAs regulate transcripts in mouse oocytes. Nature 2008;453:539–43.

[13] Ghildiyal M, Seitz H, Horwich MD, Li C, Du T, Lee S, et al. Endogenous siRNAs derived from transposons and mRNAs in *Drosophila* somatic cells. Science 2008;320:1077–81.

[14] Meister G. Argonaute proteins: functional insights and emerging roles. Nat Rev Genet 2013;14:447–59.

[15] Lee RC, Feinbaum RL, Ambros V. The *C. elegans* heterochronic gene lin-4 encodes small RNAs with antisense complementarity to lin-14. Cell 1993;75:843–54.

[16] Finnegan EF, Pasquinelli AE. MicroRNA biogenesis: regulating the regulators. Crit Rev Biochem Mol Biol 2013;48:51–68.

[17] Pasquinelli AE. MicroRNAs and their targets: recognition, regulation and an emerging reciprocal relationship. Nat Rev Genet 2012;13:271–82.

[18] Kim VN, Han J, Siomi MC. Biogenesis of small RNAs in animals. Nat Rev Mol Cell Biol 2009;10:126–39.

[19] Borchert GM, Lanier W, Davidson BL. RNA polymerase III transcribes human microRNAs. Nat Struct Mol Biol 2006;13:1097–101.

[20] Winter J, Jung S, Keller S, Gregory RI, Diederichs S. Many roads to maturity: microRNA biogenesis pathways and their regulation. Nat Cell Biol 2009;11:228–34.

[21] Krol J, Loedige I, Filipowicz W. The widespread regulation of microRNA biogenesis, function and decay. Nat Rev Genet 2010;11:597–610 [Nature Publishing Group, a division of Macmillan Publishers Limited. All Rights Reserved].

[22] Bushati N, Cohen SM. MicroRNA functions. Ann Rev Cell Dev Biol 2007;23: 175–205.

[23] Bartel DP. MicroRNAs: target recognition and regulatory functions. Cell 2009;136:215–33.

[24] Huntzinger E, Izaurralde E. Gene silencing by microRNAs: contributions of translational repression and mRNA decay. Nat Rev Genet 2011;12:99–110.

[25] Aalto AP, Pasquinelli AE. Small non-coding RNAs mount a silent revolution in gene expression. Curr Opin Cell Biol 2012;24:333–40.

[26] Axtell MJ, Westholm JO, Lai EC. Vive la différence: biogenesis and evolution of microRNAs in plants and animals. Genome Biol 2011;12:221.

[27] Wyckoff GJ, Wang W, Wu CI. Rapid evolution of male reproductive genes in the descent of man. Nature 2000;403:304–9.

[28] Haerty W, Jagadeeshan S, Kulathinal RJ, Wong A, Ravi Ram K, Sirot LK, et al. Evolution in the fast lane: rapidly evolving sex-related genes in *Drosophila*. Genetics 2007;177:1321–35.

[29] Swanson WJ, Vacquier VD. The rapid evolution of reproductive proteins. Nat Rev Genet 2002;3:137–44.

[30] Chi SW, Zang JB, Mele A, Darnell RB. Argonaute HITS-CLIP decodes microRNA–mRNA interaction maps. Nature 2009;460:479–86.

[31] Hafner M, Landthaler M, Burger L, Khorshid M, Hausser J, Berninger P, et al. Transcriptome-wide identification of RNA-binding protein and microRNA target sites by PAR-CLIP. Cell 2010;141:129–41.

[32] Keene JD, Komisarow JM, Friedersdorf MB. RIP-Chip: the isolation and identification of mRNAs, microRNAs and protein components of ribonucleoprotein complexes from cell extracts. Nat Protoc 2006;1:302–7.

[33] Cambronne XA, Shen R, Auer PL, Goodman RH. Capturing microRNA targets using an RNA-induced silencing complex (RISC)-trap approach. Proc Natl Acad Sci USA 2012;109:20473–8.

[34] Huang C-CJ, Yao HHC. Inactivation of Dicer1 in steroidogenic factor 1-positive cells reveals tissue-specific requirement for Dicer1 in adrenal, testis, and ovary. BMC Dev Biol 2010;10:66.

[35] Papaioannou MD, Pitetti J-L, Ro S, Park C, Aubry F, Schaad O, et al. SC Dicer is essential for spermatogenesis in mice. Dev Biol 2009;326:250–9.

[36] Kim G-J, Georg I, Scherthan H, Merkenschlager M, Guillou F, Scherer G, et al. Dicer is required for SC function and survival. Int J Dev Biol 2010;54: 867–75.

[37] Maatouk DM, Loveland KL, McManus MT, Moore K, Harfe BD. Dicer1 is required for differentiation of the mouse male germline. Biol Reprod 2008;79:696–703.

[38] Hayashi K, Chuva de Sousa Lopes SM, Kaneda M, Tang F, Hajkova P, Lao K, et al. MicroRNA biogenesis is required for mouse primordial germ cell development and spermatogenesis. PLoS One 2008;3:e1738.

[39] Liu D, Li L, Fu H, Li S, Li J. Inactivation of Dicer1 has a severe cumulative impact on the formation of mature germ cells in mouse testes. Biochem Biophys Res Commun 2012;422:114–20.

[40] Romero Y, Meikar O, Papaioannou M. Dicer1 depletion in male germ cells leads to infertility due to cumulative meiotic and spermiogenic defects. PLoS One 2011;6.

[41] Greenlee AR, Shiao M-S, Snyder E, Buaas FW, Gu T, Stearns TM, et al. Deregulated sex chromosome gene expression with male germ cell-specific loss of Dicer1. PLoS One 2012;7:e46359.

[42] Wu Q, Song R, Ortogero N, Zheng H, Evanoff R, Small CL, et al. The RNase III enzyme DROSHA is essential for microRNA production and spermatogenesis. J Biol Chem 2012;287:25173–90.

[43] Korhonen HM, Meikar O, Yadav RP, Papaioannou MD, Romero Y, Da Ros M, et al. Dicer is required for haploid male germ cell differentiation in mice. PLoS One 2011;6:e24821.

[44] De Pietri Tonelli D, Pulvers JN, Haffner C, Murchison EP, Hannon GJ, Huttner WB. miRNAs are essential for survival and differentiation of newborn neurons but not for expansion of neural progenitors during early neurogenesis in the mouse embryonic neocortex. Development 2008;135:3911–21.

[45] Sharpe RM, McKinnell C, Kivlin C, Fisher JS. Proliferation and functional maturation of SCs, and their relevance to disorders of testis function in adulthood. Reproduction 2003;125:769–84.

[46] Rao MK, Pham J, Imam JS, MacLean JA, Murali D, Furuta Y, et al. Tissue-specific RNAi reveals that WT1 expression in nurse cells controls germ cell survival and spermatogenesis. Genes Dev 2006;20:147–52.

[47] Clark AM, Garland KK, Russell LD. Desert hedgehog (Dhh) gene is required in the mouse testis for formation of adult-type Leydig cells and normal development of peritubular cells and seminiferous tubules. Biol Reprod 2000;63:1825–38.

[48] Hemendinger RA, Gores P, Blacksten L, Harley V, Halberstadt C. Identification of a specific SC marker, Sox9, for use in transplantation. Cell Transplant 2002;11:499–505.

[49] Song H-W, Wilkinson MF. Transcriptional control of spermatogonial maintenance and differentiation. Semin Cell Dev Biol 2014.

[50] Papaioannou MD, Lagarrigue M, Vejnar CE, Rolland AD, Kühne F, Aubry F, et al. Loss of Dicer in SCs has a major impact on the testicular proteome of mice. Mol Cell Proteomics 2011;10: M900587MCP200.
[51] Vogl AW, Vaid KS, Guttman JA. The SC cytoskeleton. Adv Exp Med Biol 2008;636:186–211.
[52] Tam OH, Aravin AA, Stein P, Girard A, Murchison EP, Cheloufi S, et al. Pseudogene-derived small interfering RNAs regulate gene expression in mouse oocytes. Nature 2008;453:534–8.
[53] Song R, Hennig GW, Wu Q, Jose C, Zheng H, Yan W. Male germ cells express abundant endogenous siRNAs. Proc Natl Acad Sci USA 2011;108:13159–64.
[54] Chong MMW, Zhang G, Cheloufi S, Neubert TA, Hannon GJ, Littman DR. Canonical and alternate functions of the microRNA biogenesis machinery. Gene Dev 2010;24:1951–60.
[55] Gromak N, Dienstbier M, Macias S, Plass M, Eyras E, Cáceres JF, et al. Drosha regulates gene expression independently of RNA cleavage function. Cell Rep 2013;5:1499–510.
[56] González-González E, López-Casas PP, del Mazo J. The expression patterns of genes involved in the RNAi pathways are tissue-dependent and differ in the germ and somatic cells of mouse testis. Biochim Biophys Acta 2008;1779:306–11.
[57] Barad O, Meiri E, Avniel A, Aharonov R, Barzilai A, Bentwich I, et al. MicroRNA expression detected by oligonucleotide microarrays: system establishment and expression profiling in human tissues. Genome Res 2004;14:2486–94.
[58] Ro S, Park C, Sanders KM, McCarrey JR, Yan W. Cloning and expression profiling of testis-expressed microRNAs. Dev Biol 2007;311:592–602.
[59] Yan N, Lu Y, Sun H, Tao D, Zhang S, Liu W, et al. A microarray for microRNA profiling in mouse testis tissues. Reproduction 2007;134:73–9.
[60] Yan N, Lu Y, Sun H, Qiu W, Tao D, Liu Y, et al. Microarray profiling of microRNAs expressed in testis tissues of developing primates. J Assist Reprod Genet 2009;26: 179–86.
[61] Luo L, Ye L, Liu G, Shao G, Zheng R, Ren Z, et al. Microarray-based approach identifies differentially expressed microRNAs in porcine sexually immature and mature testes. PLoS One 2010;5:e11744.
[62] Buchold GM, Coarfa C, Kim J, Milosavljevic A, Gunaratne PH, Matzuk MM. Analysis of microRNA expression in the prepubertal testis. PLoS One 2010;5:e15317.
[63] Wu W, Qin Y, Li Z, Dong J, Dai J, Lu C, et al. Genome-wide microRNA expression profiling in idiopathic non-obstructive azoospermia: significant up-regulation of miR-141, miR-429 and miR-7-1-3p. Hum Reprod 2013;28:1827–36.
[64] Guo X, Su B, Zhou Z, Sha J. Rapid evolution of mammalian X-linked testis microRNAs. BMC Genomics 2009;10:97.
[65] Wang X, Zhang J. Rapid evolution of mammalian X-linked testis-expressed homeobox genes. Genetics 2004;167:879–88.
[66] Niu A, Wang Y, Zhang H, Liao C, Wang J, Zhang R, et al. Rapid evolution and copy number variation of primate RHOXF2, an X-linked homeobox gene involved in male reproduction and possibly brain function. BMC Evol Biol 2011;11:298.
[67] Soumillon M, Necsulea A, Weier M, Brawand D, Zhang X, Gu H, et al. Cellular source and mechanisms of high transcriptome complexity in the mammalian testis. Cell Rep 2013;3:2179–90.

[68] Lercher MJ, Urrutia AO, Hurst LD. Evidence that the human X chromosome is enriched for male-specific but not female-specific genes. Mol Biol Evol 2003;20:1113–6.
[69] Torgerson DG, Singh RS. Enhanced adaptive evolution of sperm-expressed genes on the mammalian X chromosome. Heredity 2006;96:39–44.
[70] Ortogero N, Hennig GW, Langille C, Ro S, McCarrey JR, Yan W. Computer-assisted annotation of murine SC small RNA transcriptome. Biol Reprod 2013;88:3.
[71] Panneerdoss S, Chang Y-F, Buddavarapu KC, Chen H-I H, Shetty G, Wang H, et al. Androgen-responsive microRNAs in mouse SCs. PLoS One 2012;7:e41146.
[72] Liang X, Zhou D, Wei C, Luo H, Liu J, Fu R, et al. MicroRNA-34c enhances murine male germ cell apoptosis through targeting ATF1. PLoS One 2012;7:e33861.
[73] Bao J, Li D, Wang L, Wu J, Hu Y, Wang Z, et al. MicroRNA-449 and microRNA-34b/c function redundantly in murine testes by targeting E2F transcription factor-retinoblastoma protein (E2F-pRb) pathway. J Biol Chem 2012;287:21686–98.
[74] Chang Y-F, Lee-Chang JS, Imam JS, Buddavarapu KC, Subaran SS, Sinha-Hikim AP, et al. Interaction between microRNAs and actin-associated protein Arpc5 regulates translational suppression during male germ cell differentiation. Proc Natl Acad Sci USA 2012;109:5750–5.
[75] Wainwright EN, Jorgensen JS, Kim Y, Truong V, Bagheri-Fam S, Davidson T, et al. SOX9 regulates microRNA miR-202-5p/3p expression during mouse testis differentiation. Biol Reprod 2013;89:34.
[76] Mazaud-Guittot S, Meugnier E, Pesenti S, Wu X, Vidal H, Gow A, et al. Claudin 11 deficiency in mice results in loss of the SC epithelial phenotype in the testis. Biol Reprod 2010;82:202–13.
[77] Pitetti J-L, Calvel P, Zimmermann C, Conne B, Papaioannou MD, Aubry F, et al. An essential role for insulin and IGF1 receptors in regulating SC proliferation, testis size, and FSH action in mice. Mol Endocrinol 2013;27:814–27.
[78] Peláez N, Carthew RW. Biological robustness and the role of microRNAs: a network perspective. Curr Top Dev Biol 2012;99:237–55.
[79] Tong M-H, Mitchell D, Evanoff R, Griswold MD. Expression of Mirlet7 family microRNAs in response to retinoic acid-induced spermatogonial differentiation in mice. Biol Reprod 2011;85:189–97.
[80] Tong M-H, Mitchell DA, McGowan SD, Evanoff R, Griswold MD. Two miRNA clusters, Mir-17-92 (Mirc1) and Mir-106b-25 (Mirc3), are involved in the regulation of spermatogonial differentiation in mice. Biol Reprod 2012;86:72.
[81] Niu Z, Goodyear SM, Rao S, Wu X, Tobias JW, Avarbock MR, et al. MicroRNA-21 regulates the self-renewal of mouse spermatogonial stem cells. Proc Natl Acad Sci USA 2011;108:12740–5.
[82] Bouhallier F, Allioli N, Lavial F, Chalmel F, Perrard M-H, Durand P, et al. Role of miR-34c microRNA in the late steps of spermatogenesis. RNA 2010;16:720–31.
[83] Liu W-M, Pang RTK, Chiu PCN, Wong BPC, Lao K, Lee K-F, et al. Sperm-borne microRNA-34c is required for the first cleavage division in mouse. Proc Natl Acad Sci USA 2012;109:490–4.
[84] Rakoczy J, Fernandez-Valverde SL, Glazov EA, Wainwright EN, Sato T, Takada S, et al. MicroRNAs-140-5p/140-3p modulate Leydig cell numbers in the developing mouse testis. Biol Reprod 2013;88:143.
[85] Huszar JM, Payne CJ. MicroRNA 146 (Mir146) modulates spermatogonial differentiation by retinoic acid in mice. Biol Reprod 2013;88:15.

[86] Yang Q-E, Racicot KE, Kaucher AV, Oatley MJ, Oatley JM. MicroRNAs 221 and 222 regulate the undifferentiated state in mammalian male germ cells. Development 2013;140:280–90.
[87] Tian H, Cao Y-X, Zhang X-S, Liao W-P, Yi Y-H, Lian J, et al. The targeting and functions of miRNA-383 are mediated by FMRP during spermatogenesis. Cell Death Dis 2013;4:e617.
[88] Dai L, Tsai-Morris C-H, Sato H, Villar J, Kang J-H, Zhang J, et al. Testis-specific miRNA-469 up-regulated in gonadotropin-regulated testicular RNA helicase (GRTH/DDX25)-null mice silences transition protein 2 and protamine 2 messages at sites within coding region: implications of its role in germ cell development. J Biol Chem 2011;286:44306–18.
[89] Mruk DD, Cheng CY. Sertoli–Sertoli and Sertoli–germ cell interactions and their significance in germ cell movement in the seminiferous epithelium during spermatogenesis. Endocr Rev 2004;25:747–806.
[90] Nicholls PK, Harrison CA, Walton KL, McLachlan RI, O'Donnell L, Stanton PG. Hormonal regulation of SC micro-RNAs at spermiation. Endocrinology 2011;152:1670–83.
[91] McLachlan RI, O'Donnell L, Meachem SJ, Stanton PG, de Kretser DM, Pratis K, et al. Identification of specific sites of hormonal regulation in spermatogenesis in rats, monkeys, and man. Recent Prog Horm Res 2002;57:149–79.
[92] O'Donnell L, Nicholls PK, O'Bryan MK, McLachlan RI, Stanton PG. Spermiation: the process of sperm release. Spermatogenesis 2011;1:14–35.
[93] Denison FC, Smith LB, Muckett PJ, O'Hara L, Carling D, Woods A. LKB1 is an essential regulator of spermatozoa release during spermiation in the mammalian testis. PLoS One 2011;6:e28306.
[94] Justine OS, Peter NK, Rob MI, Criag HA, Peter SG. Activin A-responsive microRNAs in differentiated rat SCs. In: The annual scientific meeting of the Endocrine Society of Australia and the Society for Reproductive Biology. ESA-SRB, Sydney; 2013.
[95] Mendis SHS, Meachem SJ, Sarraj MA, Loveland KL. Activin A balances Sertoli and germ cell proliferation in the fetal mouse testis. Biol Reprod 2011;84:379–91.
[96] Nicholls PK, Stanton PG, Chen JL, Olcorn JS, Haverfield JT, Qian H, et al. Activin signaling regulates SC differentiation and function. Endocrinology 2012;153:6065–77.
[97] Liu T, Cheng W, Gao Y, Wang H, Liu Z. Microarray analysis of microRNA expression patterns in the semen of infertile men with semen abnormalities. Mol Med Rep 2012;6:535–42.
[98] Wang C, Yang C, Chen X, Yao B, Yang C, Zhu C, et al. Altered profile of seminal plasma microRNAs in the molecular diagnosis of male infertility. Clin Chem 2011;57:1722–31.
[99] Lian J, Zhang X, Tian H, Liang N, Wang Y, Liang C, et al. Altered microRNA expression in patients with non-obstructive azoospermia. Reprod Biol Endocrinol 2009;7:13.
[100] Liu L, Lian J, Zhang H, Tian H, Liang M, Yin M, et al. MicroRNA-302a sensitizes testicular embryonal carcinoma cells to cisplatin-induced cell death. J Cell Physiol 2013;228:2294–304.

Biochemistry of Sertoli cell/germ cell junctions, germ cell transport, and spermiation in the seminiferous epithelium

12

C. Yan Cheng and Dolores D. Mruk
The Mary M. Wohlford Laboratory for Male Contraceptive Research, Center for Biomedical Research, Population Council, New York, NY

I. Introduction

Mammalian spermatogenesis that takes place in the testis is comprised of four interrelated but discrete cellular events: (i) self-renewal of spermatogonial stem cells (SSCs) at the stem cell niche via mitosis, (ii) mitotic renewal of spermatogonia and differentiation of type A to type B spermatogonia, (iii) meiosis that transforms spermatocytes to spermatids, and (iv) development of spermatids via spermiogenesis and the eventual release of sperm at spermiation [1–8]. Furthermore, these processes are tightly regulated by hormones at the hypothalamic–pituitary–testicular axis involving follicle-stimulating hormone, testosterone, and estrogen [9–14]. However, developing germ cells, in particular preleptotene spermatocytes and spermatids, have to be transported across the blood–testis barrier (BTB) and the adluminal compartment of the seminiferous epithelium, respectively, during the epithelial cycle of spermatogenesis [7,15,16]. Without the timely transport of developing germ cells residing near the basement membrane in the seminiferous epithelium to the adluminal compartment, so that fully developed spermatids can line up at the edge of the tubule lumen to prepare for their release at spermiation, spermatogenesis is disrupted, leading to infertility. While the transport of germ cells across the seminiferous epithelium has long been known, the biology of germ cell transport, in particular its biochemistry and molecular biology, remained poorly understood until the 1990s, when advances in the field were made.

The transport of germ cells across the BTB and/or the seminiferous epithelium involves continuous and cyclic breakdown and reassembly of junctions at the Sertoli cell/Sertoli cell and Sertoli cell/germ cell interface [1,7,16–18]. Furthermore, more than 300, 30, or 50 million spermatozoa are produced each day from the testes of men, mice, and rats, respectively, starting at approximately 12 years of age (for men), 35 day postpartum (dpp; in mice), and 45 dpp (in rats), and these levels remain stable well into the adulthood until death [19–21]. Thus, an

Sertoli Cell Biology. DOI: http://dx.doi.org/10.1016/B978-0-12-417047-6.00012-0

enormous number of germ cells are produced via tightly coordinated and regulated mechanism(s), also, the turnover and remodeling of junctions in the seminiferous epithelium are extensive [3,22,23].

In this chapter, we provide the latest update on the biochemistry, cell biology, and molecular biology of the transport of germ cells across the seminiferous epithelium. We also provide two models on germ cell transport. One model focuses on the transport of preleptotene spermatocytes, which are connected in clones via intercellular bridges across the BTB at stage VIII of the epithelial cycle in the rat testis and the mechanism(s) by which the immunological barrier can be maintained during the extensive turnover of cell junctions. The other model focuses on the transport of spermatids across the seminiferous epithelium in the adluminal compartment during spermiogenesis and the release of sperm into the tubule lumen at spermiation. Many outstanding questions remain about the biology of germ cell transport, and specific areas that deserve attention in future studies are highlighted in corresponding sections below. We hope that this information will serve as the framework for investigators in the years to come.

II. Cell junctions and their restructuring during the epithelial cycle in the testis

A. Background

The types of junctions in the testis have some similarity to junctions in other epithelia, such as the gap junction (GJ), desmosome, and hemidesmosome (Table 12.1) [24–28]. Interestingly, the testis has a unique junction, known as the ectoplasmic specialization (ES), that is not found in other epithelia and/or endothelia (Figure 12.1). Furthermore, the BTB is a unique blood–tissue barrier that is strikingly different from other blood–tissue barriers (Figure 12.1). Also focal adhesion complex (FAC; also known as “focal contact”), an actin-based cell/extracellular matrix (ECM) junction restricted to the cell–matrix/basal laminin interface, is absent from the testis. Proteins that are restricted to FAC in other epithelia or endothelia are found at the Sertoli cell/Sertoli cell and Sertoli cell/germ cell interface, creating the hybrid junction type ES.

B. Types of cell junctions in the testis

Like other epithelia, cell junctions in the testis are broadly divided into two types, F-actin or vimentin-based intermediate filament, depending on the cytoskeleton that is being used for attachment of the adhesion protein complexes. Also, cell junctions can be found at the cell/cell or cell/ECM interface (or basal lamina). As noted in Table 12.1, junctions that use F-actin for attachment at the cell/cell interface consist of the (i) tight (occluding) junction; (ii) ES, a testis-specific adherens junction (AJ); and (iii) GJ. As noted above, there is no cell/ECM junction (known as “basement membrane”) in the

Table 12.1 Junctions and adhesion protein complexes in the seminiferous epithelium of adult testes[a]

	Location	Adhesion protein complex
Anchoring junction		
Apical ES	Sertoli cell–spermatid (steps 8–19)	α6β1-integrin–laminin α3β3γ3 N-cadherin-β-catenin Nectin-2/3-afadin JAM-C-ZO-1 CAR-ZO-1 ICAM-2-afadin
Adherens junction (AJ)	Sertoli cell–spermatocyte/ spermatogonium	E-cadherin/β-catenin N-cadherin/β-catenin β1-integrin
Desmosome	Sertoli cell–spermatocyte/ spermatogonium	Desmoglein–desmocollin
	Sertoli cell–spermatid (steps 1–7)	Desmoglein–desmocollin
Hemidesmosome	Sertoli cell–basement membrane	β1-integrin/laminin α2
Communicating junction		
Gap junction	Sertoli cell–spermatid (steps 1–8)	Connexin 43–plakophilin-2
	Sertoli cell–spermatocyte/ spermatogonium	Connexin 43–plakophilin-2
Blood–testis barrier		
Tight junction	Sertoli–Sertoli cell	Occludin-ZO-1 N-cadherin-β-catenin JAM-A-ZO-1 JAM-B-ZO-1 CAR-ZO-1 Claudin-ZO-1
Basal ES	Sertoli–Sertoli cell	N-cadherin-β-catenin Nectin-2-afadin JAM-A-ZO-1 CAR-ZO-1 ICAM-1-ZO-1
Desmosome	Sertoli–Sertoli cell	Desmoglein-2-desmocollin-2
Gap junction	Sertoli–Sertoli cell	Connexin 43-plakophilin-2

[a]This table was prepared based on earlier studies in the rat testis and it was compiled and updated based on several recent reviews [7,15,79]. Apical ES, basal ES, tight junction, and gap junction are using actin for their attachment; desmosome and hemidesmosome are using intermediate filament for their attachment. Focal contact (or focal adhesion complex), a cell-matrix anchoring junction type using actin for its attachment is *not* found in the testis. Abbreviations: CAR, coxsackievirus and adenovirus receptor; JAM, junctional adhesion molecule; ZO-1, zonula occludens-1.

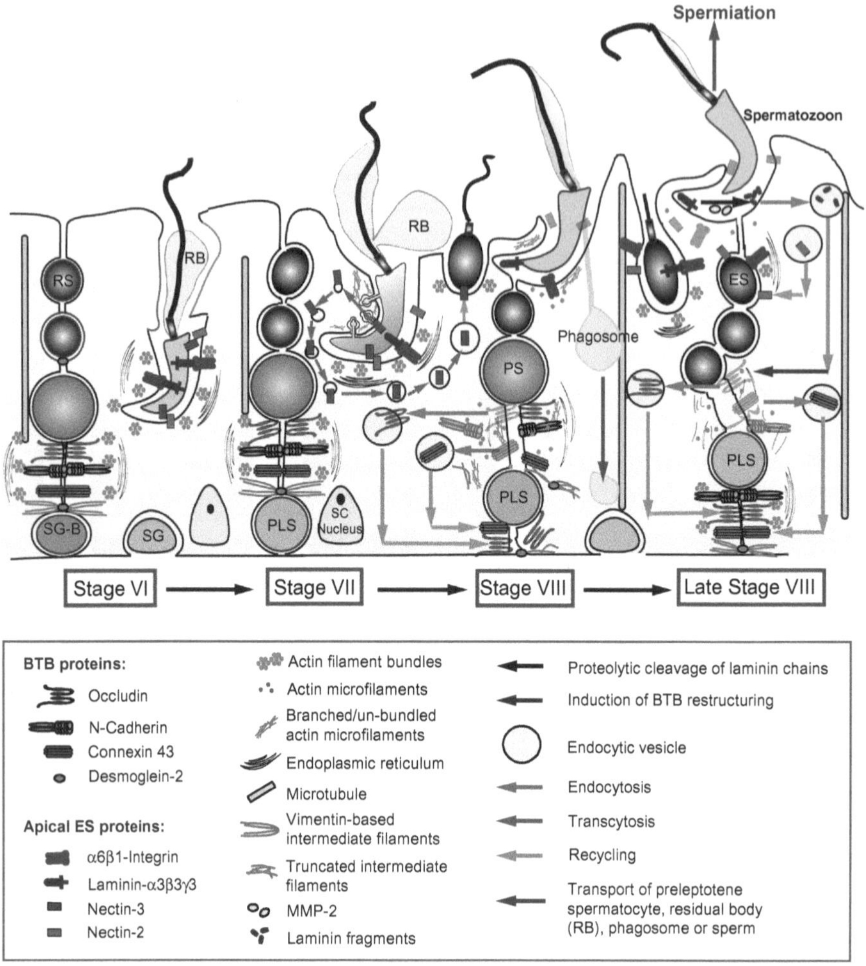

Figure 12.1 A schematic drawing that illustrates changes at the cell−cell interface during the epithelial cycle of spermatogenesis. This figure illustrates the spatial relationship between the Sertoli cell (SC) and germ cells at different stages of development in the seminiferous epithelium during the epithelial cycle, such as from stages VI to VIII, in the adult rat testis. Also illustrated are changes at the Sertoli cell−cell and the Sertoli−germ cell interface at these stages of the epithelial cycle. These changes are necessary to facilitate: (i) the transport of germ cells such as spermatids across the adluminal compartment from stages VI to VIII and also preleptotene spermatocytes (PLS) across the BTB at stage VII−VIII, (ii) endocytic vesicle-mediated recycling of proteins, and (iii) formation of residual body (RB) and its transport from the edge of the tubule lumen to the basal compartment for degradation. At stage VI, integrity of the BTB and the apical ES (ectoplasmic specialization) are maintained by the corresponding adhesion protein complexes. However, at stage VII, apical ES at the concave (ventral) side of the spermatid head undergoes extensive endocytic vesicle-mediated protein trafficking, forming an ultrastructure designated apical TBC (tubulobulbar complex), so that old apical ES proteins (e.g., nectins, integrins) can be recycled to assemble new

testis, wherein the basement membrane is a modified form of an ECM [29,30]. Also, AJ is absent in the seminiferous epithelium, where it is replaced by ES, which is restricted to the Sertoli cell/spermatid interface (steps 8–19) in the apical compartment and is known as "apical ES." Once apical ES appears, it replaces desmosome and the GJ, becoming the only anchoring device until it is degenerated in stage VIII of the epithelial cycle to facilitate spermiation [31–35]. ES is also found at the Sertoli cell/Sertoli cell interface of the BTB known as the "basal ES," which, unlike apical ES, coexists with tight junction (TJ) or GJ. Together with desmosome, TJ and GJ constitute the BTB, which also divides the seminiferous epithelium into the basal and adluminal (apical) compartments [1,36–38] (Figure 12.1). AJ is only detected between Sertoli cells and spermatogonia in the stem cell niche. In the testis, the junction that uses vimentin-based intermediate filament to tether its adhesion protein complexes at the cell/cell interface is desmosome; in contrast, hemidesmosome is found at the Sertoli cell/basement membrane interface [27,28,39] (Table 12.1). As noted in Table 12.1, the composition of the constituent proteins, in particular the integral membrane proteins and the underlying adaptors and regulatory proteins that constitute the regulatory adhesion protein complexes, has been better defined, and this list keeps expanding. These findings also show that there are highly organized signaling pathways downstream of these adhesion protein complexes to regulate adhesion function at the corresponding cell junctions [1,40–43], suggesting that many functional studies can now be designed to explore the role of these adhesion protein complexes in spermatogenesis.

C. *Functions of cell junctions in the testis*

Figure 12.1 shows the relative distribution of several types of cell junctions (Table 12.1) at the Sertoli cell/Sertoli cell or Sertoli cell/germ cell interface in the seminiferous epithelium. The BTB physically divides the seminiferous epithelium into the basal and adluminal (apical) compartments (Figure 12.1). The BTB thus creates a unique microenvironment for meiosis I/II and for postmeiotic spermatid development, which take place in the adluminal compartment.

◀ apical ES when step 8 spermatids are formed at stage VIII of the epithelial cycle. At stage VIII, BTB also undergoes restructuring in which actin filament bundles at the basal ES are induced to undergo reorganization in which microfilaments become unbundled via the action of Arp2/3 complex (actin-related protein 2/3), and similar changes also take place at the apical ES. This thus facilitates the transport of preleptotene spermatocytes across the BTB and to prepare elongated spermatids for their release at spermiation that takes place at late stage VIII of the cycle. Furthermore, laminin chains at the apical ES also undergo cleavage induced by MMP-2 (matrix metalloprotease 2) so that biologically active laminin fragments induce BTB restructuring, thereby coordinating the events of spermiation and BTB restructuring that take place simultaneously at the opposite ends of the seminiferous epithelium. Abbreviations: SG-B, spermatogonium, type B; SG, spermatogonium; RS, round spermatid; ES, elongating/elongated spermatid; RB, residual body; PS, pachytene spermatocyte; PLS, preleptotene spermatocyte.

Similar to other blood–tissue barriers [44], such as the blood–brain barrier [45,46], the blood–retina barrier [47,48], the blood–epididymal barrier [36,49], and the gut barrier [50–53], TJ, which located at the BTB between adjacent Sertoli cells near the basement membrane, restricts the paracellular and transcellular transport of substances, including nutrients, electrolytes, ions, and water, across the barrier [1,36,54,55]. TJ at the BTB together with the basal ES also confers Sertoli cell polarity. For instance, the Sertoli cell nuclei are located near the basement membrane, and organelles, such as phagosomes, that contain residual bodies under degradation/digestion are also located near the basement membrane.

The basal ES is equipped with two arrays of actin filament bundles that are sandwiched between cisternae of endoplasmic reticulum and the apposing Sertoli cell plasma membranes. These arrays of actin microfilament bundles thus confer the BTB with unusual strength, making it one of the tightest blood–tissue barriers in the mammalian body. However, the BTB undergoes cyclic restructuring during the epithelial cycle of spermatogenesis in which the old BTB located above the preleptotene spermatocytes being transported across the immunological barrier is disassembled, while the new BTB behind these germ cells is rapidly reassembled [7,56]. Thus, the immunological barrier is maintained when preleptotene spermatocytes connected in clones via intercellular bridges are transported across the BTB.

Yet the underlying mechanism(s), in particular the regulatory molecules involved that coordinate these events, remain unknown. Studies using various biochemical assays to assess changes in protein endocytosis, transcytosis, recycling, and degradation have shown that testosterone promotes endocytosed integral membrane proteins at the Sertoli cell BTB to the pathway of transcytosis and recycling. Thus, internalized proteins can be effectively used to assemble new BTB from the disassembled old BTB, whereas cytokines (e.g. transforming growth factor [TGF] β2 and tumor necrosis factor α) promote endocytosed Sertoli cell BTB proteins to endosome-mediated protein degradation for the degeneration of old BTB [57–59]. Other studies have shown that intercellular adhesion molecule 1 (ICAM-1) and soluble ICAM-1 [60,61], mammalian target of rapamycin complex 1 (mTORC1) and mTORC2 [62–64], as well as p-FAK-Tyr^{407} and p-FAK-Tyr^{397} [65,66] have antagonistic effects on Sertoli cell-TJ permeability barrier function, showing that these molecules can act as molecular switches that can turn the barrier function on or off during the transport of preleptotene spermatocytes across the BTB, such as the apical or basal region of the spermatocyte. A hypothetical model is proposed based on these findings that can be used as the framework for future investigation of the involvement of these molecules in regulating the timely opening and closing of the BTB during the transport of preleptotene spermatocytes at the site.

However, apical ES at the Sertoli cell/spermatid interface confers spermatid adhesion. Apical ES also confers spermatid polarity that is similar to the function of basal ES at the BTB. Studies have shown that apical ES is perhaps one of the strongest anchoring junctions in the mammalian body and was at least twice as strong as desmosome when the force required to tear apart apical ES was compared to that of desmosome [67]. Desmosome is the anchoring junction that confers unusual adhesive strength to the skin [68,69].

Apical ES also confers polarity to developing spermatids during spermiogenesis so that the heads of spermatids are pointing toward the basement membrane (Figure 12.1). Such an orderly arrangement of spermatids allows the maximum number of developing spermatids to be packed into the seminiferous epithelium. As such, millions of germ cells can be produced in the microenvironment of the seminiferous tubules, which are about 300 m long and packed in each testis (testis weight is approximately 24 g) of men [70] versus 20 m per testis in adult rats (testis weight of approximately 1.6 g) [71], showing the physiological significance of polarity during spermatogenesis [72,73].

Studies have shown that this unique function of ES that confers Sertoli cell and spermatid polarity at the basal and apical ES, respectively, is due to the presence of three polarity protein complexes: (i) the partition-defective protein (Par)-based [74,75], (ii) Crumbs (CRB)-based [74], and (iii) Scribble-based [76] polarity complexes [72,73,77]. The mechanism by which these three polarity protein modules elicit cell polarity is based on their ability to recruit their specific protein partners. For instance, the Par-based and CRB-based protein complexes coexist in a microdomain in either the Sertoli cell epithelium or germ cells. These two protein modules and the Scribble-based polarity protein module are mutually exclusive [78–80]. Because each of these polarity protein complexes recruits its own protein-binding partners via their unique adaptor proteins, each module creates its own protein complexes, causing the formation of apico-basal polarity in an epithelium [78–80].

Furthermore, cell junctions, in particular GJ, at the Sertoli cell/cell and Sertoli cell/germ cell interfaces also provide crucial communications between adjacent cells in the seminiferous epithelium so that these cells can be coordinated in their development or signaling function during different stages of the epithelial cycle [25,81]. Studies using a fluorescence-based functional assay, such as fluorescence recovery after photobleaching, have identified the transfer of a fluorescence tag across the GJ at the Sertoli cell/Sertoli cell interface [63,82]. These data also show that signaling biomolecules can be transported across the GJ [26,81,83]. In this way, cellular events that are pertinent to spermatogenesis can be tightly coordinated between Sertoli and germ cells during the epithelial cycle of spermatogenesis. In fact, studies have shown that GJ communications between Sertoli cells at the BTB are disrupted when the Sertoli cell epithelium is exposed to the environmental toxicants bisphenol A [82] and perfluorooctanesulfonate [84]. It is known that GJ at the Sertoli cell BTB also provides crucial communications between different junction types, such as TJ, basal ES, and desmosome, so that these junctions can be coordinated during the transport of preleptotene spermatocytes across the BTB at stage VIII of the epithelial cycle [82,85].

While there are no ultrastructures that are analogous to FAC, an actin-based cell-matrix anchoring junction found in other epithelia, in the testis, hemidesmosome, an intermediate filament-based cell-matrix anchoring junction, is found in the testis at the interface of Sertoli cells and the basement membrane [29,30]. However, there are virtually no reports in the literature by investigators who have probed the functional role of hemidesmosome in spermatogenesis. Also, the constituent proteins of hemidesmosome in the testis are virtually unknown, except that laminin-α2 chain and β1-integrin are two known components of the hemidesmosome [86]. Furthermore, an earlier report showed that a knockdown of β1-integrin at the hemidesmosome by RNAi can impede the Sertoli cell tight junction permeability barrier function [86]. This observation prompted us to postulate the presence of functional crosstalk between the BTB and the hemidesmosome, which are intimately related to one another morphologically (Figure 12.1). This hypothesis is also supported by studies in other epithelia, such as in the skin, where hemidesmosomal proteins are known to regulate keratinocyte movement during wound healing [87]. While Sertoli cells are not motile cells *per se* in the seminiferous epithelium during the epithelial cycle, Sertoli cells undergo extensive morphological changes, in particular at the Sertoli cell/germ cell interface, during spermatogenesis, such as the transport of preleptotene spermatocytes across the immunological barrier. Thus, hemidesmosomal proteins may be involved in BTB restructuring during the epithelial cycle, as supported by the finding that a loss of β1-integrin at the hemidesmosome impedes Sertoli cell tight junction permeability barrier function [86]. Nonetheless, it is obvious that more intensive investigation is needed to probe the function of hemidesmosome in spermatogenesis, in particular its role in germ cell transport in the seminiferous epithelium.

III. Ectoplasmic specialization

ES was first observed at the BTB in the early 1970s, when an array of actin filament bundles that lie perpendicular to the Sertoli cell plasma membrane was identified [88,89]. These bundles of actin microfilaments are sandwiched between the cisternae of endoplasmic reticulum and the apposing Sertoli cell plasma membranes, found on both sides of the adjacent Sertoli cells that create the BTB [88]. The name "ectoplasmic specialization" was coined in 1977 to designate this testis-specific ultrastructure, which is also found at the Sertoli cell/spermatid (steps 8–19 spermatids in the rat testis) interface in the adluminal compartment and the Sertoli cell/ Sertoli cell interface at the BTB, which are designated apical and basal ES, respectively [17,31,37]. Apical and basal ES are indistinguishable from one another by electron microscopy, except that apical ES has only one array of actin filament bundles because no visible ultrastructure is found in the elongating/elongated spermatid, whereas basal ES is composed of two arrays of actin filament bundles (Figure 12.1) [18,32,33]. Furthermore, once apical ES appears in step 8 spermatids at the Sertoli cell/spermatid cell interface, it is the only anchoring device that anchors spermatids to the seminiferous epithelium, replacing GJ and desmosome [31,39].

Apical ES persists from step 8 through step 19 spermatids until it begins to undergo degeneration in late stage VII, beginning at the concave (or ventral) side of spermatids. Apical ES forms an ultrastructure that is analogous to giant endocytic vesicles and can be detected by fluorescence microscopy, representing the site of extensive protein endocytosis to be used for transcytosis and recycling of integral membrane proteins. This site is likely where the cellular events are initiated that lead to the elimination of cytoplasmic debris resulting from spermiogenesis and known as "apical tubulobulbar complex" (apical TBC) [16,79,90–92]. By stage VIII, degeneration of apical ES spreads also to the convex (or dorsal) side of the spermatid head in preparation for the release of sperms at spermiation [8].

Unlike apical ES, basal ES coexists with TJ and GJ [18,93], which together with desmosome create the BTB between adjacent Sertoli cells near the basement membrane in the rat testis [1,36,55]. Basal TBC is also detected at the Sertoli cell/Sertoli cell interface at the BTB [38]. Similar to apical TBC, basal TBC represents endocytic vesicles used for transcytosis and recycling of proteins at the BTB to maintain the integrity of the immunological barrier during the transport of preleptotene spermatocytes connected in clones at the BTB, such that old BTB proteins above the spermatocyte in transit can be recycled to assemble the new BTB behind the germ cell (Figure 12.1) [59,79,94].

In this chapter, we do not provide details on the morphology of the ES because this subject has been reviewed elsewhere [1,7,16,18,24,42,79]. The ultrastructural and functional features of the various junction types at the BTB are described only briefly here because this subject has also been extensively reviewed [1,7,15,24,25,42,55,95]. Instead, we focus our discussion on the latest findings and concepts based on studies that have important implications for guiding future studies.

IV. Spermatid transport and spermiation

A. Background

During spermatogenesis, in addition to the transport of preleptotene spermatocytes across the BTB, round spermatids that are transformed from secondary spermatids during meiosis are also transported up and down across the seminiferous epithelium during the epithelial cycle until elongated spermatids line up at the edge of the tubule lumen to prepare for their release at spermiation in the mammalian testis [2,3,96,97] (Figure 12.1). The biology of spermatid transport and spermiation is discussed here collectively because spermiation is the final step of spermatid transport during spermatogenesis.

In the rat testis, spermiation takes place at late stage VIII of the epithelial cycle [3,31,98]. However, the preparation of spermiation begins at stage VII, when step 19 spermatids are transported and aligned near the adluminal edge of the tubule lumen [99], which is supported by the actin-based cytoskeleton (Figure 12.2). Because the subject of spermiation has been reviewed in two excellent articles [8,16], only a brief summary is given here. The discussion is focused on the regulation of this process based on findings that have not been carefully evaluated before. We also provide a likely hypothetical model, shown in Figure 12.3, that is based on these findings. This model is likely to be updated when more data are available in the future.

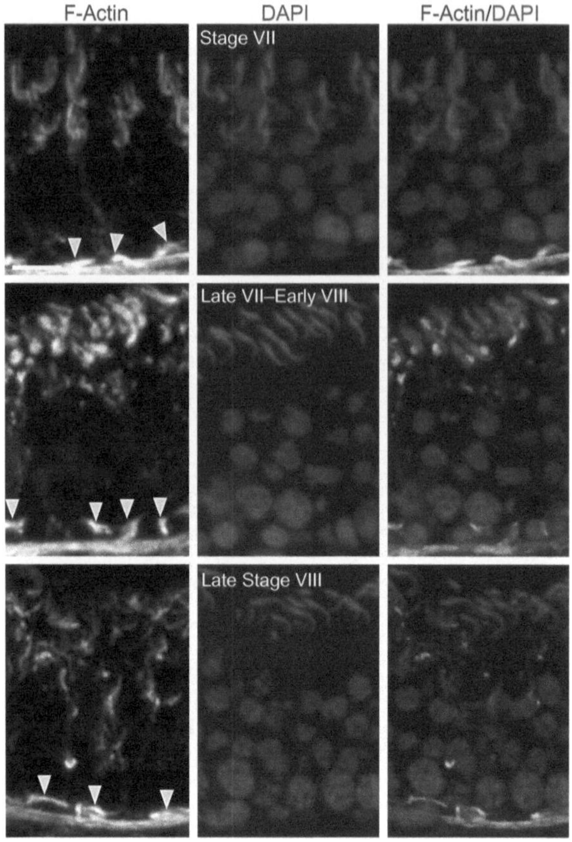

Figure 12.2 Changes in the organization of F-actin at the apical and basal ectoplasmic specialization (ES) during the epithelial cycle at stage VII–VIII in the rat testis. F-Actin (red fluorescence, labeled by rhodamine phalloidin) is detected at the basal ES of the BTB (annotated by yellow arrowheads) and also tunica propria (such as peritubular myoid cells and endothelial cells of the lymphatic vessel) behind the basal ES in stage VII–VIII of the epithelial cycle. F-Actin is also detected prominently surrounding the heads of elongated spermatids at the apical ES, and most notably found at the concave (ventral) side of the spermatid head in late stage VII–early VIII. However, F-actin is rapidly reorganized at late stage VIII to prepare for the release of spermatids (to be transformed to spermatozoa at late stage VIII) at spermiation, becoming unbundled and eventually replaced by G-actin, and thus F-actin staining rapidly diminished surrounding the spermatid heads. However, step 8 spermatids that arise at stage VIII of the cycle are equipped with newly formed apical ES, and actin filament bundles are associated with apical ES in these spermatids; as such, F-actin staining is detected at late stage VIII tubules in these newly formed step 8 spermatids as noted herein. Scale bar = 50 μm, applies to all other micrographs. (For interpretation of the references to color in this figure legend, the reader is referred to the web version of this book.)

B. Cascade of cellular events at the Sertoli cell/spermatid interface at spermiation

In the rat testis, stages VII and VIII last for 56 and 29.1 h, respectively [42,100], showing that it takes approximately 3 days to prepare for spermiation. The release of spermatozoa into the tubule occurs in late stage VIII, during the last few hours of this stage. Spermiation involves degeneration of the apical ES, which is largely the result of disintegration of the actin filament bundles in the Sertoli cells that lie perpendicular to the Sertoli cell plasma membrane and are sandwiched between cisternae of endoplasmic reticulum and the apposing Sertoli cell/spermatid plasma membranes [8].

Degeneration of apical ES is first noted at the concave (ventral) side of spermatid heads in late stage VII of the epithelial cycle, in which actin microfilament bundles undergo reorganization, creating an ultrastructure known as apical TBC [16,101,102]. In some species, such as in humans, that have spatulate spermatids, TBC forms at the tip of the spermatid head [103]. Thus, apical TBC is first found in the area where apical ES initiates its degeneration. Basically, apical TBC consists of giant endocytic vesicles that are being used for the elimination of cytoplasm from the head region of late spermatids [90], and it is readily detectable by fluorescence microscopy using specific markers of the TBC [16].

The notion that TBC is an ultrastructure containing endocytic vesicles is supported by the finding that clathrin [104]—which is an integrated component of the endosome that is necessary for protein endocytosis, transcytosis, and recycling—is highly expressed at the TBC [16,105]. Furthermore, N-WASP (neuronal Wiskott-Aldrich syndrome protein) and cortactin are also detected at the apical TBC [104] which are the upstream regulators of the actin-related protein 2/3 (Arp2/3) protein complex. The Arp2/3 complex, once activated by N-WASP and cortactin, is known to induce branched actin polymerization. Thus, the barbed-end protein nucleation and polymerization activity mediated by the Arp2/3 complex in an existing microfilament is effective for converting microfilaments from a bundled to a debundled/branched configuration [65,106]. Thus, N-WASP and cortactin work in concert with epidermal growth factor receptor pathway substrate 8 (Eps8) and Arp3, which are also expressed stage specifically at the apical ES on the ventral (concave) side of the spermatid head [107,108]. The findings that these proteins co-localize at the apical TBC (i.e., the degenerating apical ES) are consistent with the notion that these proteins confer plasticity to the apical TBC to facilitate endocytosis [79,109].

In this context, it is of interest to note that the expression of dynamin II and III is also upregulated at the apical ES/apical TBC [110,111]. These proteins are known to be involved in pinching off endocytic vesicles from plasma membranes [112]. In short, apical TBC at the concave side of the spermatid head is composed of the necessary machinery to support the cellular events that are pertinent to endocytic vesicle-mediated protein trafficking.

In stage VIII, degeneration of the apical ES is no longer restricted to the concave side of the spermatid head. This degeneration of apical ES rapidly spreads across the entire spermatid head, causing defragmentation of the actin filament bundles at the dorsal side (convex) of the spermatid head as well [8,16]. This disintegration of

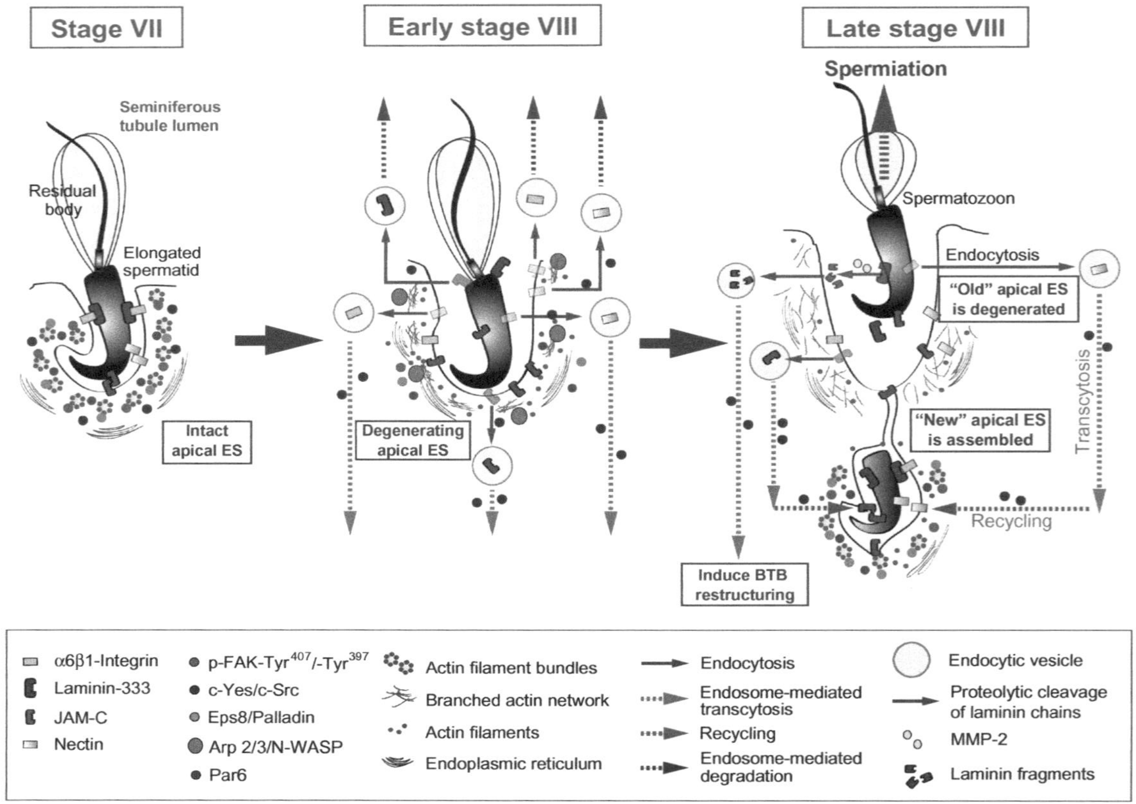

Figure 12.3 A schematic drawing that illustrates the likely changes at the apical ES associated with the events of spermiation in the rat testis. Left panel illustrates the intact apical ES which is conferred by bundles of actin filaments surrounding the head of a step 19 spermatid in a stage

apical ES at late stage VIII is facilitated, at least in part, by the spatiotemporal expression of matrix metalloproteinase-2 (MMP-2) at the apical ES, just prior to spermiation [113].

In addition to being used to eliminate unwanted cytoplasmic substances of the spermatid head that are likely to be digested via endosome-mediated degradation [90], apical TBC is an ultrastructure used for protein endocytosis and, possibly, recycling [91]. Thus, old apical ES proteins (e.g., nectin-2, nectin-3, and α6β1-integrin) can be recycled for the assembly of new apical ES when step 8 spermatids first appear in stage VIII of the epithelial cycle [79]. Figure 12.3 briefly summarizes the likely cascade of cellular events that is pertinent to spermiation based on findings in the rat testis.

However, many outstanding questions remain unanswered. For instance, what is the molecule (or molecules) that trigger the initiation of endocytosis at the apical TBC and the timely expression of MMP-2? Does this process involve the tightly aligned spermatid 19 spermatids near the tubule lumen in which spermatids send the necessary signal(s) via GJ to the Sertoli cells? Or is this largely a Sertoli cell-dependent cellular event via a local paracrine/autocrine regulatory loop? Furthermore, the molecule(s) that trigger the spatiotemporal expression of Eps8, Arp3, and other actin regulatory proteins necessary to perform the timely reorganization of actin microfilaments at the site remain to be identified. Nonetheless, the model shown in Figure 12.3 serves as a framework for the design of functional experiments in future studies.

C. The apical ES/BTB/basement membrane axis

Among the cell adhesion protein complexes known to confer spermatid adhesion at the apical ES—namely nectin-2/-3-afadin complex [114–116], cadherins/catenins [117,118], and α6β1-integrin–laminin-α3/β3/γ3 [113,119–123]—the best studied

◀ VII tubule. These actin microfilaments assume the bundled configuration are supported by actin barbed end capping/bundling protein Eps8 and actin bundling protein palladin. Apical ES integrity is also supported by nonreceptor protein kinases c-Yes, c-Src and p-FAK-Tyr^{397} and -Tyr^{407}. These actin filament bundles also provide proper attachment sites for adhesion protein complexes at the apical ES, such as integrin–laminin, JAM-C-ZO-1, and nectin–afadin. However, in early stage VIII of the epithelial cycle (middle panel), the Arp2/3/N-WASP protein complex is activated, mediated by changes in the spatiotemporal expression of p-FAK-Tyr^{397} and -Tyr^{407}, actin filament bundles undergo extensive reorganization, assuming an unbundled and branched configuration. Thus, cell adhesion proteins no longer confer adhesion of spermatids onto the Sertoli cell. These changes in actin organization also favor endocytic vesicle-mediated protein trafficking, and with the likely involvement of c-Yes/c-Src, old apical ES proteins either undergo endosome-mediated degradation or undergo transcytosis and recycling (see late stage VIII in the right panel), so that these proteins can be reused to assemble apical ES for the newly formed step 8 spermatids at stage VIII of the epithelial cycle. Furthermore, laminin chains at the apical ES also undergo cleavage induced by MMP-2 (matrix metalloprotease 2), generating biologically active laminin fragments, which are being used to induced BTB restructuring. See text for additional details.

adhesion complex is the integrin–laminin-based protein complex [35,124,125]. For instance, laminin chains are exclusively expressed by late spermatids, whereas integrins are expressed by Sertoli cells which form a bona fide adhesion complex [113,119,123]. It is of interest to note that the integrin–laminin-based adhesion protein complex is usually restricted to the cell/ECM interface called focal adhesion complex (FAC). FAC, also known as focal contact, uses actin-based cytoskeleton for its attachment in virtually all other epithelia, and it is crucial to facilitate cell migration [126,127]. However, FAC is absent in the seminiferous epithelium at the Sertoli cell-basement membrane interface.

Thus, apical ES is an atypical and hybrid AJ because it is composed of proteins that are usually restricted to FAC (e.g., laminins and integrins), TJ (e.g., JAM-C and CAR), and GJ (e.g., connexin 43) instead of AJ-specific proteins [32–35]. Studies have shown that at late stage VIII, when spermiation occurs, this event is facilitated by proteolytic cleavage of the laminin chains by MMP-2, which results in the loss of a putative binding partner for $\alpha6\beta1$-integrin in Sertoli cells, causing disassembly of the apical ES at the Sertoli cell/spermatid interface [86,113,128] (Figure 12.3).

Studies in other epithelia have shown that peptides cleaved from laminin chains that contain specific stretches of sequences are biologically active fragments that are known to regulate morphogenesis, wound healing, immune response, tumorigenesis, and cell junction function [129–133]. Thus, it was not entirely unexpected when domain IV of laminin-$\gamma3$ chain or domain I of laminin-$\beta3$ chain was overexpressed in Sertoli cells (or the addition of purified recombinant protein to Sertoli cell epithelium cultured on bicameral units), both peptides were found to be capable of perturbing the Sertoli cell/TJ permeability barrier. This effect was mediated by an impairment of the distribution of BTB-associated integral membrane proteins (e.g., occludin and N-cadherin) at the Sertoli cell/cell interface, causing these proteins to become mislocalized and moved from the cell surface into the cell cytosol [86], thereby destabilizing the Sertoli cell/TJ barrier.

However, domain I of laminin-$\gamma3$ chain has no effects [86]. Collectively, these findings show that only specific stretches of amino acid sequences of the laminin chain at the apical ES are biologically active peptides. A subsequent study mapped a stretch of sequence located in domain IV of laminin-$\gamma3$ chain, designated F5 peptide, which contains 50 amino acid residues that potently perturbs Sertoli cell tight junction barrier function both *in vitro* and *in vivo* [134]. These findings also demonstrate the presence of a physiological link between the apical ES and the BTB, showing that the events of spermiation and BTB restructuring that occur simultaneously at stage III of the epithelial cycle to facilitate the release of sperm and the transport of preleptotene spermatocytes across the immunological barrier, respectively, are tightly coordinated via a paracrine/autocrine-based regulatory axis across the epithelium [2,40,79]. Studies using a toxicant model of Sertoli cell injury in mice also support the presence of this apical ES–BTB axis because phthalate-induced spermatid release from the epithelium was found to perturb the BTB integrity via the generation of laminin-$\gamma3$ fragments mediated by MMP-2 [128,135,136].

In addition to affecting BTB function, these biologically active fragments perturb hemidesmosome function via their effects on downregulating β1-integrin expression at the hemidesmosome. β1-integrin is not found at the Sertoli cell/Sertoli cell interface and there are no spermatids in these cultures; thus, no apical ES is present, but β1-integrin is found at the Sertoli-basement membrane interface, at the hemidesmosome [86]. Furthermore, a knockdown of β1-integrin by RNAi also perturbs the Sertoli cell/TJ permeability barrier [86].

These data demonstrate the presence of a functional loop between the BTB and hemidesmosome in which a disruption of hemidesmosome function can further potentiate BTB restructuring. The presence of a physiological link between the BTB and the basement membrane in which hemidesmosome is found is also supported by the earlier finding that a disruption of collagen function in the basement membrane, such as through the use of an anticollagen antibody, could perturb the Sertoli cell/TJ permeability barrier [137]. A subsequent study showed that the noncollagenous 1 domain in collagen IV chain, when possibly cleaved via the action of MMP-9 at the microenvironment of the BTB, indeed possesses the biological activity required to perturb the Sertoli cell tight junction function, which is mediated by changes in the distribution of BTB-associated proteins (e.g., CAR, ZO-1, and N-cadherin) at the Sertoli cell/Sertoli cell interface [138].

Collectively, these findings demonstrate unequivocally the presence of a local functional axis, known as the apical ES-BTB-basement membrane in the testis that is used to coordinate cellular events that take place across the seminiferous epithelium during the epithelial cycle. This axis, for instance, is crucial in regulating the release of spermatozoa that are differentiated from step 19 spermatids at spermiation. Nonetheless, many questions remain to be addressed in future studies. For instance, does the biologically active peptide(s) released at the apical ES potentiate further apical ES break down in addition to acting on the BTB near the basement membrane? Does the laminin-peptide-induced BTB disruption send feedback signal to the apical ES after the transport of preleptotene spermatocytes across the barrier? If such a signal exists, what is its identity?

D. *Regulation of spermatid transport and sperm release at spermiation*

While morphological details that are pertinent to the release of sperm at spermiation during stage VIII of the epithelial cycle in rat and mouse testes have known since the 1950s [3,97,100,139,140], the underlying molecular mechanism(s) that regulate this cellular event remains largely unexplored until the early 2000s. Here, we provide a summary of these findings. We also highlight areas of research that deserve attention by investigators and provide a hypothetical model of the regulation of spermiation based on our current knowledge.

1. Cytoskeleton

As noted above, the transport of germ cells, in particular spermatids, across the seminiferous epithelium in the adluminal compartment during spermiogenesis is tightly regulated by the actin- and tubulin-based cytoskeleton. At spermiation, spermatozoa transformed from step 19 spermatids at late stage VIII of the epithelial cycle are transported from the edge of the tubule lumen and emptied into the lumen so that spermatozoa enter the epididymis for further development (Figure 12.3). Thus, spermatozoa are the cargoes that are transported by the vehicle (i.e., actin microfilaments) to the tubule lumen utilizing a track (i.e., polarized microtubules) that involves motor proteins in a process that is analogous to the transport of spermatids across the adluminal compartment during spermiogenesis [141].

This concept is supported by the finding that polarized microtubules are tightly associated with apical ES, are located adjacent to the actin filament bundles/cisternae of endoplasmic reticulum complex, and are parallel to the long axis of the Sertoli cell. This placement is consistent with the notion that polarized microtubules serve as the track for the directional transport of spermatids during spermiogenesis [142] (Figure 12.3). It is noted that microtubules are intrinsically polar cytoskeletal elements with their plus and minus ends directed basally and apically in the Sertoli cell, respectively [143–146].

Also, the motor proteins dynein and kinesin that are found in the testis [147–149] are also associated with microtubules and move toward the minus and plus ends of a microtubule, respectively [148], to provide directional transport of cargoes, namely spermatids and spermatozoa. Interestingly, while dynein is a minus end-directed motor protein, when it forms a complex with dynactin (an adaptor protein), it can also move to the plus end [150]. This finding shows that depending on the nature of specific adaptor proteins, spermatids can be transported back and forth across the seminiferous epithelium during the epithelial cycle efficiently via spatiotemporal expression of dynactin to partner with dynein to provide directional transport of spermatids. This process may even be necessary to fine-tune the alignment of spermatids to prepare for spermiation.

In addition to working in concert with microtubules to transport step 19 spermatids into the tubule lumen while they are transformed into spermatozoa, changes in the organization of actin microfilaments at the apical ES are also necessary to facilitate several crucial cellular events that are pertinent to spermiation. These include: (i) phagocytosis of residual bodies to eliminate unwanted cellular debris and cytosolic organelles; (ii) endocytic vesicle-mediated trafficking events that are pertinent to endocytosis, transcytosis, and recycling of old apical ES proteins for the assembly of new apical ES; and (iii) mediation of integrin-based signaling function by biologically active laminin fragments to modulate Sertoli cell BTB function. These three cellular events require extensive re-organization of actin filament bundles at the apical ES. For instance, biologically active laminin fragment that mediates its effects via the inside-out as well as the outside-in signaling pathway to transduce signaling function downstream to modulate Sertoli cell tight junction permeability barrier function [134]. These changes require tightly coordinated reorganization of F-actin, and also the involvement of microtubules and small GTPases. Once spermatozoa begin to empty into the tubule lumen at late stage VIII, the expression of F-actin and of α- and β-tubulin rapidly diminishes because these cytoskeletons are no longer necessary. However, it is of interest that F-actin is highly expressed surrounding the region of newly appearing step 8 spermatids at the future acrosomal site, which is also the area of the apical ES that undergoes rapid assembly (Figure 12.2).

2. Focal adhesion kinase

In virtually all epithelia, focal adhesion kinase (FAK) is the downstream signal transducer that mediates signaling function of integrin-based receptors at the cell–ECM interface known as FAC [151–154] via inside-out, outside-in, or inside-outside-in mechanisms [155,156], depending on the ligand, usually laminins, collagens, or their fragments. In the seminiferous epithelium of adult rat testes, however, no ultrastructure analogous to FAC (an actin-based cell–ECM anchoring junction) is found at the Sertoli cell/basement membrane interface, except for the intermediate filament-based cell/ECM anchoring junction called "hemidesmosome" [7,15,29,30].

Interestingly, two phosphorylated (activated) forms of FAK, p-FAK-Tyr^{397} and p-FAK-Tyr^{576}, are restrictively expressed at the apical ES (not found at the basal ES/BTB) and localized exclusively on the convex (dorsal) side of spermatid heads, displaying spatiotemporal expression during the epithelial cycle at the Sertoli cell/spermatid interface prominently from stage VI to stage VIII, until shortly before spermiation at late stage VIII when their expression is considerably diminished [157,158]. These findings suggest that p-FAK-Tyr^{397} may be crucial for step 19 spermatid adhesion at the apical ES because its upregulation is needed until late in stage VIII, when spermiation occurs.

A recent study that overexpressed p-FAK-Tyr^{397} in the testis *in vivo* using a phosphomimetic mutant of p-FAK-Y397E, in which Tyr^{397} is converted to Glu^{397} via site-directed mutagenesis to make the mutant constitutively active [66], supported the notion that p-FAK-Tyr^{397} is crucial for spermatid adhesion at the apical ES [159]. For instance, transfection of rat testes *in vivo* with the p-FAK-Y397E mutant at 30% efficiency was found to cause defects in spermiation by which step 19 spermatids failed to be transported to the luminal edge of the adluminal compartment but remained tightly anchored onto the Sertoli cells at the apical ES, deep inside the seminiferous epithelium, at stage VIII [159]. In fact, many step 19 spermatids were still visible in the seminiferous epithelium in stage IX tubules [159].

Detailed analysis on the phenotypes of these testes transfected with the p-FAK-Y397E mutant has shown that overexpression of p-FAK-Tyr^{397} leads to persistence of F-actin at the apical ES until late in stage VIII [159]. This persistence of actin filament bundles at the apical ES that fail to be debundled and converted to a branched/unbundled configuration to facilitate spermatid transport and the occurrence of spermiation is due to the upregulation of the actin barbed-end capping and bundling protein Eps8 and upregulation of actin cross-linking/bundling protein palladin. These changes are concomitant with a mislocalization of Arp3, which thereby perturbs the proper reorganization of F-actin from a bundled to unbundled configuration in response to the epithelial cycle [159]. Thus, cell adhesion proteins nectin-2 and nectin-3 are also retained at the apical ES at late stage VIII, when they should have been downregulated, and their persistent presence at the apical ES thereby impedes spermatid transport and spermiation [159].

A study has also detected p-FAK-Tyr^{407} at the apical ES; however, unlike the two other phosphorylated forms, p-FAK-Tyr^{407} is localized most prominently to the concave (ventral) side of the spermatid head, weakly expressed at the convex side, and is expressed at the basal ES at the BTB [66]. Furthermore, p-FAK-Tyr^{407} colocalizes with F-actin at the apical ES, and a study using a biochemical assay has shown that p-FAK-Tyr^{407} stimulates actin polymerization kinetics, which is mediated by the recruitment of N-WASP to Arp3, thereby leading to actin nucleation and polymerization [66]. Because the Arp2/3 complex is a barbed-end actin nucleation and polymerization inducer when activated by N-WASP [106,160,161], its recruitment by p-FAK-Tyr^{407} to the apical ES site can effectively reorganize actin filament bundles at the apical ES, creating a branched actin network at the apical ES, thereby destabilizing apical ES. These observations show that p-FAK-Tyr^{407} acts as an actin polymerization signaling switch at the apical ES by altering the status of actin filament bundles at the site, converting F-actin from "bundled to unbundled/branched configuration" (and vice versa by working in concert with the actin bundling proteins Eps8 and palladin) in response to the stages of the epithelial cycle.

In summary, these findings demonstrate the crucial involvement of p-FAK-Tyr^{397} and -Tyr^{407} on spermatid transport and the release of sperm at spermiation. It is likely that activated FAK exerts its effects by modifying the organization of F-actin at the apical ES, which also involves the spatiotemporal expression of Eps8, palladin, and Arp3. These changes thus affect the distribution of cell adhesion proteins (e.g., nectin-2 and nectin-3) at the apical ES. Future studies should focus on the mechanism(s) by which p-FAK regulates the spatiotemporal expression of actin regulatory proteins, such as Eps8, palladin, and Arp3, at the apical ES. For instance, future research should determine whether p-FAK-Tyr^{397} recruits other signaling molecule(s) to the site to affect spatiotemporal expression of these regulators.

3. Polarity proteins

Polarity proteins that regulate apico-basal polarity during development were first reported in *Drosophila melanogaster* and *Caenorhabditis elegans* [162–165]. Subsequent studies have shown that these proteins, in addition to regulating polarity, regulate other cellular events, including cell cycle progression, adhesion, and cytoskeleton, as well as tumorigenesis and metastasis [73,164,166–168]. Polarity proteins are classified into three functionally distinctive modules: the Par (partitioning-defective)-based, CRB (Crumbs)-based, and Scribble-based complexes [78,80,169,170].

Subsequent studies have shown that these proteins (or their analogs) are also found in mammalian cells and they are well conserved between species, from flies and worms to rodents and humans [73,78,168,171,172]. The Par-based polarity complex is composed of cytosolic proteins Par3, Par6, Cdc42, and the Ser/Thr kinase atypical protein kinase C (aPKC), whereas the CRB-based complex is composed of the integral membrane protein CRB3 and the cytoplasmic scaffolding proteins PALS1 (protein associated with Lin-7 1) and PATJ (PALS1 associated tight junction protein). Both of these protein complexes are localized at the apical region of a cell epithelium, closest to the TJ [73,80,168]. However, the Scribble-based complex is constituted of the cytoplasmic proteins Scribble, Discs large (Dlg), and Lethal giant larvae (Lgl), which are located at the basolateral region of a cell, below the AJ plaque in a cell epithelium [77,80,173,174].

Each of these three protein modules recruits its own binding partner proteins, usually scaffolding proteins, adaptors, and/or nonreceptor protein (or lipid) kinases and phosphatases, forming a giant protein complex module.

Scribble-based proteins display mutual exclusivity with the Par- and CRB-based protein complexes in an epithelium. The result is apico-basal polarity in the epithelium [78,80], including the seminiferous epithelium in the rat testis, conferring Sertoli cell and spermatid polarity [73]. Studies have shown that Scribble-based proteins are tumor suppressors in both flies and mammals [175,176], a finding that has sparked the interest of investigators in probing the role of polarity proteins in cancer therapy.

In the testis, step 8–19 spermatids display restrictive polarity in which the head of the spermatid is pointed toward the basement membrane during spermiogenesis (Figure 12.1). Furthermore, Sertoli cells in the seminiferous epithelium display strict polarity, during different stages of the epithelial cycle in which multiple cellular events that take place in the Sertoli cell. For instance, Sertoli cell nuclei are restrictively localized near the basement membrane in the seminiferous epithelium throughout the epithelial cycle [73,96,100]. Lysosomal degradation of phagosomes following the engulfment of residual bodies at stage VIII takes place at stage IX of the cycle, and phagosomes are restrictively found near the basal region of the Sertoli cells [177], supporting the concept of Sertoli cell polarity. These observations also support the notion that polarity proteins are involved in conferring these phenotypes.

Yet the role of polarity proteins in spermatogenesis remained unknown until a study showed the presence of polarity proteins in the testis as well as their crucial role in spermatid polarity [74]. In rats treated with either adjudin [1-(2,4-dichlorobenzyl)-1H-indazole-3-carbohydrazide; 50 mg/kg body weight by gavage] [74] or cadmium (3 mg/kg body weight intraperitoneally) [178], spermatid polarity was found to be disrupted. For instance, spermatid heads were deviated from their intended orientation of pointing toward the basemene membrane. Instead, spermatid heads were deviated by 90–180° from the basement membrane. More important, spermatids that had defects in polarity were eventually depleted from the epithelium and emptied into the tubule lumen, implying a functional correlation between polarity and cell adhesion.

Indeed, studies that silenced polarity proteins, such as Par3 and Par6, using RNA interference (RNAi) and that used co-immunoprecipitation to assess changes in protein–protein interactions have shown that the Par-based proteins are crucial for proper distribution of adhesion proteins, such as JAM-A, ZO-1, and N-cadherin, at the Sertoli cell/Sertoli cell interface [74]. Par6 also stabilizes the JAM-based adhesion proteins at the apical ES by recruiting Pals1 to bind to JAM-C. However, during adjudin-induced spermatid loss that mimics spermiation, Src kinase (e.g., c-Src) becomes tightly associated with the Par6/Pals1 complex, pulling it away from JAM-C and thereby destabilizing spermatid adhesion that leads to spermatid loss from the epithelium [74]. These findings demonstrate the crucial role of polarity proteins in spermiation.

Subsequent studies also showed that polarity proteins, such as Par5 (also known as 14-3-3) and Cdc42 (a component of the Par3/6-based polarity protein), regulate protein distribution at the ES via their effects on the kinetics of protein endocytosis, including TGF-β3-induced enhancement of protein endocytosis in Sertoli cells [58,75]. These latter findings support the notion that polarity proteins can also stabilize proteins at the apical ES. For instance, when polarity protein expression is downregulated, such as when it is mediated by spatiotemporal expression during the epithelial cycle, spermatid adhesion is perturbed, such as at late stage VIII, and this perturbation contributes to the apical ES degeneration to facilitate spermiation. This hypothesis is indeed supported by the downregulation of Par6 at the apical ES in late VIII of the epithelial cycle [74]. In addition, a triple knockdown of Scribble, Lgl2, and Dlg1 by RNAi in the testis *in vivo* is known to perturb spermatid polarity at the apical ES, most prominently in stage VIII tubules, concomitantly with a disorganization of F-actin at the apical ES, leading to a perturbation of the distribution and downregulation of laminin-γ3 chain in step 19 spermatids [76]. These changes thus destabilize spermatid adhesion and polarity, contributing to a premature loss of spermatids from the seminiferous epithelium, a process that is analogous to spermiation. Collectively, these findings clearly demonstrate the essential role of polarity proteins in spermiation as depicted in the hypothetical model in Figure 12.3.

4. Endocytic vesicle-mediated trafficking of proteins

Endocytic vesicle-mediated protein trafficking is emerging as one of the most critical cellular events in all mammalian cells and epithelia. These proteins regulate a wide range of events, including cell signaling, cell movement, recycling of proteins, immune response, cell and tissue homeostasis, embryogenesis, development, spermatogenesis, and pathogenesis of disease (such as tumorigenesis and metastasis) [179–188].

The involvement of endocytosis in spermatogenesis and in particular spermiation was first noted in the late 1970s, when the giant endocytic vesicles known as TBC were detected at the degenerating apical ES when they were examined by electron microscopy. These ultrastructures were thought to be involved in the elimination of cytoplasmic debris and/or recycling of membrane structures of spermatids [90,101]. However, the significance of endocytic vesicles in protein trafficking and signaling function was not known until the early 2000s [8,16,40,43,79,189].

It is now known that the proteins that are necessary for endocytosis, transcytosis, and recycling (e.g., clathrin, caveolin, small GTPases, dynamins, cortactin, N-WASP, Eps8, and Arp3) are spatiotemporally expressed at the ventral (concave) side of spermatid head, at the apical TBC, the site of initial apical ES degeneration

[7,104,107–109,111,112,190]. Physiologically, endosome-mediated trafficking is essential to spermatogenesis because it is physiologically impossible for the fixed number of Sertoli cells [191]—approximately 4 and 30 million in each testis of mice [192] and rats [193], respectively—to support almost countless numbers of spermatids that arise from meiosis.

Thus, spermatid proteins, such as those at the plasma membrane of spermatids that confer adhesion at the Sertoli cell/spermatid interface, can be endocytosed, transcytosed, and recycled instead of having new proteins constantly made via *de novo* synthesis. It is likely that the events of endocytic vesicle-mediated protein trafficking that occur at the apical TBC/degenerating apical ES at the Sertoli cell/spermatid interface are analogous to the Sertoli cell/Sertoli cell interface (see Section V.B.) because the ultrastructures of apical and basal ES are virtually indistinguishable (Figure 12.1). This concept is also supported by the timely appearance of step 8 spermatids at stage VIII of the epithelial cycle so that endocytic vesicles containing old apical ES proteins that form in late stage VII to early stage VIII of the cycle at the apical TBC can be recycled to assemble new apical ES at the spermatid/Sertoli cell interface (Figure 12.1). Furthermore, endocytic vesicles may also function in endosomal signaling by carrying fragments of biologically active laminin chains at the degenerating apical ES to induce BTB restructuring at the basal ES/BTB [86,134]. This possibility is also supported by the finding that the events of endocytosis and recycling require approximately 1–2 h to complete based on biochemical assays in Sertoli cells [58,59,75,194].

Because the duration of stage VIII is approximately 29.1 h in the rat testis [12,100,195], fragments generated at the apical ES located near the edge of the seminiferous tubule can biologically coordinate BTB restructuring at the basal ES near the basement membrane via endosomal signaling, which involves endocytosis and transcytosis of biologically active laminin fragments. In short, biologically active laminin peptides that are generated at the apical ES can be rapidly shuffled to the basal ES/BTB via endosomal signaling to exert their biological function.

In fact, this signaling mechanism can also be used to coordinate the timely transport of spermatids back and forth across the seminiferous epithelium during the epithelial cycle so that signals can be coordinated by spermatids during spermiogenesis, a process that is similar to that used by other epithelia [186,196].

The signal(s) that trigger and/or coordinate the transport of spermatids across the seminiferous epithelium during spermiogenesis remain to be determined, but studies of membrane trafficking suggest the involvement of nonreceptor protein tyrosine kinases [197]. Indeed, studies have shown that members of the Src family of kinases, such as c-Yes and c-Src, are involved in endocytic vesicle-mediated trafficking because a knockdown of c-Yes by RNAi was shown to impede protein endocytosis in Sertoli cells [198,199]. This possibility must be carefully evaluated in future studies.

Collectively, these data thus support the emerging concept that endocytic vesicle is crucial to spermatid transport across the seminiferous epithelium and for spermiation. These data also support the role of endosomal signaling in coordinating the cellular events that take place across the seminiferous epithelium, such as spermiation and BTB restructuring. In this context, it is of interest that in addition to their role in recycling, endosomes are vital organelles for intracellular degradation of proteins. In fact, studies in other epithelia have shown that endocytosis and autophagy may have overlapping functions in protein degradation [200], so that energy from apoptotic or diseased germ cells can be recycled to sustain Sertoli cell function [201]. Thus, endosomes can also be used as the pivotal organelles for the degradation of unwanted cellular debris generated in the seminiferous epithelium during spermatogenesis. While this is a tempting concept, this is an emerging area of research that deserves more attention in the future.

E. Phagocytosis

1. Background

Phagocytosis was first reported in 1882 by Elie Metchnikoff [202]. In this study, macrophages were found to have the ability to seek out and eliminate invading infectious agents, such as bacteria; repair tissue damage by eliminating cellular debris; and mediate inflammatory response [203]. During phagocytosis, solid particles larger than 0.5 μm (e.g., bacteria) are engulfed by phagocytes (e.g., macrophages). The particles form internal phagosomes, which are then lysed by lysosomes of the phagocytes [204]. Phagocytosis that targets larger solid particles, such as bacteria and germ cells, is triggered by receptor–ligand interaction between the phagocytes (e.g., macrophages and Sertoli cells) and the engulfing particles (e.g., bacteria, cellular debris, and defective/diseased germ cells).

Studies in macrophages have identified two phagocytic receptors: (i) the Fcγ receptor (FcγR) that recognizes particles opsonized by antibodies and (ii) the complement receptor 3 (CR3; also known as integrin$\alpha_M\beta_2$, CD11b/CD18, and Mac1) that recognizes particles coated with the complement molecule C3b [204]. In short, macrophages extend membrane processes to engulf IgG-opsonized beads [205]. The term "zipper model" [206] is used to refer to the concept that IgG-coated surface acts as a railway track to guide the extending membrane (pseudopod) to wrap around the particle for engulfment [204,207]. A subsequent study has shown that IgG coating is not sufficient to trigger phagocytosis because objects of certain shapes are not suitable for engulfment regardless of the presence of IgG coating [208].

In short, phagocytosis is a crucial cellular event for maintaining tissue homeostasis by which phagocytes engulf and digest external pathogens, apoptotic cells, or dead cells; remove cellular debris in healthy and pathological conditions, such as inflammation; and maintain the immune defense system [207].

2. Phagocytosis in the testis

In the seminiferous epithelium of mammalian testes, Sertoli cells were first reported in 1901 to have the ability to phagocytose the residual cytoplasmic bodies, the organelles that contain cytoplasmic debris derived from developing spermatids during spermiogenesis [177,209]. The phagocytic activity of Sertoli cells in eliminating the residual bodies and cellular debris from the seminiferous epithelium during spermatogenesis is physiologically important to maintain the homeostasis of the testis. In addition, the timely removal of residual bodies from step 19 spermatids at stage

VIII is crucial for the release of mature spermatozoa at spermiation that takes place in late stage VIII of the epithelial cycle in the rat testis because sperms found at the tubule lumen are free of residual bodies [3,7,8,42,96]. In fact, the removal of residual bodies from step 19 spermatids may trigger a yet-to-be-identified signaling cascade of events that initiates spermiation, which is a subject of interest for future investigation. Without this timely removal of residual bodies by Sertoli cells via phagocytosis, immature, defective sperm that are found in the epididymis can possibly be tagged by ubiquitinylation, and it is known that ubiquitinated (i.e., defective) sperm is phagocytosed by epididymal epithelial cells [210–212].

But this viewpoint has been challenged because phagocytosis of sperm by the epididymal epithelial cells was not detected in earlier classical histological and ultrastructural analyses, and it is known that sperm heads are difficult to digest intracellularly [213]. Nonetheless, a failure to eliminate residual bodies as well as apoptotic germ cells from the seminiferous epithelium leads to male infertility [214]. Studies have shown that phagocytosed residual bodies engulfed at stage VIII of the epithelial cycle in the rat testis that form phagosomes migrate swiftly from the apex to the base of Sertoli cells at stage IX of the cycle to be fused with lysosomes for complete degradation near the basement membrane, usually within hours, and this degradation is completed by stage X [177,215,216] (Figure 12.4). However, the molecular mechanism(s) underlying these events remain largely unexplored. It is

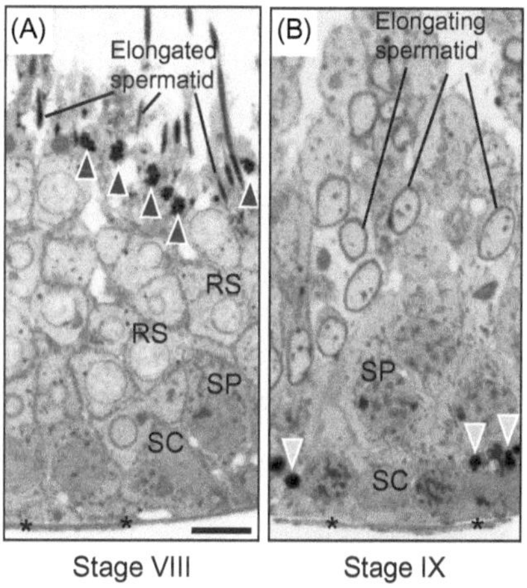

Figure 12.4 Phagocytosis of residual bodies by the Sertoli cell and the transport of phagosomes for degradation during the epithelial cycle. (A) Cross-section of a stage VIII tubule illustrating residual bodies arise from step 19 spermatids are being engulfed by the Sertoli cell (annotated by red arrowheads) via phagocytosis (see text for details), which will then form phagosomes. (B) Cross-section of a stage IX tubule. Phagosomes formed at stage VIII are rapidly transported to the basal compartment near the basement membrane of the tunica propria (tunica propria is annotated by asterisks) so that residual bodies in these phagosomes (annotated by yellow arrowheads) can be degraded via proteolysis at stage IX. Scale bar, 40 μm, applies to the other micrograph. Abbreviations: SC, Sertoli cell nucleus; SP, primary spermatocyte; RS, round spermatid.

likely that annexins [214], integrins, and GTPases in particular Rho GTPases [217,218]), and nonreceptor protein kinases [219] are involved in the engulfment of residual bodies/apoptotic germ cells, intracellular trafficking, and the eventual degradation of phagosomes, which is an area that deserves attention in future studies.

In addition to residual bodies, Sertoli cells phagocytose germ cells, in particular apoptotic germ cells [220–225] in the testis. In fact more than 75% of developing germ cells undergo apoptosis during spermatogenesis [226–228], and these germ cells must be removed by Sertoli cells via phagocytosis. A recent report has shown that apoptotic germ cells that are phagocytosed by Sertoli cells also serve as a crucial energy source for these nursery cells during spermatogenesis [201] because the BTB created by adjacent Sertoli cells limits the amount of nutrients available to Sertoli cells from the systemic circulation to support developing germ cells.

Thus, the phagocytic activity of Sertoli cells in eliminating germ cells is physiologically critical to maintain the homeostasis of spermatogenesis because there is a fixed number of Sertoli cells in the adult testis, indicating the limited capacity of Sertoli cells to support developing germ cells [229]. For instance, Sertoli cells cease to proliferate by days 15–17 in rodents [191,230] and after puberty by approximately 12 years of age in humans [191]. Sertoli cell numbers are approximately 4 million in mice [192] and 30 million in rats [193].

Studies have shown that the ratio of Sertoli cells to germ cells is maintained at approximately 1:30–50 in the rat testis [231,232], so that many spermatogenic cells undergo apoptosis during spermatogenesis [233,234] to avoid overwhelming Sertoli cell capacity in the seminiferous epithelium. However, unlike macrophages, apoptotic germ cells, residual bodies, and cellular debris in the seminiferous epithelium that are to be phagocytosed are not opsonized by antibodies or coated with complement molecule C3b so that they can be marked for disposal via phagocytosis by Sertoli cells because the BTB limits the passage of these molecules (e.g., IgG and C3b) from the systemic circulation to enter the adluminal compartment. Also, FcγR and CR3 have not been shown to be the phagocytic receptors in Sertoli cells. Instead, studies have shown that phagocytosis of residual bodies and/or apoptotic germ cells by Sertoli cells is regulated by two likely receptor–ligand-dependent pathways:

1. *Tyro3, Axl, and Mer (TAM) receptor–growth arrest-specific gene 6 (Gas6)/Protein S encoded by Pros1 gene (Pros1) ligand phagocytic pathway.* Phagocytes—including macrophages, dendritic cells, and natural killer cells in the immune system; osteoclasts in bone; retinal pigment epithelial cells in retina; and Sertoli cells in testes—are known to express the TAM family of receptor protein tyrosine kinases [235–240]. The TAM family is comprised of three members: (i) Tyro3 (also called Brt, Dtk, Etk, Rse, Sky, or Tif), (ii) Axl (Ark, Tyro 7, or Ufo), and (iii) Mer (c-Mer, Eyk, Mertk, Nyk, or Tyro 12) [236,239,241,242]. TAM receptors are pleiotropic negative regulators of Toll-like receptor and cytokine receptor signaling, most notably in dendritic cells, involving in innate immune response [243].

   Two ligands for TAM receptors, Gas6 and Pros1, have been identified with *in vitro* approaches [244]. For instance, Pros1 is a plasma glycoprotein and a negative regulator of blood coagulation [245,246]. Interestingly, its anticoagulant activity is independent of TAM receptors [247,248]. It is highly expressed by mouse and human activated T cells and inhibits dendritic cell function [249].

Sertoli cells express all three TAM receptors and the two ligands Gas6 and Pros1 [237]. In triple-knockout (KO) mice, the deletion of all three TAM family receptors produces signs of spermatogenesis failure in young adult males, which leads to an increase in apoptotic germ cells, followed by complete arrest of spermatogenesis without differentiating germ cells beyond spermatogonia in the seminiferous epithelium of adult KO mouse testes, such as by 30 weeks of age [237,250]. Sertoli cells isolated from the triple KO mice lose the ability to bind to the apoptotic germ cells without detectable phagocytosis [251]. Furthermore, the lack of TAM in Leydig cells appears not to perturb Leydig cell function [250], even though these cells express Mer [252].

It is of interest that either single or various combination of double KO of Tyro3, Axl, and Mer, including mutant mice of heterozygous in Axl and homozygous in Tyro3 and Mer (Tyr3$^{-/-}$, Axl$^{+/-}$, and Mer$^{-/-}$) are fertile and display normal spermatogenesis [237], showing that the presence of a single member of the TAM family is sufficient to maintain Sertoli cell phagocytic activity. A study showed that Leydig cells also express Mer, but not Tyro 3 or Axl, unlike Sertoli cells, which predominantly express Tyro 3 and Axl with moderate expression of Mer [252].

Furthermore, spermatogonia, but not advanced germ cells, have been shown to express Gas6, which serves as the ligand for the TAM receptors [252]. Gas6 is also a member of the vitamin K-dependent protein family and is related to Pros1, which is an anticoagulation factor that plays a role in thrombosis [253]. Both Gas6 and protein S (e.g., Pros1) are putative ligands that stimulate phagocytosis of rat retinal pigment epithelial cells [254]. Furthermore, Gas6 is known to induce phagocytic activity of Sertoli cells by as much as fivefold [251], demonstrating its functional role as a ligand for Tyro-3-induced phagocytosis.

Collectively, these findings suggest that other ligand(s) for TAM receptors remain to be identified in spermatocytes and, in particular, spermatids. Also, the downstream signaling molecules that mediate the TAM receptor remain to be identified. A recent report has shown that c-Src is involved in the endocytic vesicle-mediated protein degradation in Sertoli cells [94]. It would be of interest for future studies to establish the physiological link between c-Src and the TAM receptor/Gas6 phagocytic pathway.

2. *Class B scavenger receptor type I (SR-BI) receptor—phosphatidylserine (PS) ligand phagocytic pathway.* Studies have shown that SR-BI, also a receptor for high-density lipoprotein [255], is a phagocytic receptor that is restrictively expressed by Sertoli cells in the seminiferous epithelium of rat testes [224,256,257]. SR-BI also displays high affinity for acidic liposomes and other apoptotic mammalian cells, such as Chinese hamster ovary cells and thymocytes [255,258,259].

The use of an anti-SR-BI antibody via microinjection in the testis *in vivo* has been shown to block Sertoli cell phagocytosis of apoptotic germ cells [224], showing that SR-BI is a putative phagocytic receptor. Apoptosis of germ cells is manifested by DNA fragmentation, chromatin condensation, and externalization of PS [222,260]. PS is one of the most well-characterized phagocytosis markers [261]. SR-BI has been shown to bind to PS with high affinity, but not to other phospholipids [256]. SR-BI is also highly expressed by polarized Sertoli cells in both the apical and basal regions [257]. These results show that SR-BI serves as the phagocytic receptor in both compartments of the seminiferous epithelium. Furthermore, the use of annexin V, a known binding protein for PS, can block phagocytosis, as shown in studies both *in vitro* and *in vivo* [214] that have confirmed the involvement of PS-mediated phagocytosis. Collectively, these data support the notion that SR-BI and PS expressed by phagocytic Sertoli cells and apoptotic germ cells, respectively, create the receptor/ligand pathway for phagocytosis in the testis.

3. Remarks

It is likely that the TAM receptor Gas6/Pros1 and the SR-BI/PS pathways are connected downstream via a common signaling molecule(s) or pathway(s). If both pathways operated independently, the triple-KO mutants of Tyro 3, AxI, and Mer would not lead to spermatogenetic arrest because the SR-BI/PS phagocytic pathway can supersede the lost function of the TAM receptors to maintain spermatogenesis, at least to a certain extent, and tubules would not be devoid of all spermatogenic cells by 30 weeks of age [237,250]. Thus, identification of the common downstream signaling molecule(s)/pathway(s) that regulate Sertoli cell phagocytosis is of great interest in the field.

Phagocytosis involves extensive remodeling of Sertoli cell morphology for the engulfment of residual bodies and/or apoptotic germ cells. There is extensive involvement in phagocytosis of the actin- and tubulin-based cytoskeletons and their regulatory proteins, in particular the actin barbed-end nucleation protein Arp2/3 protein complex, and actin-bundling proteins (e.g., Eps8, palladin) [40,79], as demonstrated in studies during phagocytosis in macrophages [204,262]. Furthermore, while Gas6 and Pros1 are the putative ligands of the TAM tyrosine kinase receptors that elicit phagocytosis, Gas6 is only weakly expressed in spermatogonia but not in spermatocytes and spermatids [252]. Thus, the putative ligand(s) for TAM receptors in spermatocytes and spermatids remain to be identified.

The rationale for challenging Gas6 and Pros1 as the putative ligands for TAM receptors in Sertoli cells includes findings in which $Gas^{-/-}$ mutant mice were fertile and apparently normal without a distinctive pathology, except that these mice had platelet dysfunction and were resistant to experimentally induced thrombosis [263]. The function of Pros1 in the testis is not known because $Pros1^{-/-}$ mice died between E15.5 and E17.5 due to embryonic lethal coagulopathy, manifested by massive hemorrhages [247]. Thus, a testis-specific KO model of Pros1 would be helpful to address its function in phagocytosis in the testis. In short, future studies that can better understand the regulation of phagocytosis are crucial to delineate the biology of spermiation because these two cellular events are highly coordinated at stage VIII of the epithelial cycle. For instance, a failure of phagocytosis can impede spermiation and spermatogenesis.

V. Transport of preleptotene spermatocytes at the BTB

Background—the BTB

The BTB in the testis is one of the tightest blood–tissue barriers in the mammalian body [1,36,44,55]. The concept of BTB was discovered in the nineteenth century, when dyes injected into rodents were shown to be prevented from reaching the brain and the seminiferous tubules [1,264]. In short, the BTB limits paracellular (i.e., between Sertoli cells) and transcellular (i.e., across Sertoli cells) transport of substances. The adluminal compartment behind the BTB is a specialized microenvironment that is significantly different from the interstitial space and the systemic circulation. This difference is reflected by the significant differences in protein concentration and the relative levels of electrolytes, ions, and other substances found in the fluids of seminiferous tubules versus interstitial fluids and blood plasma or serum [1,264,265].

The BTB provides a unique microenvironment in the adluminal (apical) compartment of the seminiferous epithelium that is composed of Sertoli and germ cells only, and it is also considered to contribute, at least in part, to the immune-privileged status of the testis [1,266,267] by preventing autoantigens residing in late spermatocytes and postmeiotic germ cells from being recognized by the host immune cells [268]. However, this immunological barrier can be leaky in that many germ cell autoantigens also reside in spermatogonia and early spermatocytes that are found outside the BTB [269,270]. Collectively, these findings show that perhaps the immunosuppressive biomolecules secreted by Sertoli cells [271–274] are more important in conferring immune privilege status to the testis.

Nonetheless, the BTB creates a unique microenvironment for meiosis and post-meiotic spermatid development because a delay in BTB assembly by treatment of neonatal rats with diethylstilbestrol (a synthetic nonsteroidal estrogen) delays the onset of spermatogenesis in which type B spermatogonia that are transformed to spermatocytes fail to enter meiosis but undergo degeneration instead [275]. In short, the absence of a functional BTB can impede meiosis and spermatogenesis. This concept is supported by studies in which functional spermatogonia failed to differentiate into spermatocytes in rats following treatment with a high dose of adjudin that led to an irreversible disruption of the BTB [276]. Collectively, these data suggest the possibility that the BTB may also be used to protect SSCs and spermatogonia that reside outside the BTB from the biomolecules that are likely released from developing germ cells in particular postmeiotic spermatids. Nonetheless, these findings illustrate the physiological significance of the BTB in spermatogenesis.

A. Preleptotene spermatocyte transport at the BTB

While the BTB is one of the tightest blood–tissue barriers, at stage VIII of the epithelial cycle, when spermiation takes place in the adluminal compartment, the BTB also undergoes remodeling to facilitate the transport of preleptotene spermatocytes across the BTB [1,37,96]. Germ cells, including preleptotene spermatocytes, are connected in clones via intercellular bridges [277–279], also known as tunneling nanotubes, in other mammalian cells [280–283] to transport signals and organelles (e.g., ribosomes and DNA/RNA-binding proteins) between these cells [279,284–286] to synchronize cellular events during the epithelial cycle. A unique mechanism must be in place at the BTB to facilitate preleptotene spermatocyte transport to avoid a massive disruption of the immunological barrier. If a disruption of the BTB takes place during spermatogenesis, it perturbs spermatogenesis.

Based on morphological studies in the 1970s, it was speculated that BTB integrity could only be maintained if new BTB was assembled behind spermatocytes connected in clones that were being transported across the immunological barrier prior to the disassembly of the old BTB located above the preleptotene spermatocytes (reviewed in [7,79]). This concept is supported by studies that used lanthanum, an electron-dense salt, to track the relative location of BTB surrounding preleptotene spermatocytes at the immunological site [59,287]. Indeed, a functional new BTB was detected behind a preleptotene spermatocyte while the old BTB was gradually remodeled or disassembled based on the relative distribution of lanthanum around the spermatocyte. This concept was supported by the use of biochemical assays of protein endocytosis and recycling [59] in addition to morphological data [287]. A study that used confocal microscopy has elegantly confirmed this earlier concept in the rodent testis [56]. However, the precise molecular mechanism(s) and the molecules involved remained elusive for many years. Below is a critical discussion and evaluation of this concept based on these findings. A hypothetical model is depicted in Figure 12.5.

B. A biochemical model of preleptotene spermatocyte transport at the BTB

As outlined in Figure 12.5, the rapid restructuring of the BTB that allows the timely transport of preleptotene spermatocytes across the BTB is regulated via tightly coordinated, but differentially regulated, events of endocytic vesicle-mediated protein trafficking that take place below and above the transiting spermatocytes. Studies *in vitro* and *in vivo* have shown that testosterone and cytokines (e.g., TNFα, TGF-β2/-β3, and interleukin (IL)-1α) that are known to promote [288–293] and perturb [294–299] BTB integrity, respectively, enhance the kinetics of protein endocytosis at the Sertoli cell BTB [59,194]. However, testosterone promotes recycling of integral membrane proteins (e.g., occludin and N-cadherin) at the BTB back to the Sertoli cell surface, whereas cytokines (e.g., TGF-β2) enhance endosome-mediated degradation of endocytosed proteins [57,59].

In addition, other studies have shown that endocytic vesicle-mediated protein trafficking at the Sertoli cell BTB requires the active participation of polarity proteins. For instance, TGF-β3-induced enhancement of protein endocytosis requires the presence of Cdc42 (a component of the Par3/6-based polarity protein complex at the ES [74]) because overexpression of a dominant-negative mutant of Cdc42 in Sertoli cells is known to abolish the TGF-β3-mediated enhancement in protein endocytosis at the Sertoli cell BTB [58]. Furthermore, 14-3-3 (also known as Par5) plays a role in protein endocytosis at the Sertoli cell BTB because a knockdown of 14-3-3 by RNAi induces a significant increase in the kinetics of endocytosis of JAM-A and N-cadherin, two putative integral membrane proteins at the BTB [75]. These findings are important because they demonstrate that testosterone, which promotes BTB integrity, facilitates recycling of integral membrane proteins from the old BTB site located behind the preleptotene spermatocytes for the assembly of the new BTB at the apical region of the spermatocytes (Figure 12.5).

However, cytokines (e.g., TNFα and TGF-β2/-β3) that enhance protein endocytosis are found to promote endosome-mediated degradation of nonfunctional/old integral membrane proteins from the old BTB to facilitate its disassembly (Figure 12.5). In short, it is through the tightly regulated changes in the levels of testosterone (or its receptor) and cytokines at the microenvironment of the BTB plus the microdomains surrounding the preleptotene spermatocytes in transit at the site that the tightly regulated biochemical events of protein endocytosis, transcytosis, and recycling can provide an effective mechanism to promote spermatocyte transport while the integrity of the immunological barrier is maintained (Figure 12.5).

A study of two nonreceptor protein tyrosine kinases at the BTB—c-Yes and c-Src—by RNAi coupled with biochemical assays of protein endocytosis, transcytosis, and recycling has further supported the concept depicted in Figure 12.5 [94]. It is shown that the silencing of c-Yes impedes endocytosis of integral membrane proteins at the BTB, whereas a knockdown of c-Src perturbs intracellular protein degradation [94]. It is likely that c-Yes and c-Src are working in concert with testosterone and cytokines to provide a delicate system to coordinate the events of endocytic vesicle-mediated protein trafficking to affect the transport of preleptotene spermatocytes across the BTB as depicted in Figure 12.5.

It is of interest that the BTB is constituted by actin-based TJ and GJ that coexist with basal ES as well as intermediate filament-based desmosome, as shown in Figure 12.5. Studies have shown that there is a physiological necessity for multiple coexisting junction types at the BTB to safeguard its integrity during the transport of preleptotene spermatocytes. As noted in Figure 12.5, proteins from the old BTB are recycled to assemble the developing new BTB above and behind the spermatocyte, respectively. If the BTB is composed of TJ alone, as in other blood–tissue barriers, the degeneration of TJ at the old BTB would lead to immunological barrier disruption before the new BTB could be functionally established. It has been shown that other adhesion protein complexes at the basal ES and, perhaps desmosome, can supersede transiently the lost TJ function to maintain the barrier's integrity via a unique engagement/disengagement mechanism [300].

It is noted that in stages other than VII–VIII, TJ (e.g., occludin-ZO-1) and basal ES (e.g., N-cadherin-α/γ-catenin) physically interact (i.e., engage) with each other to maintain BTB integrity [300] (Figure 12.5, left panel). However, ZO-1 and α- or γ-catenin are disengaged and no longer structurally interact with each other when rats are exposed to adjudin [300], which is known to induce BTB remodeling that mimics stage VIII tubules [276,301,302]. Thus, the disruption of old TJ, as mediated by recycling of occludin and ZO-1 to assemble new BTB via endocytosis and transcytosis, does not interfere with the adhesion function of N-cadherin-α- or γ-catenin at the basal ES, which could supersede the temporarily lost function of TJ at the BTB (Figure 12.5, middle panel). Subsequent studies have shown that such coordination of multiple junction types at the BTB is mediated by GJ [82,85]. Thus, the coexisting junction types at the BTB are crucial to the transport of preleptotene spermatocytes at stage VIII of the epithelial cycle to maintain the BTB integrity, as illustrated in Figure 12.5.

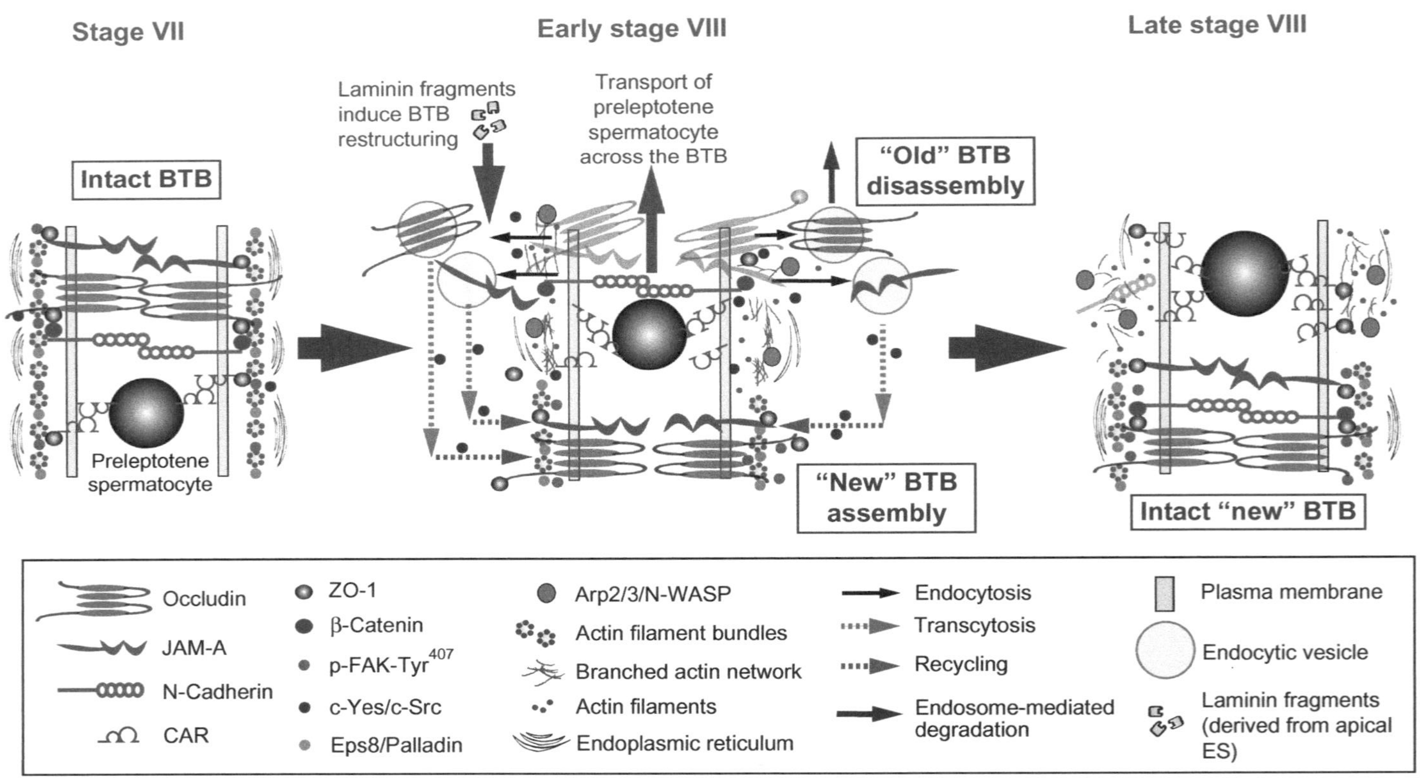

Figure 12.5 A schematic drawing that illustrates the likely events at the BTB/basal ES during the transport of preleptotene spermatocytes across the immunological barrier in the rat testis. Left panel illustrates the intact BTB at stage VII of the epithelial cycle in which actin filament bundles

C. The role of actin- and tubulin-based cytoskeleton in the transport of preleptotene spermatocytes at the BTB

The model depicted Figure 12.5 is based on findings in the field, but many outstanding questions remain to be addressed. For instance, it is likely that the transport of endocytic vesicles between the apical and basal regions of preleptotene spermatocytes at the BTB requires the active participation of actin- and tubulin-based cytoskeletons. The mechanism(s) by which these cytoskeletons are regulated remain to be explored.

A study has shown that p-FAK-Tyr^{407}, the phosphorylated and activated form of FAK at the BTB, is the likely regulator of actin-based cytoskeleton to confer BTB integrity in the rat testis [66]. For instance, overexpression of the p-FAK-Tyr^{407} phosphomimetic mutant p-FAK Y407E (i.e., making p-FAK-Tyr^{407} constitutively active) in Sertoli cells at the BTB makes the Sertoli cell tight junction permeability barrier tighter by promoting BTB integrity, which is mediated by changes in the kinetics of actin polymerization [66]. Findings from this report support the notion that p-FAK-Tyr^{407} serves as a molecular switch that modulates F-actin dynamics at the BTB to confer changes in its organization that favor the bundled configuration of actin microfilaments at the BTB. This concept is important because protein endocytosis, transcytosis, and recycling are cytoskeleton-dependent cellular events. Indeed, overexpression of FAK Y407E phosphomimetic mutant modulates the organization of actin microfilaments in Sertoli cells. This process also alters the localization of cell adhesion proteins at the Sertoli cell/cell interface, such as claudin-11 (a TJ-integral membrane protein) and ZO-1 (a TJ adaptor protein), in which claudin-11 and ZO-1 are restrictively localized to the Sertoli cell/cell interface to strengthen the TJ barrier [66].

These findings thus support the concept that cytoskeletons are intimately related to the events of endocytic vesicle-mediated trafficking in the BTB microenvironment during the transit of preleptotene spermatocytes at the immunological barrier.

◀ at the basal ES of the BTB are maintained via the intricate action of actin bundling proteins Eps8 and palladin, mediated by p-FAK-Tyr^{407} and nonreceptor protein tyrosine kinases c-Yes and c-Src. However, in early stage VIII of the epithelial cycle (middle panel), biologically active laminin fragments derived from the apical ES exert their effects at the BTB, most likely via their actions on p-FAK-Tyr^{407}, which activates the branched actin polymerization protein complex, Arp2/3-N-WASP, inducing reorganization of the actin filament bundles at the basal ES/BTB, converting actin microfilaments to an unbundled/branched configuration, thereby destabilizing adhesion protein complexes (e.g., occludin-ZO-1, JAM-A-ZO-1, N-cadherin-β-catenin) at the site. Unbundled actin filaments, coupled with the actions of c-Yes/c-Src and p-FAK-Tyr^{407}, also facilitate endocytic vesicle-mediated protein trafficking events, which induce endocytosis, transcytosis, and recycling of old BTB proteins above the preleptotene spermatocyte in transit to assemble a new BTB behind the spermatocyte. Thus, the disassembly of the old BTB that facilitates the transport preleptotene spermatocyte across the BTB does not lead to a disruption of the immunological barrier, even transiently, because a new BTB is simultaneously assembled as shown in the right panel. See text for additional details.

Nonetheless, much research is needed in this area, such as the identification of the functional role of actin and/or tubulin regulatory proteins at the BTB so that the molecular mechanism(s) underlying the events of preleptotene spermatocyte transport can be fully elucidated.

D. Involvement of mammalian target of rapamycin complex 1 (mTORC1) and mTORC2 in preleptotene spermatocyte transport at the BTB

mTOR is a Ser/Thr protein kinase. It is the mammalian homolog of the protein first identified in yeast, in which rapamycin, an antibiotic, was shown to inhibit the activity of TOR in cells [303–305]. mTOR is found in virtually all mammalian cells and regulates a wide spectrum of cellular events, including actin organization, cell metabolism, survival, autophagy, diabetes, obesity, and tumorigenesis [62,306–311]. To execute its cellular function, mTOR must be associated with either regulatory-associated protein of mTOR (Raptor) or rapamycin-insensitive companion of mTOR (Rictor) together with other binding partners to form the mTORC1 and mTORC2 signaling complexes, respectively [62,310,312].

Studies have shown that mTORC1 and mTORC2, which are expressed by both Sertoli and germ cells in the testis, display antagonistic effects on the Sertoli cell tight junction permeability barrier function both *in vitro* and *in vivo.* mTORC1 induces Sertoli cell BTB disruption [64], whereas mTORC2 promotes Sertoli cell BTB integrity [63]. For instance, a knockdown of ribosomal protein S6 (rpS6), the downstream signaling regulator of mTORC1, promotes the Sertoli cell tight junction permeability barrier function [64], showing that its expression in the BTB microenvironment is used to disrupt BTB function. This hypothesis is indeed supported by an *in vivo* study that knocked down rpS6 by transfecting rat testes using shRNA [64]. For instance, a transient loss of rpS6 promotes the recruitment of BTB integral membrane proteins, such as occludin and claudin-11, to the BTB, which is consistent with findings of the RNAi study *in vitro* [64]. More important, this effect is mediated by changes in the organization of F-actin in Sertoli cells that constitute the BTB [64].

However, a knockdown of Rictor, the key binding partner of mTOR that creates the mTORC2 signaling complex, perturbs the Sertoli cell tight junction permeability barrier function [63], which is in sharp contrast to the knockdown of rpS6 that promotes Sertoli cell tight junction barrier function after mTORC1 function is disrupted [64], demonstrating that the expression of mTORC2/Rictor is crucial to promote BTB integrity. Furthermore, the observation that mTORC2 promotes BTB integrity has also been reproduced in studies *in vivo* via a knockdown of Rictor in the testis [63]. More important, the transient loss of Rictor also induces a significant downregulation of the expression of the GJ protein connexin 43 (Cx43) [63], showing that in addition to changes in the organization of F-actin at the BTB, GJ function may be altered. This notion is supported by the finding that the transient loss of Rictor function by RNAi that perturbs the mTORC2 signaling complex function indeed disrupts Sertoli cell/cell GJ communication based on a functional assay [63].

Collectively, these findings demonstrate the likely involvement of mTORC1 and mTORC2 in coordinating the timely transport of preleptotene spermatocytes across the BTB at stage VIII of the epithelial cycle. However, what remains to be addressed in future studies is to identify the upstream biomolecule(s) that regulate the spatiotemporal expression of Raptor and Rictor, as well as their corresponding binding partners, such as rpS6 (for Raptor) and PKC-α (for Rictor), which in turn maintain the homeostasis of the mTORC1 and mTORC2 function in the microenvironment of the BTB. It is plausible that testosterone and cytokines (such as TNFα, TGF-β2/-β3, and IL-1α) are the upstream biomolecules that regulate the action of mTORC1 and mTORC2 at the Sertoli cell BTB during the epithelial cycle of spermatogenesis. However, this possibility remains to be investigated.

VI. Concluding remarks and future perspectives

As discussed above, we have provided an update on the biochemistry of germ cell transport in the seminiferous epithelium during the epithelial cycle of spermatogenesis. In each topic discussed above, area(s) that deserve attention in future studies have been highlighted. To avoid redundancy, the questions that remain to be addressed are not repeated here. It is also noted that we published our first report on this topic based on a series of biochemical experiments [313] showing that the assembly of Sertoli cell/germ cell junctions—the initial cellular event of germ cell transport—appeared to be more than a morphological phenomenon and, instead, were well-defined biochemical changes involving activation of proteases and protease inhibitors. The field has since advanced appreciably. In addition to our laboratory, many investigators have contributed significantly to understanding the cellular events that take place at the Sertoli cell/cell and Sertoli cell/germ cell interface that are pertinent to the biology of germ cell transport during spermatogenesis.

These findings have also uncovered some unique opportunities to develop nonhormonal male contraceptives, which are urgently needed, in particular in developing countries [314–316]. For instance, the unique adhesion junction at the Sertoli cell/cell and Sertoli cell/spermatid interface, namely the basal and apical ES, respectively, is a novel target for male contraception. Studies have shown that adjudin exerts its effects primarily at the apical ES [317–322]. Adjudin and its analog gamendazole [323,324] have been actively investigated as a potential nonhormonal contraceptive.

However, much work is needed to better understand apical ES and the molecules necessary to regulate germ cell transport, such as the Arp2/3 complex, Eps8, palladin, FAK, and others. For instance, research needs to identify the mechanism(s) and the molecules that regulate changes in F-actin organization at the ES, in particular the rapid conversion between bundled and unbundled/branched configuration, because any mishap in this process leads to unwanted spermatid depletion from the epithelium or retention in the epithelium and these spermatids will be removed by Sertoli cells via phagocytosis. Both scenarios lead to infertility.

Also, research is needed on the molecule(s), in particular transcription factors, that regulate/modulate the spatiotemporal expression of the Arp2/3 complex, Eps8, palladin, FAK and its activated/phosphorylated forms (e.g., p-FAK-Tyr^{397} and p-FAK-Tyr^{407}) at the apical versus the basal ES during the epithelial cycle. This information will better determine the molecular mechanism(s) that regulate germ cell transport in the testis.

Acknowledgments

This work was supported by grants from the National Institutes of Health, NICHD HD029990 Project 5 to CYC; R01 HD056034 to CYC; R03 HD061401 to DDM.

We are indebted to the former and current members of the Cheng laboratory who contributed their time and effort to better understand the biology of germ cell transport in the seminiferous epithelium. Without the dedication and commitment of this group of young investigators, this review could not have been written. Some of their studies along with those from many other investigators in the field were cited, discussed, and carefully evaluated here. We are also indebted to many investigators who are not reproductive biologists, but their findings have contributed significantly to our understanding of the biology of germ cell transport during spermatogenesis. Also, due to space limits, much earlier work could not be cited. However, every effort was made to cite reviews of specific topics in the field so that many valuable and important references are covered and they can be found in these reviews. We thus apologize to our colleagues if their work has not been cited here and deeply regret this oversight.

References

[1] Cheng CY, Mruk DD. The blood–testis barrier and its implication in male contraception. Pharmacol Rev 2012;64:16–64.

[2] Cheng CY, Wong EWP, Yan HHN, Mruk DD. Regulation of spermatogenesis in the microenvironment of the seminiferous epithelium: new insights and advances. Mol Cell Endocrinol 2010;315:49–56.

[3] de Kretser DM, Kerr JB. The cytology of the testis. In: Knobil E, Neill JB, Ewing LL, Greenwald GS, Markert CL, Pfaff DW, editors. The physiology of reproduction, Vol. 1. New York, NY: Raven Press; 1988. p. 837–932.

[4] de Rooij DG. The spermatogonial stem cell niche. Microsc Res Tech 2009;72:580–5.

[5] Ehmcke J, Schlatt S. A revised model for spermatogonial expansion in man: lessons from non-human primates. Reproduction 2006;132:673–80.

[6] Hermo L, Pelletier RM, Cyr DG, Smith CE. Surfing the wave, cycle, life history, and genes/proteins expressed by testicular germ cells. Part 1: Background to spermatogenesis, spermatogonia, and spermatocytes. Microsc Res Tech 2010;73:241–78.

[7] Mruk DD, Cheng CY. Sertoli–Sertoli and Sertoli–germ cell interactions and their significance in germ cell movement in the seminiferous epithelium during spermatogenesis. Endocr Rev 2004;25:747–806.

[8] O'Donnell L, Nicholls PK, O'Bryan MK, McLachlan RI, Stanton PG. Spermiation: the process of sperm release. Spermatogenesis 2011;1:14–35.

[9] Carreau S, Hess RA. Oestrogens and spermatogenesis. Philos Trans R Soc Lond B Biol Sci 2010;365:1517–35.

[10] Carreau S, Wolczynski S, Galeraud-Denis I. Aromatase, estrogens and human male reproduction. Phil Trans R Soc Lond B Biol Sci 2010;365:1571–9.

[11] O'Donnell L, Meachem SJ, Stanton PG, McLachlan RI. Endocrine regulation of spermatogenesis. In: Neill JD, editor. Physiology of reproduction. 3rd ed. Amsterdam: Elsevier; 2006. p. 1017–69.

[12] O'Donnell L, Robertson KM, Jones ME, Simpson ER. Estrogen and spermatogenesis. Endocr Rev 2001;22:289–318.

[13] Sharpe RM. Regulation of spermatogenesis. In: Knobil E, Neill JD, editors. The physiology of reproduction. New York, NY: Raven Press; 1994. p. 1363–434.

[14] Winters SJ, Moore JP. Paracrine control of gonadotrophs. Semin Reprod Med 2007;25:379–87.

[15] Cheng CY, Mruk DD. Cell junction dynamics in the testis: Sertoli–germ cell interactions and male contraceptive development. Physiol Rev 2002;82:825–74.

[16] Vogl AW, Young JS, Du M. New insights into roles of tubulobulbar complexes in sperm release and turnover of blood–testis barrier. Int Rev Cell Mol Biol 2013;303:319–55.

[17] Russell LD, Peterson RN. Sertoli cell junctions: morphological and functional correlates. Int Rev Cytol 1985;94:177–211.

[18] Vogl AW, Vaid KS, Guttman JA. The Sertoli cell cytoskeleton. Adv Exp Med Biol 2008;636:186–211.

[19] Amann RP, Howards SS. Daily spermatozoal production and epididymal spermatozoal reserves of the human male. J Urol 1980;124:211–5.

[20] Auharek SA, Franca LR. Postnatal testis development, Sertoli cell proliferation and number of different spermatogonial types in C57BL/6J mice made transiently hypo and hyperthyroidic during the neonatal period. J Anat 2010;216:577–88.

[21] Johnson L, Petty CS, Neaves WB. A comparative study of daily sperm production and testicular composition in humans and rats. Biol Reprod 1980;22:1233–43.

[22] Hermo L, Pelletier RM, Cyr DG, Smith CE. Surfing the wave, cycle, life history, and genes/proteins expressed by testicular germ cells. Part 5: Intercellular junctions and contacts between germ cells and Sertoli cells and their regulatory interactions, testicular cholesterol, and genes/proteins associated with more than one germ cell generation. Microsc Res Tech 2010;73:409–94.

[23] Lie PPY, Cheng CY, Mruk DD. Coordinating cellular events during spermatogenesis: a biochemical model. Trends Biochem Sci 2009;34:366–73.

[24] Byers S, Pelletier RM, Suarez-Quain C. Sertoli cell junctions and the seminiferous epithelium barrier. In: Russell LD, Griswold MD, editors. The Sertoli cell. Clearwater, FL: Cache River Press; 1993. p. 431–46.

[25] Enders G. Sertoli–Sertoli and Sertoli–germ cell communications. In: Russell L, Griswold M, editors. The Sertoli cell. Clearwater, FL: Cache River Press; 1993. p. 447–60.

[26] Li MWM, Mruk DD, Cheng CY. Gap junctions and blood–tissue barriers. Adv Exp Med Biol 2012;763:260–80.

[27] Lie PPY, Cheng CY, Mruk DD. The biology of the desmosome-like junction: a versatile anchoring junction and signal transducer in the seminiferous epithelium. Int Rev Cell Mol Biol 2011;286:223–69.

[28] Mruk DD, Cheng CY. Desmosomes in the testis. Moving into an unchartered territory. Spermatogenesis 2011;1:47–51.

[29] Dym M. Basement membrane regulation of Sertoli cells. Endocr Rev 1994;15:102–15.
[30] Siu MKY, Cheng CY. Dynamic cross-talk between cells and the extracellular matrix in the testis. BioEssays 2004;26:978–92.
[31] Russell LD. Observations on rat Sertoli ectoplasmic ('junctional') specializations in their association with germ cells of the rat testis. Tissue Cell 1977;9:475–98.
[32] Toyama Y, Maekawa M, Yuasa S. Ectoplasmic specializations in the Sertoli cells: new vistas based on genetic defects and testicular toxicology. Anat Sci Int 2003;78:1–16.
[33] Vogl A, Pfeiffer D, Mulholland D, Kimel G, Guttman J. Unique and multifunctional adhesion junctions in the testis: ectoplasmic specializations. Arch Histol Cytol 2000;63:1–15.
[34] Wong EWP, Mruk DD, Cheng CY. Biology and regulation of ectoplasmic specialization, an atypical adherens junction type, in the testis. Biochem Biophys Acta 2008;1778:692–708.
[35] Yan HHN, Mruk DD, Lee WM, Cheng CY. Ectoplasmic specialization: a friend or a foe of spermatogenesis? BioEssays 2007;29:36–48.
[36] Franca LR, Auharek SA, Hess RA, Dufour JM, Hinton BT. Blood–tissue barriers: morphofunctional and immunological aspects of the blood–testis and blood–epididymal barriers. Adv Exp Med Biol 2012;763:237–59.
[37] Russell LD. Movement of spermatocytes from the basal to the adluminal compartment of the rat testis. Am J Anat 1977;148:313–28.
[38] Russell LD. Observations on the inter-relationships of Sertoli cells at the level of the blood–testis barrier: evidence for formation and resorption of Sertoli–Sertoli tubulobulbar complexes during the spermatogenic cycle of the rat. Am J Anat 1979;155:259–79.
[39] Russell LD. Desmosome-like junctions between Sertoli and germ cells in the rat testis. Am J Anat 1977;148:301–12.
[40] Cheng CY, Mruk DD. An intracellular trafficking pathway in the seminiferous epithelium regulating spermatogenesis: a biochemical and molecular perspective. Crit Rev Biochem Mol Biol 2009;44:245–63.
[41] Lie PPY, Cheng CY, Mruk DD. The biology of interleukin-1: emerging concepts in the regulation of the actin cytoskeleton and cell junction dynamics. Cell Mol Life Sci 2012;69:487–500.
[42] Mruk DD, Silvestrini B, Cheng CY. Anchoring junctions as drug targets: role in contraceptive development. Pharmacol Rev 2008;60:146–80.
[43] Su WH, Mruk DD, Cheng CY. Regulation of actin dynamics and protein trafficking during spermatogenesis—insights into a complex process. Crit Rev Biochem Mol Biol 2013;48:153–72.
[44] Cheng CYE. Biology and regulation of blood–tissue barriers. *Austin, TX: Landes Bioscience/Springer Science + Business Media, LLC*; 2012. p. 1–361.
[45] Ashraf T, Kis O, Banerjee N, Bendayan R. Drug transporters at brain barriers: expression and regulation by neurological disorders. Adv Exp Med Biol 2012;763:20–69.
[46] Easton AS. Regulation of permeability across the blood–brain barrier. Adv Exp Med Biol 2012;763:1–19.
[47] Campbell M, Humphries P. The blood–retina barrier: tight junctions and barrier modulation. Adv Exp Med Biol 2012;763:70–84.
[48] Hosoya KI, Tachikawa M. The inner blood–retinal barrier: molecular structure and transport biology. Adv Exp Med Biol 2012;763:85–104.
[49] Dube E, Cyr DG. The blood–epididymis barrier and human male fertility. Adv Exp Med Biol 2012;763:218–36.

[50] Groschwitz KR, Hogan SP. Intestinal barrier function: molecular regulation and disease pathogenesis. J Allergy Clin Immunol 2009;124:3–20.
[51] Ivanov AI. Structure and regulation of intestinal epithelial tight junctions. Adv Exp Med Biol 2012;763:132–48.
[52] Nalle SC, Turner JR. Endothelial and epithelial barriers in graft-verus-host disease. Adv Exp Med Biol 2012;763:105–31.
[53] Turner JR. Intestinal mucosal barrier function in health and disease. Nat Rev Immunol 2009;9:799–809.
[54] Mital P, Hinton BT, Dufour JM. The blood–testis and blood–epididymis barriers are more than just their tight junctions. Biol Reprod 2011;84:851–8.
[55] Pelletier RM. The blood–testis barrier: the junctional permeability, the proteins and the lipids. Prog Histochem Cytochem 2011;46:49–127.
[56] Smith BE, Braun RE. Germ cell migration across Sertoli cell tight junctions. Science 2012;338:798–802.
[57] Su L, Mruk DD, Lee WM, Cheng CY. Differential effects of testosterone and TGF-β3 on endocytic vesicle-mediated protein trafficking events at the blood–testis barrier. Exp Cell Res 2010;316:2945–60.
[58] Wong EWP, Mruk DD, Lee WM, Cheng CY. Regulation of blood–testis barrier dynamics by TGF-β3 is a Cdc42-dependent protein trafficking event. Proc Natl Acad Sci USA 2010;107:11399–404.
[59] Yan HHN, Mruk DD, Lee WM, Cheng CY. Blood–testis barrier dynamics are regulated by testosterone and cytokines via their differential effects on the kinetics of protein endocytosis and recycling in Sertoli cells. FASEB J 2008;22:1945–59.
[60] Xiao X, Cheng CY, Mruk DD. Intercellular adhesion molecule (ICAM)-1 is a regulator of blood–testis barrier function. J Cell Sci 2012;125:5677–89.
[61] Xiao X, Mruk DD, Cheng CY. Intercellular adhesion molecules (ICAMs) and spermatogenesis. Hum Reprod Update 2013;19:167–86.
[62] Mok KW, Mruk DD, Cheng CY. Regulation of blood–testis barrier (BTB) dynamics during spermatogenesis via the "yin" and "yang" effects of the mammalian target of rapamycin complex 1 (mTORC1) and mTORC2. Int Rev Cell Mol Biol 2013;301:291–358.
[63] Mok KW, Mruk DD, Lee WM, Cheng CY. Rictor/mTORC2 regulates blood–testis barrier dynamics via its effects on gap junction communications and actin filament network. FASEB J 2013;27:1137–52.
[64] Mok KW, Mruk DD, Silvestrini B, Cheng CY. rpS6 regulates blood-testis barrier dynamics by affecting F-actin organization and protein recruitment. Endocrinology 2012;153:5036–48.
[65] Cheng CY, Lie PPY, Wong EWP, Mruk DD. Focal adhesion kinase and actin regulatory/binding proteins that modulate F-actin organization at the tissue barrier. Lession from the testis. Tissue Barriers 2013;1:e24252. Available from: http://dx.doi.org/doi:10.4161/tisb.24252.
[66] Lie PPY, Mruk DD, Mok KW, Su L, Lee WM, Cheng CY. Focal adhesion kinase-Tyr^{407} and -Tyr^{397} exhibit antagonistic effects on blood–testis barrier dynamics in the rat. Proc Natl Acad Sci USA 2012;109:12562–7.
[67] Wolski KM, Perrault C, Tran-Son-Tay R, Cameron DF. Strength measurement of the Sertoli–spermatid junctional complex. J Androl 2005;26:354–9.
[68] Green KJ, Getsios S, Troyanovsky S, Godsel LM. Intercellular junction assembly, dynamics, and homeostasis. Cold Spring Harb Perspect Biol 2010;2:a000125.
[69] Green KJ, Simpson CL. Desmosomes: new perspectives on a classic. J Invest Dermatol 2007;127:2499–515.

[70] Johnson L, Petty CS, Neaves WB. Age-related variation in seminiferous tubules in men. A stereologic evaluation. J Androl 1986;7:316–22.
[71] Wing TY, Christensen AK. Morphometric studies on rat seminiferous tubules. Am J Anat 1983;165:13–25.
[72] Cheng CY, Wong EWP, Lie PPY, Mruk DD, Xiao X, Li MWM, et al. Polarity proteins and actin regulatory proteins are unlikely partners that regulate cell adhesion in the seminiferous epithelium during spermatogenesis. Histol Histopathol 2011;26:1465–74.
[73] Wong EWP, Cheng CY. Polarity proteins and cell–cell interactions in the testis. Int Rev Cell Mol Biol 2009;278:309–53.
[74] Wong EWP, Mruk DD, Lee WM, Cheng CY. Par3/Par6 polarity complex coordinates apical ectoplasmic specialization and blood–testis barrier restructuring during spermatogenesis. Proc Natl Acad Sci USA 2008;105:9657–62.
[75] Wong EWP, Sun S, Li MWM, Lee WM, Cheng CY. 14-3-3 protein regulates cell adhesion in the seminiferous epithelium of rat testes. Endocrinology 2009;150:4713–23.
[76] Su WH, Wong EWP, Mruk DD, Cheng CY. The Scribble/Lgl/Dlg polarity protein complex is a regulator of blood–testis barrier dynamics and spermatid polarity during spermatogenesis. Endocrinology 2012;153:6041–53.
[77] Su WH, Mruk DD, Wong EWP, Lui WY, Cheng CY. Polarity protein complex Scribble/Lgl/Dlg and epithelial cell barriers. Adv Exp Med Biol 2012;763:149–70.
[78] Assemat E, Bazellieres E, Pallesi-Pocachard E, Le Bivic A, Massey-Harroche D. Polarity complex proteins. Biochim Biophys Acta 2008;1778:614–30.
[79] Cheng CY, Mruk DD. A local autocrine axis in the testes that regulates spermatogenesis. Nat Rev Endocrinol 2010;6:380–95.
[80] Iden S, Collard JG. Crosstalk between small GTPases and polarity proteins in cell polarization. Nature Rev Mol Cell Biol 2008;9:846–59.
[81] Pointis G, Gilleron J, Carette D, Segretain D. Physiological and physiopathological aspects of connexins and communicating gap junctions in spermatogenesis. Philos Trans R Soc Lond B Biol Sci 2010;365:1607–20.
[82] Li MWM, Mruk DD, Lee WM, Cheng CY. Connexin 43 is critical to maintain the homeostasis of blood–testis barrier via its effects on tight junction reassembly. Proc Natl Acad Sci USA 2010;107:17998–8003.
[83] Pointis G, Gilleron J, Carette D, Segretain D. Testicular connexin 43, a precocious molecular target for the effect of environmental toxicants on male fertility. Spermatogenesis 2011;1:303–17.
[84] Wan HT, Mruk DD, Wong CKC, Cheng CY. Perfluorooctanesulfonate (PFOS) perturbs male rat Sertoli cell blood–testis barrier function by affecting F-actin organization via p-FAK-Tyr^{407}—an *in vitro* study. Endocrinology 2014;155:249–62.
[85] Li MWM, Mruk DD, Lee WM, Cheng CY. Connexin 43 and plakophilin-2 as a protein complex that regulates blood–testis barrier dynamics. Proc Natl Acad Sci USA 2009;106:10213–8.
[86] Yan HHN, Mruk DD, Wong EWP, Lee WM, Cheng CY. An autocrine axis in the testis that coordinates spermiation and blood–testis barrier restructuring during spermatogenesis. Proc Natl Acad Sci USA 2008;105:8950–5.
[87] Tsuruta D, Hashimoto T, Hamill KJ, Jones JC. Hemidesmosomes and focal contact proteins: functions and cross-talk in keratinocytes, bullous diseases and wound healing. J Dermatol Sci 2011;62:1–7.
[88] Fawcett D. Ultrastructure and function of the Sertoli cell. In: Hamilton D, Greep R, editors. Handbook of physiology. Washington, DC: American Physiological Society; 1975. p. 21–5.

[89] Fawcett DW, Leak LV, Heidger PM. Electron microscopic observations on the structural components of the blood–testis barrier. J Reprod Fertil 1970;(Suppl. 10): 105–22.
[90] Russell LD. Spermatid–Sertoli tubulobulbar complexes as devices for elimination of cytoplasm from the head region in late spermatids of the rat. Anat Rec 1979;194:233–46.
[91] Young JS, Takai Y, Kojic KL, Vogl AW. Internalization of adhesion junction proteins and their association with recycling endosome marker proteins in rat seminiferous epithelium. Reproduction 2012;143:347–57.
[92] Young JS, Vogl AW. Focal adhesion proteins zyxin and vinculin are co-distributed at tubulobulbar complexes. Spermatogenesis 2012;2:63–8.
[93] Yan HHN, Mruk DD, Lee WM, Cheng CY. Cross-talk between tight and anchoring junctions—lesson from the testis. Adv Exp Med Biol 2008;636:234–54.
[94] Xiao X, Mruk DD, Wong EWP, Lee WM, Han D, Wong CKC, et al. Differential effects of c-Src and c-Yes on the endocytic vesicle-mediated trafficking events at the Sertoli cell blood–testis barrier. Am J Physiol Endocrinol Metab 2014 (in press).
[95] Byers S, Jegou B, MacCalman C, Blaschuk O. Sertoli cell adhesion molecules and the collective organization of the testis. In: Russell LD, Griswold MD, editors. The Sertoli cell. Clearwater, FL: Cache River Press; 1993. p. 461–76.
[96] Hess RA, de Franca LR. Spermatogenesis and cycle of the seminiferous epithelium. Adv Exp Med Biol 2008;636:1–15.
[97] Parvinen M. Regulation of the seminiferous epithelium. Endocr Rev 1982;3:404–17.
[98] Leblond C, Clermont Y. Definition of the stages of the cycle of the seminiferous epithelium in the rat. Ann NY Acad Sci 1952;55:548–73.
[99] Russell L. Role in spermiation. In: Russell L, Griswold M, editors. The Sertoli cell. Clearwater, FL: Cache River Press; 1993. p. 269–303.
[100] Clermont Y. Kinetics of spermatogenesis in mammals: seminiferous epithelium cycle and spermatogonial renewal. Physiol Rev 1972;52:198–235.
[101] Russell LD. Further observations on tubulobulbar complexes formed by late spermatids and Sertoli cells in the rat testis. Anat Rec 1979;194:213–32.
[102] Russell LD, Clermont Y. Anchoring device between Sertoli cells and late spermatids in rat seminiferous tubules. Anat Rec 1976;185:259–78.
[103] Russell LD, Malone JP. A study of Sertoli–spermatid tubulobulbar complexes in selected mammals. Tissue Cell 1980;12:263–85.
[104] Young JS, Guttman JA, Vaid KS, Vogl AW. Cortactin (CTTN), N-WASP (WASL), and clathrin (CLTC) are present at podosome-like tubulobulbar complexes in the rat testis. Biol Reprod 2009;80:153–61.
[105] Xiao X, Wong EWP, Lie PPY, Mruk DD, Wong CKC, Cheng CY. Cytokines, polarity proteins and endosomal protein trafficking and signaling—the Sertoli cell blood–testis barrier *in vitro* as a study model. Methods Enzymol 2014;534:181–94.
[106] Cheng CY, Mruk DD. Regulation of spermiogenesis, spermiation and blood–testis barrier dynamics: novel insights from studies on Eps8 and Arp3. Biochem J 2011;435:553–62.
[107] Lie PPY, Chan AYN, Mruk DD, Lee WM, Cheng CY. Restricted Arp3 expression in the testis prevents blood–testis barrier disruption during junction restructuring at spermatogenesis. Proc Natl Acad Sci USA 2010;107:11411–6.
[108] Lie PPY, Mruk DD, Lee WM, Cheng CY. Epidermal growth factor receptor pathway substrate 8 (Eps8) is a novel regulator of cell adhesion and the blood–testis barrier integrity in the seminiferous epithelium. FASEB J 2009;23:2555–67.

[109] Young JS, Guttman JA, Vaid KS, Vogl AW. Tubulobulbar complexes are intercellular podosome-like structures that internalize intact intercellular junctions during epithelial remodeling events in the rat testis. Biol Reprod 2009;80:162–74.

[110] Lie PPY, Xia W, Wang CQF, Mruk DD, Yan HHN, Wong CH, et al. Dynamin II interacts with the cadherin- and occludin-based protein complexes at the blood–testis barrier in adult rat testes. J Endocrinol 2006;191:571–86.

[111] Vaid KS, Guttman JA, Babyak N, Deng W, McNiven MA, Mochizuki N, et al. The role of dynamin 3 in the testis. J Cell Physiol 2007;210:644–54.

[112] Lie PPY, Mruk DD, Cheng CY. Dynamins, spermatogenesis and contraceptive development. Immun Endocr Metab Agents Med Chem 2008;8:51–8.

[113] Siu MKY, Cheng CY. Interactions of proteases, protease inhibitors, and the β1 integrin/laminin γ3 protein complex in the regulation of ectoplasmic specialization dynamics in the rat testis. Biol Reprod 2004;70:945–64.

[114] Inagaki M, Irie K, Ishizaki H, Tanaka-Okamoto M, Miyoshi J, Takai Y. Role of cell adhesion molecule nectin-3 in spermatid development. Genes Cells 2006;11:1125–32.

[115] Ozaki-Kuroda K, Nakanishi H, Ohta H, Tanaka H, Kurihara H, Mueller S, et al. Nectin couples cell–cell adhesion and the actin scaffold at heterotypic testicular junctions. Curr Biol 2002;12:1145–50.

[116] Takai Y, Ikeda W, Ogita H, Rikitake Y. The immunoglobulin-like cell adhesion molecule nectin and its associated protein afadin. Annu Rev Cell Dev Biol 2008;24:309–42.

[117] Johnson K, Boekelheide K. Dynamic testicular adhesion junctions are immunologically unique. II. Localization of classic cadherins in rat testis. Biol Reprod 2002;66:992–1000.

[118] Lee NPY, Mruk DD, Lee WM, Cheng CY. Is the cadherin/catenin complex a functional unit of cell–cell–actin-based adherens junctions (AJ) in the rat testis? Biol Reprod 2003;68:489–508.

[119] Koch M, Olson P, Albus A, Jin W, Hunter D, Brunken W, et al. Characterization and expression of the laminin γ3 chain: a novel, non-basement membrane-associated, laminin chain. J Cell Biol 1999;145:605–18.

[120] Palombi F, Salanova M, Tarone G, Farini D, Stefanini M. Distribution of β1 integrin subunit in rat seminiferous epithelium. Biol Reprod 1992;47:1173–82.

[121] Salanova M, Ricci G, Boitani C, Stefanini M, De Grossi S, Palombi F. Junctional contacts between Sertoli cells in normal and aspermatogenic rat seminiferous epithelium contain α6β1 integrins, and their formation is controlled by follicle-stimulating hormone. Biol Reprod 1998;58:371–8.

[122] Salanova M, Stefanini M, De Curtis I, Palombi F. Integrin receptor α6β1 is localized at specific sites of cell-to-cell contact in rat seminiferous epithelium. Biol Reprod 1995;52:79–87.

[123] Yan HHN, Cheng CY. Laminin α3 forms a complex with β3 and γ3 chains that serves as the ligand for α6β1-integrin at the apical ectoplasmic specialization in adult rat testes. J Biol Chem 2006;281:17286–303.

[124] Mulholland DJ, Dedhar S, Vogl AW. Rat seminiferous epithelium contains a unique junction (ectoplasmic specialization) with signaling properties both of cell/cell and cell/matrix junctions. Biol Reprod 2001;64:396–407.

[125] Siu MKY, Cheng CY. Extracellular matrix: recent advances on its role in junction dynamics in the seminiferous epithelium during spermatogenesis. Biol Reprod 2004;71:375–91.

[126] Kuo JC. Mechanotransduction at focal adhesions: integrating cytoskeletal mechanics in migrating cells. J Cell Mol Med 2013;17:704–12.
[127] Wehrle-Haller B. Assembly and disassembly of cell matrix adhesions. Curr Opin Cell Biol 2012;24:569–81.
[128] Yao PL, Lin YC, Richburg JH. Mono-(2-ethylhexyl) phthalate-induced disruption of junctional complexes in the seminiferous epithelium of the rodent testis is mediated by MMP2. Biol Reprod 2010;82:516–27.
[129] Ekblom P, Lonai P, Talts JF. Expression and biological role of laminin-1. Matrix Biol 2003;22:35–47.
[130] Hamasaki H, Koga K, Aoki M, Hamasaki M, Koshikawa N, Seiki M, et al. Expression of laminin 5-γ2 chain in cutaneous squamouos cell carcinoma and its role in tumour invasion. Br J Cancer 2011;105:824–32.
[131] Monboisse JC, Oudart JB, Ramont L, Brassart-Pasco S, Maquart FX. Matrikines from basement membrane collagens: a new anti-cancer stragegy. Biochim Biophys Acta 2014;1840:2589–98.
[132] Okazaki I, Suzuki N, Nishi N, Utani A, Matsuura H, Shinkai H, et al. Identification of biologically active sequences in the laminin α 4 chain G domain. J Biol Chem 2002;277:37070–8.
[133] Senyurek I, Kempf WE, Klein G, Maurer A, Kalbacher H, Schafer L, et al. Processing of laminin α chains generates peptides invovled in wound healing and host defense. J Innate Immun 2014; PMID:24458132.
[134] Su L, Mruk DD, Lie PPY, Silvestrini B, Cheng CY. A peptide derived from laminin-γ3 reversibly impairs spermatogenesis in rats. Nat Commun 2012;3:1185. Available from: http://dx.doi.org/doi:10.1038/ncomms2171.
[135] Mazaud-Guittot S. Dissecting the phthalate-induced Sertoli cell injury: the fragile balance of proteases and their inhibitors. Biol Reprod 2011;85:1091–3.
[136] Yao PL, Lin YC, Richburg JH. TNFα-mediated disruption of spermatogenesis in response to Sertoli cell injury in rodents is partially regulated by MMP2. Biol Reprod 2009;80:581–9.
[137] Siu MKY, Lee WM, Cheng CY. The interplay of collagen IV, tumor necrosis factor-α, gelatinase B (matrix metalloprotease-9), and tissue inhibitor of metalloprotease-1 in the basal lamina regulates Sertoli cell-tight junction dynamics in the rat testis. Endocrinology 2003;144:371–87.
[138] Wong EWP, Cheng CY. NC1 domain of collagen α3(IV) derived from the basement membrane regulates Sertoli cell blood testis barrier dynamics. Spermatogenesis 2013;3: e25465. Available from: http://dx.doi.org/10.4161/spmg.25465. [PMID:23885308; PMCID:PMC23710226].
[139] Clermont Y, Leblond CP. Spermiogenesis of man, monkey, ram and other mammals as shown by the "periodic acid-Schiff" technique. Am J Anat 1955;96: 229–53.
[140] Leblond CP, Clermont Y. Spermiogenesis of rat, mouse, hamster, and guinea pig as revealed by the periodic acid-fuchin sulfurous acid technique. Am J Anat 1952;90:167–210.
[141] Redenbach DM, Vogl AW. Microtubule polarity in Sertoli cells: a model for microtubule-based spermatid transport. Eur J Cell Biol 1991;54:277–90.
[142] Redenbach D, Boekelheide K, Vogl A. Binding between mammalian spermatid–ectoplasmic specialization complexes and microtubules. Eur J Cell Biol 1992;59: 433–48.

[143] Redenbach DM, Boekelheide K. Microtubules are oriented with their minus-ends directed apically before tight junction formation in rat Sertoli cells. Eur J Cell Biol 1994;65:246–57.
[144] Redenbach DM, Hall ES, Boekelheide K. Distribution of Sertoli cell microtubules, microtubule-dependent motors, and the Golgi apparatus before and after tight junction formation in developing rat testis. Microsc Res Tech 1995;32:504–19.
[145] Tang EI, Mruk DD, Cheng CY. MAP/microtubule affinity-regulating kinases, microtubule dynamics, and spermatogenesis. J Endocrinol 2013;217:R13–23.
[146] Wang F, Zhang Q, Cao J, Huang Q, Zhu X. The microtubule plus end-binding protein EB1 is involvevd in Sertoli cell plasticity in testicular seminiferous tubules. Exp Cell Res 2008;314:213–26.
[147] Guttman JA, Kimel GH, Vogl AW. Dynein and plus-end microtubule-dependent motors are associated with specialized Sertoli cell junction plaques (ectoplasmic specializations). J Cell Sci 2000;113:2167–76.
[148] Hall ES, Eveleth J, Jiang C, Radenbach DM, Boekelheide K. Distribution of the microtubule-dependent motors cytoplasmic dynein and kinesin in rat testis. Biol Reprod 1992;46:817–28.
[149] Vaid KS, Guttman JA, Singaraja RR, Vogl AW. A kinesin is present at unique Sertoli/spermatid adherens junctions in rat and mouse testes. Biol Reprod 2007;77:1037–48.
[150] Kardon JR, Vale RD. Regulators of the cytoplasmic dynein motor. Nat Rev Mol Cell Biol 2009;10:854–65.
[151] Boutros T, Chevet E, Metrakos P. Mitogen-activated protein (MAP) kinase/MAP kinase phosphatase regulation: roles in cell growth, death, and cancer. Pharmacol Rev 2008;60:261–310.
[152] Cheng CY, Mruk DD. Regulation of blood–testis barrier dynamics by focal adhesion kinase (FAK). An unexpected turn of events. Cell Cycle 2009;8:3493–9.
[153] Lechertier T, Hodivala-Dilke K. Focal adhesion kinase and tumour angiogenesis. J Pathol 2012;226:404–12.
[154] Pentassuglia L, Sawyer DB. ErbB/integrin singaling interactions in regulation of myocardial cell–cell and cell–matrix interactions. Biochim Biophys Acta 2013;1833:909–16.
[155] Hu P, Luo BH. Integrin bi-directional signaling across the plasam membrane. J Cell Physiol 2013;228:306–12.
[156] Shen B, Delaney MK, Du X. Inside-out, outside-in, and inside-outside-in: G protein signaling in integrin-mediated cell adhesion, spreading, and retraction. Curr Opin Cell Biol 2012;24:600–6.
[157] Beardsley A, Robertson DM, O'Donnell L. A complex containing α6β1-integrin and phosphorylated focal adhesion kinase between Sertoli cells and elongated spermatids during spermatid release from the seminiferous epithelium. J Endocrinol 2006;190:759–70.
[158] Siu MKY, Mruk DD, Lee WM, Cheng CY. Adhering junction dynamics in the testis are regulated by an interplay of β1-integrin and focal adhesion complex (FAC)-associated proteins. Endocrinology 2003;144:2141–63.
[159] Wan HT, Mruk DD, Li SYT, Mok KW, Lee WM, Wong CKC, et al. p-FAK-Tyr^{397} regulates spermatid adhesion in the rat testis via its effects on F-actin organization at the ectoplasmic specialization. Am J Physiol Endocrinol Metab 2013;305:E687–99.
[160] Bruianek LE, Soderling SH. Under lock and key: spatiotemporal regulation of WASP family proteins coordinates separate dynamic cellular processes. Semin Cell Dev Biol 2013;24:258–66.

[161] Suetsugu S. Activation of nucleation promoting factors for directional actin filament elongation: allosteric regulation and multimerization on the membrane. Semin Cell Dev Biol 2013;24:267–71.
[162] Betschinger J, Knobloich JA. Dare to be different: asymmetric cell division in *Drosophila*, *C. elegans* and vertebrates. Curr Biol 2004;14:R674–85.
[163] Gonczy P. Mechanisms of asymmetric cell division: flies and worms pave the way. Nat Rev Mol Cell Biol 2008;9:355–66.
[164] Noatynska A, Tavernier N, Gotta M, Pintard L. Coordinating cell polarity and cell cycle progression: what can be learn from flies and worms? Open Biol 2013;3:130083. Available from: http://dx.doi.org/doi:10.1098/rsob.130083.
[165] Thompson BJ. Cell polarity: models and mechanisms from yeast, worms and flies. Development 2013;140:13–21.
[166] Bazzoun D, Lelievre S, Talhouk R. Polarity proteins as regulators of cell junction complexes: Implications for breast cancer. Pharmacol Ther 2013;138:418–27.
[167] Goldenring JR. A central role for vesicle trafficking in epithelial neoplasia: intracellular highways to carcinogenesis. Nat Rev Cancer 2013;13:813–20.
[168] Goldstein B, Macara IG. The PAR proteins: fundamental players in animal cell polarization. Dev Cell 2007;13:609–22.
[169] Goehring NW, Grill SW. Cell polarity: mechanochemical patterning. Trends Cell Biol 2013;23:72–80.
[170] Tepass U. The apical polarity protein network in Drosophila epithelial cells: regulation of polarity, junctions, morplhogenesis, cell growth, and survival. Annu Rev Cell Dev Biol 2012;28:655–85.
[171] Apodaca G, Gallo LI, Bryant DM. Role of membrane traffic in the generation of epithelial cell asymmetry. Nat Cell Biol 2012;14:1235–343.
[172] Wan G, Corfas G, Stone JS. Inner ear supporting cells: rethinking the silent majority. Semin Cell Dev Biol 2013;24:448–59.
[173] Elsum I, Yates L, Humbert PO, Richardson HE. The Scribble-Dlg-Lgl polarity module in development and cancer: from flies to man. Essays Biochem 2012;53: 141–68.
[174] Humbert PO, Dow LE, Russell SM. The Scribble and Par complexes in polarity and migration: friends or foes? Trends Cell Biol 2006;16:622–30.
[175] Ellenbroek SI, Iden S, Collard JG. Cell polarity proteins and cancer. Semin Cancer Biol 2012;22:208–15.
[176] Lee M, Vasioukhin V. Cell polarity and cancer—cell and tissue polarity as a non-canonical tumor suppressor. J Cell Sci 2008;121:1141–50.
[177] Clermont Y, Morales C, Hermo L. Endocytic activities of Sertoli cells in the rat. Ann NY Acad Sci 1987;513:1–15.
[178] Cheng CY, Wong EWP, Lie PPY, Li MWM, Su L, Siu ER, et al. Environmental toxicants and male reproductive function. Spermatogenesis 2011;1:2–13.
[179] Cosker KE, Segal RA. Neuronal signaling through endocytosis. Cold Spring Harb Perspect Biol 2014;6: [pii,a020669].
[180] Hu CT, Wu JR, Wu WS. The role of endosomal signaling triggered by metastatic growth factors in tumor progression. Cell Signal 2013;25:1539–45.
[181] Mahmutefendic H, Zagtorac GB, Tomas MI, Groettrup M, Momburg F, Lucin P. Endosomal trafficking of open Major Histocompatibility Class I conformers—implications for presentation of endocytosed antigens. Mol Immunol 2013;55:149–52.
[182] Mellman I, Yarden Y. Endocytosis and cancer. Cold Spring Harb Perspect Biol 2013;5:a016949.

[183] Mruk DD, Cheng CY. The myotubularin family of lipid phosphatases in disease and in spermatogenesis. Biochem J 2011;433:253–62.

[184] Mruk DD, Xiao X, Lydka M, Li MWM, Bilinska B, Cheng CY. Intercellular adhesion molecule 1: recent findings and new concepts involved in mammalian spermatogenesis. Semin Cell Dev Biol 2014;29:43–54.

[185] Rainero E, Norman JC. Late endosomal and lysomal trafficking during integrin-mediated cell migration and invasion: cell matrix receptors are trafficked through the late endosomal pathway in a way that dictates how cells migrate. Bioessays 2013;35:523–32.

[186] Seaman MN. The retromer complex—endosomal protein recycling and beyond. J Cell Sci 2012;125:4693–702.

[187] Taguchi T. Emerging roles of recycling endosomes. J Biochem 2013;153:505–10.

[188] Wu LG, Hamid E, Shin W, Chiang HC. Exocytosis and endocytosis: modes, function, and coupling mechanisms. Annu Rev Physiol 2014;76:301–31.

[189] Upadhyay RD, Kumar AV, Ganeshan M, Balasinor NH. Tubulobulbar complex: cytoskeletal remodeling to release spermatozoa. Reprod Biol Endocrinol 2012;10:27. Available from: http://dx.doi.org/doi:10.1186/1477-7827-10-27.

[190] Mruk D, Lau A, Conway A. Crosstalk between Rab GTPases and cell junctions. Contraception 2005;72:280–90.

[191] Sharpe RM, McKinnell C, Kivlin C, Fisher JS. Proliferation and functional maturation of Sertoli cells, and their relevance to disorders of testis function in adulthood. Reproduction 2003;125:769–84.

[192] Auharek SA, Avelar GF, Lara NLM, Sharpe RM, Franca LR. Sertoli cell numbers and spermatogenic efficency are increased in inducible nitric oxide synthase (iNOS) mutant mice. Int J Androl 2011;34:e621–9.

[193] Berndtson WE, Thompson TL. Changing relationships between testis size, Sertoli cell number and spermatogenesis in Sprague-Dawley rats. J Androl 1990;11:429–35.

[194] Lie PPY, Cheng CY, Mruk DD. Interleukin-1α is a regulator of the blood–testis barrier. FASEB J 2011;25:1244–53.

[195] Clermont Y, Harvey SC. Duration of the cycle of the seminiferous epithelium of normal, hypophysectomized and hypophysectomized hormone treated albino rats. Endocrinology 1965;76:80–9.

[196] Miaczynska M, Bar-Sagi D. Signaling endosomes: seeing is believing. Curr Opin Cell Biol 2010;22:535–40.

[197] Miaczynska M. Effects of membrane trafficking on signaling by receptor tyrosine kinases. Cold Spring Harb Perspect Biol 2013;5:a009035.

[198] Xiao X, Mruk DD, Cheng CY. c-Yes regulates cell adhesion at the apical ectoplasmic specialization blood–testis barrier axis via its effects on protein recruitment and distribution. Am J Physiol Endocrinol Metab 2013;304:E145–59.

[199] Xiao X, Mruk DD, Lee WM, Cheng CY. c-Yes regulates cell adhesion at the blood–testis barrier and the apical ectoplasmic specialization in the seminiferous epithelium of rat testes. Int J Biochem Cell Biol 2011;43:651–65.

[200] Lamb CA, Dooley HC, Tooze SA. Endocytosis and autophagy: shared machinery for degradation. Bioessays 2013;35:34–45.

[201] Xiong W, Wang H, Wu HM, Chen Y, Han D. Apoptotic spermatogenic cells can be energy sources for Sertoli cells. Reproduction 2009;137:469–79.

[202] Gordon S. Elie Metchnikoff: father of natural immunity. Eur J Immunol 2008;38:3257–64.

[203] Murray PJ, Wynn TA. Protective and pathogenic functions of macrophage subsets. Nat Rev Immunol 2011;11:723–37.

[204] Rougerie P, Miskolci V, Cox D. Generation of membrane structures during phagocytosis and chemotaxis of macrophages: role and regulation of the actin cytoskeleton. Immunol Rev 2013;256:222–39.

[205] Griffin FMJ, Griffin JA, Silvestein SC. Studies on the mechanism of phagocytosis. II. The interacction of macrophages with anti-immunoglobulin G-coated bone marrow-derived lymphocytes. J Exp Med 1976;144:788–809.

[206] Griffin FM, Silverstein SC. Segmental response of the macrophage plasma membrane to a phagocytic stimulus. J Exp Med 1974;139:323–36.

[207] Aderem A, Underhill DM. Mechanisms of phagocytosis in macrophages. Annu Rev Immunol 1999;17:593–623.

[208] Champion JA, Mitragotri S. Role of target geometry in phagocytosis. Proc Natl Acad Sci USA 2006;103:4930–4.

[209] Regaud C. Etudes sur la structure des tube seminiferes et sur la spermatogenese chez les mammiferes. Arch Anat Microsc Morphol Exp 1901;4(101–156):231–380.

[210] Quintar AA, Nishioka S, Torres AI, Maldonado CA, Aoki A. The initial segment of the rat epididymis is able to uptake immature germ cells shed by testicular damage. Cell Tissues Organs 2011;193:170–83.

[211] Ramos-Ibeas P, Pericuesta E, Fernandez-Gonzalez R, Ramirez MA, Gutierrez-Adan A. Most regions of mouse epididymis are able to phagocytose immature germ cells. Reproduction 2013;146:481–9.

[212] Sutovsky P, Moreno R, Ramalho-Santos J, Dominko T, Thompson WE, Schatten G. A putative, ubiquitin-dependent epididymal mechanism for the recognition and elimination of defective spermatozoa in mammalian epididymis. J Cell Sci 2001;114:1665–75.

[213] Cooper TG, Yeung CH, Jones R, Orgebin-Crist MC, Robaire B. Rebuttal of a role for the epididymis in sperm quality control by phagocytosis of defective sperm. J Cell Sci 2002;115:5–7.

[214] Maeda Y, Shiratsuchi A, Namiki M, Nakanishi Y. Inhibition of sperm production in mice by annexin V microinjection into seminiferous tubules: possible etiology of phagocytic clearance of apoptotic spermatogenic ccells and male infertility. Cell Death Differ 2002;9:742–9.

[215] Morales C, Clermont Y, Hermo L. Nature and function of endocytosis in Sertoli cells of the rat. Am J Anat 1985;173:203–17.

[216] Morales C, Clermont Y, Nadler NJ. Cyclic endocytic activity and kinetics of lysosomes in Sertoli cells of the rat: a morphometric analysis. Biol Reprod 1986;34:207–18.

[217] Niedergang F, Chavrier P. Regulation of phagocytosis by Rho GTPases. Curr Top Microbiol Immunol 2005;291:43–60.

[218] Sayedyahossein S, Dagnino L. Integrins and small GTPases as modulators of phagocytosis. Int Rev Cell Mol Biol 2013;302:321–54.

[219] Park H, Ishihara D, Cox D. Regulation of tyrosine phosphorylation in macrophage phagocytosis and chemotaxis. Arch Biochem Biophys 2011;510:101–11.

[220] Chemes H. The phagocytic function of Sertoli cells: a morphological, biochemical, and endocrinological study of lysosomes and acid phosphatase localization in the rat testis. Endocrinology 1986;119:1673–81.

[221] Miething A. Germ-cell death during prespermatogenesis in the testis of the golden hamster. Cell Tissue Res 1992;267:583–90.

[222] Mizuno K, Shiratsuchi A, Masamune Y, Nakanishi Y. The role of Sertoli cells in the differentiation and exclusion of rat testicular germ cells in primary culture. Cell Death Differ 1996;3:119–23.

[223] Pineau C, Le Magueresse B, Courtens JL, Jegou B. Study *in vitro* of the phagocytic function of Sertoli cells in the rat. Cell Tissue Res 1991;264:589–98.

[224] Shiratsuchi A, Kawasaki Y, Ikemoto M, Arai H, Nakanishi Y. Role of class B scavenger receptor type I in phagocytosis of apoptotic rat spermatogenic cells by Sertoli cells. J Biol Chem 1999;274:5901–8.

[225] Wang H, Wang H, Xiong W, Chen Y, Ma Q, Ma J, et al. Evaluation on the phagocytosis of apoptotic spermatogenic cells by Sertoli cells *in vitro* through detecting lipid droplet formation by Oil Red O staining. Reproduction 2006;132:485–92.

[226] Huckins C. The morphology and kinetics of spermatogonial degeneration in normal adult rats: an analysis using a simplified classification of the germinal epithelium. Anat Rec 1978;190:905–26.

[227] Johnson L, Petty CS, Neaves WB. Further quantification of human spermatogenesis: germ cell loss during postprophase of meiosis and its relationship to daily sperm production. Biol Reprod 1983;29:207–15.

[228] Oakberg EF. A description of spermiogenesis in the mouse and its use in analysis of the cycle of the seminiferous epithelium and germ cell renewal. Am J Anat 1956;99:391–413.

[229] Orth JM, Gunsalus GL, Lamperti AA. Evidence from Sertoli cell-depleted rats indicates that spermatid number in adults depends on numbers of Sertoli cells produced during perinatal development. Endocrinology 1988;122:787–94.

[230] Orth JM. Proliferation of Sertoli cells in fetal and postnatal rats: a quantitative autoradiographic study. Anat Rec 1982;203:485–92.

[231] Weber JE, Russell LD, Wong V, Peterson RN. Three dimensional reconstruction of a rat stage V Sertoli cell: II. Morphometry of Sertoli–Sertoli and Sertoli–germ cell relationships. Am J Anat 1983;167:163–79.

[232] Wong V, Russell LD. Three-dimensional reconstruction of a rat stage V Sertoli cell: I. Methods, basic configuration, and dimensions. Am J Anat 1983;167:143–61.

[233] Billig H, Furuta I, Rivier C, Tapanainen J, Parvinen M, Hsueh AJ. Apoptosis in testis germ cells: developmental changes in gonadotropin dependence and localization to selective tubule stages. Endocrinology 1995;136:5–12.

[234] Shaha C. Estrogens and spermatogenesis. Adv Exp Med Biol 2008;636:42–64.

[235] Caraux A, Lu Q, Feernandez N, Riou S, di Santo JP, Raulet DH, et al. Natural killer cell differntiation driven by Tyro3 receptor tyrosine kinases. Nat Immunol 2006;7:747–54.

[236] Lemke G, Lu Q. Macrophage regulation by Tyro 3 family receptors. Curr Opin Immunol 2003;15:31–6.

[237] Lu Q, Gore M, Zhang Q, Camenisch T, Boast S, Casagranda F, et al. Tyro-3 family receptors are essential regulators of mammalian spermatogenesis. Nature 1999; 398:723–8.

[238] Lu Q, Lemke G. Homeostatic regulation of the immune system by receptor tyrosin kinases of the Tyro 3 family. Science 2001;293:306–11.

[239] Lu Q, Li Q, Lu Q. Regulation of phagocytosis by TAM receptors and their ligands. Front Biol 2010;5:227–37.

[240] Prasad D, Rothlin CV, Burrola P, Burstyn-Cohen T, Lu Q, Garcia de Frutos P, et al. TAM receptor function in the retinal pigment epithelium. Mol Cell Neurosci 2006;33:96–108.

[241] Crosier KE, Crosier PS. new insights into the control of cell growth: the role of the Axl family. Pathology 1997;29:131–5.
[242] Lemke G, Rothlin CV. Immunobiology of the TAMP receptors. Nat Rev Immunol 2008;8:327–36.
[243] Rothlin CV, Ghosh S, Zuniga EI, Oldstone MB, Lemke G. TAM receptors are pleiotropic inhibitors of the innate immune response. Cell 2007;131:1124–36.
[244] Stitt TN, Conn G, Gore M, Lai C, Bruno J, Radziejewski C, et al. The anticoagulation factor protein S and its relative, Gas6, are ligands for the Tyro3/Axl family of receptor tyrosine kinases. Cell 1995;80:661–70.
[245] DiScipio RG, Davie EW. Characterization of protein S, a gamma-carboxyglutamic acid containing protein from bovine and human plasma. Biochemistry 1979;18: 899–904.
[246] Heeb MJ, Mesters RM, Tans G, Rosing J, Griffin JH. Binding of protein S to factor Va associated with inhibition of prothrombinase that is independent of activated protein C. J Biol Chem 1993;268:2872–7.
[247] Burstyn-Cohen T, Heeb MJ, Lemke G. Lack of protein S in mice causes embryonic lethal coagulapathy and vascular dysgenesis. J Clin Invest 2009;119:2942–53.
[248] Dahlback B. The tale of protein S anbd C4b-binding protein, a story of affection. Thromb Haemost 2007;98:90–6.
[249] Carrera Silva EA, Chan PY, Joannas L, Errasti AE, Gagliani N, Bosurgi L, et al. T cell-pderived protein S engages TAM receptor signaling in dendritic cells to control the magnitude of the immune response. Immunity 2013;39:169–70.
[250] Chen Y, Wang H, Qi N, Wu H, Xiong W, Ma J, et al. Functions of TAM RTKs in regulating spermatogenesis and male fertility in mice. Reproduction 2009;138:655–66.
[251] Xiong W, Chen Y, Wang H, Wang H, Wu H, Lu Q, et al. Gas6 and the Tyro 3 receptor tyrosine kinase subfamily regulate the phagocytic function of Sertoli cells. Reproduction 2008;135:77–87.
[252] Wang H, Chen Y, Ge Y, Ma P, Ma Q, Ma J, et al. Immunoexpression of Tyro 3 family receptors—Ayro 3, Axl, and Mer—and their ligand Gas6 in postnatal developing mouse testis. J Histochem Cytochem 2005;53:1355–64.
[253] Manfioletti G, brancolini C, Avanzi G, Schneider C. The protein encoded by a growth arrest specific gene (gas6) is a new member of the vitamin K-dependent proteins related to protein S, a negative coregulator in the blood coagulation cascade. Mol Cell Biol 1993;13:4976–85.
[254] Hall MO, Obin MS, Heeb MJ, Burgess BL, Abrams TA. Both protein S and Gas6 stimulate outer segment phagocytosis by cultured rat retinal pigment epithelial cells. Exp Eye Res 2005;81:581–91.
[255] Murao K, Terpstra V, Green SR, Kondratenko N, Steinberg D, Quehenberger O. Characterization of CLA-1, a human homologue of rodent scavenger receptor BI, as a receptor for high density lipoprotein and apoptotic thymocytes. J Biol Chem 1997;272:17551–7.
[256] Kawasaki Y, Nakagawa A, Nagaosa K, Shiratsuchi A, Nakanishi Y. Phosphatidylserine binding of class B scavenger receptor type 1, a phagocytosis receptor of testicular Sertoli cells. J Biol Chem 2002;277:27559–66.
[257] Nakanishi Y, Shiratsuchi A. Phagocytic removal of apoptotic spermatogenic cells by Sertoli cells: mechanisms and consequences. Bill Pharm Bull 2004;27:13–6.
[258] Fukasawa M, Adachi H, Hirota K, Tsujimoto M, Arai H, Inoue K. SRB1, a class B scavenger receptor, recognizes both negatively charged liposomes and apoptotic cells. Exp Cell Res 1996;222:246–50.

[259] Rigotti A, Acton SL, Krieger M. The class B scavenger receptors SR-BI and CD36 are receptors for anionic phospholipids. J Biol Chem 1995;270: 16221–4.
[260] Shiratsuchi A, Umeda M, Ohba Y, Nakanishi Y. Recognition of phosphatidylserine on the surface of apoptotic spermastogenic cells and subsequent phagocytosis by Sertoli cells of the rat. J Biol Chem 1997;272:2354–8.
[261] Schlegel RA, Williamson P. Phosphatidylserine, a death knell. Cell Death Differ 2001;8:551–63.
[262] Niedergang F, Chavrier P. Signaling and membrane dynamics during phagocytosis: many roads lead to the phagos(R)ome. Curr Opin Cell Biol 2004;16: 422–8.
[263] Angelillo-Scherrer A, Garcia de Frutos P, Aparicio C, Melis E, Savi P, Lupu F, et al. Deficiency or inhibition of Gas6 causes platelet dysfunction and protects mice against thrombosis. Nat Med 2001;7:215–21.
[264] Setchell BP. Blood–testis barrier, functional and transport proteins and spermatogenesis. Adv Exp Med Biol 2008;636:212–33.
[265] Setchell BP, Waites GMB. The blood–testis barrier. In: Hamilton DW, Greep RO, editors. The handbook of physiology section 7, Vol. V: male reproductive system. Washington, DC: American Physiological Society; 1975. p. 143–72.
[266] Kaur G, Thompson LA, Dufour JM. Sertoli cells—immunological sentinels of spermatogenesis. Sem Cell Dev Biol 2014;30:36–44.
[267] Meinhardt A, Hedger MP. Immunological, paracrine and endocrine aspects of testicular immune privilege. Mol Cell Endocrinol 2011;335:60–8.
[268] Cheng YH, Wong EWP, Cheng CY. Cancer/testis (CT) antigens, carcinogenesis and spermatogenesis. Spermatogenesis 2011;1:209–20.
[269] Mahi-Brown CA, Yule TD, Tung KS. Evidence for active immunological regulation in prevention of testicular autoimmune disease independent of the blood–testis barrier. Am J Reprod Immunol Microbiol 1988;16:165–70.
[270] Yule TD, Montoya GD, Russell LD, Williams TM, Tung KSK. Autoantigenic germ cells exist outside the blood testis barrier. J Immunol 1988;141:1161–7.
[271] Dufour JM, Dass B, Halley KR, Korbutt GS, Dixon DE, Rajotte RV. Sertoli cell line lacks the immunoprotective properties associated with primary Sertoli cells. Cell Transplant 2008;17:525–34.
[272] Dufour JM, Gores P, Hemendinger R, Emerich DF, Halberstadt CR. Transgenic Sertoli cells as a vehicle for gene therapy. Cell Transplant 2004;13:1–6.
[273] Selawry HP, Cameron DF. Sertoli cell-enriched fractions in successful islet cell transplantation. Cell Transplant 1993;2:123–9.
[274] Shamekh R, El-Badri NS, Saporta S, Pascual C, Sanberg PR, Cameron DF. Sertoli cells induce systemic donor-specific tolerance in xenogenic transplantation model. Cell Transplant 2006;15:45–53.
[275] Toyama Y, Ohkawa M, Oku R, Maekawa M, Yuasa S. Neonatally administered diethylstilbestrol retards the development of the blood–testis barrier in the rat. J Androl 2001;22:413–23.
[276] Mok KW, Mruk DD, Lee WM, Cheng CY. Spermatogonial stem cells alone are not sufficient to re-initiate spermatogenesis in the rat testis following adjudin-induced infertility. Int J Androl 2012;35:86–101.
[277] Fawcett DW. Intercellular bridges. Exp Cell Res 1961;8:174–87.
[278] Hamer G, Roepers-Gajadien HL, Gademan IS, Kal HB, De Rooij DG. Intercellular bridges and apoptosis in clones of male germ cells. Int J Androl 2003;26:348–53.

[279] Ventela S, Toppari J, Parvinen M. Intercellular organelle traffic through cytoplasmic bridges in early spermatids of the rat: mechanisms of haploid gene product sharing. Mol Biol Cell 2003;14:2768–80.
[280] Gerdes HH, Carvalho RN. Intercellular transfer mediated by tunneling nanotubes. Curr Opin Cell Biol 2008;20:470–5.
[281] Gerdes HH, Rustom A, Wang X. Tunneling nanotubes, an emerging intercellular communication route in development. Mech Dev 2013;130:381–7.
[282] Kimura S, Hase K, Ohno H. The molecular basis of induction and formation of tunneling nanotubes. Cell Tissue Res 2013;352:67–76.
[283] Tarakanov AO, Goncharova LB. Cell–cell nanotubes: tunneling through several types of synapses. Commun Integr Biol 2009;2:359–61.
[284] Abounit S, Zurzolo C. Wiring through tunneling nanotubes—from electrical signals to organelles transfer. J Cell Sci 2012;125:1089–98.
[285] Morales CR, Wu XQ, Hecht NB. The DNA/RNA-binding protein, TB-RBP, moves from the nucleus to the cytoplasm and through intercellular bridges in male germ cells. Dev Biol 1998;201:113–23.
[286] Twiss JL, Fainzilber M. Ribosomes in axons—scrounging from neighbors? Trends Cell Biol 2009;19:236–43.
[287] Russell LD. The blood–testis barrier and its formation relative to spermatocyte maturation in the adult rat: a lanthanum tracer study. Anat Rec 1978;190:99–112.
[288] Chung NPY, Cheng CY. Is cadmium chloride-induced inter-Sertoli tight junction permeability barrier disruption a suitable *in vitro* model to study the events of junction disassembly during spermatogenesis in the rat testis? Endocrinology 2001;142:1878–88.
[289] Janecki A, Jakubowiak A, Steinberger A. Effect of cadmium chloride on transepithelial electrical resistance of Sertoli cell monolayers in two-compartment cultures—a new model for toxicological investigations of the "blood–testis" barrier *in vitro*. Toxicol Appl Pharmacol 1992;112:51–7.
[290] Meng J, Holdcraft RW, Shima JE, Griswold MD, Braun RE. Androgens regulate the permeability of the blood–testis barrier. Proc Natl Acad Sci USA 2005;102:16670–96.
[291] Siu ER, Wong EWP, Mruk DD, Porto CS, Cheng CY. Focal adhesion kinase is a blood–testis barrier regulator. Proc Natl Acad Sci USA 2009;106:9298–303.
[292] Wang RS, Yeh S, Chen LM, Lin HY, Zhang C, Ni J, et al. Androgen receptor in Sertoli cell is essential for germ cell nursery and junction complex formation in mouse testes. Endocrinology 2006;147:5624–33.
[293] Wang RS, Yeh S, Tzeng CR, Chang C. Androgen receptor roles in spermatogenesis and fertility: lessons from testicular cell-specific androgen receptor knockout mice. Endocr Rev 2009;30:119–32.
[294] Li MWM, Xia W, Mruk DD, Wang CQF, Yan HHY, Siu MKY, et al. TNFα reversibly disrupts the blood–testis barrier and impairs Sertoli-germ cell adhesion in the seminiferous epithelium of adult rat testes. J Endocrinol 2006;190:313–29.
[295] Lui WY, Lee WM, Cheng CY. Transforming growth factor-β3 perturbs the inter-Sertoli tight junction permeability barrier *in vitro* possibly mediated via its effects on occludin, zonula occludens-1, and claudin-11. Endocrinology 2001;142:1865–77.
[296] Lui WY, Lee WM, Cheng CY. Transforming growth factor-β3 regulates the dynamics of Sertoli cell tight junctions via the p38 mitogen-activated protein kinase pathway. Biol Reprod 2003;68:1597–612.

[297] Lui WY, Wong CH, Mruk DD, Cheng CY. TGF-β3 regulates the blood–testis barrier dynamics via the p38 mitogen activated protein (MAP) kinase pathway: an *in vivo* study. Endocrinology 2003;144:1139–42.

[298] Wong CH, Mruk DD, Lui WY, Cheng CY. Regulation of blood–testis barrier dynamics: an *in vivo* study. J Cell Sci 2004;117:783–98.

[299] Xia W, Mruk DD, Lee WM, Cheng CY. Differential interactions between transforming growth factor-β3/TβR1, TAB1, and CD2AP disrupt blood–testis barrier and Sertoli–germ cell adhesion. J Biol Chem 2006;281:16799–813.

[300] Yan HHN, Cheng CY. Blood–testis barrier dynamics are regulated by an engagement/disengagement mechanism between tight and adherens junctions via peripheral adaptors. Proc Natl Acad Sci USA 2005;102:11722–7.

[301] Su L, Cheng CY, Mruk DD. Adjudin-mediated Sertoli-germ cell junction disassembly affects Sertoli cell barrier function *in vitro* and *in vivo*. Int J Biochem Cell Biol 2010;42:1864–75.

[302] Xia W, Cheng CY. TGF-β3 regulates anchoring junction dynamics in the seminiferous epithelium of the rat testis via the Ras/ERK signaling pathway: an *in vivo* study. Dev Biol 2005;280:321–43.

[303] Brown EJ, Albers MW, Shin TB, Ichikawa K, Keith CT, Lane WS, et al. A mammalian protein targeted by G1-arresting rapamycin-receptor complex. Nature 1994;369:756–8.

[304] Chiu MI, Katz H, Berlin V. RAPT1, a mammalian homolog of yeast Tor, interqacts with the FKBP12/rapamycin complex. Proc Natl Acad Sci USA 1994;91:12574–8.

[305] Sabatini DM, Erdjument-Bromage H, Lui M, Tempst P, Snyder SH. RAFT1: a mammalian protein that binds to FKBP12 in a rapamycin-dependent fashion and is homologous to yeast TORs. Cell 1994;78:35–43.

[306] Francipane MG, Lagasse E. mTOR pathway in colorectal cancer: an update. Oncotarget 2014;5:49–66.

[307] Goodman CA. The role of mTORC1 in regulating protein synthesis and skeletal muscle mass in response to various mechanical stimuli. Rev Physiol Biochem Pharmacol 2014;166:43–95.

[308] Husseinzadeh N, Husseinzadeh HD. mTOR inhibitors and their clinical application in cervical, endometrial and ovarian cancers: a critical review. Gynecol Oncol 2014;133:375–81.

[309] Inoki K. mTOR signaling in autophagy regulation in the kidney. Semin Nephrol 2014;34:2–8.

[310] Laplante M, Sabatini DM. mTOR signaling in growth control and disease. Cell 2012;49:274–93.

[311] Martinet W, De Loof H, de Meyer GR. mTOR inhibition: a promising strategy for stabilization of atherosclerotic plaques. Atherosclerosis 2014;233:601–7.

[312] Dazert E, Hall MN. mTOR signaling in disease. Curr Opin Cell Biol 2011;23:744–55.

[313] Mruk DD, Zhu LJ, Silvestrini B, Lee WM, Cheng CY. Interactions of proteases and protease inhibitors in Sertoli-germ cell cocultures preceding the formation of specialized Sertoli-germ cell junctions *in vitro*. J Androl 1997;18:612–22.

[314] Campo-Engelstein L. Contraceptive justice: why we need a male pill. Virtual Mentor 2012;14:146–51.

[315] Cheng CY, Mruk DD. Why do we need male contraceptives? Spermatogenesis 2013;3:e24098.

[316] Dorman E, Bishai D. Demand for male contraception. Expert Rev Pharmacoecon Outcomes Res 2012;12:605–13.

[317] Chen YM, Lee NPY, Mruk DD, Lee WM, Cheng CY. Fer kinase/Fer *T* and adherens junction dynamics in the testis: an *in vitro* and *in vivo* study. Biol Reprod 2003;69:656–72.

[318] Cheng CY, Mruk DD, Silvestrini B, Bonanomi M, Wong CH, Siu MKY, et al. AF-2364 [1-(2,4-dichlorobenzyl)-*1H*-indazole-3-carbohydrazide] is a potential male contraceptive: a review of recent data. Contraception 2005;72:251–61.

[319] Cheng CY, Silvestrini B, Grima J, Mo MY, Zhu LJ, Johansson E, et al. Two new male contraceptives exert their effects by depleting germ cells prematurely from the testis. Biol Reprod 2001;65:449–61.

[320] Mruk DD, Cheng CY. Cell–cell interactions at the ectoplasmic specialization in the testis. Trends Endocrinol Metab 2004;15:439–47.

[321] Mruk DD, Cheng CY. Delivering non-hormonal contraceptives to men: advances and obstacles. Trends Biotechnol 2008;26:90–9.

[322] Wolski KM, Mruk DD, Cameron DF. The Sertoli-spermatid junctional complex adhesion strength is affected *in vitro* by adjudin. J Androl 2006;27:790–4.

[323] Tash JS, Attardi B, Hild SA, Chakrasali R, Jakkaraj SR, Georg GI. A novel potent indazole carboxylic acid derivative blocks spermatogenesis and is contraceptive in rats after a single oral dose. Biol Reprod 2008;78:1127–38.

[324] Tash JS, Chakrasali R, Jakkaraj SR, Hughes J, Smith SK, Hornbaker K, et al. Gamendazole, an orally active indazole carboxylic acid male contraceptive agent, targets HSP90AB1 (HSP90BHETA) and EEF1A1 (eEF1A), and stimulates *Il1a* transcription in rat Sertoli cells. Biol Reprod 2008;78:1139–52.

Sertoli cell structure and function in anamniote vertebrates

13

Luiz R. França[a,], Rafael H. Nóbrega[b], Roberto D.V.S. Morais[b], Luiz H. De Castro Assis[b] and Rüdiger W. Schulz[b,*]*
[a]Department of Morphology, Institute of Biological Sciences, Federal University of Minas Gerais, Belo Horizonte-MG, Brazil, [b]Department Biology, Science Faculty, Utrecht University, Utrecht, The Netherlands

I. Introduction

The "raison d'être" of Sertoli cells in the adult testis is to facilitate the survival and development of male germ cells so that spermatogenesis can provide spermatozoa in sufficient numbers to guarantee fertility. This function is conserved across vertebrates. However, taxonomic groups at the basis of the phylum—fish and amphibians collectively referred to as anamniote vertebrates—differ from the amniote vertebrates in having the cystic type of spermatogenesis. The main differences between cystic and noncystic spermatogenesis are summarized in Figure 13.1. In cystic spermatogenesis, Sertoli cell structure and function have a number of characteristics that are very interesting from a comparative point of view and provide new insights into Sertoli cell physiology. This chapter focuses on specific aspects of Sertoli cells in cystic spermatogenesis that are relevant for understanding important Sertoli cell functions in vertebrates in general.

Testicular anatomy in amphibians and fish (except for sharks, which do not have steroidogenic Leydig cells in the interstitial compartment) follows the general vertebrate pattern of two compartments, germinative and interstitial, that are separated from each other by a basement membrane and peritubular myoid cells [1]. The interstitial compartment contains steroid hormone-producing Leydig cells, blood vessels, macrophages and mast cells, neural and connective tissue elements that are continuous with the tunica albuginea, and the testis organ wall. The (intra)tubular compartment houses the germinal epithelium that holds two cell types, the somatic Sertoli cells and the germ cells. Via the pituitary gonadotropic hormones, follicle-stimulating hormone (FSH), and luteinizing hormone (LH), interstitial steroidogenic Leydig cells as well as Sertoli cells, expressing the receptors for FSH and LH, are integrated into endocrine regulatory circuits governing puberty and adult functioning of the testis. The survival and development of germ cells *in vivo* depends on their close and continuous interaction with Sertoli cells so that the Sertoli cell number limits the spermatogenic capacity of a testis [2]. This limitation renders

*Communicating authors: de França and Schulz

Sertoli Cell Biology. DOI: http://dx.doi.org/10.1016/B978-0-12-417047-6.00013-2

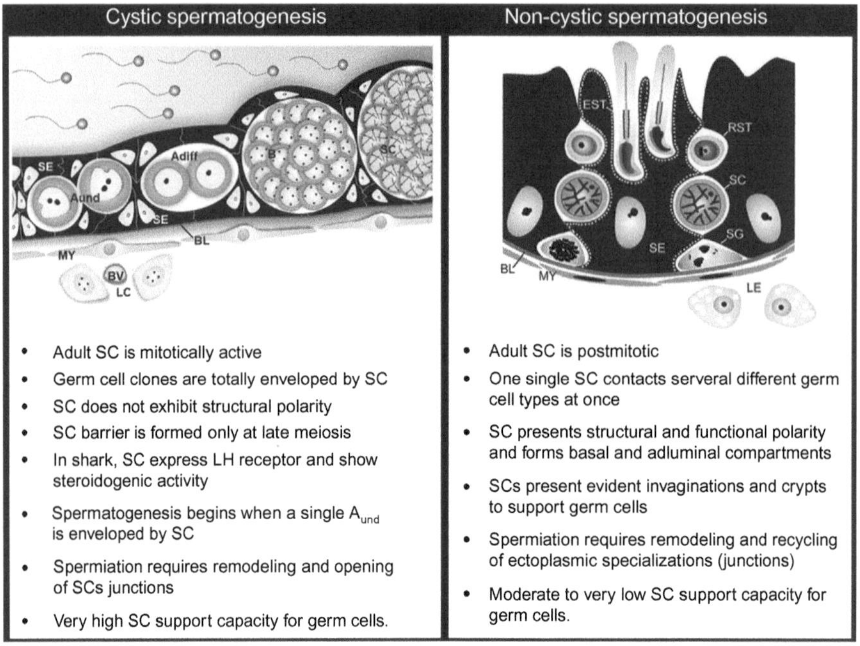

Figure 13.1 Schematic representation of the main differences between Sertoli cells in cystic and noncystic spermatogenesis. A_{diff}, type A differentiated spermatogonia; A_{und}, type A undifferentiated spermatogonia; B, type B spermatogonia; BL, basal lamina; BV, blood vessel; EST, elongated spermatid; LE or LC, Leydig cells; MY, peritubular myoid cells; RST, round spermatid; SC, spermatocytes; SE, Sertoli cell; SG, spermatogonia. Part of the figure was published previously [1].

determination of Sertoli cell number one of the most important aspects of testis physiology, and we discuss the occurrence and regulation of Sertoli cell proliferation in cystic spermatogenesis. Sertoli cell proliferation is accompanied by progressive morphological and functional differentiation that is discussed, along with information on the regulation of these processes.

II. Sertoli cell proliferation

A. Development of existing spermatogenic cysts

The manners of Sertoli cell proliferation differ between amniote and anamniote vertebrates, reflecting a major difference in the Sertoli cell/germ cell relationship.

In birds and mammals, Sertoli cells proliferate until puberty so that in the adult testis, a given number of resident, postmitotic Sertoli cells support successive waves of spermatogenesis. During these waves, a given Sertoli cell supports several different developmental stages of germ cells at a time. Typically, at the basolateral

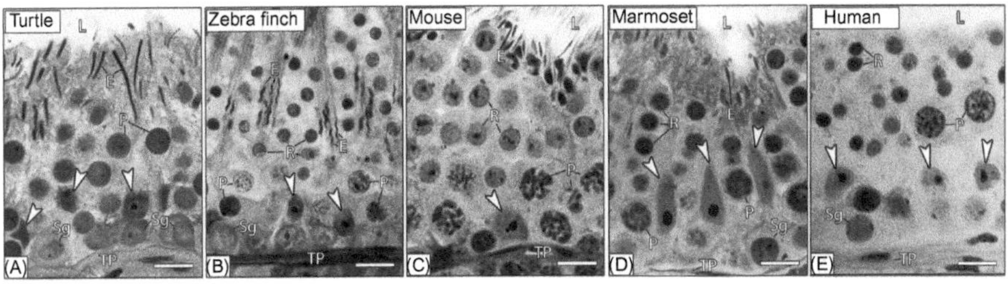

Figure 13.2 Seminiferous tubule cross sections of noncystic spermatogenesis in a turtle (A), bird, zebra finch (B), mouse (C), marmoset (D), and human (E), showing Sertoli cell nuclei (arrowhead), spermatogonia (Sg), pachytene spermatocytes (P), and round (R) and elongated (E) spermatids. Also shown are the tunica propria (TP) and lumen (L). In this mode of spermatogenesis, one single Sertoli cell contacts several different germ cell types at the same time, and the Sertoli cell barrier delineates two compartments in the seminiferous epithelium: the basal compartment in contact with the tunica propria, where spermatogonia and early spermatocytes are located; and the adluminal compartment containing spermatocytes and spermatids. Bars = 20 μm.

surface (basal compartment), mammalian Sertoli cells contact spermatogonia and early spermatocytes. Inside the blood–testis barrier, lateral parts contact spermatocytes and early (round) spermatids, while apical/adluminal parts contact late (elongating) spermatids (Figure 13.2 and Chapter 1).

In anamniote vertebrates (fishes and amphibians), in contrast, the functional unit of the germinal epithelium is the spermatogenic cyst (Figure 13.3). This cyst consists initially of a single spermatogonium enveloped by the cytoplasmic extensions of Sertoli cells. When this initial cyst enters the spermatogenic process, both the cyst-forming Sertoli cells and germ cells proliferate and differentiate. A species-specific number of Sertoli cells is associated with cysts at specific stages of spermatogenesis, thereby establishing predictable Sertoli cell/germ cell ratios for a given stage of germ cell development [2–5]. Because Sertoli cells associated with growing cysts are positive in proliferation assays, at least part of the increase in Sertoli cell number per cyst is attributed to the division of these cells [4]. The major increase in cyst volume and Sertoli cell number per cyst occurs during the mitotic expansion of spermatogonia [2,4,5]. Therefore, unlike in amniote vertebrates, Sertoli cells are not postmitotic cells but proliferate in the anamniote testis. The situation is complicated somewhat by the fact that a Sertoli cell can contact more than one germ cell clone on the two sides of a cytoplasmic extension (Figure 13.4), although this ability is less complex than the typical situation of Sertoli cells in adult birds and mammals (Figure 13.2).

It seems that in all vertebrate species, testis function requires a functional Sertoli cell barrier and the interaction between the morphological and physiological barrier components to efficiently regulate the entry and exit of molecules [6]. This immunological barrier leads to an immune-privileged site that protects germ cells from immunological destruction [6]. The Sertoli cell barrier function is relatively well studied in mammals but still requires careful investigation in lower vertebrates. In fish, Sertoli cell proliferation stops when the germ cell clone completes meiosis and

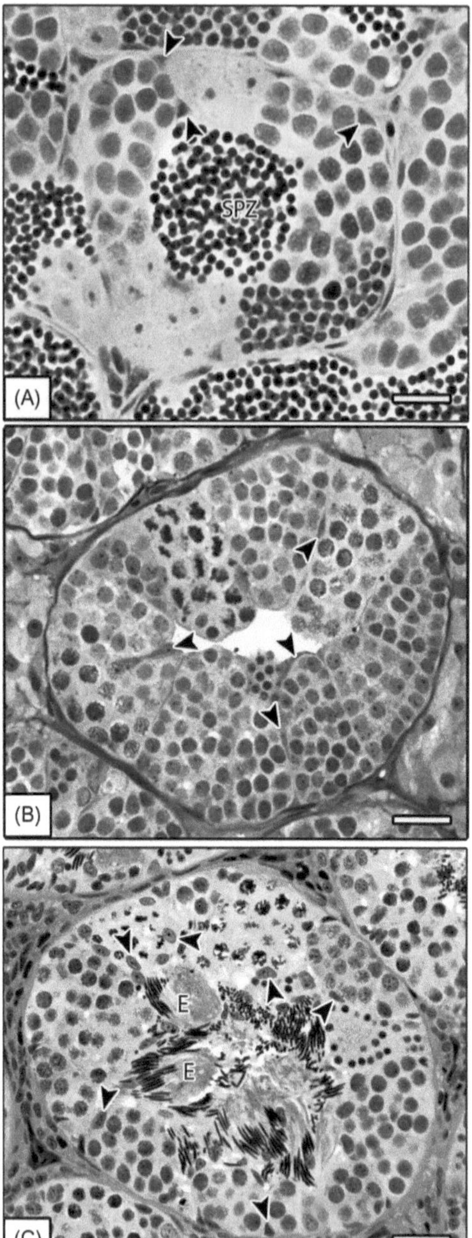

Figure 13.3 Seminiferous tubule cross sections showing cystic spermatogenesis in different phases of development and enveloped by Sertoli cells (arrowhead) in zebrafish (A), Nile tilapia (B), and amphibian bullfrogs (C). The zebrafish tubular lumen is usually filled with sperm (SPZ), while the elongate spermatids (E) cysts are open in bullfrogs but are still contacting SCs. Bars = 100 μm.

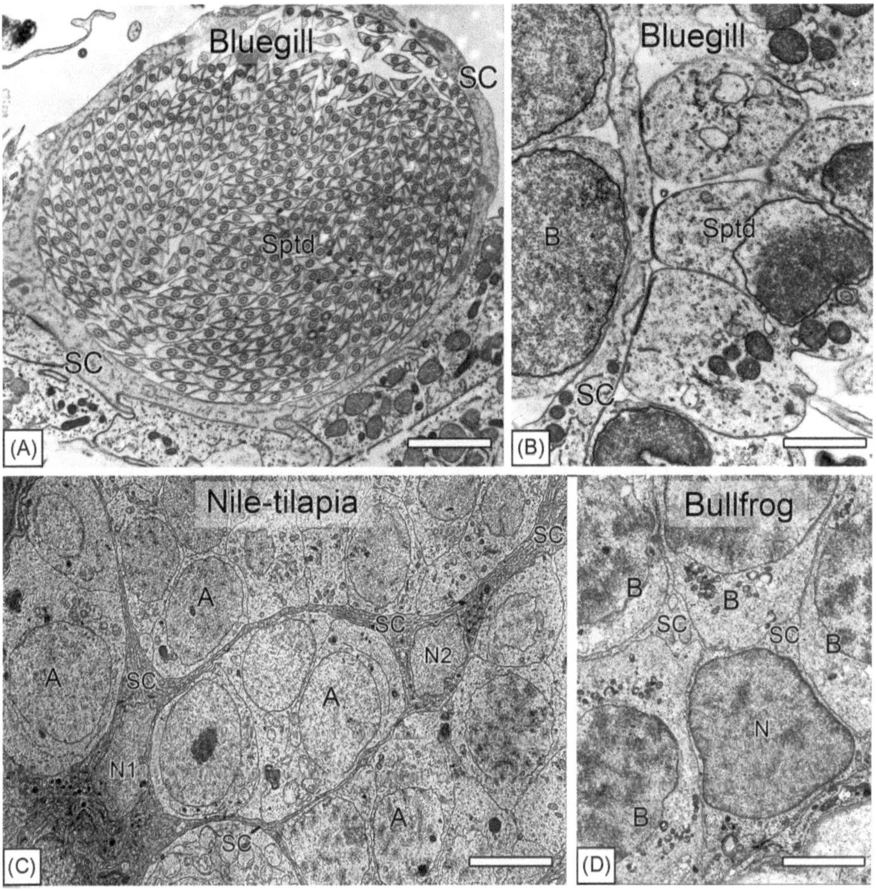

Figure 13.4 Seminiferous epithelium ultrastructure in fish and amphibians showing Sertoli cells (in purple) and germ cells. Strikingly, in the bluegill fish (A), a Sertoli cell is enveloping hundreds of spermatids (Sptd). This figure also shows that one Sertoli cell is able to contact different cysts, which can be observed in the bluegill (B), where Sertoli cell cytoplasm is facing type B spermatogonia (B) and Sptd. In the Nile tilapia (C), one Sertoli cell is apparently facing different type A spermatogonial (A) cysts. Sertoli cell cytoplasmic projections (D) are also observed between different germ cells in a bullfrog type B spermatogonial cyst. N, Sertoli cell nucleus. Bars: A = 2 μm; B = 5 μm; C = 5 μm; D = 3 μm. (For interpretation of the references to color in this figure legend, the reader is referred to the web version of this book.)

enters spermiogenesis, at which time tight junctions are also established among the Sertoli cells [2,4,5] (Figure 13.5). A similar trend has been observed in studies in bullfrogs in our laboratory. Because meiotic germ cells in fish are apparently not shielded from the vascular system, a tight Sertoli cell barrier seems not to be necessary for meiosis in fish. However, our knowledge of the barrier structure and function in fish is rather fragmentary. For instance, as shown for zebrafish

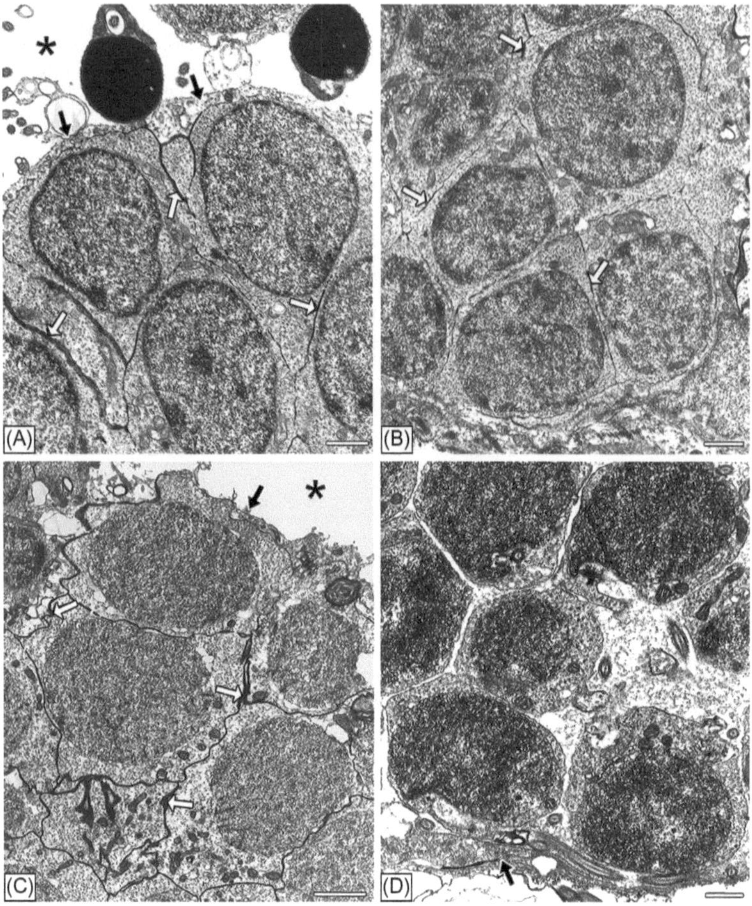

Figure 13.5 Ultrastructure of zebrafish testis using lanthanum as a tracer to investigate the Sertoli cell barrier. In cystic spermatogenesis, the barrier is not observed in cysts containing type B spermatogonia (A) or primary spermatocytes cysts at the initial stages of meiosis (leptotene/zygotene) (B and C) because lanthanum is present between the germ cells (white arrows in A–C). (D) Lanthanum is found at the level of tight junctions between Sertoli cells in spermatid cysts (black arrow) and are not present between the germ cells and in the lumen (asterisks in A and C). The black arrows in A and C indicate the Sertoli cell cytoplasm enveloping the cyst. Bars in A and B = 1 μm, C = 2 μm, D = 0.5 μm. The figure was published previously [5].

(Figure 13.5), even without a functional barrier, lanthanum, a tracer used to investigate the barrier efficacy, is never found in the tubular lumen. With respect to proliferation activity, however, Sertoli cells seem to behave in a similar way throughout vertebrates, given that Sertoli cells stop proliferating when tight junctions have been established. In this way, a predictable spermatid/Sertoli cell ratio is established. Interestingly, this ratio decreases during vertebrate evolution, being around

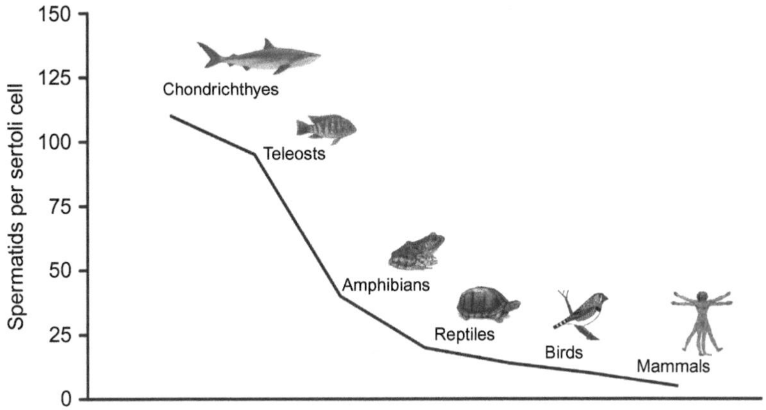

Figure 13.6 Number of spermatids per Sertoli cell, based on the available literature, for different vertebrate groups. This figure illustrates a progressive decrease in Sertoli cell efficiency during evolution.

100 in fish but decreasing to <10 in most mammalian species, reaching about 4 in humans (Figure 13.6).

After completion of spermiogenesis, the cyst opens to release sperm into the tubular lumen (spermiation). The seminiferous epithelium in anamniote vertebrates is composed of spermatogenic cysts in which different germ cell clones (i.e., descendants of different stem cells, usually in different stages of development) are taken care of by Sertoli cells enveloping the clone until spermiation. This finding seems to imply that an individual Sertoli cell is not a resident but rather a transient cell in the anamniote testis (see section III D).

B. Generation of new spermatogenic cysts—Sertoli cell progenitors

In addition to the Sertoli cell proliferation that accompanies the development of spermatogenic cysts, another mode of Sertoli cell proliferation takes place in the fish testis in the context of the production of new spermatogenic cysts. Because germ cells cannot survive unless they are in close contact with Sertoli cells, we hypothesize that the first new Sertoli cells are produced, thereby creating a niche into which a newly formed, single type A_{und} (type A undifferentiated) spermatogonium, a spermatogonial stem cell (SSC) candidate can be recruited. This assumption explains the observation made in the seasonally reproducing Atlantic salmon, for example, that at the beginning of the testis growth phase, groups of Sertoli cells appear that are not (yet) in contact with germ cells or several Sertoli cells are grouped around a single germ cell (Figure 13.7). In this setting, it seems possible that similar to recent observations in mice [7], Sertoli cells produce a paracrine factor, glial cell-lined derived neurotrophic factor (GDNF) that increases the number of SSCs in the vicinity of these Sertoli cells, for example, by stimulating SSC

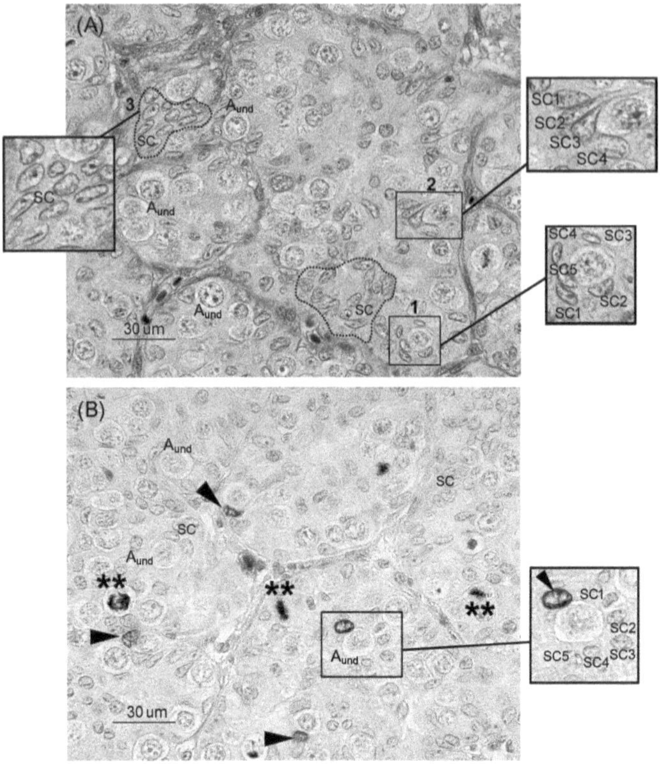

Figure 13.7 Sertoli cell groups in Atlantic salmon testis at the beginning of the seasonal testis growth phase. (A) **Squares (1,2)** show several Sertoli cells, indicated by numbered Sertoli cell nuclei that are grouped around a single type A undifferentiated spermatogonium ($\mathbf{A_{und}}$). Areas delineated by **dashed lines** (e.g., **3**) show Sertoli cells apparently not (yet) in contact with germ cells (possibly containing Sertoli cell progenitor cells). Staining: Hematoxylin and eosin. (B) Immunocytochemical detection of a proliferation marker [phosphorylated histone H3 (pH3)] on another section of the same testis shown in A. pH3-positive Sertoli cell nuclei (**arrowheads**) indicate proliferation of Sertoli cells at the beginning of testis growth. The square highlights several Sertoli cells (**SC1–SC5**) that are already associated with a single spermatogonium type A_{und}. Sertoli cells continue proliferating, potentially providing niche space for new, single type A_{und} spermatogonia (cyst formation). Double asterisks indicate pH3-positive single type A_{und} spermatogonia.

self-renewal divisions or by attracting SSCs from other areas. Interestingly, in rainbow trout, Gdnf and its receptor have been detected in spermatogonia type A_{und}, suggesting that this factor can also function in an autocrine loop [8]. Studies in adult zebrafish have revealed a regulatory link (via thyroid hormone and a growth factor of the Igf family; see below) between the proliferation of Sertoli cells and of single type A_{und} spermatogonia [9]. Therefore, our current model of cyst formation assumes that Sertoli cells are first generated to provide new niche space that can then be occupied by single type A_{und} spermatogonia.

New cysts are produced at different stages of the life cycle, and the intensity of cyst production varies with the reproductive strategy of a species, in particular in seasonally reproducing species. Regardless of the reproductive strategy, cyst generation takes place in all species between completion of male sex differentiation and puberty. This slow growth populates the prepubertal testis that often grows not more than allometrically with cysts containing mainly type A_{und} spermatogonia. Because many fish species display lifelong growth, a certain level of (allometric) testis growth associated with the generation of new cysts and Sertoli cells continues after puberty.

In seasonally reproducing species, large (e.g., 50-fold) changes in testis weight are recorded [10,11]. These changes reflect the production of new spermatogenic cysts associated with a relatively slow growth of the testis at the beginning of a seasonal cycle. This growth becomes rapid when the cysts go through spermatogenesis, and germ cells and the associated Sertoli cells proliferate and differentiate until the cysts open to release spermatozoa into the tubular lumen. At a certain point in time, the generation of new cysts ceases and existing cysts continue to develop and complete spermatogenesis, such that the fully mature testis in these species typically shows spermatogenic tubuli with lumina filled with spermatozoa and a small number of quiescent cysts with a single type A_{und} spermatogonium, representing the start-up reservoir for the next season's spermatogenic wave.

The testis weight built up during spermatogenic activity declines in a stepwise way. The first step occurs during completion of spermiogenesis, when Sertoli cells remove residual bodies while the production of new cysts stops. The second step of testis weight decrease is due to the use of sperm during the spawning season. Testis weight finally returns to low levels when residual spermatozoa are phagocytized by Sertoli cells after completion of the spawning season in preparation for the start of the next seasonal growth period. Thus, in seasonally reproducing species, Sertoli cell proliferation for the formation of new cysts is activated at the beginning of the seasonal testicular growth phase and is turned off to initiate the completion of the ongoing spermatogenic wave. In contrast, many species living in (sub)tropical zones display continuous spermatogenesis after puberty; prominent examples used as experimental models are medaka, tilapia, and zebrafish. In these species, Sertoli cell proliferation in the context of the formation of new cysts is observed continuously after puberty.

We hypothesize that the new Sertoli cells required for the generation of new spermatogenic cysts are derived from a Sertoli cell precursor population that potentially has stem cell characteristics. There is no formal proof for this hypothesis yet, but circumstantial evidence suggests the presence of an undifferentiated population of somatic cells in the testis. One line of evidence is the long-term capacity to produce new Sertoli cells during successive reproductive cycles. Another line of evidence is the fully functional sex reversal in adults, for example, in female goldfish with androgen-induced female-to-male sex change [12], in estrogen-induced male-to-female sex change in medaka [13], or in the natural sexual plasticity in sequentially hermaphroditic species [14]. These observations suggest that an undifferentiated somatic precursor cell population can produce cells that have a male (Sertoli) or female (granulosa) cell fate, depending on the environment in which they are situated. Clearly, germ cell-supporting somatic cells (or their precursors) in fish show an astonishing developmental plasticity in terms of sexual fate and proliferation capacity.

C. Intratesticular sites of Sertoli cell proliferation

Where does Sertoli cell proliferation in the context of cyst production take place in the testis? Depending on the species, cysts with a single type A_{und} spermatogonium, among which are SSCs, are distributed apparently at random throughout the germinal epithelium or are restricted to the periphery of the testis near the tunica albuginea [15]. In the first case, the cysts show little mobility during spermatogenesis. In the latter case, the cysts move away from the tunica as the germ cells divide and differentiate toward the region of the spermatic duct located centrally in the testis, where spermiation occurs and the cysts open to release spermatozoa. This arrangement is typical for teleost orders that are considered more evolved (atheriniformes, cyprinodontiformes, and beloniformes), while the former arrangement is found in less evolved taxonomic groups, such as in characiformes, cypriniformes, and salmoniformes [16]. Intermediate situations exist in perciformes (e.g., tilapia, *Oreochromis niloticus* [17]) or some pleuronectiformes (e.g. sole, *Solea senegalensis* [18]), where cysts with single type A_{und} spermatogonia have a preferred, but not exclusive, location close to the tunica albuginea. Close inspection of the Atlantic cod *Gadus morhua* revealed that new cysts are formed in a germinative zone in the periphery of the testicular parenchyma, which is divided into several lobes arranged around an efferent duct [11]. This structure results in a zonation of the lobes because early stages of development take place in the periphery, while advanced stages occur close to the efferent duct. The setting is based on appositional growth that is fuelled by the cyst-generating activity of the germinative zone rather than through movement of developing cysts. Hence, positional cues for the formation of new cysts exist in certain species, such that progenitor germ and Sertoli cells are likely to be found close to the tunica in the periphery of the testis. In the bluehad wrasse *Thalassoma bifasciatum*, it is possible that cellular elements from the tunica can differentiate into Sertoli cell progenitors [19].

However, in species in which the cysts with single type A_{und} spermatogonia (i.e., the potential site for the production of new cysts) are distributed throughout the testis, such as zebrafish, the intratubular location of these cysts is not random. As in rodents [20–22] and other mammals, such as horses [23] and peccaries [24], single type A_{und} spermatogonia in zebrafish preferentially locate to areas of the seminiferous tubules near blood vessels and the interstitium [25].

D. Regulation of Sertoli cell proliferation

The two modes of Sertoli cell proliferation, one in the context of the production of new spermatogenic cysts (mode 1) and the other in the context of the growth and development of existing cysts (mode 2), are regulated differently (Figure 13.8). This conclusion is based on the observation that these modes can take place

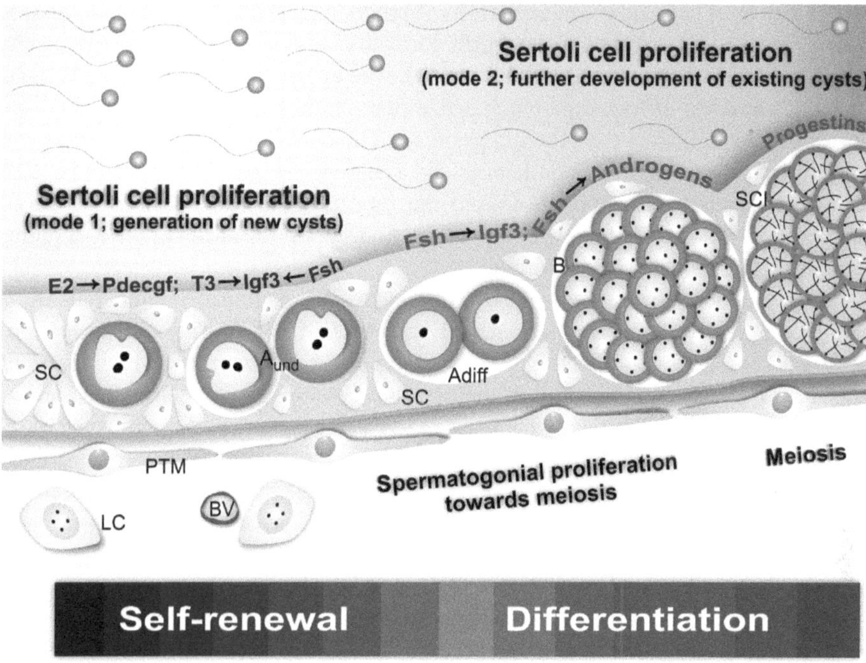

Figure 13.8 Schematic representation of Sertoli cell proliferation in relation to endocrine and paracrine regulation of fish spermatogenesis. Generation of new cysts occurs by **mode 1** of Sertoli cell proliferation, while **mode 2** accompanies the development of existing cysts. Proliferation in **mode 1** can occur when Sertoli cells are not (yet) in contact with germ cells or when multiple Sertoli cells contact a single type A undifferentiated spermatogonium (**A_{und}**). In zebrafish, cysts with single type A_{und} spermatogonia are preferentially found close to the interstitial area (Leydig cells [LC]) and blood vessels [25] (BVs). We hypothesize that mode 1 of Sertoli cell proliferation increases niche space into which single type A_{und} spermatogonia can be recruited for the generation of new cysts. **Mode 1** is stimulated by thyroid hormone, in part by increases in the production and release of Sertoli cell-derived insulin-like growth factor 3 [9] (**Igf3**). Because *igf3* gene expression is strongly stimulated by follicle-stimulating hormone, this gonadotropin may also be relevant to mode 1, as is the case for estrogens via stimulation of Sertoli cell production of Pdecgf [26]. **Mode 2** of Sertoli cell proliferation accompanies the development of existing cysts from the expansion of the spermatogonial population until the end of meiosis/beginning of spermiogenesis and stops when tight junctions are formed between Sertoli cells in teleost testes [2,4,5]. During spermatogonial proliferation toward meiosis, FSH stimulates Sertoli cell Igf3 and LC androgen production, which both promote cyst development. In the meiotic phase, progestins might be involved in **mode 2**. Sertoli cell (**SC**), type A undifferentiated spermatogonia (**A_{und}**), type A differentiated spermatogonia (**A_{diff}**), type B spermatogonia (**B**), primary spermatocyte (**SCI**), Leydig cell (**LC**), peritubular myoid cell (**PTM**), and blood vessel (**BV**).

independently of each other. In seasonally reproducing species, such as the Atlantic salmon, Sertoli cell proliferation for cyst formation (mode 1) starts earlier than the second mode of Sertoli cell proliferation. The latter mode commences once cysts that have started to differentiate reach a size that requires increasing the Sertoli cell number to accommodate the growing germ cell clones. When mode 2 of Sertoli cell proliferation starts, mode 1 is still ongoing. Later in the season, when mode 1 of Sertoli cell proliferation stops at around the time that the testis reaches its maximum weight, mode 2 of Sertoli cell proliferation continues in the context of completing the development of existing cysts.

The production of new cysts also requires the production of single type A_{und} spermatogonia derived from self-renewal divisions of SSCs. The self-renewal of SSCs is observed when they divide in a specific niche environment, while displacement from the niche would result in the tipping of the signaling balance toward differentiating cell division (see Chapter 4). SSCs are among the single type A_{und} spermatogonia that are present in all vertebrates; however, in mice, paired or aligned undifferentiated spermatogonia can also retain stem cell potency [27]. In the cystic type of spermatogenesis, SSCs are not in contact with the basement membrane and are enveloped by Sertoli cells, which are thought to contribute importantly to the niche characteristics. To survive, a newly formed, single type A_{und} spermatogonium must recruit its own Sertoli cell support to form a new spermatogenic cyst. As shown in Figure 13.7, Sertoli cells proliferate excessively at the beginning of the testicular growth phase in the seasonally reproducing Atlantic salmon, which transiently leads to a situation in which a single type A_{und} spermatogonium is either surrounded by several Sertoli cells that nevertheless keep proliferating, or in which Sertoli cells form groups that are not in contact with germ cells (yet).

What is the available information regarding the regulation of mode 1 of Sertoli cell proliferation? Studies in different fish species have shown that a low dose of estrogen stimulates SSC self-renewal (eel [26]; huchen [28]; medaka [29]). In eel, this effect involved estrogen receptor-dependent expression of a Sertoli cell-derived growth factor [26]. Although it has not been studied directly, we believe that this stimulatory effect of low estrogen doses is associated with the production of new spermatogenic cysts which must involve mode 1 of Sertoli cell proliferation.

Thyroid hormones exert well-known (inhibitory) effects on Sertoli cell proliferation in mammals [30]. Studies in zebrafish showed, however, that in contrast to findings in rodents, triiodothyronine (T3) increased the mitotic index of Sertoli cells, in particular of Sertoli cells not associated with germ cells (yet) or associated with type A spermatogonia. Moreover, proliferation of type A_{und} spermatogonia was stimulated as well. Jointly, these observations suggest that T3 increases the available niche space by stimulating mode 1 of Sertoli cell proliferation, resulting in the production of new spermatogenic cysts containing a single type A_{und} spermatogonium [9]. T3-stimulated proliferation involves the Igf signaling system because the T3 effect on proliferation is abrogated fully (A_{und}) or partially (Sertoli cell) by an Igf receptor inhibitor. Interestingly, fish gonads express the *igf3* gene, a new Igf family member [31] that may have arisen from the *igf1* gene in a neofunctionalization event in the context of the teleost-specific whole genome duplication.

Thyroid hormone receptors are expressed by Sertoli cells in zebrafish [9], as is Igf3 protein/mRNA (see below), and exposure to T3 increased testicular *igf3* mRNA levels so that T3-stimulated proliferation of spermatogonia and (part of) Sertoli cell proliferation are mediated by Sertoli cell-derived Igf3.

A point not studied yet in this concept is that, as mentioned above, we believe that mode 1 of Sertoli cell proliferation is fueled by a Sertoli cell progenitor or stem cell population. Therefore, we must postulate a niche for Sertoli cell progenitor cells as well. Signaling molecules that are potentially relevant to the SSC niche and are derived from other somatic elements (e.g., myoid, Leydig, or endothelial cells) but also from Sertoli cell progenitors themselves might be relevant to a potential Sertoli cell niche in the fish testis.

The *igf3* gene provides a link to the next aspect of regulation of Sertoli cell proliferation, the role of gonadotropins, given that FSH is a more potent stimulator of *igf3* gene expression than T3 in the zebrafish testis [32]; stimulatory effects of FSH on *igf3* mRNA have also been described for other species (e.g., rainbow trout [33]).

With regard to the biological activity of piscine gonadotropins, cellular localization data show that Leydig cells express the receptors for FSH and LH, which both stimulate steroidogenesis directly, while Sertoli cell functions are predominantly regulated by FSH [1]. Consequently, an important difference for the situation in higher vertebrates is that FSH in fish is also a potent steroidogenic hormone, while also regulating Sertoli cell activities. Most information on circulating FSH levels in male fish is available from salmonid species [34–36]; these species have annual reproductive cycles or their reproduction occurs only once in a lifetime. Transiently elevated FSH blood levels typically are associated with the period of spermatogonial proliferation at the beginning of the testis growth phase. FSH blood levels increase again when spermiation becomes increasingly prominent, but they decrease before the actual spawning season commences, at which time prominent increases in plasma LH levels occur.

In mammals, FSH is an important regulator of Sertoli cell proliferation and adult Sertoli cell function. Many of these functions were uncovered by studies that analyzed experimentally induced or spontaneous FSHR loss-of-function mutations in animal models or humans [37].

What do we know about the possible roles of FSH as a direct regulator of Sertoli cell activities in fish? With respect to Sertoli cell proliferation, it is tempting to speculate that elevated FSH plasma levels at the start of spermatogonial proliferation are important. In support of this concept, we found that incubating adult zebrafish testis tissue with recombinant zebrafish FSH in the presence of trilostane (which inhibits the production of biologically active steroid hormones) stimulated the proliferation of Sertoli cells and of type A_{und} and type A_{diff} spermatogonia (Nóbrega, Morais, França, Schulz, Bogerd; unpublished results), suggesting that not only can new cysts be formed but also that the further differentiation of existing cysts is stimulated. Experiments in prepubertal male sea bass that were injected with recombinant FSH point in the same direction. Next to plasma androgen levels, testis growth was stimulated, along with Sertoli cell and germ cell proliferation; postmeiotic stages of development were reached 23 days after injection [38]. In adult African catfish, surgical removal of one testis induced increased androgen release and *fshr* mRNA levels in the contralateral, remaining testis, which were associated with increased proliferation activity of Sertoli cells and spermatogonia type A_{und}, that is compatible with activated FSH signaling [39].

As mentioned above, FSH is a potent steroidogenic hormone in fish and induces the testicular release of 11-ketotestosterone (11KT), the main androgen in fish [40,41]. In a tissue culture system with prepubertal eel testis (which is rich in type A spermatogonia), incubation with 11KT induced full spermatogenesis, such that all stages of germ cell development were present [42]. This stimulatory effect of 11KT has been observed in other species as well (e.g., Japanese huchen [28] and zebrafish [43]). Although the effects of FSH/androgen on mode 2 of Sertoli cell proliferation, which accompanies the growth of differentiating cysts, have not been studied directly yet, we predict that FSH/androgen stimulation of germ cell proliferation toward advanced spermatogonial generations will also stimulate mode 2 of Sertoli cell proliferation.

A progestin is relevant to spermatogenesis in fish. There are two peaks of circulating 17α,20β-dihydroxy-4-pregnen-3-one (DHP) during the reproductive cycle of salmonid fish—a prominent peak in the spawning season and a small peak during the progression of spermatogonial proliferation and entry into meiosis [44–46]. Studies in Japanese eel showed that DHP stimulated spermatogonial DNA synthesis and was required to initiate meiosis [47]. A strong, stimulatory effect of DHP on spermatogenesis was also reported in adult zebrafish [48]. In this study, spermatogenesis was first interrupted by estrogen-induced androgen insufficiency and was then restarted by exposure to DHP, which induced proliferation of type A and B spermatogonia and entry into meiosis.

In an individual spermatogenic cyst, the Sertoli cell number increases steadily during the mitotic phase and levels off during meiosis to reach its maximum value upon completion of meiosis and the start of spermiogenesis, when tight junctions are also established between neighboring Sertoli cells [2,5]. We have explained that estrogens, thyroid hormones, and FSH can stimulate mode 1 of Sertoli cell proliferation, while FSH, androgens, and progestins can stimulate mode 2 of Sertoli cell proliferation (Figure 13.8), which stops when spermiogenesis starts. Leaving the level of a single cyst towards the tissue level in seasonally reproducing species, mode 1 of Sertoli cell proliferation (i.e., the production of new spermatogenic cysts) stops at a certain moment, after which mode 2 continues until all cysts have entered spermiogenesis. When spermiation follows, these species typically show spermatogenic tubules filled with spermatozoa and a limited number of quiescent cysts with single type A_{und} spermatogonia scattered along the basement membrane and awaiting the start of the subsequent spermatogenic wave [1]. Studies in Atlantic salmon showed that the cessation of the production of new cysts was associated with a decrease in the expression of *fshbeta* mRNA in the pituitary gland [49]. Because FSH plasma levels decrease in salmonids at this stage [35], one possible explanation for the cessation of the production of new cysts is reduced FSH stimulation. However, it is also possible that FSH receptor expression decreases in Sertoli cells in association with type A_{und} spermatogonia in testes that have reached their seasonal maximum weight, which might bear some similarity to the stage-dependent (i.e., induced by the specific germ cell composition) changes occurring in FSH receptor mRNA levels in adult rat testis [50].

III. Sertoli cell functions

After discussing different aspects of Sertoli cell proliferation, we now review Sertoli cell functions, starting with the function of these cells as paracrine communicators that translate endocrine signals into locally acting signals, and potentially also respond to signals from the germ cells.

A. Paracrine relay station

With respect to germ cell signaling to Sertoli cells, experimental evidence has yet to be published in fish. However, ongoing ultrastructural work in our laboratory showed that a single Sertoli cell can be in contact with different germ cell clones in quite different stages of spermatogenesis (Figure 13.5). We believe that Sertoli cells in this situation receive signals from the two different germ cell clones and respond appropriately. In rats, the density of undifferentiated spermatogonia modulates growth factor production by Sertoli cells [51], and the changing germ cell composition during the epithelial cycle is associated with changes in FSH receptor expression [50]. In zebrafish, the transforming growth factor (Tgf) β family member anti-Müllerian hormone (Amh) is produced by Sertoli cells in contact with type A spermatogonia. However, *amh* gene expression is downregulated when Sertoli cells contact type B spermatogonia or at later stages of germ cell development [43]. As mentioned above, there is no direct experimental evidence available yet to show that germ cells modulate Sertoli cell functions in fish, so we turn to the opposite direction of information flow, Sertoli cell signaling to germ cells.

When we discussed the regulation of Sertoli cell proliferation, we introduced the concept of the major endocrine input received by Sertoli cells through FSH and via the steroidogenic activity of (in fish) FSH and LH. All three types of sex steroids modulate Sertoli cell functions. In some selected cases, (part of) the molecular mechanisms involved have been elucidated, and these findings are summarized below.

With respect to the early stages of spermatogenesis, we referred to the stimulatory effects of low estrogen levels and thyroid hormone on the production of single type A_{und} spermatogonia. For thyroid hormone, this stimulation also involved Igf3. The effect of low estrogen levels is mediated by a nuclear estrogen receptor located in Sertoli cells that triggers the expression of platelet-derived endothelial cell growth factor (Pdecgf); recombinant Pdecgf mimics estrogen's effects and was designated in eel as an SSC renewal factor [26].

Although thyroid hormone is stimulatory, FSH has a much stronger effect on upregulating *igf3* mRNA in zebrafish Sertoli cells (Nobrega, Morais, de Waal, França, Schulz, Bogerd, unpublished results). These ongoing studies also show that zebrafish Igf3 stimulates the mitotic indices of A_{und} and A_{diff} spermatogonia without modulating androgen production. Still, FSH does have clear steroidogenic potency, and sex steroids have clear effects on Sertoli cell functions. As in other vertebrates, the androgen receptor is expressed by Sertoli cells but not by germ cells, and in zebrafish in particular, by Sertoli cells in contact with type A spermatogonia [52]. In general, the stimulatory effect of androgens on spermatogenesis and Sertoli cell proliferation has already been discussed, while little is known yet about the molecular mechanisms involved in fish. One exception is a finding in eel that androgen-stimulated Inhba (activin) production is involved in the stimulation of spermatogonial proliferation toward meiosis [53].

FSH-driven androgen production also provides a cross-link to thyroid hormone effects. In zebrafish, thyroid hormone receptors are not only expressed by Sertoli cells but also by Leydig cells, and thyroid hormone potentiates the stimulatory effect of FSH on androgen production [9]. In this way, thyroid hormone in the presence of low FSH levels stimulates the production of new spermatogenic cysts (as discussed earlier), while in combination with FSH, differentiation-inducing signals prevail in the context of the upregulation of Igf3 production and the potentiation of steroidogenic effects.

Also relevant to FSH is the Tgf β family member Amh. Fish do not have Müllerian ducts, and Amh probably has evolutionarily older functions in teleosts. Amh inhibited the onset of gonadotropin- or androgen-stimulated spermatogenesis in prepubertal Japanese eel [54]. Later work in adult zebrafish showed that Amh inhibited FSH-stimulated androgen production by downregulating the expression of steroidogenesis-related genes, resulting in reduced androgen release. Amh also inhibited androgen-supported spermatogenesis by blocking the differentiation of type A spermatogonia, resulting in an accumulation of type A_{und} spermatogonia [43]. Finally, FSH downregulated *amh* mRNA levels in adult zebrafish testis. Taken together, these observations place FSH in a central position with respect to the switch to differentiation—FSH downregulates Amh expression in Sertoli cells, which would otherwise inhibit differentiation of type A_{und} spermatogonia and compromise steroidogenesis. At the same time, FSH directly stimulates Leydig cell androgen production, which is potentiated by thyroid hormone. FSH also stimulates the expression of Igf3, which promotes proliferation of Sertoli cells and of type A spermatogonia. Thus, the biological activities of FSH coordinate signaling systems to increase the number of spermatogenic cysts and then support their progress through the mitotic phase of spermatogenesis.

Other research, again based on the eel model, revealed a signaling system regulated by the fish-specific progestin that becomes relevant when germ cells approach meiosis. DHP, which is typically also found in male fish, induced trypsin expression [55] and taurine biosynthesis [56] in Sertoli cells. Trypsin then stimulated germ cells to express a solute carrier gene (*slc6a6*) that functions as a taurine transporter so that elevated taurine levels in germ cells could trigger germ cell entry into meiosis [57].

B. Spermiation

In most fish species, spermatogenic cysts open after completion of spermiogenesis, so that the lumen of the opening cyst becomes continuous with the lumen of the spermatogenic tubule. The close contact between germ cells and Sertoli cells is broken and spermatozoa are released. This process is called "spermiation." In mammals, spermiation takes place at a specific stage of the cycle of the germinal epithelium that is known to be particularly dependent on androgens, and androgen signaling is required for the disintegration of the ectoplasmic specializations that connect late spermatids and Sertoli cells [58,59].

Experimental data on cellular and molecular mechanisms operating in Sertoli cells during spermiation in fish are limited. Data in Atlantic salmon indicate that spermiation is associated with elevated plasma androgen levels; the mitotic, meiotic, and most of the spermiogenic phases take place in the presence of plasma androgen levels (15–20 ng/ml) that are clearly higher than the levels found in immature fish (<2 ng/ml). However, the transition to large-scale spermiation, which leads to the typical histological picture of mature testes, in which the tubular lumen is filled with (and serves as a storage place for) spermatozoa, is associated with a second step of plasma androgen level increase to >40 ng/ml [49]. However, a direct role of androgens in spermiation has not been demonstrated yet. In fish, the highest levels of circulating progestins are observed in fully mature males in seasonally reproducing species. Progestins like DHP or 17α,20β,21-trihydroxy-4-pregnen-3-one (20β-S) are believed to induce or advance spermiation in salmonid and cyprinid fishes [60], increasing testicular fluid production [61,62], and stimulating spermatozoa motility [63,64] by altering the pH and fluidity of the seminal fluid [65].

In a number of taxonomic groups of fish (e.g., opheliidae [66], scorpaenidae [67], bleniidae [68], and soleidae [18]), spermiation takes place shortly after the completion of meiosis and spermatogenesis is referred to as "semicystic" [66]. In bullfrogs, cysts also open early (Figure 13.3), but spermatids remain in close contact with Sertoli cells until the completion of spermiogenesis; in this species, opening of the cyst and spermiation are separate events. In fishes with semicystic spermatogenesis, round spermatids complete spermiogenesis while in a free state in the tubular lumen. The Sertoli cells that formed the cysts are transformed into a single-layered epithelium lining the tubular lumen, and these Sertoli cells may still provide support to spermatids. Spermiogenesis in fish is less complex than in tetrapod vertebrates, as suggested by the absence of an acrosome in teleost spermatozoa and by its short duration (e.g., 2 days in zebrafish [5]). Direct, individual contact between Sertoli cells and spermatids during spermiogenesis may not be available to all members of spermatid clones in teleosts in general because the clone size usually amounts to hundreds of cells while only the peripheral layer(s) seem to have easily access to direct Sertoli cell contact.

C. Phagocytosis of apoptotic germ cells and removal of residual sperm

Fish Sertoli cells are astonishing with regard to their efficiency as phagocytotic cells, which is part of the normal physiological function of Sertoli cells. Electron microscopy studies have demonstrated Sertoli cell phagocytic activity and shown vesicles with hydrolytic enzymes, lysosomes, or phagosomes with degenerating spermatids and spermatozoa. In addition, residual bodies were observed in the cytoplasm of Sertoli cells. Acid phosphatase in Sertoli cell lysosomes is a marker of phagocytic activity in fish and rodents [69,70]. Altogether, these morphological features indicate the capability of Sertoli cells to remove residual bodies during spermiogenesis, apoptotic germ cells (in fish only 30–40% loss compared to the expected number [3,5,17]), and residual sperm after spermiation. These characteristics have been reported for the Sertoli cells of several teleost species, especially after spermiation and at the end of a spawning season [71–73].

In many fishes, the tubular lumen is an important sperm storage site. At the end of a spawning season, residual spermatozoa can remain in the tubular lumen. These spermatozoa are phagocytized by the Sertoli cells that form the epithelial lining of spermatogenic tubules, an aspect of Sertoli cell physiology that is absent in higher vertebrates in which derivatives of mesonephric tissue form efferent duct and sperm storage tissues. Interestingly, the clearing of residual sperm is completed before the next wave of spermatogenesis can start. The timing of these processes differs locally because neighboring tubules can have rather different activities. For example, phagocytosis of residual sperm was ongoing in one tubule in which spermatogonia seemed to be quiescent, while spermatogonial proliferation resumed in an adjacent tubule in which residual sperm were not visible [11,74]. This finding suggests that local signaling, potentially originating from Sertoli cells still occupied with phagocytosis, can suppress the resumption of spermatogonial proliferation in the vicinity. No information is available about the endocrine regulation (if it does exist) of the phagocytotic activity of Sertoli cells in fish. In rats, phagocytosis mediated by acid phosphatase in Sertoli cells is not hormone dependent [69]. In humans, however, testicular acid phosphatase gene expression is upregulated by androgens and downregulated by estrogens [75].

D. Fate after completion of cyst development

The lifelong generation of new Sertoli cells along with the production and differentiation of spermatogenic cysts supports the hypothesis that Sertoli cells are at least partially lost when a cyst has completed its development, in particular because seasonally reproducing species show large increases and then similarly large decreases in testis weight after the spawning season before embarking on the next season's growth phase.

So far, clear evidence for Sertoli cell apoptosis is only available for sharks. In the shark testis, developing cysts migrate during development through the testis from the dorsal to the ventral surface, where cysts open to release sperm into an efferent duct system while the cyst-forming Sertoli cells become apoptotic [76]. There are no results available in teleosts or amphibians yet. However, it is conceivable that during spermiation, some Sertoli cells are lost, perhaps especially those Sertoli cells that had a mainly adluminal position and did not contact another germ cell clone. Other Sertoli cells—perhaps preferentially those that had extended contact areas with the basement membrane—can become integrated into the epithelial lining of the spermatogenic tubule, while Sertoli cells that do contact another germ cell clone may continue to support this clone [19].

IV. Concluding remarks

From a comparative point of view, there are, in our opinion, two main differences worth pinpointing between anamniote and amniote Sertoli cells. One difference is the astonishing plasticity of the numbers and stages of differentiation of Sertoli cells in the adult anamniote testis. The other difference is the considerably higher germ cell-supporting capacity. With respect to the first point, it will be exciting to investigate further the presence, characteristics, and regulation of Sertoli cell precursor cells and to search for formal proof for the presence of a true somatic stem cell population in the anamniote testis. The potential to provide additional Sertoli cells via mode 1 and the tailored increase in Sertoli cell number during the progress of cystic spermatogenesis via mode 2 of Sertoli cell proliferation may be part of the "secret" of the high capacity of anamniote Sertoli cells to support germ cells. However, several other structural and functional aspects of Sertoli cell/germ cell interaction probably contribute to this high efficiency and await clarification and comparative analysis in higher vertebrates. For both issues, we wish to stress the equal importance of developing technical approaches on the molecular, physiological, and morphological levels.

Technical help from Paulo Henrique de Almeida Campos-Junior in the preparation of figures is highly appreciated. Financial support from FAPEMIG and CNPq are fully appreciated.

References

[1] Schulz RW, de França LR, Lareyre JJ, LeGac F, Chiarini-Garcia H, Nóbrega RH, et al. Spermatogenesis in fish. Gen Comp Endocrinol 2010;165:390–411.

[2] Matta SL, Vilela DA, Godinho HP, França LR. The goitrogen 6-n-propyl-2-thiouracil (PTU) given during testis development increases Sertoli and germ cell numbers per cyst in fish: the tilapia (*Oreochromis niloticus*) model. Endocrinology 2002;143:970–8.

[3] Billard R. La spermatogenèse de *Poecilia reticulata* I. Estimation du nombre de générations goniales et rendement de la spermatogenèse. Ann Biol Anim Bioch Biophys 1969;9:251–71.

[4] Schulz RW, Menting S, Bogerd J, de França LR, Vilela DAR, Godinho HP. Sertoli cell proliferation in the adult testis: evidence from two fish species belonging to different orders. Biol Reprod 2005;73:891–8.

[5] Leal MC, Cardoso ER, Nóbrega RH, Batlouni SR, Bogerd J, de França LR, et al. Histological and stereological evaluation of zebrafish (*Danio rerio*) spermatogenesis with an emphasis on spermatogonial generations. Biol Reprod 2009;81:177–87.

[6] França LR, Auharek SA, Hess RA, Dufour JM, Hinton BT. Morphofunctional and immunological aspects of the blood–testis and blood–epididymal barriers. In: Cheng CY, editor. Biology and regulation of blood–tissue barriers. New York, NY, USA: Landes Bioscience and Springer Science+Business Media; 2012. p. 237–59.

[7] Dovere L, Fera S, Grasso M, Lamberti D, Gargioli C, Muciaccia B, et al. The niche-derived glial cell line-derived neurotrophic factor (GDNF) induces migration of mouse spermatogonial stem/progenitor cells. PLoS One 2013;8:e59431.

[8] Nakajima S, Hayashi M, Kouguchi T, Yamaguchi K, Miwa M, Yoshizaki G. Expression patterns of *gdnf* and *gfrα1* in rainbow trout testis. Gene Expr Patterns 2014;14:111–20.

[9] Morais RD, Nóbrega RH, Gómez-González NE, Schmidt R, Bogerd J, de França LR, et al. Thyroid hormone stimulates the proliferation of Sertoli cells and single type A spermatogonia in adult zebrafish (*Danio rerio*) testis. Endocrinology 2013;154:4365–76.

[10] Billard R, Breton B. Rhythm of reproduction in teleost fish. In: Thorpe JE, editor. Rhythmic activity of fishes. New York, NY: Academic Press; 1978. p. 31–53.

[11] Almeida FFL, Kristoffersen C, Taranger GL, Schulz RW. Spermatogenesis in Atlantic cod (*Gadus morhua*): a novel model of cystic germ cell development. Biol Reprod 2008;78:27–34.

[12] Kobayashi M, Aida K, Stacey NE. Induction of testis development by implantation of 11-ketotestosterone in female goldfish. Zool Sci 1991;8:389–93.

[13] Shibata N, Hamaguchi S. Evidence for the sexual bipotentiality of spermatogonia in the fish, *Oryzias latipes*. J Exp Zool 1988;245:71–7.

[14] Kobayashi Y, Nakamura M, Sunobe T, Usami T, Kobayashi T, Manabe H, et al. Sex change in the Gobiid fish is mediated through rapid switching of gonadotropin receptors from ovarian to testicular portion or vice versa. Endocrinology 2009;150:1503–11.

[15] Grier HJ. Cellular organization of the testis and spermatogenesis in fishes. Am Zool 1981;21:345–57.

[16] Parenti LR, Grier HJ. Evolution and phylogeny of gonad morphology in bony fishes. Integr Comp Biol 2004;44:333–48.

[17] Vilela DAR, Silva SGB, Peixoto MTD, Godinho HP, de França LR. Spermatogenesis in teleost: insights from the Nile tilapia (*Oreochromis niloticus*) model. Fish Physiol Biochem 2003;28:187–90.

[18] García-López Á, Martínez-Rodríguez G, Sarasquete C. Male reproductive system in Senegalese sole *Solea senegalensis* (Kaup): anatomy, histology and histochemistry. Histol Histopathol 2005;20:1179–89.

[19] Koulish S, Kramer CR, Grier HJ. Organization of the male gonad in a protogynous fish, *Thalassoma bifasciatum* (Teleostei: Labridae). J Morphol 2002;254:292–311.

[20] Chiarini-Garcia H, Raymer AM, Russell LD. Non-random distribution of spermatogonia in rats: evidence of niches in the seminiferous tubules. Reproduction 2003;126:669–80.

[21] Yoshida S, Sukeno M, Nabeshima Y. A vasculature-associated niche for undifferentiated spermatogonia in the mouse testis. Science 2007;317:1722–6.

[22] de Rooij DG, Griswold MD. Questions about spermatogonia posed and answered since 2000. J Androl 2012;33:1085–95.

[23] Costa GM, Avelar GF, Rezende-Neto JV, Campos-Junior PH, Lacerda SM, Andrade BS, et al. Spermatogonial stem cell markers and niche in equids. PLoS One 2012;7(8):e44091.

[24] Campos-Junior PH, Costa GM, Lacerda SM, Rezende-Neto JV, de Paula AM, Hofmann MC, et al. The spermatogonial stem cell niche in the collared peccary (*Tayassu tajacu*). Biol Reprod 2012;86(5):155 1–10.

[25] Nóbrega RH, Greebe CD, van de Kant H, Bogerd J, de França LR, Schulz RW. Spermatogonial stem cell niche and spermatogonial stem cell transplantation in zebrafish. PLoS One 2010;5(9):e12808.

[26] Miura T, Ohta T, Miura CI, Yamauchi K. Complementary deoxyribonucleic acid cloning of spermatogonial stem cell renewal factor. Endocrinology 2003;144:5504–10.

[27] Yoshida S. Elucidating the identity and behavior of spermatogenic stem cells in the mouse testis. Reproduction 2012;144:293–302.

[28] Amer MA, Miura T, Miura C, Yamauchi K. Involvement of sex steroid hormones in the early stages of spermatogenesis in Japanese huchen (*Hucho perryi*). Biol Reprod 2001;65:1057–66.

[29] Song M, Gutzeit HO. Effect of 17α-ethynylestradiol on germ cell proliferation in organ and primary culture of medaka (*Oryzias latipes*) testis. Dev Growth Differ 2003;45:327–37.

[30] Cooke PS, Holsberger DR, França LR. Thyroid hormone regulation of Sertoli cell development. In: Skinner MK, Griswold MD, editors. The Sertoli cell biology. Amsterdam, The Netherlands: Elsevier Science; 2005. p. 217–26.

[31] Wang DS, Jiao B, Hu C, Huang X, Liu Z, Cheng CH. Discovery of a gonad-specific IGF subtype in teleost. Biochem Biophys Res Commun 2008;367:336–41.

[32] de Waal PP. Hormonal regulation of spermatogenesis in zebrafish [Ph.D. thesis]. Utrecht University, the Netherlands; 2009. ISBN: 978.90.393.5143.7.

[33] Sambroni E, Lareyre JJ, Le Gac F. Fsh controls gene expression in fish both independently of and through steroid mediation. PLoS One 2013;8(10):e76684.

[34] Campbell B, Dickey JT, Swanson P. Endocrine changes during onset of puberty in male spring Chinook salmon, *Oncorhynchus tshawytscha*. Biol Reprod 2003;69:2109–17.

[35] Gomez JM, Weil C, Ollitrault M, Lebail PY, Breton B, LeGac F. Growth hormone (GH) and gonadotropin subunit gene expression and pituitary and plasma changes during spermatogenesis and oogenesis in rainbow trout (*Oncorhynchus mykiss*). Gen Comp Endocrinol 1999;113:413–28.

[36] Prat F, Sumpter JP, Tyler CR. Validation of radioimmunoassays for two salmon gonadotropins (GTH I and GTH II) and their plasma concentrations throughout the reproductive cycle in male and female rainbow trout (*Oncorhynchus mykiss*). Biol Reprod 1996;54:1375–82.

[37] Huhtaniemi IT, Themmen AP. Mutations in human gonadotropin and gonadotropin receptor genes. Endocrine 2005;26:207–17.

[38] Mazón MJ, Gómez A, Yilmaz O, Carrillo M, Zanuy S. Administration of follicle-stimulating hormone *in vivo* triggers testicular recrudescence of juvenile European sea bass (*Dicentrarchus labrax*). Biol Reprod 2014;90(1):6 1–10.

[39] Schulz RW, van Dijk W, Chaves-Pozo E, García-López A, de França LR, Bogerd J. Sertoli cell proliferation in the adult testis is induced by unilateral gonadectomy in African catfish. Gen Comp Endocrinol 2012;177:160–7.

[40] Idler DR, Bitners II, Schmidt PJ. 11-Ketotestosterone: an androgen for sockeye salmon. Can J Biochem Physiol 1961;39:1737–42.

[41] Borg B. Androgens in teleost fish. Comp Biochem Physiol 1994;109C:219–45.

[42] Miura T, Yamauchi K, Takahashi H, Nagahama Y. Hormonal induction of all stages of spermatogenesis in vitro in the male Japanese eel (*Anguilla japonica*). Proc Natl Acad Sci USA 1991;88:5774–8.

[43] Skaar KS, Nóbrega RH, Magaraki A, Olsen LC, Schulz RW, Male R. Proteolytically activated, recombinant anti-Müllerian hormone inhibits androgen secretion, proliferation, and differentiation of spermatogonia in adult zebrafish testis organ cultures. Endocrinology 2011;152:3527–40.

[44] Dépêche J, Sire O. In vitro metabolism of progesterone and 17α-hydroxyprogesterone in the testis of the rainbow trout, *Salmo gairdneri* Rich., at different stages of spermatogenesis. Reprod Nutr Develop 1982;22:427–38.

[45] Scott AP, Sumpter JP. Seasonal variations in testicular germ cell stages and in plasma concentrations of sex steroids in male rainbow trout (*Salmo gairdneri*) maturing at 2 years old. Gen Comp Endocrinol 1989;73:46–58.

[46] Vizziano D, Le Gac F, Fostier A. Effect of 17β-estradiol, testosterone, and 11-ketotestosterone on 17α,20β-dihydroxy-4-pregnen-3-one production in the rainbow trout testis. Gen Comp Endocrinol 1996;104:179–88.
[47] Miura T, Higuchi M, Ozaki Y, Ohta T, Miura C. Progestin is an essential factor for the initiation of the meiosis in spermatogenetic cells of the eel. Proc Natl Acad Sci USA 2006;103:7333–8.
[48] Chen SX, Bogerd J, Schoonen NE, Martijn J, de Waal PP, Schulz RW. A progestin (17α,20β-dihydroxy-4-pregnen-3-one) stimulates early stages of spermatogenesis in zebrafish. Gen Comp Endocrinol 2013;185(1–9).
[49] Melo MC, Andersson E, Fjelldal PG, Bogerd J, de França LR, Taranger GL, et al. Salinity and photoperiod modulate pubertal development in Atlantic salmon (*Salmo salar*). J Endocrinol 2014;220:319–32.
[50] Kliesch S, Penttila TL, Gromoll J, Saunders PTK, Nieschlag E, Parvinen M. FSH receptor messenger-RNA is expressed stage-dependently during rat spermatogenesis. Mol Cell Endocrinol 1992;84:R45–9.
[51] Johnston DS, Olivas E, DiCandeloro P, Wright WW. Stage-specific changes in GDNF expression by rat Sertoli cells: a possible regulator of the replication and differentiation of stem spermatogonia. Biol Reprod 2011;85:763–9.
[52] de Waal PP, Wang DS, Nijenhuis WA, Schulz RW, Bogerd J. Functional characterization and expression analysis of the androgen receptor in zebrafish (*Danio rerio*) testis. Reproduction 2008;136:225–34.
[53] Miura T, Miura C, Yamauchi K, Nagahama Y. Human recombinant activin induces proliferation of spermatogonia *in vitro* in the Japanese eel (*Anguilla japonica*). Fisheries Sci 1995;61:434–7.
[54] Miura T, Miura C, Konda Y, Yamauchi K. Spermatogenesis-preventing substance in Japanese eel. Development 2002;129:2689–97.
[55] Miura C, Ohta T, Ozaki Y, Tanaka H, Miura T. Trypsin is a multifunctional factor in spermatogenesis. Proc Natl Acad Sci USA 2009;106:20972–7.
[56] Higuchi M, Celino FT, Tamai A, Miura C, Miura T. The synthesis and role of taurine in the Japanese eel testis. Amino Acids 2012;43:773–81.
[57] Higuchi M, Miura C, Iwai T, Miura T. Trypsin regulates meiotic initiation in the Japanese eel (*Anguilla japonica*) by promoting the uptake of taurine into germ cells during spermatogenesis. Biol Reprod 2013;89(3):58 1–9.
[58] O'Donnell L, Stanton PG, Bartles JR, Robertson DM. Sertoli cell ectoplasmic specializations in the seminiferous epithelium of the testosterone-suppressed adult rat. Biol Reprod 2000;63:99–108.
[59] Cheng CY, Wong EW, Yan HH, Mruk DD. Regulation of spermatogenesis in the microenvironment of the seminiferous epithelium: new insights and advances. Mol Cell Endocrinol 2010;315:49–56.
[60] Ueda H, Kambegawa A, Nagahama Y. Involvement of gonadotrophin and steroid hormones in spermiation in the amago salmon, *Oncorhynchus rhodurus*, and goldfish, *Carassius auratus*. Gen Comp Endocrinol 1985;59:24–30.
[61] Baynes SM, Scott AP. Seasonal variations in parameters of milt production and in plasma concentration of sex steroids of male rainbow trout (*Salmo gairdneri*). Gen Comp Endocrinol 1985;57:150–60.
[62] Yueh WS, Chang CF. 17α,20β,21-trihydroxy-4-pregnen-3-one and 17α,20β-dihydroxy-4-pregnen-3-one stimulated spermiation in protandrous black porgy, *Acanthopagrus schlegeli*. Fish Physiol Biochem 1997;17:187–93.

[63] Miura T, Yamauchi K, Takahashi H, Nagahama Y. The role of hormones in the acquisition of sperm motility in salmonid fish. J Exp Zool 1992;261:359–63.
[64] Tubbs C, Thomas P. Functional characteristics of membrane progestin receptor alpha (mPRalpha) subtypes: a review with new data showing mPRalpha expression in seatrout sperm and its association with sperm motility. Steroids 2008;73:935–41.
[65] Scott AP, Sumpter JP, Stacey N. The role of the maturation-inducing steroid, 17,20β-dihydroxypregn-4-en-3-one, in male fishes: a review. J Fish Biol 2010;76: 183–224.
[66] Mattei X. Peculiarities in the organization of testis of *Ophidion sp.* (Pisces: Teleostei). Evidence for two types of spermatogenesis in teleost fish. J Fish Biol 1993;43:931–7.
[67] Muñoz M, Casadevall M, Bonet S. Testicular structure and semicystic spermatogenesis in a specialized ovuliparous species: *Scorpaena notata* (Pisces, Scorpaenidae). Acta Zool 2002;83:213–9.
[68] Lahnsteiner F, Patzner RA. Spermiogenesis and structure of mature spermatozoa in bleniid fishes (Pisces, Bleniidae). J Submicr Cytol Pathol 1990;22:565–76.
[69] Chemes H. The phagocitic function of Sertoli cells: a morphological, biochemical and endocrinological study of lysosomes and acid phosphatase localization in the rat testis. Endocrinology 1986;119:1673–81.
[70] Porawski M, Wassermann GF, Achaval M. Localization of acid phosphatase activity in the testis of two teleostean species (*Oreochromis niloticus* and *Odonthestes perugiae*). Braz J Biol 2004;64:853–8.
[71] van Vuren JHJ, Soley JT. Some ultrastructural observations of Leydig and Sertoli cells in the testis of *Tilapia rendalli* following induced testicular recrudenscence. J Morphol 1990;206:57–63.
[72] Lo Nostro FL, Grier H, Meijide FJ, Guerrero GA. Ultrastructure of the testis in *Synbranchus marmoratus* (Teleostei, Synbranchidae): the germinal compartment. Tissue Cell 2003;35:121–32.
[73] Chung EY, Yang YC, Kang HW, Choi KH, Jun JC, Lee KY. Ultrastructure of germ cells and the functions of Leydig cells and Sertoli cells associated with spermatogenesis in *Pampus argenteus* (Teleostei: Perciformes: Stromateidae). Zool Studies 2010;49:39–50.
[74] Schulz R. Serum levels of 11-oxotestosterone in male and 17β-estradiol in female rainbow trout (*Salmo gairdneri*) during the first reproductive cycle. Gen Comp Endocrinol 1984;56:111–20.
[75] Yousef GM, Diamandis M, Jung K, Diamandis EP. Molecular cloning of a novel human acid phosphatase gene (ACPT) that is highly expressed in the testis. Genomics 2001;74:385–95.
[76] McClusky LM. The caspase-dependent apoptosis gradient in the testis of the blue shark, *Prionace glauca*. Reproduction 2013;145:297–310.

Adult Sertoli cell differentiation status in humans

14

Jenna T. Haverfield[a,b], *Peter G. Stanton*[a,c] *and Sarah J. Meachem*[a,b]
[a]Monash Institute of Medical Research-Prince Henry's Institute, Clayton, Victoria, Australia, [b]Department of Anatomy and Developmental Biology, Monash University, Victoria, Australia, [c]Department of Biochemistry, Monash University, Victoria, Australia

I. Introduction and scope of the chapter

Successful spermatogenesis during adult life is dependent upon the establishment of a normal complement of fully differentiated Sertoli cells; this complement develops at around puberty. The transition from an undifferentiated to a differentiated Sertoli cell population involves a series of well-defined structural and functional changes that are essential to support and cater to the individual needs of each germ cell type. This complex, stepwise cascade of transformations includes changes in proliferative activity, junctional protein organization, blood−testis barrier (BTB) permeability, protein expression, and morphology. To the extent that is known, these changes are largely regulated by a combination of endocrine and paracrine extracellular ligands presented to the Sertoli cell during fetal, postnatal, prepubertal, and adult life *in vivo*. Disturbances to these endocrine and paracrine signaling pathways directly affect Sertoli cell differentiation and thus spermatogenesis.

Since the nineteenth century, it has been believed that once adult Sertoli cells complete differentiation, their differentiation state is fixed for life, a feature called "terminal" differentiation. However, research since the early 2000s, largely from seasonal-breeding Djungarian hamsters and some limited data from humans, reveal that the adult Sertoli cell population *in vivo* is capable of dedifferentiation in settings of impaired spermatogenesis and therefore may not be a homogenous terminally differentiated population. These adult Sertoli cells feature an intermediate state of differentiation, in that they display at least one hallmark characteristic of prepubertal undifferentiated Sertoli cells. While the hamster model has been useful, to advance our understanding of adult Sertoli cell differentiation and its impact on spermatogenesis, access to more clinical samples as well as traditional laboratory species that feature breeding systems that are more similar to those of humans, such as rats and mice, are needed.

This chapter discusses how the adult Sertoli cell population develops, briefly describes some of the factors that regulate this population, and outlines the differences between undifferentiated and differentiated Sertoli cells. The aim is to use findings from animals and, where possible, to include data from humans to build an

Sertoli Cell Biology. DOI: http://dx.doi.org/10.1016/B978-0-12-417047-6.00014-4

understanding of Sertoli cells' biological, and perhaps their unrealized therapeutic, potential. To understand the adult Sertoli cell population, one has to understand how the Sertoli cell population transforms from an undifferentiated to a differentiated state during normal testicular development. In this chapter, changes in the status of Sertoli cell differentiation are traced primarily from postnatal life and, where relevant from the fetus, to adulthood, with a focus on puberty, where the major switch in differentiation has been reported.

Finally, it is necessary to define what this chapter defines as "puberty." Generally speaking, puberty is defined in males as the time when sperm is first produced, which occurs around 10–13 years of age in humans and 35–45 days postpartum (dpp) in rodents [1]. Puberty is underpinned by the reactivation of the hypothalamic–pituitary–testicular axis as a result of pulsatile gonadotropin-releasing hormone (GnRH) secretion [1a,2]. However, from a scientific point of view, puberty is much more complex than this definition because it is accompanied by many biological, sociological, and psychological changes that occur across a period of time and is not just one event. In this chapter, "puberty" refers to the attainment of a full complement of Sertoli cells, which occurs at 15–21 dpp in rats [3,4]; 12–20 dpp in mice [5–7]; 38–47 months of age in rhesus monkeys, commonly referred to as the early pubertal period [8]; and 10–13 years of age in boys.

II. Development of the adult Sertoli cell population

The foundation of normal testicular function and, thus, fertility in adult life is entirely dependent on the development of a normal, fully differentiated Sertoli cell population, which occurs as a result of developmental changes during fetal, postnatal, and pubertal life. These developmental processes involve a series of proliferation and differentiation events that are regulated by a complex network of genetic mechanisms as well as hormonal messengers acting through endocrine, paracrine, and autocrine pathways. These regulatory factors can act separately and/or synergistically at either a distinct point during Sertoli cell development, or they can have broader, long-lasting effects throughout Sertoli cell development. To give the reader an overview of key events that define Sertoli cell development, the following sections will examine the proliferation of the Sertoli cell population in prepubertal life, the transition to a differentiated Sertoli cell population around pubertal life, and the ability of differentiated Sertoli cells to dedifferentiate during adult life.

A. *Proliferation in prepubertal life*

During postnatal life, Sertoli cells undergo extensive proliferative activity, which is essential for the expansion of the testis cords and to produce the final number of Sertoli cells in adulthood (Figure 14.1A and B) (reviewed in [9]). Because each Sertoli cell supports a fixed number of germ cells, the number of adult Sertoli cells quantitatively determines sperm output, with decreases and increases in Sertoli cell number resulting in a relatively proportionate increase or decrease in spermatids, respectively [10,11].

Human and monkey Sertoli cells proliferate throughout postnatal life; however, proliferation is not continuous in the way it appears to be in rodents, and it is separated by a period of quiescence (Figure 14.1A) (reviewed in [9,12]). Bromodeoxyuridine (BrdU) incorporation studies in rhesus monkeys show that the highest rates of Sertoli cell proliferation occur during infancy, at around 4–5 months of age, and then again around puberty, resulting in a final number of Sertoli cells at 38–47 months of age [8,12,13]. ^{3}H-thymidine and BrdU incorporation is not possible in humans due to ethical and practical constraints; nonetheless, stereological analyses show that the total number of Sertoli cells increases both during the first 3 months of life and during puberty (10–18 years) [14].

In rodents, ^{3}H-thymidine incorporation studies show that Sertoli cells commence proliferation around 14–16 days postcoitum (dpc) [4,6], with maximum proliferative activity occurring at 20 dpc in rats [4] and 16 dpc in mice [6]. Sertoli cells continue to proliferate after birth throughout postnatal life; however, their proliferation rate progressively decreases, with no mitotic activity detectable after 15–21 dpp in rats [3,4] and 12–20 dpp in mice [5–7]. However, stereological assessments have shown in mice that Sertoli cell numbers increase well into adulthood [15]. Loss of Sertoli cell proliferative activity around puberty is discussed in detail later.

III. Proliferation and differentiation around puberty

Puberty represents an important time in the development of the Sertoli cell population. Undifferentiated Sertoli cells phenotypically transform into a differentiated population, acquiring the functions necessary to support and cater to the individual needs of four to five generations of germs cells, each with different structural and nutritional needs (reviewed in [16]). Transformations in proliferation activity, junctional protein organization, BTB permeability, protein expression, and morphology (Table 14.1 and Figure 14.1) are instrumental to adult male reproductive health; failure to undergo proper pubertal differentiation is associated with the manifestation of many developmental disorders falling under the umbrella of testicular dysgenesis syndrome [19], including hypospadias [20], cryptorchidism [21], and germ cell cancer [22] (for a detailed review, see [9]). Each of the phenotypic changes that occurs around puberty are similar across mammalian species and are hallmark features used to diagnose the differentiation status of Sertoli cells in clinical settings and animal models (Tables 14.1 and 14.2). The status of Sertoli cell differentiation can be useful during investigations of idiopathic primary seminiferous tubule failure, which accounts for more than half of all male infertility cases [32,33].

From this point forward, pubertal Sertoli cell differentiation is distinct from mesenchymal-to-epithelial differentiation, which occurs during fetal life. References to undifferentiated and differentiated Sertoli cells in this chapter will pertain only to pubertal differentiation and, thus, pre- and postpubertal Sertoli cells,

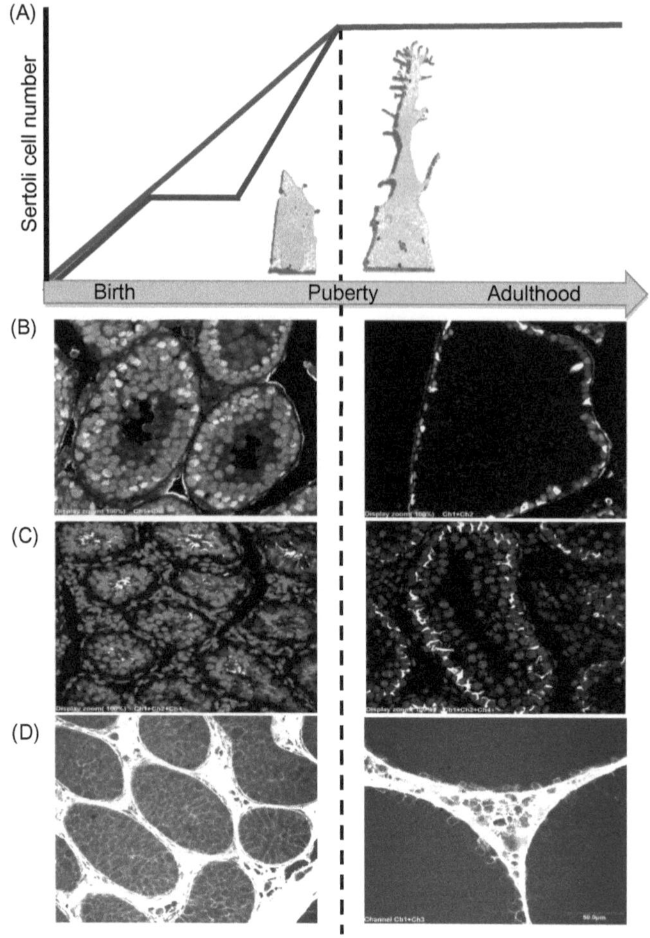

Figure 14.1 Conceptual representation of the development of a Sertoli cell population in humans and rodents. (A) Sertoli cells proliferate during fetal, postnatal, and prepubertal life. In humans (blue line), a period of quiescence for 8 years occurs during this developmental period. In rodents, the two proliferative phases appear as one occurring without an intervening quiescent period (red line). At puberty, Sertoli cell proliferation is thought to cease for the remainder of adulthood, and as a result, adult Sertoli cells are considered terminally differentiated. Puberty also marks the point at which the Sertoli cells acquires other structural and functional changes, including organization of Sertoli cell junction proteins (C) and formation of the BTB (D). In hamsters (B) Prepubertal Sertoli cells are still proliferative (left panel: red, proliferating cell nuclear antigen positive cells; green, Sertoli, Leydig, and peritubular myoid cells (GATA4 positive); and yellow, proliferating Sertoli cells (colocalization of red and green markers)). After puberty (right panel), no colocalization can be observed. (C) Prepubertal Sertoli cells feature disorganized Sertoli cell junctions throughout the seminiferous epithelium (left panel: claudin-11, green; cell nuclei, blue). After puberty (right panel), Sertoli cell junctions acquire an organized phenotype by forming a scalloped pattern positioned along the basal aspect of the epithelium. Analysis of junction organization is a qualitative technique, but recently quantified in humans (23). (D) The BTB has not formed between prepubertal Sertoli cells (left panel), and, as a result, a biotin tracer (green) injected into testes can permeate the seminiferous epithelium. The BTB forms at puberty (right panel), and, as a result, the biotin tracer cannot permeate beyond the

Table 14.1 Morphological and functional differences between undifferentiated and differentiated Sertoli cells

Undifferentiated	Differentiated
Proliferative ability	
Proliferative • PCNA positive • Ki-67 positive • PH3 positive • EdU positive	Non proliferative • PCNA negative • Ki-67 negative • PH3 negative • EdU negative
Sertoli cell junctions	
• Claudin-11 absent or disorganized • Connexin-43 absent or disorganized • JAM-A absent or disorganized	• Claudin-11 organized • Connexin-43 organized • JAM-A organized
Morphology	
• Uniform small shape nuclei • Round nucleoli • Large nucleus:cytoplasm ratio • Minimal cytoplasmic processes	• Irregular large shape nuclei • Tripartite nucleoli • Small nucleus:cytoplasm ratio • Elaborate cytoplasmic processes
Protein expression	
• CK18 present • AMH present • M2A present • Arom present • NCAM present • Inhibin bodies present • GATA1 absent • AR absent • $p27^{kip1}$ absent • Laminin $\alpha 5$ absent	• CK18 absent • AMH absent • M2A absent • Arom absent • NCAM absent • Inhibin bodies absent • GATA1 present • AR present • $p27^{kip1}$ present • Laminin $\alpha 5$ present

Morphological differences can be distinguished by light or electron microscopy. Markers for Sertoli cell junctions, proliferative ability, and protein expression can be detected immunohistochemically and visualized using light or fluorescent microscopy. JAM-A, Junction adhesion molecule-A; PCNA, Proliferating cell nuclear antigen; PH3, Phosphatase histone 3; EdU, 5-ethynyl-2′-deoxyuridine; CK18, cytokeratin 18; AMH, anti-Müllerian hormone; Arom, aromatase; NCAM, neural cell adhesion molecule; AR, androgen receptor; M2A, M2A antigen. *Note*: While the Sertoli cell AR is considered to be a marker of pubertal Sertoli cell differentiation, AR protein expression has been reported to turn on as early as 5 dpp in rats [18]. Therefore, one should exert caution when diagnosing Sertoli cell differentiation status based on this marker alone.
Adapted from Brehm and Steger [17] and Sharpe et al. [9].

◀ inter-Sertoli cell tight junctions positioned above the preleptotene spermatocytes. (For interpretation of the references to color in this figure legend, the reader is referred to the web version of this book.)

Table 14.2 Compilation of data available from human adult Sertoli cells *in vivo* as assessed either through histology or specific validated immunoprobes associated with Sertoli cell differentiation

	Proliferative ability	Junction organization	BTB function	Protein expression	Morphology	Differentiation status
Normospermic	**No** PCNA −ve [22]	**Organized** [22,23,24,144]	**Impermeable** [25]	**CK18 −ve** [22,26] **AR +ve** [22,27] **AMH − ve** [28]	**Single nucleolus; irregular large nucleus; cytoplasmic processes abundant** [29,145]	**Differentiated**
Subfertile (hormone suppression)	**Yes** PCNA +ve (1.7 ± 0.6%) [22]	**Organized** [22]	**Impermeable** [25]	**AR −ve** [22]	Not assessed	**Mixed phenotype**
Sertoli cell only	**No** PCNA −ve [26]	**Disorganized** [23,144]	**Permeable** [25]	**CK18 +ve AMH +ve** [26,28,30]	**Oval nuclei, nontripartite, poor in organelles** [25,145]	**Mixed phenotype**
Germ cell arrest	Not assessed	**Disorganized** [23]	**Permeable** [25]	**CK18 +ve AMH +ve** [26,28,30]	**Tripartite nucleoli; abundant organelles** [25]	**Mixed phenotype**
Azoospermic	**Yes** PCNA +ve (7%) [26]	**Disorganized** [25]	**Permeable** [25]	**CK18** +ve [28] **AMH** +ve [28]	**Tripartite nucleoli; Uniformed small nucleus** [145]	**Mixed phenotype**
Oligospermia	**Yes** [22]	**Organized** [25]	**Impermeable** [25]	Not assessed	**Tripartite nucleoli; abundant organelles** [25]	**Possibly differentiated**
Testicular cancer	**Yes** PCNA+ [146,147]	**Disorganized** [22,24]	Not assessed	CK18 +ve [146,148]	**Tripartite nucleoli; minimal cytoplasmic processes** [149]	**Mixed phenotype**
Hypogonadotropic hypogonadal			**Permeable** [29,31]			

Enables comparison between limited data available linking Sertoli cell differentiation to different human fertility phenotypes (−ve=negative; + ve=positive).

respectively. In this section, we briefly discuss the role of the germ cells in initiating this differentiation process, and in the subsequent sections we discuss the key phenotypic changes occurring in Sertoli cells during differentiation focussing on proliferative ability, Sertoli cell junctions, protein expression, and morphology.

Regulation. The role of hormones and germ cells as regulators (inducers) of Sertoli cell differentiation probably results from the timely rise in pituitary hormones and the appearance of meiotic and postmeiotic germ cells in the epithelium at around puberty. Broadly speaking, there is ample evidence that hormones—such as follicle-stimulating hormone (FSH), androgen, and thyroid hormone (TH)—regulate one or more of the hallmark characteristics of Sertoli cell differentiation. With respect to germ cells influencing the induction of Sertoli cell differentiation, early reports showing that germ cells can modify Sertoli cell function emerged from culture experiments, in which removal of germ cells via hypotonic shock treatment decreased the secretion of the Sertoli cell-specific proteins androgen-binding protein [34–36], transferrin [35,36], and inhibin [37] and upon germ cell replacement, protein secretion was restored. Selective destruction of pachytene spermatocytes and round spermatids *in vivo* resulted in a 30–50% decrease in the incorporation of ^{35}S-methionine into proteins secreted at stages VI–VIII of the seminiferous cycle [38]. Moreover, stage-specific spermatocytes and spermatids secrete particular growth factors that, in turn, regulate Sertoli cell functions *in vitro*, such as inhibin production and Sertoli cell junction protein expression of claudin-11, occludin, and ZO-1 [39]. The first piece of *in vivo* evidence that stage-specific germ cells may influence Sertoli cell differentiation emerged in a study showing that the reappearance of steps 4–6 round spermatids coincides with full BTB function [40].

A. Proliferative ability

One of the key features of differentiated Sertoli cells is a loss of proliferation. This concept was first proposed by Enrico Sertoli himself [41] and was subsequently supported by rodent studies in which limited mitotic activity was detectable after 15–21 dpp in rats [3,4] and 12–20 dpp in mice [5–7]. The loss of proliferative activity is considered the central foundation of Sertoli cell differentiation and has led to classification of adult Sertoli cells as terminally differentiated (for a review, see [9]). "Terminal differentiation" is a term given to many other cells in the body that irreversibly lose their ability to proliferate, including cardiomyocytes [42], neurons [43], and skeletal muscle cells [44]. The mechanisms that underpin the loss of proliferative ability not only in Sertoli cells but also in all other so-called terminally differentiated cells in the mammalian body are not well understood [45]. Nonetheless, research since the early 2000s shows that the changes in the proliferative ability of Sertoli cells are in part regulated by a combination of endocrine and paracrine factors, each of which is discussed. It is important to note, however, that the classification of Sertoli cells and other cells within the body as terminally differentiated is probably inappropriate because, at the very least, a subset of these cells is in cell cycle and may retain the residual capacity to proliferate.

Regulation. FSH is important for the establishment of the Sertoli cell population and is one of the major supporters of their proliferation prior to puberty (reviewed in [16]). Exit from the cell cycle in rodent Sertoli cells is not underpinned by FSH because serum FSH levels and the expression of FSH receptors continue to rise during early postpubertal development [46,47]. Moreover, while cultured postnatal Sertoli cells respond mitogenically to FSH, postpubertal Sertoli cells do not [48]. Instead, Sertoli cell proliferation at puberty is limited by a series of other signals, including TH, testosterone, retinoic acid, and inhibin, which induce a coordinated, progressive association of cyclins and specific cyclin-dependent kinase inhibitors, including $p27^{kip1}$ and $p21^{Cip1}$ [49–51].

TH plays a major role in terminating Sertoli cell proliferation at puberty and, therefore, in the establishment of a normal complement of Sertoli cells in adulthood [52,53]. Postnatally induced hypothyroidism in rats extends the period of Sertoli cell proliferation beyond that of normal rats to 30 dpp [54]. Conversely, induced hyperthyroidism results in a precocious inhibition of mitotic activity at 12 dpp in rats [55]. The effect of TH inhibition on Sertoli cell proliferation is manifested during early postnatal life and relates to the period in which the rates of Sertoli cell proliferation decline. Serum levels of TH increase from just above the level of detection on 5 dpp to peak on 15 dpp in mice [56], whereas TH receptor levels in Sertoli cells decrease from 5 dpp until they are barely detectable on 20 dpp in rats, humans, and mice, respectively [57–59]. This declining period of Sertoli cell proliferation is coincident with an increase in androgen receptor (AR) expression, commencing around 5 dpp in rats [18], and may indicate that testosterone plays an inhibitory role in Sertoli cell proliferation. Both TH and testosterone independently reduce ^{3}H-thymidine incorporation and, therefore, Sertoli cell proliferation in 9 dpp rat Sertoli cell cultures; however, combined treatment does not result in further reductions [49]. This lack of effect of combined treatment may be due to the fact that TH and testosterone both alter the expression of the cell cycle inhibitors $p21^{Cip1}$ and $p27^{Kip1}$ via similar mechanisms [49,51]. Nonetheless, TH has a strong stimulatory effect on AR mRNA expression in 5 and 20 dpp rat Sertoli cells [60], and, conversely, suppression of TH decreases LH mRNA in the pituitary gland [61]. One could postulate that the decrease in Sertoli cell proliferation caused by reduced TH may involve secondary effects on the AR signalling pathway. Whether similar regulatory mechanisms involving TH and testosterone occur in primates remains debatable because expression of AR in primates occurs as a relatively late event during juvenile quiescence [62].

Retinoic acid is another important extracellular ligand that is required for the cessation of Sertoli cell proliferation at puberty. Earlier work using 6 dpp rat Sertoli cell cultures showed that retinoic acid treatment significantly reduces ^{3}H-thymidine incorporation after only 2 days of treatment, with proliferation suppressed to approximately 20% of control values following 8 days of treatment [49]. More recent work showed that retinoic acid antagonizes activin-induced rat Sertoli cell proliferation *in vitro* [63]. Activin A is a potent Sertoli cell mitogen during fetal, neonatal, and prepubertal life, and levels of activin A decline at puberty [46]. Prolonged activin A exposure at pathological levels after puberty is implicated in the development of Sertoli cell tumors [64].

Generally speaking, cell proliferation and differentiation are typically considered to be mutually exclusive events, such that when cells start to differentiate, they lose proliferative capacity by transcriptional repression of cell cycle genes and coordinate expression of specialized sets of genes, leading to dramatic changes in cellular structures and functions. However, the proliferation and differentiation events of Sertoli cells are not distinct, and, in fact, the mechanisms underpinning both of these events functionally interact. For example, several transgenic mouse models with targeted deletions for junction proteins feature prolonged Sertoli cell proliferation beyond puberty, including claudin-11 [65], occludin [66], and connexin-43 [67].

B. Sertoli cell junctions

Another feature of differentiated Sertoli cells at puberty is the appearance and organization of Sertoli cell junction proteins, resulting in the formation of a functional BTB (Table 14.1 and Figures 14.1 and 14.2). BTB formation at puberty is a progressive process, and the exact point at which the BTB becomes fully functional is not clear. Traditional BTB function analyses in rats and mice using hypertonic perfusion experiments and lanthanum electron opaque tracer permeability studies suggest that initial BTB formation occurs between 15 and 19 dpp, coincident with the appearance of pachytene spermatocytes [68–72]. In comparison, studies using a fluorescent biotin tracer of small molecular weight (<0.6 kDa) showed that the BTB is formed by 25 dpp in mice [73]. Moreover, alternative approaches using penetration of Cr-EDTA or albumin show that the BTB does not attain the "tightness" of the adult barrier until 44 dpp [70,74]. These differences in the timing of BTB formation are likely attributable to differences in tracer type and size. While the initial appearance of junction proteins occurs at around 12–15 dpp, they feature only diffuse localization within the seminiferous epithelium. Complete organization and localization of junctional complexes at the BTB in mice occurs at around 25 dpp for connexin-43 and ZO-1 [72] and 20 dpp for claudin-3 [73]. The timing of Sertoli cell junction organization and BTB formation is much later than the cessation of Sertoli cell proliferation. From these findings, one could ask, do the majority of Sertoli cells have to cease proliferation for the Sertoli cell junctions and BTB to form?

Regulation. The initial formation of Sertoli cell junctions and BTB function at puberty is dependent upon gonadotropin and testosterone stimulation. The data largely stem from genetically modified mice and natural alterations in gonadotropins in men. To date, no studies have directly assessed the importance of these hormones in BTB formation at puberty in rats. Rats have largely been used to study the hormonal regulation of adult BTB function, which is discussed in the section on differentiation in adult life.

Gonadotropin stimulation is essential for BTB formation in mice. In *hpg* mice lacking gonadotropins due to a natural GnRH gene mutation, the BTB does not form, as shown by the permeability of the fluorescent biotin (0.6 kDa) tracer into the seminiferous epithelium [75] and incomplete inter-Sertoli cell junctions [76]. While claudin-11 protein is still present in the testes of *hpg* mice, it is located away from the BTB in adluminal regions of Sertoli cells [75]. In comparison, claudin-3 is

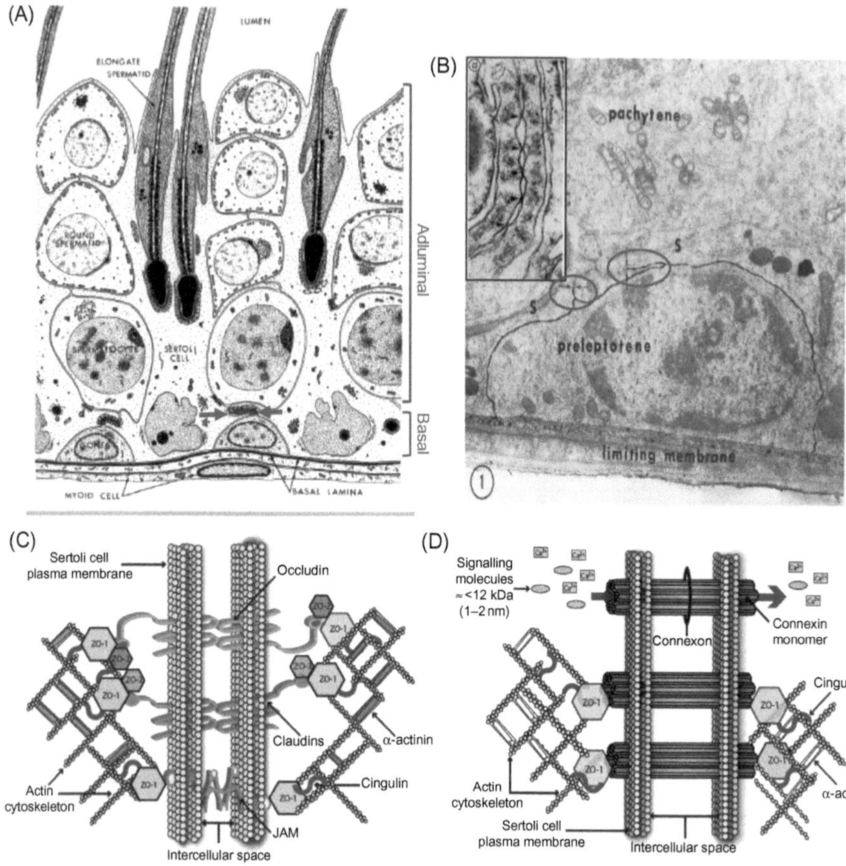

Figure 14.2 Diagrammatic and microscopic representations of Sertoli cell junctions and the BTB. (A) Schematic illustration of a cross section of the seminiferous epithelium demonstrating the position of the inter-Sertoli cell junctional complex constituting the BTB (red arrows). The BTB compartmentalizes the seminiferous epithelium in basal and adluminal regions (red lines and text), sequestering spermatocytes and round and elongate spermatids in the adluminal compartment inside the BTB. (B) Electron micrograph of a Sertoli–Sertoli intercellular tight junction (red circles). Lanthanum tracer is seen permeating around a stage VIII preleptotene spermatocyte between adjacent Sertoli cells (S). Tight junctions meet above this preleptotene spermatocyte. The furthermost upward penetration of lanthanum is indicated by opposing arrows inside the red circles. Lanthanum is not seen permeating a nearby pachytene spermatocyte. Insert in top left corner (129,000 magnification) shows that the tight junction is comprised of numerous kissing points along the length of the Sertoli cell plasma membrane. (C) Structure of tight junction and associated proteins found between Sertoli cells forming the BTB. Adjacent Sertoli cells form tight junctions with each other. The three main tight junction proteins include the claudin family and occludin, which are tetraspan proteins, and the JAM family, whose members have a single transmembrane domain. Claudins and occludin are structurally similar in that they span the membrane four times and keep their N- and C-termini within the cell. JAMs are structurally different in that they dimerize with an adjacent junctional adhesion molecule,

not detectable in *hpg* mice, suggesting a role for the endocrine regulation of the translocation of germ cells through the BTB in mice [75,77,78]. This possibility is further supported by data from Sertoli cell-specific AR-deficient mice (SCARKO) showing that claudin-3 is androgen dependent [73]. Data exist for humans that lack gonadotropins; electron microscopy observations show that hypogonadotropic hypogonadal men do not feature a mature junction structure and are permeable to lanthanum [29,31]. Gonadotropin treatment of these men resulted in the formation of junctions based on assessment by electron microscopy [29].

No known studies have assessed the role of FSH alone in the development of Sertoli cell junction organization and BTB formation at puberty *in vivo*. Nevertheless, more is known about the relative contributions of androgens. SCARKO mice feature delayed and incomplete BTB formation, whereby many tubules at 35 dpp feature diverse areas of the epithelium that are permeable to lanthanum [72], and at 56 dpp, many tubules are permeable to biotin (0.6 kDa) [73]. This process is likely underpinned by the lack of obvious junctional kissing points in 90% of seminiferous tubules in these mice [79]. Similarly, SCARKO mice feature at least a 10-day delay in the organization of ZO-1 and connexin-43, and their arrangement is still not comparable to that of wild-type littermates at 35 dpp [72]. The compromised BTB function in SCARKO mice likely contributes to the elevated testicular immunoglobulin levels; increase in interstitial macrophages, neutrophils, plasma cells, and eosinophils; and presence of antibodies against germ cell antigens in their serum [79]. Interestingly, the studies of Meng and Willems both showed variations in the permeability of the BTB throughout seminiferous tubule cross sections [72,73], suggesting possible stage-specific differences, although these studies did not assess such differences. The observations that androgens regulate BTB formation in the SCARKO mice extended earlier findings in *Tfm* mice in which the BTB is defective in adulthood [80] and are supported by work in rats in which *in utero* exposure to flutamide suppressed claudin-11 protein expression at puberty and adulthood [81]. Moreover, testosterone increases claudin-11 mRNA, protein localization, and tight junction function in 20 dpp rat Sertoli cells *in vitro* [82].

TH also plays a role in the organization of Sertoli cell junction proteins and the formation of the BTB at puberty. The effect of TH may be mediated via interactions with testosterone, as discussed above and below. Triiodothyronine administration between 0 and 15 dpp to rats results in an abnormal cytoplasmic localization of connexin-43 [83] and PTU-induced (TH antagonist) neonatal hypothyroidism

◀ forming a U-shaped domain in the extracellular space. All of these proteins are connected to the Sertoli cell cytoskeleton (actin and α actinin) via the peripheral adaptor proteins ZOs and cingulins. (D) Structure of gap junctions between two opposing cell membranes. Connexin proteins form hexamers called "connexons" that generate pores within the cell membranes, allowing for free diffusion of small molecules between cells. Gap junctions are linked to the actin cytoskeleton via adaptor proteins, such as ZO-1. (For interpretation of the references to color in this figure legend, the reader is referred to the web version of this book.)
With permission from (A) Russell et al. [102] and (B) Dym and Faucet [150] and Russell [151].

results in an approximately threefold reduction in connexin-43 immunoreactivity [84] as well as altered localization away from the BTB [85]. While TH has not been shown to directly regulate tight junction proteins, such as claudin-11, at the BTB, connexin-43 is essential to maintain BTB homeostasis by the regulation of tight junction reassembly [86]. In addition to endocrine regulation of junction organization and BTB formation, retinoic acid is also important. RARαknockout mice feature a delay in the appearance and localization of Sertoli cell junctions and incomplete BTB formation, as shown by adluminal germ cell shrinkage to hypertonic solution [87].

Retinoic acid has recently been identified as an important regulator of Sertoli cell tight junction mRNA expression, protein localization, and function. Although the data stem from *in vitro* observations, retinoic acid can initiate tight junction formation in 9 dpp rat Sertoli cells, which do not feature tight junctions under normal conditions [63]. Moreover, stimulation of 20 dpp rat Sertoli cells with retinoid signaling resulted in a 2.3-fold increase in transepithelial electrical resistance, a real-time measurement of tight junction function, and a significant increase in claudin-11 and ZO-1 mRNA expression (2.1- and 1.9-fold, respectively) [63]. In addition, *in vivo* work shows that retinoic acid regulates junction organization and BTB function in a stage-specific manner (stages VII–XII) as a result of its downstream effects on stage-dependent gene expression [88].

C. *Protein expression*

A third feature of differentiated Sertoli cells is a switch in the expression of Sertoli cell-specific proteins. The timely expression of these proteins is used as a marker of Sertoli cell differentiation *in vivo* through immunohistochemical staining (Tables 14.1 and 14.2). Adult Sertoli cells characteristically express AR, GATA1, laminin α5, and $p27^{kip1}$ and express low to no levels of anti-Müllerian hormone (AMH), cytokeratin 18, M2A antigen (now recognized as podoplanin [89]), aromatase, inhibin bodies, and neural cell adhesion molecule [9,17,90] (Table 14.1). While AR is widely used as a marker of differentiated Sertoli cells, its reliability as an accurate pubertal differentiation marker of Sertoli cells must be used with some caution because many studies have shown that AR protein can be detected immunohistochemically much earlier than the pubertal period, at 5 dpp in mice [91] and rats [18,92] and at 4–8 years in humans [27]. In comparison, other factors, such as Wilms' tumor gene, GATA4, vimentin, and inhibin α are expressed consistently from fetal life to adulthood and are routinely used as stable Sertoli cell markers in immunohistochemical analyses [9,17].

Regulation. Interestingly, data from the SCARKO and ARKO mice show that androgen action is not required for immunoexpression of AMH, $p27^{kip1}$, GATA1, and sulfated glycoprotein-2 [93]. However, androgens appear to be important for the switch in protein expression at puberty because patients with complete androgen insensitivity syndrome exhibit undifferentiated Sertoli cell characteristics, such as postpubertal persistence of AMH expression [94–96]. Likewise, in cases of testicular dysgenesis resulting from estrogen and antiandrogen exposure during fetal life,

Sertoli cells express a variety of undifferentiated markers, including M2A antigen, inhibin bodies, and AMH [20,21,97].

In addition to the role that TH plays in reducing Sertoli cell proliferation, reports show that TH also regulates the timely differentiation of Sertoli cells [54,55,98]. However, direct *in vivo* evidence for regulation by TH of changes in the expression of specific Sertoli cell differentiation proteins has not yet been reported. Nevertheless, TH receptors are expressed maximally during fetal and prepubertal periods of development, suggesting that TH is important during this time [58]. TH receptors are present on Sertoli cells and on germ cells up to mid-pachytene spermatocytes, suggesting that TH affects other events at puberty, such as the initiation of meiosis [57]. However, *in vitro* data support a role for TH in the regulation of the change in Sertoli cell protein expression because TH induces AR expression [60] and suppresses AMH expression [99] in immature rat Sertoli cells. FSH also creates an additive effect in combination with TH on AR and AMH expression [60,99], highlighting an important role for gonadotropins in the switch in protein expression. Indeed, other studies have shown that both FSH and testosterone regulate the expression of another protein marker of differentiation, GATA1 [100].

D. Morphology

There are many differences in nuclear and cytoplasmic morphology between undifferentiated and differentiated Sertoli cells (Table 14.1). These morphological changes are readily distinguished by electron microscopy; however, nuclear changes can also be visualized using light microscopy coupled with optimum fixation using Bouin's solution.

One of the most distinctive features of a differentiated Sertoli cell is the large irregular shaped nucleus with deep nuclear envelope invaginations and the tripartite nucleolus [101] (Table 14.1 and Figure 14.1A). Only the large central component of the tripartite nucleolus is the functional nucleolus; the two bodies flanking this nucleolus are satellite karyosomes thought to represent the centromeric region of the chromosomes, and they have little involvement in RNA synthesis [101,102]. This nuclear profile differs from that of the undifferentiated Sertoli cells, which feature an elongated-to-oval-shaped nucleus with a smooth nuclear envelope and a single nucleolus [103] (Table 14.1). Other distinctive features of the differentiated Sertoli cell relate to the cytoplasm. The differentiated Sertoli cell grows markedly in size to reach a total volume of 6000 μm^3 and a surface area of 12,000 μm^3 in rats and, as a result, features a much larger cytoplasm-to-nucleus ratio then its undifferentiated form [102] (Table 14.1). The differentiated Sertoli cell also features extensive cytoplasmic projections that encircle the germ cells and a marked increase in the number of organelles, particularly smooth endoplasmic reticulum and mitochondria, which is thought to relate to the high metabolic activity and lipid and steroid synthesis of the differentiated Sertoli cell [101]. The presence of differentiated Sertoli cells, as assessed using these nuclear and cytoplasmic transformations in rats, occurs later than the other features of Sertoli cell differentiation, as discussed above, at around 35 dpp; prior to this, the Sertoli cell display morphological features that are in between those of undifferentiated and differentiated Sertoli cells [103].

Regulation. It appears that TH plays an important role in the morphological differentiation of Sertoli cells because induced hypothyroidism delays morphological differentiation [54]. Moreover, seasonal breeders show marked reductions in the number of endoplasmic reticulum, mitochondria, and Golgi complexes as well as a loss of the nucleolus during testicular regression, highlighting an important role for gonadotropin regulation of Sertoli cell morphological differentiation [104].

IV. Differentiation in adult life

Sertoli cell differentiation has been considered a unidirectional event; once Sertoli cells have acquired their phenotypic transformations around puberty, they are thought to remain a stable, terminally differentiated population for the remainder of adult life with the inability to revert back to their undifferentiated state. Technically speaking, their classification as terminally differentiated refers to their inability to proliferate in the adult testis; however, this classification is commonly used to imply that all markers of differentiation are fixed and unmodifiable. Research since the early 2000s challenges the terminal classification of Sertoli cell differentiation and shows that in particular experimental settings, adult Sertoli cells can dedifferentiate, meaning that they reach full differentiation in adulthood but are capable of reverting to a less differentiated form, displaying a mixed phenotype of undifferentiated and differentiated characteristics, by gaining proliferative ability, losing BTB function, and changing their protein expression (Figure 14.3) (reviewed in [16]). Similarly, other cell types traditionally considered terminally differentiated, such as human epidermal keratinocytes [105] and myocytes [45,106,107], can revert to a dedifferentiated state. These discoveries are due in part to the development of newer technical approaches, including well-validated immune probes and modern stereological methods, and the use of seasonal breeders, Djungarian hamsters, to study adult cell proliferation and differentiation.

In this section, we discuss evidence largely from the Djungarian hamsters as well as *in vitro* data that overturn the view that adult Sertoli cells are terminally differentiated using proliferative ability, Sertoli cell junction organization, BTB permeability, and other expression protein markers of differentiation. What is not discussed is the role that metabolism has in cell cycle progression. Differentiation requires energy and the energy transport system. Thus, how may it play a role in the changing differentiation status of adult Sertoli cells? Suffice it to say that the metabolic features of adult Sertoli cells remain high based on their glycolytic activity and lipid and steroid synthesis as they progress from an undifferentiated, proliferative state to the differentiated, nonproliferative state, a feature not usually observed in nonproliferative cells.

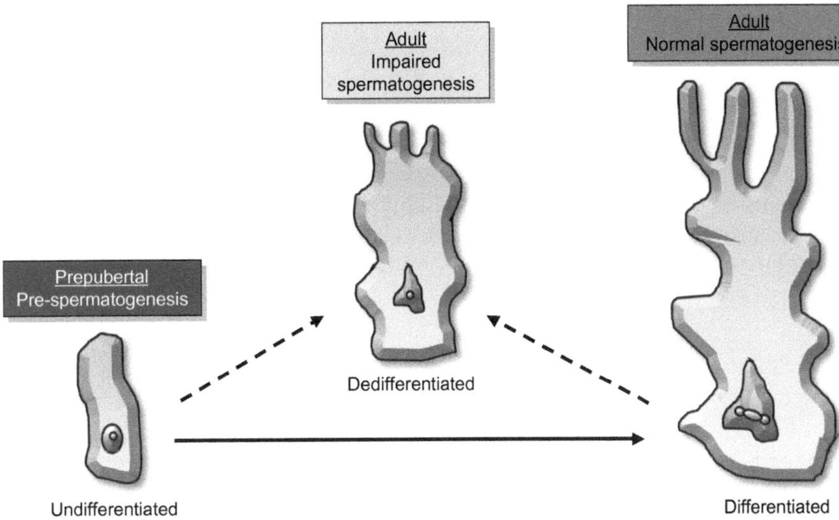

Figure 14.3 Working model of Sertoli cell differentiation. The traditional Sertoli cell differentiation pathway is considered a unidirectional event; undifferentiated prepubertal Sertoli cells are phenotypically transformed into a differentiated population at around puberty. This population is considered stable for the remainder of adult life in that its differentiation state is fixed and unmodifiable. A growing body of evidence shows that in settings of impaired spermatogenesis, adult Sertoli cells exhibit one or more features of undifferentiated Sertoli cells, termed dedifferentiated Sertoli cells. Studies show that Sertoli cells may never reach their differentiated form (as in the case of developmental disorders, such as testicular dysgenesis syndrome) or that in unique endocrine settings, Sertoli cells are capable of reverting back to a dedifferentiated (less differentiated) form.

The concept that Sertoli cells are permanently quiescent throughout adult life stems from work in the nineteenth century by Enrico Sertoli, who reported "branched" cells do not divide in the adult testis [41]. Sertoli's work was subsequently supported by the finding that no overt Sertoli cell proliferation occurs in normal [3,108] or hormonally manipulated adult rats [109] and that these cells do not degenerate after hypophysectomy [110,111]. Importantly, more recent investigations show that normal adult Sertoli cells may retain the residual capacity to proliferate. Human and mouse adult Sertoli cells resume proliferation *in vitro* after 4 [112] and 20 [113] days, respectively, when cultured in the presence of serum. The addition of activin in a similar Sertoli cell culture experiment in rats resulted in a more potent induction of proliferation (approximately threefold) in as little as 24 h [114]. These findings suggest that adult Sertoli cells maintain the residual capacity to proliferate, yet to this day, no known reports of observations of Sertoli cells proliferating in the testes of healthy men, adult primates, or rats *in vivo* are available. However, observations in adult Sertoli cells in unique endocrine settings *in vivo* show directly that adult Sertoli cells can be induced to proliferate, as evidenced by an induction of the percentage of Sertoli cells that were proliferating cell nuclear antigen (PCNA) reactive ([115], reviewed in [16]). This finding was in accord with this laboratory's earlier work [116] that provided the first compelling evidence using best-practice stereological methods that the adult Sertoli cell population is modifiable in response to a changing endocrine environment in seasonal-breeding Djungarian hamsters [115,116]. This animal displays 30–40% reductions in adult Sertoli cell number during the "short-day" period, which features low serum gonadotropins and essentially a spermatogonial-only testis [116]. The addition of exogenous FSH during this short-day period normalizes the Sertoli cell population within a week [116] and induces Sertoli cell proliferation, promotes Sertoli cell junction organization and BTB function, alters the expression of proteins such as Gata1, and reinitiates spermatogenesis [115,117]. These data from Djungarian hamsters support earlier findings of similar fluctuations in Sertoli cell number across other photoperiod-sensitive breeders, such as stallions [118], golden hamsters [119], and rams [120,121]. However, these findings were largely dismissed due to the use of older stereological techniques and the belief that the ability of adult Sertoli cells to alter their differentiation status may just be unique to seasonal breeders.

To date, humans are the only nonseasonal breeding species to demonstrate changes in proliferative potential *in vivo* following endocrine stimulation; 12 weeks of gonadotropin suppression in men resulted in $1.7 \pm 0.6\%$ PCNA-positive adult Sertoli cells [22]. In addition, Sertoli cells were observed to be Ki-67 and TUNEL positive, suggesting that Sertoli cells were in cell cycle (nonterminally differentiated and in G0 phase). There is some evidence to imply that at least some of the component of the Sertoli cell population is nonfunctional based on what appears to be a diminished AR protein expression associated with PCNA-reactive Sertoli cells [22]. The limiting nature of working with fixed human testis tissue makes dissecting the mechanisms regulating this change in cellular kinetics a great scientific challenge. Therefore, this research needs to be furthered in appropriate and accessible animal models.

While earlier work showed that no proliferative activity or change in Sertoli cell number occurs in adult rats or mice [3–6,109], these studies were conducted using less sensitive techniques to detect Sertoli cell proliferation and older stereological approaches to enumerate Sertoli cell numbers. To be fair, even modern stereological techniques in rats have not shown significant changes in the number of adult Sertoli cells following chronic [122] or acute hormone suppression [123]; the same is true in humans [124,125]. However, this possibility cannot be ruled out, even when modern stereological techniques are used, small fluctuations in the size of the Sertoli cell population are masked, and the use of other methods will be required to confirm whether the Sertoli cell population is modifiable. There are data to support reduced Sertoli cell numbers in men following vasectomy [126], in patients with impaired spermatogenesis [127], and during the aging process [128]. Whether these findings are due to failure of the Sertoli cell population to reach its full complement remained unknown in these reports [127,128]; however, this failure of Sertoli cells would not explain changes in the Sertoli cell population following vasectomy [126].

Normal adult wild-type mouse data show that the Sertoli cell population is not stable after puberty and, in fact, continues to rise well into adulthood between 30 and 70 days of age [15]. In these mice, 3% of the Sertoli cell population is in cell cycle as assessed by incorporation of EdU and measured by flow cytometry [129]. Thus, taken together, laboratory mice may be a suitable model to further our understanding of the regulation of adult Sertoli cell proliferation and differentiation. Data from hamsters and humans show that hormones play a role in the status of adult Sertoli cell differentiation, with FSH having an important role in modulating the size of the adult Sertoli cell and its differentiation status, at least in hamsters. Although a very different model, inhibin alpha knockout mice displays Sertoli cell tumorigenesis, whereas in combination with the deletion of the FSH-β subunit gene, tumor formation is suppressed in adult mice [64]. These mice are an important model that may shed light on the factors that regulate normal and dysregulated Sertoli cell differentiation and function.

It seems likely that such dynamic cells within the highly active environment of the seminiferous epithelium would maintain the capacity to regenerate or undergo repair. While other terminally differentiated cell types, such as keratinocytes and erythrocytes, that lose their nucleus during the differentiation process are obviously irreversibly growth arrested [107], adult Sertoli cells retain their nucleus and differentially express an unusual suite of genes and proteins associated with differentiation, proliferation, and DNA repair that are not characteristic of terminally differentiated cells. First, adult Sertoli cells express ID proteins, helix-loop-helix proteins that inhibit differentiation and promote proliferation [130,131]. Moreover, manipulation of ID expression in adult Sertoli cells enables them to reenter the cell cycle [132]. Adult Sertoli cells also express genes involved in inhibition of the cell cycle, such as the cyclin-dependent kinase inhibitor 1B (CDKN1B/$p27^{kip1}$), which arrests cells in the G1 phase of the cell cycle [133]. Adult Sertoli cells that resume proliferation in culture feature a 70% reduction in CDKN1B/$p27^{kip1}$ expression and a 50% increase in ID2 expression [113]. Moreover, activin-induced proliferation of Sertoli cells increases mRNA levels of the cell cycle regulator cyclin D2 by 9.2-fold within 18 h [114]. Adult Sertoli cells also express a number of DNA repair proteins, including XRCC5 (KU86), XTCC6 (KU70), PARP1, and phosphorylated XRCC1; these cells also express TP53BP1 before irradiation and phosphorylated ATM following irradiation [113,133–137]. These DNA repair proteins are not expressed in other somatic cells in the testis, including Leydig cells and peritubular myoid cells, and, indeed, adult Sertoli cells can use these proteins to repair DNA double strand breaks [113]. The ability of adult Sertoli cells to proliferate under endocrine stimulation in seasonal breeders and humans and their expression factors associated with cell proliferation, differentiation, and DNA suggest that differentiated adult Sertoli cells may maintain the residual capacity to proliferate and regenerate. In addition, given that in hamsters, upon FSH stimulation, the majority of Sertoli cells are in cell cycle as assessed by PCNA reactivity also implies that perhaps most, if not all, Sertoli cells have the capacity to regain proliferative potential.

In addition to proliferative ability, features of adult Sertoli cell differentiation, such as the organization of junctional proteins and BTB function, are not stable throughout adult life. While the junction organization and BTB formation at puberty (discussed above) require endocrine hormones, they also require continued endocrine stimulation throughout adult life to maintain their structure and function.

Administration of the GnRH antagonist acyline to adult rats after 75 dpp for 8 weeks diminishes occludin protein expression; alters claudin-11, ZO-1, and JAM-A protein organization; and impairs BTB function with biotin tracer permeation throughout the seminiferous epithelium [138]. Significant recoveries in protein organization and BTB impermeability required 7 days of treatment with human chorionic gonadotropin and FSH; however, FSH alone was not sufficient for junction structure and function [135]. These findings support the need for androgen stimulation to stimulate claudin-11 protein localization and function, as seen *in vitro* [81]. Furthermore, data show that during reinitiation of spermatogenesis following hormonal suppression, meiosis can occur when the BTB is permeable, and BTB permeability relates to the presence of particular meiotic and postmeiotic germ cell types in the rat epithelium. These findings challenge our current understanding of BTB function as being either open or closed, and, instead, show that the BTB is selectively permeable to molecular weights of up to 70 kDa during reinitiation of spermatocyte development and early round spermatid development. These findings also raise the question of whether germ cells regulate the BTB and, thus, Sertoli cell differentiation [40].

Moreover, the seasonal decreases in gonadotropins in adult short-day Djungarian hamsters result in a disorganized BTB in terms of the localization of particular tight junction and associated proteins (115); a non functional BTB as assessed by permeability to biotin (0.6 kDa) [117]. Unpublished data from our laboratory show that a chronic hormone suppression regime in men [139] also impairs Sertoli cell junction organization. This finding is in contrast to published data in which no change in junction proteins was shown following short-term hormone suppression in men; however, longer suppression was not studied [140]. These data provide direct evidence that endocrine perturbations can alter the adult Sertoli cell differentiation state based on the use of junction organization and BTB function as markers.

In addition to showing that adult Sertoli cells can dedifferentiate, studies using genetic reprogramming experiments have shown that adult Sertoli cells can transdifferentiate. Deletion of the *Dmrt1* transcription factor in adult mouse Sertoli cells activates *Foxl2* and reprograms differentiated Sertoli cells into granulosa cells [141]. Therefore, antagonism between the sex-specific transcriptional regulators *Dmrt1* and *Foxl2* is a key to maintaining Sertoli cell fate and differentiation throughout adult life [141]. Importantly, the sex steroid environment was altered in these mice, with elevated levels of aromatase protein and reduced androgens, showing that the endocrine milieu is central to maintaining the Sertoli cell adult differentiation state [141]. This result supports earlier work in which long-term estrogen therapy to patients who had experienced sex reversal transformed the nuclear and cytoplasmic morphology of adult Sertoli cells into that of its undifferentiated form [142].

V. Human Sertoli cell differentiation and pathology

Healthy men that present as normospermic based in the immune markers of differentiation have not shown evidence of undifferentiated Sertoli cells [22,23]. Indeed, the literature provides ample evidence of altered adult Sertoli cell differentiation states in various pathologies of clinical testicular tissue, including Sertoli cell tumors [19], testicular intraepithelial neoplasia [24], azoospermia [28,143], and varying degrees of impaired spermatogenesis [23,25,28,143] (Table 14.2). Haverfield and colleagues have recently shown that claudin-11, a marker of the tight junction, and the gap junction protein connexin-43 display altered spatial patterns of organization in men with primary seminiferous tubule failure compared with normospermic controls. However, surprisingly, no differences in claudin-11 and connexion-43 protein organization were observed between tubules with meiotic germ cells present and those with no germ cells. The data imply two important pieces of information pertaining to adult human Sertoli cell differentiation: (i) an important relationship exists between the organization of the Sertoli cell junctional complex and sustained spermatogenesis and (ii) the degree of spermatogenic impairment may not be associated with the extent of adult Sertoli cell differentiation, at least in the case of the BTB and in primary spermatogenic failure.

In healthy men with suppressed gonadotropins resulting in severe oligospermia, adult Sertoli cells have been shown to dedifferentiate [22]. However, in diseased states, the situation is not as clear-cut; the presence of adult Sertoli cells with undifferentiated characteristics may be a result of a failure to differentiate during pubertal life or of dedifferentiation during in adult life (Figure 14.3).

VI. Future perspectives

At this time, the adult Sertoli cell population has greater biological and perhaps clinical potential that we ever envisioned in the early 2000s. The subject of adult Sertoli cell differentiation status needs to be readdressed as a result of the compelling data presented. It has been realized in a number of species that adult Sertoli cells are not in fact terminally differentiated and their population is not fixed and unmodifiable, which is similar to other cells of the body. In this chapter, we have attempted to review what we know and what we do not know about adult Sertoli cell differentiation. However, given the newness of the topic, we anticipate that what is not known is likely to be the place where critical leaps in knowledge will be created to understand the unrealized potential of the adult Sertoli cell population. The growing body of evidence opens up new avenues of thinking and research and raises interesting questions, including:

1. To what extent can adult undifferentiated Sertoli cells support germ cell development? Is this relevant for fertility?
2. Do all Sertoli cells in the testis retain the ability to alter their differentiation state or is this a characteristic of a subset, perhaps the stem-Sertoli cell-like progenitor cells?
3. What hormones influence adult Sertoli cell differentiation? What keeps the adult Sertoli cell population differentiated?
4. How tight does tge BTB have to be to support spermatogenesis?
5. How are proliferation, BTB function, protein expression, and morphology related and regulated?

References

[1] McCarrey J. Development of the germ cell. In: Desjardins C, Ewing LL, editors. Cell and molecular biology of the testis. New York, NY: Oxford University Press; 1993. p. 58–89.

[1a] McCullagh DR. Dual endocrine activity of the testes. Science 1932;76:19–20.

[2] Ebling FJP. The neuroendocrine timing of puberty. Reprod Camb Engl 2005;129: 675–83.

[3] Steinberger A, Steinberger E. Replication pattern of Sertoli cells in maturing rat testis *in vivo* and in organ culture. Biol Reprod 1971;4:84–7.

[4] Orth JM. Proliferation of Sertoli cells in fetal and postnatal rats: a quantitative autoradiographic study. Anat Rec 1982;203:485–92.

[5] Kluin PM, Kramer MF, Rooij DG. Proliferation of spermatogonia and Sertoli cells in maturing mice. Anat Embryol (Berl.) 1984;169:73–8.

[6] Vergouwen RPFA, Jacobs SGPM, Huiskamp R, Davids JaG, de Rooij DG. Proliferative activity of gonocytes, Sertoli cells and interstitial cells during testicular development in mice. J Reprod Fertil 1991;93:233–43.

[7] Baker PJ, O'Shaughnessy PJ. Role of gonadotrophins in regulating numbers of Leydig and Sertoli cells during fetal and postnatal development in mice. Reproduction 2001; 122:227–34.

[8] Simorangkir DR, Ramaswamy S, Marshall GR, Roslund R, Plant TM. Sertoli cell differentiation in rhesus monkey (*Macaca mulatta*) is an early event in puberty and precedes attainment of the adult complement of undifferentiated spermatogonia. Reproduction 2012;143:513–22.

[9] Sharpe RM, McKinnell C, Kivlin C, Fisher JS. Proliferation and functional maturation of Sertoli cells, and their relevance to disorders of testis function in adulthood. Reproduction 2003;125:769–84.

[10] Meachem SJ, McLachlan RI, Kretser DM, de Robertson DM, Wreford NG. Neonatal exposure of rats to recombinant follicle stimulating hormone increases adult Sertoli and spermatogenic cell numbers. Biol Reprod 1996;54:36–44.

[11] Orth JM, Gunsalus GL, Lamperti AA. Evidence from Sertoli cell depleted rats indicates that spermatid number in adults depends on numbers of Sertoli cells produced during perinatal development. Endocrinology 1988;122:787–94.

[12] Plant TM, Ramaswamy S, Simorangkir D, Marshall GR. Postnatal and pubertal development of the rhesus monkey (*Macaca mulatta*) testis. Ann N Y Acad Sci 2005; 1061:149–62.

[13] Simorangkir DR, Marshall GR, Plant TM. Sertoli cell proliferation during prepubertal development in the rhesus monkey (*Macaca mulatta*) is maximal during infancy when gonadotropin secretion is robust. J Clin Endocrinol Metab 2003;88:4984–9.

[14] Cortes D, Müller J, Skakkebæk NE. Proliferation of Sertoli cells during development of the human testis assessed by stereological methods. Int J Androl 1987;10:589–96.

[15] Hazra R, Corcoran L, Robson M, McTavish KJ, Upton D, Handelsman DJ, et al. Temporal role of Sertoli cell androgen receptor expression in spermatogenic development. Mol Endocrinol Baltim Md 2013;27:12–24.

[16] Tarulli GA, Stanton PG, Meachem SJ. Is the adult Sertoli cell terminally differentiated? Biol Reprod 2012;87(13):1–11.

[17] Brehm R, Steger K. Regulation of Sertoli cell and germ cell differentation. Adv Anat Embryol Cell Biol 2005;181:1–93.

[18] You L, Sar M. Androgen receptor expression in the testes and epididymides of prenatal and postnatal Sprague-Dawley rats. Endocrine 1998;9:253–61.

[19] Brehm R, Rey R, Kliesch S, Steger K, Marks A, Bergmann M. Mitotic activity of Sertoli cells in adult human testis: an immunohistochemical study to characterize Sertoli cells in testicular cords from patients showing testicular dysgenesis syndrome. Anat Embryol (Berl.) 2006;211:223–36.

[20] Raĭgorodskaia NI, Bolotova NV, Morozov DA, Sedova LN, Zakharova NB, Ivanenko IL, et al. Sexual development of boys with hypospadias. Urol Mosc Russ 2013;1999: 88–92.

[21] Nistal M, Gonzalez-Peramato P, De Miguel MP. Sertoli cell dedifferentiation in human cryptorchidism and gender reassignment shows similarities between fetal environmental and adult medical treatment estrogen and antiandrogen exposure. Reprod Toxicol 2013; 42:172–9.

[22] Tarulli GA, Stanton PG, Loveland KL, Meyts ER-D, McLachlan RI, Meachem SJ. A survey of Sertoli cell differentiation in men after gonadotropin suppression and in testicular cancer. Spermatogenesis 2013;3:e24014.

[23] Haverfield JT, Meachem SJ, O'Bryan MK, McLachlan RI, Stanton PG. Claudin-11 and connexin-43 display altered spatial patterns of organization in men with primary seminiferous tubule failure compared with controls. Fertil Steril 2013;100(3):658–66. Available from: http://dx.doi.org/10.1016/j.fertnstert.2013.04.034 [Epub May 23, 2013].

[24] Fink C, Weigel R, Fink L, Wilhelm J, Kliesch S, Zeiler M, et al. Claudin-11 is overexpressed and dislocated from the blood–testis barrier in Sertoli cells associated with testicular intraepithelial neoplasia in men. Histochem Cell Biol 2009;131:755–64.

[25] Cavicchia JC, Sacerdote FL, Ortiz L. The human blood–testis barrier in impaired spermatogenesis. Ultrastruct Pathol 1996;20:211–8.

[26] Bar-Shira Maymon B, Paz G, Elliott DJ, Hammel I, Kleiman SE, Yogev L, et al. Maturation phenotype of Sertoli cells in testicular biopsies of azoospermic men. Hum Reprod 2000;15:1537–42.

[27] Chemes HE, Rey RA, Nistal M, Regadera J, Musse M, González-Peramato P, et al. Physiological androgen insensitivity of the fetal, neonatal, and early infantile testis is explained by the ontogeny of the androgen receptor expression in Sertoli cells. J Clin Endocrinol Metab 2008;93:4408–12.

[28] Steger K, Rey R, Kliesch S, Louis F, Schleicher G, Bergmann M. Immunohistochemical detection of immature Sertoli cell markers in testicular tissue of infertile adult men: a preliminary study. Int J Androl 1996;19:122–8.

[29] de Kretser D, Burger H. Ultrastructural studies of the human Sertoli cell in normal men and males with hypogonadotropic hypogonadism before and after gonadotropic treatment. In: Saxena CG, Beling, HM, editors. Gonadotropins. New York; 1972.

[30] Steger K, Rey R, Louis F, Kliesch S, Behre HM, Nieschlag E, et al. Reversion of the differentiated phenotype and maturation block in Sertoli cells in pathological human testis. Hum Reprod Oxf Engl 1999;14:136–43.

[31] de Kretser DM. The fine structure of the immature human testis in hypogonadotrophic hypogonadism. Virchows Arch B 1968;1:283–96.

[32] McLachlan RI, Rajpert-De Meyts E, Hoei-Hansen CE, de Kretser DM, Skakkebaek NE. Histological evaluation of the human testis—approaches to optimizing the clinical value of the assessment: mini review. Hum Reprod Oxf Engl 2007;22:2–16.

[33] McLachlan RI, O'Bryan MK. State of the art for genetic testing of infertile men. J Clin Endocrinol Metab 2010;95:1013–24.

[34] Galdieri M, Monaco L, Stefanini M. Secretion of androgen binding protein by Sertoli cells is influenced by contact with germ cells. J Androl 1984;5:409–15.
[35] Le Magueresse B, Jégou B. *In vitro* effects of germ cells on the secretory activity of Sertoli cells recovered from rats of different ages. Endocrinology 1988;122:1672–80.
[36] Castellón E, Janecki A, Steinberger A. Age-dependent Sertoli cell responsiveness to germ cells *in vitro*. Int J Androl 1989;12:439–50.
[37] Allenby G, Sharpe RM, Foster PM. Changes in Sertoli cell function *in vitro* induced by nitrobenzene. Fundam Appl Toxicol Off J Soc Toxicol 1990;14:364–75.
[38] McKinnell C, Sharpe RM. The role of specific germ cell types in modulation of the secretion of androgen-regulated proteins (ARPs) by stage VI–VIII seminiferous tubules from the adult rat. Mol Cell Endocrinol 1992;83:219–31.
[39] Nicholls PK, Harrison CA, Gilchrist RB, Farnworth PG, Stanton PG. Growth differentiation factor 9 is a germ cell regulator of Sertoli cell function. Endocrinology 2009; 150:2481–90.
[40] Haverfield JT, Meachem SJ, Nicholls PK, Rainczuk KE, Simpson ER, Stanton PG. Differential permeability of the blood–testis barrier during reinitiation of spermatogenesis in adult male rats. Endocrinology 2014;. Available from: http://dx.doi.org/doi:10.1210/en.2013-1878.
[41] Sertoli E. Sulla sturttura dei canalicoli seminiferi dei testicolo. Arch Sci Med 1878;2 (107–146):267–95.
[42] Paradis AN, Gay MS, Zhang L. Binucleation of cardiomyocytes: the transition from a proliferative to a terminally differentiated state. Drug Discov Today 2013.
[43] Nouspikel T, Hanawalt PC. Terminally differentiated human neurons repair transcribed genes but display attenuated global DNA repair and modulation of repair gene expression. Mol Cell Biol 2000;20:1562–70.
[44] Fortini P, Ferretti C, Pascucci B, Narciso L, Pajalunga D, Puggioni EMR, et al. DNA damage response by single-strand breaks in terminally differentiated muscle cells and the control of muscle integrity. Cell Death Differ 2012;19:1741–9.
[45] Crescenzi M. Reactivation of the cell cycle in terminally differentiated cells. Georgetown, TX; New York, NY: Kluwer Academic/Plenum Pub.; 2003. Landes Bioscience/Eurekah.com.
[46] Barakat B, O'Connor AE, Gold E, de Kretser DM, Loveland KL. Inhibin, activin, follistatin and FSH serum levels and testicular production are highly modulated during the first spermatogenic wave in mice. Reproduction 2008;136:345–59.
[47] Rannikki AS, Zhang FP, Huhtaniemi IT. Ontogeny of follicle-stimulating hormone receptor gene expression in the rat testis and ovary. Mol Cell Endocrinol 1995;107:199–208.
[48] Boitani C, Stefanini M, Fragale A, Morena AR. Activin stimulates Sertoli cell proliferation in a defined period of rat testis development. Endocrinology 1995;136:5438–44.
[49] Buzzard JJ, Wreford NG, Morrison JR. Thyroid hormone, retinoic acid, and testosterone suppress proliferation and induce markers of differentiation in cultured rat sertoli cells. Endocrinology 2003;144:3722–31.
[50] Holsberger DR, Jirawatnotai S, Kiyokawa H, Cooke PS. Thyroid hormone regulates the cell cycle inhibitor p27Kip1 in postnatal murine Sertoli cells. Endocrinology 2003; 144:3732–8.
[51] Holsberger DR, Buchold GM, Leal MC, Kiesewetter SE, O'Brien DA, Hess RA, et al. Cell-cycle inhibitors p27Kip1 and p21Cip1 regulate murine Sertoli cell proliferation. Biol Reprod 2005;72:1429–36.
[52] Holsberger DR, Cooke PS. Understanding the role of thyroid hormone in Sertoli cell development: a mechanistic hypothesis. Cell Tissue Res 2005;322:133–40.

[53] Wagner MS, Wajner SM, Maia AL. The role of thyroid hormone in testicular development and function. J Endocrinol 2008;199:351–65.
[54] Van Haaster LH, De Jong FH, Docter R, De Rooij DG. The effect of hypothyroidism on Sertoli cell proliferation and differentiation and hormone levels during testicular development in the rat. Endocrinology 1992;131:1574–6.
[55] Van Haaster LH, de Jong FH, Docter R, de Rooij DG. High neonatal triiodothyronine levels reduce the period of Sertoli cell proliferation and accelerate tubular lumen formation in the rat testis, and increase serum inhibin levels. Endocrinology 1993;133:755–60.
[56] Hadj-Sahraoui N, Seugnet I, Ghorbel MT, Demeneix B. Hypothyroidism prolongs mitotic activity in the post-natal mouse brain. Neurosci Lett 2000;280:79–82.
[57] Buzzard JJ, Morrison JR, O'Bryan MK, Song Q, Wreford NG. Developmental expression of thyroid hormone receptors in the rat testis. Biol Reprod 2000;62:664–9.
[58] Jannini EA, Crescenzi A, Rucci N, Screponi E, Carosa E, de Matteis A, et al. Ontogenetic pattern of thyroid hormone receptor expression in the human testis. J Clin Endocrinol Metab 2000;85:3453–7.
[59] Palmero S, Maggiani S, Fugassa E. Nuclear triiodothyronine receptors in rat Sertoli cells. Mol Cell Endocrinol 1988;58:253–6.
[60] Arambepola NK, Bunick D, Cooke PS. Thyroid hormone effects on androgen receptor messenger RNA expression in rat Sertoli and peritubular cells. J Endocrinol 1998;156:43–50.
[61] Rao JN, Liang JY, Chakraborti P, Feng P. Effect of thyroid hormone on the development and gene expression of hormone receptors in rat testes *in vivo*. J Endocrinol Invest 2003;26:435–43.
[62] McKinnell C, Saunders PT, Fraser HM, Kelnar CJ, Kivlin C, Morris KD, et al. Comparison of androgen receptor and oestrogen receptor beta immunoexpression in the testes of the common marmoset (*Callithrix jacchus*) from birth to adulthood: low androgen receptor immunoexpression in Sertoli cells during the neonatal increase in testosterone concentrations. Reprod Camb Engl 2001;122:419–29.
[63] Nicholls PK, Harrison CA, Rainczuk KE, Wayne Vogl A, Stanton PG. Retinoic acid promotes Sertoli cell differentiation and antagonises activin-induced proliferation. Mol Cell Endocrinol 2013;377:33–43.
[64] Matzuk MM, Finegold MJ, Su JG, Hsueh AJ, Bradley A. Alphainhibin is a tumour-suppressor gene with gonadal specificity in mice. Nature 1992;360:313–9.
[65] Mazaud-Guittot S, Meugnier E, Pesenti S, Wu X, Vidal H, Gow A, et al. Claudin 11 deficiency in mice results in loss of the Sertoli cell epithelial phenotype in the testis. Biol Reprod 2010;82:202–13.
[66] Saito K, O'Donnell L, McLachlan RI, Robertson DM. Spermiation failure is a major contributor to early spermatogenic suppression caused by hormone withdrawal in adult rats. Endocrinology 2000;141:2779–85.
[67] Sridharan S, Simon L, Meling DD, Cyr DG, Gutstein DE, Fishman GI, et al. Proliferation of adult sertoli cells following conditional knockout of the Gap junctional protein GJA1 (connexin 43) in mice. Biol Reprod 2007;76:804–12.
[68] Vitale R, Fawcett DW, Dym M. The normal development of the blood testis barrier and the effects of clomiphene and estrogen treatment. Anat Rec 1973;176:331–44.
[69] Bergmann M, Dierichs R. Postnatal formation of the blood–testis barrier in the rat with special reference to the initiation of meiosis. Anat Embryol (Berl.) 1983;168:269–75.

[70] Russell LD, Bartke A, Goh JC. Postnatal development of the Sertoli cell barrier, tubular lumen, and cytoskeleton of Sertoli and myoid cells in the rat, and their relationship to tubular fluid secretion and flow. Am J Anat 1989;184:179–89.
[71] Cavicchia JC, Sacerdote FL. Correlation between blood–testis barrier development and onset of the first spermatogenic wave in normal and in busulfan-treated rats: a lanthanum and freeze-fracture study. Anat Rec 1991;230:361–8.
[72] Willems A, Batlouni SR, Esnal A, Swinnen JV, Saunders PTK, Sharpe RM, et al. Selective ablation of the androgen receptor in mouse Sertoli cells affects Sertoli cell maturation, barrier formation and cytoskeletal development. PLoS One 2010;5:e14168.
[73] Meng J, Holdcraft RW, Shima JE, Griswold MD, Braun RE. Androgens regulate the permeability of the blood–testis barrier. Proc Natl Acad Sci USA 2005;102:16696–700.
[74] Setchell BP, Pöllänen P, Zupp JL. Development of the blood–testis barrier and changes in vascular permeability at puberty in rats. Int J Androl 1988;11:225–33.
[75] McCabe MJ, Allan CM, Foo CFH, Nicholls PK, McTavish KJ, Stanton PG. Androgen initiates Sertoli cell tight junction formation in the hypogonadal (hpg) mouse. Biol Reprod 2012;87:38.
[76] Myers M, Ebling FJP, Nwagwu M, Boulton R, Wadhwa K, Stewart J, et al. Atypical development of Sertoli cells and impairment of spermatogenesis in the hypogonadal (hpg) mouse. J Anat 2005;207:797–811.
[77] Komljenovic D, Sandhoff R, Teigler A, Heid H, Just WW, Gorgas K. Disruption of blood–testis barrier dynamics in ether-lipid-deficient mice. Cell Tissue Res 2009;337:281–99.
[78] Smith BE, Braun RE. Germ cell migration across Sertoli cell tight junctions. Science 2012;338:798–802.
[79] Meng J, Greenlee AR, Taub CJ, Braun RE. Sertoli cell-specific deletion of the androgen receptor compromises testicular immune privilege in mice. Biol Reprod 2011;85:254–60.
[80] Fritz IB, Lyon MF, Setchell BP. Evidence for a defective seminiferous tubule barrier in testes of Tfm and Sxr mice. J Reprod Fertil 1983;67:359–63.
[81] Florin A, Maire M, Bozec A, Hellani A, Chater S, Bars R, et al. Androgens and post-meiotic germ cells regulate claudin-11 expression in rat Sertoli cells. Endocrinology 2005;146:1532–40.
[82] Kaitu'u-Lino TJ, Sluka P, Foo CFH, Stanton PG. Claudin-11 expression and localisation is regulated by androgens in rat Sertoli cells in vitro. Reproduction 2007;133:1169–79.
[83] Marchlewska K, Kula K, Walczak-Jedrzejowska R, Kula W, Oszukowska E, Filipiak E, et al. Maturational changes in connexin 43 expression in the seminiferous tubules may depend on thyroid hormone action. Arch Med Sci 2013;9:139–45.
[84] Gilleron J, Nebout M, Scarabelli L, Senegas-Balas F, Palmero S, Segretain D, et al. A potential novel mechanism involving connexin 43 gap junction for control of sertoli cell proliferation by thyroid hormones. J Cell Physiol 2006;209:153–61.
[85] St-Pierre N, Dufresne J, Rooney AA, Cyr DG. Neonatal hypothyroidism alters the localization of gap junctional protein Connexin 43 in the testis and messenger RNA levels in the epididymis of the rat. Biol Reprod 2003;68:1232–40.
[86] Li MWM, Mruk DD, Lee WM, Cheng CY. Connexin 43 is critical to maintain the homeostasis of the blood–testis barrier via its effects on tight junction reassembly. Proc Natl Acad Sci USA 2010;107:17998–8003.

[87] Chung SSW, Sung W, Wang X, Wolgemuth DJ. Retinoic acid receptor alpha is required for synchronization of spermatogenic cycles and its absence results in progressive breakdown of the spermatogenic process. Dev Dyn Off Publ Am Assoc Anat 2004;230:754–66.

[88] Hasegawa K, Saga T. Retinoic acid signaling in Sertoli cells regulates organization of the blood–testis barrier through cyclical changes in gene expression. Development 2012;139:4347–55.

[89] Sonne SB, Herlihy AS, Hoei-Hansen CE, Nielsen JE, Almstrup K, Skakkebaek NE, et al. Identity of M2A (D2-40) antigen and gp36 (Aggrus, T1A-2, podoplanin) in human developing testis, testicular carcinoma in situ and germ-cell tumours. Virchows Arch 2006;449:200–6.

[90] Nistal M, Pastrián LG, González-Peramato P, De Miguel MP. Inhibin bodies: a new marker for immature Sertoli cells. Histopathology 2011;58:1019–27.

[91] Willems A, De Gendt K, Allemeersch J, Smith LB, Welsh M, Swinnen JV, et al. Early effects of Sertoli cell-selective androgen receptor ablation on testicular gene expression. Int J Androl 2010;33:507–17.

[92] Bremner WJ, Millar MR, Sharpe RM, Saunders PT. Immunohistochemical localization of androgen receptors in the rat testis: evidence for stage-dependent expression and regulation by androgens. Endocrinology 1994;135:1227–34.

[93] Tan KAL, De Gendt K, Atanassova N, Walker M, Sharpe RM, Saunders PTK, et al. The role of androgens in sertoli cell proliferation and functional maturation: studies in mice with total or Sertoli cell-selective ablation of the androgen receptor. Endocrinology 2005;146:2674–83.

[94] Lan K-C, Chen Y-T, Chang C, Chang Y-C, Lin H-J, Huang K-E, et al. Up-regulation of SOX9 in Sertoli cells from testiculopathic patients accounts for increasing anti-Mullerian hormone expression via impaired androgen receptor signaling. PLoS One 2013;8(10):e76303. Available from: http://dx.doi.org/10.1371/journal.pone.0076303.

[95] Rajpert-De Meyts E, Jørgensen N, Graem N, Müller J, Cate RL, Skakkebaek NE. Expression of anti-Müllerian hormone during normal and pathological gonadal development: association with differentiation of Sertoli and granulosa cells. J Clin Endocrinol Metab 1999;84:3836–44.

[96] Rey R, Mebarki F, Forest MG, Mowszowicz I, Cate RL, Morel Y, et al. Anti-Müllerian hormone in children with androgen insensitivity. J Clin Endocrinol Metab 1994;79:960–4.

[97] Wikström AM, Hoei-Hansen CE, Dunkel L, Rajpert-De Meyts E. Immunoexpression of androgen receptor and nine markers of maturation in the testes of adolescent boys with Klinefelter syndrome: evidence for degeneration of germ cells at the onset of meiosis. J Clin Endocrinol Metab 2007;92:714–9.

[98] De França LR, Hess RA, Cooke PS, Russell LD. Neonatal hypothyroidism causes delayed Sertoli cell maturation in rats treated with propylthiouracil: evidence that the Sertoli cell controls testis growth. Anat Rec 1995;242:57–69.

[99] Arambepola NK, Bunick D, Cooke PS. Thyroid hormone and follicle stimulating hormone regulate Müllerian-inhibiting substance messenger ribonucleic acid expression in cultured neonatal rat Sertoli cells. Endocrinology 1998;139:4489–95.

[100] Zhang Z, Wu AZ, Feng Z-M, Mruk D, Cheng CY, Chen C-LC. Gonadotropins, via cAMP, negatively regulate GATA-1 gene expression in testicular cells. Endocrinology 2002;143:829–36.

[101] Hess R, França L. Chapter 3—Structure of the Sertoli cell. In: Skinner MK, Griswold MD, editors. Sertoli cell biology. San Diego, CA: Elsevier Academic Press; 2005. p. 19–40.

[102] Russell LD, Hikim APS, Ettlin R. Histological and histopathological evaluation of the testis. Clearwater, FL: Cache River Press; 1990.

[103] Ramos Jr AS, Dym M. Ultrastructural differentiation of rat Sertoli cells. Biol Reprod 1979;21:909–22.

[104] Muñoz EM, Fogal T, Dominguez S, Scardapane L, Piezzi RS. Ultrastructural and morphometric study of the sertoli cell of the viscacha (*Lagostomus maximus maximus*) during the annual reproductive cycle. Anat Rec 2001;262:176–85.

[105] Sun X, Fu X, Han W, Zhao Y, Liu H, Sheng Z. Dedifferentiation of human terminally differentiating keratinocytes into their precursor cells induced by basic fibroblast growth factor. Biol Pharm Bull 2011;34:1037–45.

[106] Odelberg SJ, Kollhoff A, Keating MT. Dedifferentiation of mammalian myotubes induced by msx1. Cell 2000;103:1099–109.

[107] Sacco A, Pajalunga D, Latella, L, Siepi F, Rufini A, Crescenzi M. Cell cycle reactivation in skeletal muscle and other terminally differentiated cells; 2003.

[108] Orth JM. The role of follicle-stimulating hormone in controlling sertoli cell proliferation in testes of fetal rats. Endocrinology 1984;115:1248–55.

[109] Russell LD, Clermont Y. Degeneration of germ cells in normal, hypophysectomized and hormone treated hypophysectomized rats. Anat Rec 1977;187:347–66.

[110] Muffly KE, Nazian SJ, Cameron DF. Effects of follicle-stimulating hormone on the junction-related Sertoli cell cytoskeleton and daily sperm production in testosterone-treated hypophysectomized rats. Biol Reprod 1994;51:158–66.

[111] Murphy HD. Sertoli cell stimulation following intratesticular injections of FSH in the hypophysectomized rat. Proc Soc Exp Biol Med 1965;118:1202–5.

[112] Chui K, Trivedi A, Cheng CY, Cherbavaz DB, Dazin PF, Huynh ALT, et al. Characterization and functionality of proliferative human Sertoli cells. Cell Transplant 2011;20:619–35.

[113] Ahmed EA, Barten-van Rijbroek AD, Kal HB, Sadri-Ardekani H, Mizrak SC, van Pelt AMM, et al. Proliferative activity *in vitro* and DNA repair indicate that adult mouse and human Sertoli cells are not terminally differentiated, quiescent cells. Biol Reprod 2009;80:1084–91.

[114] Nicholls PK, Stanton PG, Chen JL, Olcorn JS, Haverfield JT, Qian H, et al. Activin signaling regulates Sertoli cell differentiation and function. Endocrinology 2012; 153:6065–77.

[115] Tarulli GA, Stanton PG, Lerchl A, Meachem SJ. Adult sertoli cells are not terminally differentiated in the Djungarian hamster: effect of FSH on proliferation and junction protein organization. Biol Reprod 2006;74:798–806.

[116] Meachem SJ, Stanton PG, Schlatt S. Follicle-stimulating hormone regulates both Sertoli cell and spermatogonial populations in the adult photoinhibited Djungarian Hamster Testis. Biol Reprod 2005;72:1187–93.

[117] Tarulli GA, Meachem SJ, Schlatt S, Stanton PG. Regulation of testicular tight junctions by gonadotrophins in the adult Djungarian hamster *in vivo*. Reproduction 2008;135:867–77.

[118] Johnson L, Varner DD, Tatum ME, Scrutchfield WL. Season but not age affects Sertoli cell number in adult stallions. Biol Reprod 1991;45:404–10.

[119] Hikim APS, Bartke A, Russell LD. Morphometric studies on hamster testes in gonadally active and inactive states: light microscope findings. Biol Reprod 1988;39:1225–37.

[120] De Reviers M, Hochereau-de Reviers M-T, Blanc MR, Brillard JP, Courot M, Pelletier J. Control of Sertoli and germ cell populations in the cock and sheep testes. Reprod Nutr Dév 1980;20:241–9.

[121] Hötzel MJ, Markey CM, Walkden-Brown SW, Blackberry MA, Martin GB. Morphometric and endocrine analyses of the effects of nutrition on the testis of mature Merino rams. J Reprod Fertil 1998;113:217–30.

[122] McLachlan RI, Wreford NG, de Kretser DM, Robertson DM. The effects of recombinant follicle-stimulating hormone on the restoration of spermatogenesis in the gonadotropin-releasing hormone-immunized adult rat. Endocrinology 1995;136:4035–43.

[123] Meachem SJ, Mclachlan RI, Stanton PG, Robertson DM, Wreford NG. FSH immunoneutralization acutely impairs spermatogonial development in normal adult rats. J Androl 1999;20:756–62 [discussion 755].

[124] Zhengwei Y, Wreford NG, Royce P, deKretser DM, McLachlan RI. Stereological evaluation of human spermatogenesis after suppression by testosterone treatment: heterologous pattern of spermatogenic impairment. J Clin Endocrinol Metab 1998;83: 1284–91.

[125] McLachlan RI, O'Donnell L, Stanton PG, Balourdos G, Frydenberg M, de Kretser DM, et al. Effects of testosterone plus medroxyprogesterone acetate on semen quality, reproductive hormones, and germ cell populations in normal young men. J Clin Endocrinol Metab 2002;87:546–56.

[126] Jarow JP, Budin RE, Dym M, Zirkin BR, Noren S, Marshall FF. Quantitative pathologic changes in the human testis after vasectomy. A controlled study. N Engl J Med 1985;313:1252–6.

[127] Hentrich A, Wolter M, Szardening-Kirchner C, Lüers GH, Bergmann M, Kliesch S, et al. Reduced numbers of Sertoli, germ, and spermatogonial stem cells in impaired spermatogenesis. Mod Pathol 2011;10:1380–9.

[128] Johnson L, Petty CS, Neaves WB. Age-related variation in seminiferous tubules in men. A stereological evaluation. J Androl 1986;7:316–22.

[129] Haverfield GT, Stanton PG, Meachem SJ. Acute inhibition of follicle-stimulating hormone suppresses Sertoli cell tumor development [Submitted].

[130] Sablitzky F, Moore A, Bromley M, Deed R, Newton J, Norton J. Stage and subcellular-specific expression of Id proteins in male germ and Sertoli cells implicates distinctive regulatory roles for Id proteins during meiosis, spermatogenesis, and Sertoli cell function. Cell Growth Differ 1998;9:1015–24.

[131] Chaudhary J, Johnson J, Kim G, Skinner MK. Hormonal regulation and differential actions of the helix-loop-helix transcriptional inhibitors of differentiation (Id1, Id2, Id3, and Id4) in Sertoli cells. Endocrinology 2001;142:1727–36.

[132] Chaudhary J, Sadler-Riggleman I, Ague JM, Skinner MK. The helix-loop-helix inhibitor of differentiation (ID) proteins induce post-mitotic terminally differentiated Sertoli cells to re-enter the cell cycle and proliferate. Biol Reprod 2005;72:1205–17.

[133] Beumer TL, Roepers-Gajadien HL, Gademan IS, van Buul PP, Gil-Gomez G, Rutgers DH, et al. The role of the tumor suppressor p53 in spermatogenesis. Cell Death Differ 1998;5:669–77.

[134] Choi EK, Lee YH, Choi YS, Kwon HM, Choi MS, Ro JY, et al. Heterogeneous expression of Ku70 in human tissues is associated with morphological and functional alterations of the nucleus. J Pathol 2002;198:121–30.

[135] Anway MD, Folmer J, Wright WW, Zirkin BR. Isolation of sertoli cells from adult rat testes: an approach to *ex vivo* studies of Sertoli cell function. Biol Reprod 2003;68: 996–1002.

[136] Hamer G, Roepers-Gajadien HL, van Duyn-Goedhart A, Gademan IS, Kal HB, van Buul PPW, et al. Function of DNA-protein kinase catalytic subunit during the early meiotic prophase without Ku70 and Ku86. Biol Reprod 2003;68:717–21.

[137] Hamer G, Kal HB, Westphal CH, Ashley T, de Rooij DG. Ataxia telangiectasia mutated expression and activation in the testis. Biol Reprod 2004;70:1206–12.

[138] McCabe MJ, Tarulli GA, Meachem SJ, Robertson DM, Smooker PM, Stanton PG. Gonadotropins regulate rat testicular tight junctions *in vivo*. Endocrinology 2010; 151:2911–22.

[139] Matthiesson KL, Stanton PG, O'Donnell L, Meachem SJ, Amory JK, Berger R, et al. Effects of testosterone and levonorgestrel combined with a 5alpha-reductase inhibitor or gonadotropin-releasing hormone antagonist on spermatogenesis and intratesticular steroid levels in normal men. J Clin Endocrinol Metab 2005;90:5647–55.

[140] Ilani N, Armanious N, Lue Y-H, Swerdloff RS, Baravarian S, Adler A, et al. Integrity of the blood–testis barrier in healthy men after suppression of spermatogenesis with testosterone and levonorgestrel. Hum Reprod Oxf Engl 2012;27:3403–11.

[141] Matson CK, Murphy MW, Sarver AL, Griswold MD, Bardwell VJ, Zarkower D. DMRT1 prevents female reprogramming in the postnatal mammalian testis. Nature 2011;476:101–4.

[142] Schulze C. Response of the human testis to long-term estrogen treatment: morphology of Sertoli cells, Leydig cells and spermatogonial stem cells. Cell Tissue Res 1988;251:31–43.

[143] Nistal M, Jimenez F, Paniagua R. Sertoli cell types in the Sertoli-cell-only syndrome: relationships between Sertoli cell morphology and aetiology. Histopathology 1990;16:173–80.

[144] Nah WH, Lee JE, Park HJ, Park NC, Gye MC. Claudin-11 expression increased in spermatogenic defect in human testes. Fertil Steril 2011;95:385–8.

[145] de Kretser DM, Kerr JB, Paulsen CA. Evaluation of the ultrastructural changes in the human sertoli cell in testicular disorders and the relationship of the changes to the levels of serum FSH. Int J Androl 1981;4:129–44.

[146] Brehm R, Marks A, Rey R, Kliesch S, Bergmann M, Steger K. Altered expression of connexins 26 and 43 in Sertoli cells in seminiferous tubules infiltrated with carcinoma-in-situ or seminoma. J Pathol 2002;197:647–53.

[147] Kratzer SS, Ulbright TM, Talerman A, Srigley JR, Roth LM, Wahle GR, et al. Large cell calcifying Sertoli cell tumor of the testis: contrasting features of six malignant and six benign tumors and a review of the literature. Am J Surg Pathol 1997;21:1271–80.

[148] Kliesch S, Behre HM, Hertle L, Bergmann M. Alteration of Sertoli cell differentiation in the presence of carcinoma in situ in human testis. J Urol 1998;160:1894–8.

[149] Janssen M, Johnston W. Anaplastic seminoma of the testis:Ultrastructural analysis of three cases. Cancer 1978;41:538–44.

[150] Dym M, Fawcett DW. The blood-testis barrier in the rat and the physiological compartmentation of the seminiferous epithelium. Biol Reprod 1970;3:308–26.

[151] Russell LD. The blood-testis barrier and its formation relative to spermatocyte maturation in the adult rat: a lanthanum tracer study. Anat Rec 1978;190:99–111.

Gene knockouts that affect Sertoli cell function 15

Wei Yan
Department of Physiology and Cell Biology, University of Nevada School of Medicine, Reno, NV

I. Introduction

Sertoli cells are functionally versatile because they not only provide physical, nutritional, and metabolic support to developing male germ cells, but they also direct and coordinate the complex spermatogenic program via endocrine (e.g., production of inhibin) and paracrine (e.g., secretion of growth factors and cytokines) mechanisms [1,2]. Consistent with the multitasking nature of Sertoli cells, the Sertoli cell transcriptome has been defined as abundant and dynamic [3,4].

The loss-of-function study remains an ideal way of defining the physiological roles of Sertoli-cell-expressed genes. Soon after the global gene knockout (KO) technology became available in the 1990s, several genes well known to be expressed in Sertoli cells and to play critical roles in hormonal signaling. These genes include follicle-stimulating hormone (FSH) receptors, androgen receptor (AR), estrogen receptors, and inhibin and were inactivated first globally and subsequently in a Sertoli-cell-specific manner (reviewed in Ref. [5]). These KO mouse models have led to a much better understanding of the effects of those endocrine signals on Sertoli cells and the spatiotemporal requirement for normal Sertoli cell function and spermatogenesis in general. That part of the work was included in the last edition of *Sertoli Cell Biology*, published in 2005.

Since then, many more global and conditional KO (cKO) mouse lines have been generated, and some display a male infertility phenotype due to compromised Sertoli cell function and, consequently, disrupted germ cell development (Tables 15.1 and 15.2). Most of the genes studied either belong to a protein family with known functions or have been functionally characterized in other cell types or organs. By analyzing the developmental defects in both Sertoli cells and spermatogenic cells, these studies dissected both primary effects due to Sertoli cell defects and secondary effects on male germ cell development.

In this chapter, I first summarize these global and cKO mouse models. By reviewing the gene ontology and phenotypes, we can clearly see that Sertoli cells have an active transcriptional program that controls the expression of a wide variety of genes that are required for almost all aspects of spermatogenesis, including spermatogonia stem cell self-renewal, mitotic proliferation of differentiated spermatogonia, meiotic

Sertoli Cell Biology. DOI: http://dx.doi.org/10.1016/B978-0-12-417047-6.00015-6

Table 15.1 Universal knockouts displaying male infertility due to Sertoli cell defects

Strain/ Genotype	Protein family	Protein encoded	Phenotype	References
Aff4$^{-/-}$	Transcription factor	AF4/FMR2 family, member 4 (AFF4)	Male sterility due to azoospermia.	[6]
Amph$^{-/-}$	Synaptic vesicle endocytosis protein	Amphiphysin	Decreased phagocytosis and phosphatidylinositol-4,5-bisphosphate-induced actin polymerization.	[7–11]
Arid4a$^{-/-}$; *Arid4b*$^{+/-}$	Transcriptional coactivator	AT-rich interaction domain 4a and 4b (ARID4A and ARID4B)	Progressive loss of male fertility due to meiotic and/or spermiogenic, accompanied by hypogonadism and seminal vesicle agenesis or hypodysplasia.	[12]
Bsg$^{-/-}$	Type I integral membrane receptor	Basigin, CD147, or Neurothelin	Male infertility characterized by azoospermia cause by compromised integrity of BTB.	[13,14]
Cldn11$^{-/-}$	Claudin tight junction protein family	Claudin protein 11	Male sterility due to Sertoli cells maturation defects, defective tight junctions between Sertoli cells and germ cells, and germ cell apoptosis.	[15]
Cnot7$^{-/-}$	Transcriptional co-factor	CCR4-NOT transcription complex, subunit 7 (CNOT7)	Male sterility due to oligo-astheno-teratozoospermia, characterized by impaired sperm motility, unsynchronized spermatid maturation, and Sertoli cell abnormalities.	[16]
Dmrt1	Transcription factor	*Doublesex* and *mab-3* related transcription factor 1	Both Sertoli cells and germ cells fail to differentiate properly after birth followed by depletion, leading to severely dysgenic testes.	[17]
Dnaja1$^{-/-}$ (i.e., *DjA1*$^{-/-}$)	Chaperone protein	DnaJ (Hsp40) homolog, subfamily A, member 1 (DANJA1)	Male sterility due to spermiogenic arrest.	[18]
Elmo1$^{-/-}$	Engulfment protein	Engulfment and cell motility 1	Disrupted seminiferous epithelium, multi-nucleated giant cells, uncleared apoptotic germ cells, and decreased sperm output.	[19]

Etv5$^{-/-}$	Transcription factor	ETS variant gene 5 (ETV5)	Spermatogonial stem cells fail to self-renew efficiently, leading to progressive germ cell depletion and male infertility.	[20]
Fads2$^{-/-}$	Enzyme	Fatty acid desaturase 2 (FADS2)	Male sterility due to defective Sertoli cell polarity and BTB, characterized by Sertoli cell vacuolization and spermiogenic arrest.	[21]
Fshr$^{-/-}$	G-protein-coupled receptor	Follicle-stimulating hormone receptor (FHSR)	Subfertility due to reduced number of spermatogenic cells.	[5,22]
Katnal1$^{1H/1H}$	Microtubule-severing protein	Katanin p60 subunit A-like1 (KATNAL1)	Male infertility due to disruptions of Sertoli cell microtubule dynamics and sloughing of spermatids from the seminiferous epithelium.	[23]
Mpf2$^{-/-}$	Enzyme	Peroxisomal multifunctional protein 2 (MFP-2)	Extensive accumulation of neutral lipids in Sertoli cells, resulting in complete testicular atrophy and male infertility.	[24]
Sox8$^{-/-}$	Transcription factor	Sex-determining region Y-box 8 (SOX8)	Adult male infertility caused by defective BTB and germ cell depletion.	[25]
Tam$^{-/-}$	Receptor tyrosine kinases	Tyro3, Axl and Mer receptor tyrosine kinases (TAM)	The testicular immune homeostasis was impaired due to compromised BTB integrity.	[26]
Tjp2 chimera	Tight junction integral membrane protein	Tight junction protein 2 (TJP2), also called zonula occluden 2 (ZO2)	*Male* infertility due to compromised BTB, characterized by increased BTB permeability, Sertoli cell vacuolization and germ cell loss.	[27]
Tlr2$^{-/-}$; *Tlr4*$^{-/-}$	Toll-like receptors	Toll-lie receptors 2 and 4 (TLR2 and TLR4)	No upregulation of TNF-α and MCP-1 expression and lack of macrophage infiltration in busulfan-induced spermatogenic cell damages *in vivo*.	[28]
Wipf3$^{-/-}$ (i.e., *Cr16*$^{-/-}$)	The Wiskott–Aldrich syndrome protein (WASP)-interacting protein (WIP) family	WAS/WASL interacting protein family, member 3 (WIPF3), also called corticosteroids and regional expression 16″ (CR16)	Male-specific sterility caused by aberrant spermiogenesis characterized by deformed head morphology.	[29]

Table 15.2 **Sertoli cell conditional knockouts displaying male infertility**

Strain/Genotype	Protein family	Protein encoded	Phenotype	References
Aldh1a1-3$^{lox/lox}$; *Amh-Cre*	Enzyme: involved in retinoid acid biosynthesis	Retinaldehyde dehydrogenases 1, 2, and 3 (RALDH1, 2, and 3)	Spermatogonia differentiation ceases at the Aal stage, failure in initiation of the first wave of spermatogenesis.	[30]
Apc$^{fl/fl}$; *Amhr2-Cre*	Tumor suppressor	Adenomatous polyposis coli protein (APC)	Premature germ cell loss and Sertoli-cell-only seminiferous tubules; loss of Sertoli cell apical extensions; defective BTB.	[31]
AR$^{lox/lox}$; *Amh-Cre*	Nuclear receptor	Androgen receptor (AR)	Azoospermia due to a spermatogenic arrest at diplotene spermatocytes stage.	[32–35]
Ctnnb1$^{fl/fl}$; *Amhr2-Cre*	Adherence junction protein	Catenin (cadherin-associated protein), beta 1 (CTNNB1)	Sertoli cell maturation defects and progressive germ cell loss, leading to male sterility. Sertoli cell tumor with partial penetrance.	[36,37]
Cx43$^{lox/lox}$; *Amh-Cre*	Gap junction protein	Connexin 43	Spermatogonial arrest or Sertoli cell-only syndrome, and male infertility.	[38]
Cyp26b1$^{fl/fl}$; *Amh-Cre*	Cytochrome P450 enzyme	Cytochrome P450 enzyme 26B1 (CYP26b1)	Loss of gonocytes.	[39]
Dicer$^{fl/fl}$; *Amh-Cre*	Enzyme: RNase III family	DICER	Defective Sertoli cell maturation and inability to properly support meiosis and spermiogenesis, leading to azoospermia and male infertility.	[40,41]
Dmrt1$^{floxed/-}$; *Dhh-Cre*	Transcription factor	*Doublesex* and *mab-3*-related transcription factor 1 (DMRT1)	Sertoli cells overproliferate and fail to differentiate completely, resulting in complete germ cell depletion by 4 weeks of age.	[42]
Elmo1$^{fl/fl}$; *Amh-Cre*	Cytoplasmic engulfment protein	Engulfment and cell motility 1 (ELMO1)	Disrupted seminiferous epithelium characterized by the presence of numerous multi-nucleated giant cells, uncleared apoptotic germ cells, and decreased sperm output.	[19]
Ggpps$^{lox/lox}$; *Amh-Cre*	Enzyme: the prenyltransferase family	Geranylgeranyl diphosphate synthase 1 (GGPPS)	BTB defects and disrupted immune response leading to depletion of spermatogonial and male infertility.	[43]

Klf4$^{fl/fl}$; *Amh-Cre*	Transcription factor	Krüppel-like factor 4 (KLF4)	Disorganized seminiferous epithelium with highly vacuolized Sertoli cell cytoplasm and impaired apical secretion of the Sertoli cell.	[44]
Lkb1$^{lox/lox}$; *Amhr2-cre*	Protein kinase	Liver kinase B1 (LKB1)	Defects in Sertoli cell polarity and testicular junction complexes, leading to progressive germ cell loss and Sertoli-cell-only tubules.	[45]
Nr5a1$^{lox/lox}$; *Amhr2-Cre*	Orphan nuclear receptor	Steroidogenic factor 1 (SF1)	Sertoli cell maturation defects leading to spermatogenic cell depletion and male infertility.	[46]
Pex5$^{fl/fl}$; *Amh-Cre*	Peroxisomal proteins	Peroxisomal biogenesis factor 5 (PEX5)	Lipid inclusions leading to progressive degeneration of the seminiferous tubules and male infertility.	[47,48]
Rara$^{lox/lox}$; *Amh-Cre*	Nuclear receptor	Retinoid acid receptor alpha (RARα)	Sertoli cells fail to initiate cyclic gene expression pattern and spermatogonia fail to differentiate. Severe germ cell depletion.	[49]
Rb$^{lox/lox}$; *Amh-Cre;*	The retinoblastoma family	Retinoblastoma protein (pRb)	Sertoli cell maturation defects, defective BTB, and Sertoli–germ cell interactions.	[50]
Rdh10$^{fl/fl}$; *Amh-Cre*$^{+}$	Retinol dehydrogenase family	Retinol dehydrogenase 10 (RDH10)	Blockage of spermatogonial differentiation in juvenile testes.	[51]
Rxrb$^{lox/lox}$; *Amh-Cre;*	Nuclear receptor	Retinoid X receptor beta (RXRB)	Failure in spermatid release (i.e., spermiation), accumulation of cholesterol esters, and subsequent testis degeneration.	[52]
Wt1$^{fl/-}$; *Amh-Cre*	Transcription factor	William's tumor protein 1 (WT1)	Disrupted testis cord formation followed by depletion of Sertoli cells and germ cells in postnatal testes, leading to severely atrophic testes and male infertility.	[53,54]
Wt1$^{fl/-}$; *ER*™-*Cre*	Transcription factor	William's tumor protein 1 (WT1)	Tamoxifen-induced *Wt1* inactivation in Sertoli cells of 8-week-old testes leads to massive loss of germ cells leading to SC-only appearance in the seminiferous tubules due to disrupted BTB and SC polarity.	[54]

entry and progression, haploid male germ cell differentiation, and spermiation (i.e., the process through which the fully developed spermatozoa are released from the seminiferous epithelium). In addition, Sertoli cells have the function of clearing dead germ cells or unwanted cytoplasmic contents of spermatids (i.e., residual bodies) during or at the end of spermiogenesis through phagocytosis or engulfment. Two types of gap junctions, which exist in between adjacent Sertoli cells as the blood−testis barrier (BTB) and in between Sertoli cells and developing spermatids as the ectoplasmic specialization (ES), represent another important function of Sertoli cells, and disruptions of these gap junctions lead to spermatogenic failure.

Despite the progress made, studies using gene inactivation, especially Sertoli-cell-specific cKO studies, remain very limited, as evidenced by the less than 40 KO lines developed (Tables 15.1 and 15.2). This small number of studies is largely due to the low efficiency of the first-generation gene-targeting technologies, which require homologous recombination-mediated gene targeting in mouse embryonic stem cells with inherent low efficiency in steps such as gene targeting and germ line transmission [55]. With more and more flox lines becoming available through the Knockout Mouse Project or next-generation genome-editing technologies, it is anticipated that most of the Sertoli-cell-expressed genes will be inactivated specifically in Sertoli cells so that their physiological roles can be defined. Therefore, at the end of this chapter, some of the latest genome-editing technologies, which will significantly improve the efficiency of gene targeting and shorten the time required for generating KO lines, are briefly discussed.

II. Genes identified as essential for normal Sertoli cell development and function through KO studies

Among the approximately 40 KO mouse lines with Sertoli cell defects (Tables 15.1 and 15.2), most have null mutations in genes encoding proteins of the following five types: enzymes, transcription factors, receptors, gap junction, and cytoplasmic signaling proteins. These proteins are involved in many critical physiological events in Sertoli cells, such as retinoid acid (RA) signaling, hormonal response, small RNA biogenesis, formation and maintenance of BTB and ES, and dynamic transcriptional programs. Although these genes represent a tiny fraction of the entire Sertoli cell transcriptome, findings from these KO studies already suggest that Sertoli cells play a critical regulatory role in male germ cell development.

A. *Enzymes*

1. *Enzymes involved in RA biosynthesis and signaling*

RA is an active metabolite of vitamin A and is synthesized by a group of specialized enzymes, including the retinaldehyde dehydrogenases 1, 2, and 3 (RALDH1, 2, and 3). RA functions by binding to their nuclear receptors, which of three isoforms, RARA, RARB, and RARG, which are either expressed in RA-producing cells or in RA target cells [56]. Through inactivation, all three genes (*Aldh1a1*, *Aldh1a2*, and *Aldh1a3*) encode RALDH1, 2, and 3, respectively, specifically in Sertoli cells. Raverdeau and colleagues reported several findings on RA action during testicular development [30]. First, Sertoli cell RA production is essential for the initiation of differentiation of Aal spermatogonia to type A1 spermatogonia and, thus, is required for the first spermatogenic cycle. Interestingly, RA of Sertoli cell origin is dispensable for the subsequent spermatogenic cycles but essential for spermiation. This study clearly demonstrated that Sertoli cells indirectly mediate RA's effects on spermatogonial differentiation, whereas the effects of RA on meiotic initiation in spermatocytes are direct.

Another enzyme critical for RA biosynthesis is retinol dehydrogenase 10 (RDH10), which is responsible for oxidation of vitamin A, a critical step for RA biosynthesis [57]. By inactivating *Rdh10* globally and conditionally, Tong and colleagues demonstrated that complete RA deficiency occurred when *Rdh10* was inactivated in both Sertoli cells and germ cells, which did not happen when *Rdh10* was only inactivated in either of the two cell types. Sertoli-cell-specific depletion of *Rdh10* showed a mild germ cell depletion phenotype [51]. These data suggest that the major source of RA in juvenile testes consists of Sertoli cells, but RA appears to be produced in germ cells as well. Interestingly, RDH10-dependent RA biosynthesis appears to be dispensable for normal spermatogenesis in adult testes. These findings are generally in agreement with those from Sertoli-cell-specific cKO of RALDH1-3 [30], indicating that Sertoli-cell-produced RA is critical for the first spermatogenic cycle because it promotes undifferentiated spermatogonia to differentiate into type A1 spermatogonia. In the subsequent spermatogenic cycles, RA is needed for meiotic entry and progression, but in a Sertoli-cell-independent manner. The catabolism of RA is mediated by a family of cytochrome P450 enzymes (CYP26A1, CYP26B1, and CYP26C1) that converts RA to more polar, inactive metabolites [58]. Inactivation of *Cyp26b1* exclusively in Sertoli cells between E15.5 and E16.5 leads to a loss of gonocytes/prospermatogonia due to early resumption of mitotic proliferation followed by entry into the meiotic prophase, suggesting that this RA-catabolizing enzyme is essential for maintaining the undifferentiated state of prospermatogonia during embryonic development [39]. Taken together, these three Sertoli-cell-specific KO studies demonstrate that RA in Sertoli cells acts as a critical regulator that affects the initiation of spermatogenesis.

2. Key upstream kinase involved in energy metabolism

Liver kinase B1 (LKB1) is a master upstream protein kinase that regulates numerous downstream kinases involved in diverse cellular functions [59]. Once activated, LKB1 can control the activities of 14 different kinases, including adenosine monophosphate-activated protein kinases (AMPKs) [60]. The LKB1–AMPK signaling cascade negatively regulates the mammalian target of rapamycin (mTOR) by both phosphorylating and activating Tuberin (TSC2)-induced Ras homolog enriched in brain-GTPase activity and by phosphorylating and inhibiting Raptor (a regulatory-associated protein of mTOR) [60]. By selectively disrupting the LKB1/AMPK/TSC1/TSC2/mTOR signaling cascade in developing Sertoli cells, *Lkb1* cKO testes display progressive germ cell loss and "Sertoli-cell-only" seminiferous tubules accompanied by defects in Sertoli cell polarity and testicular junction complexes as well as decreased activation of the MAP/microtubule affinity [45]. These findings suggest that this well-characterized signaling cascade has an essential role in regulating Sertoli cell polarity and other functions that are essential for normal spermatogenesis.

3. Enzyme mediating protein C20-prenylation

It has been reported that protein prenylation levels and the ratio of geranylgeranylated (modified with a geranylgeranyl diphosphate (GGPP) moiety) to farnesylated (modified with an farnesyl-pyrophosphate (FPP) moiety) proteins are regulated during testicular development in rats [61]. FPP and GGPP prenylate proteins that have a CaaX motif in their carbon termini, such as the RAS superfamily of GTPases [62]. GGPP synthase 1 (GGPPS) is a member of the prenyltransferase family, and it catalyzes the synthesis GGPP from FPP [63]. Inactivation of GGPPS in Sertoli cells results in constitutive activation of the proinflammatory mitogen-activated protein kinases (MAPK) and NF-κB (nuclear factor kappa-light-chain-enhancer of activated B cells) signaling pathways due to the enhanced farnesylation of H-Ras [43]. Consequently, cytokines overproduced by *Ggpps*-deficient Sertoli cells trigger spermatogonial apoptosis, and hypersecretion of chemokines leads to macrophage invasion into the seminiferous tubules, which further block spermatogonial development, resulting in Sertoli-cell-only seminiferous epithelia within the testes. This study elegantly demonstrated that protein prenylation in Sertoli cells is critical for regulating the immune response in prepubertal testes, and its disruption can have a long-lasting, negative impact on male fertility. Interestingly, patients with "Sertoli-cell-only syndrome" who have a history of mumps-virus infection often display significantly downregulated expression of GGPPS due to elevated promoter methylation. Mumps commonly affects children aged 5–9 years and can lead to permanent male sterility in certain cases due to progressive degeneration of the seminiferous epithelium and, occasionally, Sertoli-cell-only syndrome [64]. Therefore, suppressed GGPPS expression due to virus-mediated promoter methylation may represent the etiology of mumps-virus, infection-related, male infertility.

4. A cytoplasmic RNase III involved in small noncoding RNA biosynthesis

DICER is a cytoplasmic ribonuclease (RNase) III that recognizes double-stranded RNAs or precursor microRNAs and produces double-stranded RNA fragments about 20–25 base pairs long, with a two-base overhang at the 3′ end [65,66]. DICER is essential for the biogenesis of miRNAs, endogenous small interference RNAs (endo-siRNAs), and other uncharacterized small noncoding RNA species [67]. Sertoli-cell-specific KO of *Dicer* leads to aberrant Sertoli cell proliferation and differentiation during testicular development, and defective Sertoli cell maturation results in failure to properly support meiosis and spermiogenesis [40,41]. Consequently, male mice with *Dicer*-deficient Sertoli cells are infertile. The observed phenotype may not necessarily result from miRNA deficiency because *Dicer* controls biogenesis of other small noncoding RNA species (e.g., endo-siRNAs) and may have an small noncoding RNA (sncRNA)-independent role as well [67]. Overall, this study suggests that DICER is essential for the normal development of Sertoli cells and successful spermatogenesis.

B. Receptors

1. Cell surface transmembrane receptors

Signaling through integral membrane receptors is one of the most common ways for cells to communicate with their environments. Multiple surface receptor types exist on Sertoli cells that execute a wide variety of functions. These receptors include those for hormones (e.g., follicle-stimulating hormone receptor (FSHR) and luteinizing hormone receptor), for microbial ligands (TLR1-4), and for a combination of various ligands (e.g., basigins).

FSHR is a G-protein-coupled membrane receptor that binds its ligand FSH, a gonadotropin secreted by the anterior pituitary [5]. FSHR is exclusively expressed on Sertoli cells, and global inactivation of FSHR leads to compromised spermatogenesis and subfertility in mice and sterility in humans [5,22]. By analyzing testicular development in a quantitative manner, several studies have documented that FSH signaling through Sertoli cells is required for establishing the full capacity of spermatogenesis during prepubertal testicular development [5,68], and postpubertal spermatogenic capacity appears to require both FHS and testosterone [5]. Data from these KO studies are consistent with the long-standing view on FSH action on Sertoli cells, that FSH stimulates Sertoli cell proliferation and differentiation during testicular development and supports quantitative aspects of spermatogenesis during spermatogenesis in adulthood.

Basigin is a member of the immunoglobulin superfamily, with a structure related to the putative primordial form of the family [69–71]. As a type I integral membrane receptor, basigin has many ligands, including the cyclophilin (CyP) proteins CyP-A and CyP-B and certain integrins [69–71]. Interestingly, basigin KO male mice are sterile due to a complete lack of elongated spermatids and mature spermatozoa in the mutant testes [13,14]. Further functional study revealed that the integrity of the BTB in basigin-deficient mice was severely compromised, suggesting that the basigin-mediated signaling is essential for the formation and/or maintenance of the BT, and for the progression of the meiotic phase of spermatogenesis [13].

Sertoli cells are directly exposed to blood circulation and, thus, subjected to immune surveillance. It is, therefore, not surprising that Sertoli cells express Toll-like receptors (TLRs) [72]. The TLRs can trigger the innate and adaptive immune responses against invading pathogens by recognizing and binding highly conserved, pathogen-coded molecular structures termed "pathogen-associated molecular patterns" [73]. Among individual *Tlr* KOs, simultaneous inactivation of *Tlr2* and *Tlr4* leads to a compromised innate immune response of Sertoli cells to spermatogenic cell damage [28]. However, the consequences of such an alteration appear to be mild because these mice remain fertile.

2. Nuclear receptors

In addition to membrane receptors, nuclear receptors specifically bind their ligands, which are usually steroid hormones or other signaling compounds, such as RA and retinoid X [74]. Androgens are steroids that can travel via blood circulation and reach their target cells. Androgens need to diffuse into their target cells and bind to their nuclear receptors, which, once bound, act as transcriptional factors and control transcription of their target genes [75]. By selective inactivation of AR using the Cre–loxp system, several studies have demonstrated that AR signaling has an essential role in Sertoli cell function (reviewed in Ref. [76]). Mice with AR-deficient Sertoli cells are sterile due to azoospermia, and spermatogenesis is arrested at the diplotene stages of the meiotic phase [32–35]. Moreover, the BTB integrity is compromised, and many Sertoli-cell-secreted proteins known to be essential for supporting spermatogenesis are dysregulated, including specific proteases, transport proteins, and paracrine factors [32]. Analyses of compound KO mice with null mutations in FHS/FHSR and/or AR allowed in-depth dissection of the roles of FSH and AR signaling in the control of testicular development and spermatogenesis [41]. As discussed above, AR's effects on germ cell development are mediated mainly through Sertoli cells and are important for both prepubertal testicular development and postpubertal spermiogenesis.

As discussed above, RA acts through binding to its nuclear receptors, which have three isoforms, including RARA, RARB, and RARG, which are encoded by *Rara*, *Rarb*, and *Rarg*, respectively [52]. Selective inactivation of RAR genes in Sertoli cells, either individually (*Amh-Cre; Rara$^{fl/fl}$*) or jointly (triple KO), abolishes the cyclical gene expression patterns of Sertoli cells, leading to degeneration of the seminiferous epithelium [49]. The fact that global and Sertoli-cell-specific ablation of RARA leads to identical phenotypes suggests that RARA functions in a Sertoli-cell-autonomous manner. The lack of additive effects in triple-KO mice suggests that RARA is the master regulator that controls the paracrine signals required for spermatogonial differentiation and germ cell survival. RAR KOs phenocopy RADHL KOs, providing further evidence that these proteins are involved in the same signaling pathway. Interestingly, ablation of all retinoid acid receptors X (RXRs, α, β, and γ isotypes) in Sertoli cells does not recapitulate the phenotype of RAR triple KOs, suggesting that RARs and RXRs function independently [52].

C. Proteins involved in Sertoli cell-germ cell junctions and in Sertoli cell microtubule networks

One of the structural features of the seminiferous epithelium is the elegantly organized cellular association patterns among developing male germ cells at different spermatogenic phases (e.g., spermatogonia, spermatocytes, and spermatids) and Sertoli cells [77]. All spermatogenic cells of different types are anchored to the cytoplasm of Sertoli cells via various types of cellular junctions [2,78]. Between adjacent Sertoli cells lies the BTB, consisting of tight junctions, gap junctions, desmosome, and basal ES (unique actin filament bundles sandwiched in between Sertoli cell ER and plasma membrane). The blood-testis barrier (BTB), thus, physically partitions the seminiferous epithelium into two compartments, basal and adluminal [2,78]. Both spermatogonia and spermatocytes attach to Sertoli cells through desmosomes with gap junctions, whereas desmosome and gap junctions are replaced by apical ES once spermatids start to elongate [2,78]. The apical ES is also composed of actin filament bundles sandwiched in between the endoplasmic reticulum and plasma membrane of Sertoli cells. The apical ES remains the only device that anchors late spermatids to Sertoli cells, and disintegrated apical ES upon spermiation has been suggested to serve as a signal that induces BTB changes in permeability and function, which are required for temporary relaxation to allow leptotene spermatocytes to cross and enter the BTB so that spermatogenesis can occur in a continuous manner [2,78]. Gene KO studies over the past 10 years have identified many proteins, which either are the structural proteins of the BTB/ES or indirectly affect BTB formation and/or function.

1. BTB structural proteins

Claudin11 is a member of the claudin protein family, which contains approximately 24 members, and all are major constituents of tight junctions in mammals [79]. Claudin11 is specifically expressed in Sertoli cells and is an integral structural component of tight junctions in BTB [80]. Inactivation of *Cldn11* leads to defective BTB formation, disrupted Sertoli cell polarity, and detachment of Sertoli cells from the basement membrane of seminiferous tubules; consequently, *Cldn11* KO male mice are sterile [15]. In addition to having BTB defects, *Cldn11*-null Sertoli cells fail to differentiate properly and appear to undergo an epithelial-to-fibroblastic cell transformation [15].

Another BTB tight junction component is a 65-kDa protein, occludin. Global inactivation of the *Ocln* gene causes age-dependent defects in BTB and germ cell depletion [81]. Both Claudin11 and occludin are tethered to actin cytoskeleton in BTB by a group of three adaptor proteins, the zonula occludens (ZO) proteins 1, 2, and 3, also called "tight junction proteins" 1, 2, and 3, which belong to a membrane-associated guanylate-kinase-like protein superfamily [82]. *Tight junction protein* (*Tjp1*) or *2* KO mice die during embryonic development, whereas *Tjp3* inactivation causes no obvious phenotype [83,84]. The embryonic lethality of *Tjp1* or *Tjp2* KO mice precludes further study of their functions in BTB. However, chimeric mice generated using *Tjp2*-null ES cells injection into wild-type blastocysts produced BTB defects in the testis and reduced fertility [27]. Generation of *Tjp1* and *2* loxp mice will allow for further dissection of their role in BTB.

Connexin 43, also called "gap junction protein α1" (*Gja1*), is a member of the connexin protein family, which consists of structurally related transmembrane proteins involved in the formation of vertebrate gap junctions [85]. CX43 is the major component of the gap junction in testicular BTB and ES [86]. Sertoli-cell-specific deletion of *Cx43* mostly results in male infertility and smaller testes due to mitotic arrest at the spermatogonial stage, Sertoli-cell-only seminiferous epithelium and/or intratubular Sertoli cell clusters due to overproliferation and poor differentiation [38]. Thus, proper BTB gap junction formation is required for proper Sertoli cell development and spermatogonial multiplication.

2. BTB regulatory proteins

The KO studies have identified a group of proteins that are not physically located as components of the BTB, but their deficiency causes BTB defects, suggesting that these proteins have a direct or indirect regulatory role in BTB formation or function. For example, lack of AR or basigin in Sertoli cells leads to compromised BTB integrity that contributes to the defective spermatogenesis in these KO mice, as discussed above. Several additional proteins with these characteristics have been identified.

Adenomatous polyposis coli (APC), a multifunction tumor suppressor, is associated with the development of various human cancers, including colon, liver, ovarian, endometrial, and testicular cancers through the inhibition of Wnt/β-catenin signaling [87]. Interestingly, cKO of *Apc* in Sertoli cells has revealed that APC is essential for the maintenance of both BTB and ES [31]. *Apc*-deficient Sertoli cells display premature germ cell loss and defective spermatogenesis without effects on the timing and extent of Sertoli cell proliferation and differentiation or induction of tumorigenesis. *Apc*-deficient Sertoli cells lose their apical extension, leading to a lack of apical ES, and fail to form normal BTB, which correlates with aberrant expression and localization of key BTB structural proteins, including TJP1 and N-cadherin.

Similar to APC, retinoblastoma protein (pRb) is a tumor suppressor that appears to have a role in the control of Sertoli cell development [50,88,89]. Sertoli-cell-specific inactivation of the *Rb* gene causes age-dependent Sertoli cell dysfunction and infertility [50]. *Rb*-deficient Sertoli cells display increased permeability of BTB, impaired BTB remodeling, and defective Sertoli cell-germ cell interactions. The progressive loss of BTB integrity in *Rb* cKO testes suggests that *Rb* may not be required for BTB formation but is essential for BTB remodeling once the adult spermatogenic cycle is established. Given that pRb is a chromatin remodeling protein in addition to its cell cycle regulatory role [90,91], its function in BTB is likely related to the regulation of tight junction genes. Indeed, both *Cldn3* and *Ocln* are significantly downregulated in pRb cKO testes [50]. Interestingly, two members of the AT-rich interaction domain (*ARID*) family, ARID4A and ARID4B, also known as "RB-binding protein 1" (RBBP1) and "*RBBP1*-like protein 1," are reported to function in the regulation of Sertoli cell function and male fertility [12]. *Arid4a*$^{-/-}$; *Arid4b*$^{+/-}$ male mice display progressive loss of male fertility accompanied by impaired BTB, hypogonadism, and seminal vesicle agenesis/hypodysplasia. The observed phenotypes are similar to those in Sertoli-cell-specific *Ar* KO mice or *Rb* KO mice [50], supporting the notion that *Arid4a* and *Arid4b* function as transcriptional co-activators for AR and RB and are critical for the normal development and function of Sertoli cells.

Fatty acid desaturase 2 (FADS2) is a member of the FADS gene family that is responsible for the biosynthesis of omega 3- and omega 6-polyunsaturated fatty acids (PUFAs), which are essential constituents of membrane phospholipids and precursors of the eicosanoids anandamide and docosanoids [92]. Global inactivation of *Fads2* in mice abolishes the initial step in the enzymatic cascade of PUFA synthesis, and the lack of PUFAs and eicosanoids causes male sterility due to defective Sertoli cell polarity and BTB, leading to Sertoli cell vacuolization and spermiogenic arrest [21]. Therefore, PUFA-substituted membrane phospholipids have an essential role in Sertoli cell polarity and BTB.

TYRO3 protein tyrosine kinase 3 (*Tyro3*), AXL receptor tyrosine kinase (*Axl*), and c-mer proto-oncogene tyrosine kinase constitute the TAM family of receptor tyrosine kinases, which are involved in multiple cellular processes, including cell proliferation, survival, adhesion, and migration as well as inflammatory cytokine release [93]. The TAM triple-KO male mice are sterile because of oligoasthenoteratozoospermia [26]. Detailed molecular and ultrastructural characterization revealed that testicular immune homeostasis was impaired in these mutant mice and was characterized by compromised integrity of BTB and the production of autoantibodies against germ cells.

3. Proteins involved in ES

Corticosteroids and regional expression 16 (CR16; also called "WIPF3") belongs to the Wiskott–Aldrich syndrome protein (WASP)-interacting protein (WIP) family and can be regulated by glucocorticoids [94]. CR16/WIPF3 binds to the neural WASP (N-WASP), which activates the Arp2/3 complex to induce actin polymerization [95]. In the testis, CR16/WIPF3 and N-WASP are colocalized to the actin filaments (actin bundle) of the apical ES [29]. CR16/WIPF3 KO male mice display abnormal actin bundles in ES and reduced expression of N-WASP, which may cause abnormal spermiogenesis leading to deformed head morphology and male sterility [29]. These studies suggested that normal actin bundle and ES are required for normal late spermiogenesis and, thus, male fertility.

4. Proteins involved in Sertoli cell microtubule network

Microtubule networks display dynamic changes at different stages of the spermatogenic cycle. Microtubule remodeling events in Sertoli cells have been demonstrated to be critical in Sertoli cell architecture, secretion of tubule fluid, intracellular vesicle transport, and translocation of spermatids during maturation [96]. Katanin p60 subunit A-like1 (KATNAL1) is an ATPase and may have a microtubule-severing role similar to katanin p60 [23,97]. Loss of function of KATNAL1 leads to male infertility due to disruptions of Sertoli cell microtubule dynamics and depletion of spermatids from the seminiferous epithelium [23]. KATNAL1 appears to be localized to all segments of the Sertoli cell microtubule network from basal to apical regions, whereas in the absence of KATNAL1, the microtubule network is confined to the basal region, suggesting disruptions in microtubule remodeling in Sertoli cells [23]. Therefore, KATNAL1 plays an essential role in spermatogenesis by regulating Sertoli cell microtubule remodeling.

D. Transcription factors and transcriptional coregulators

Versatility requires that Sertoli cells have an active transcription program so that various gene sets needed for multitasking can be synthesized in a timely manner. Because one transcription factor can control multiple genes, inactivation of one transcription factor may affect the expression of numerous genes that are usually functionally related, thus causing functional disruptions. For example, nuclear receptors mostly function as transcriptional factors. Once bound by their specific ligands (e.g., steroid hormones), they undergo conformational changes and, together with their coregulators, they act on the promoter region of their target genes to control transcription. Several transcription factors have been inactivated either globally or specifically in Sertoli cells, and this inactivation has led to Sertoli cell defects.

WT1 is a member of the Wilms' tumor protein family, and its global KO leads to complete gonadal agenesis, which precludes further analyses of its role in postnatal testicular development [53]. By crossing the *Amh-Cre* deletor line with *Wt1* loxp mice, investigators generated a Sertoli-cell-specific *Wt1* cKO line [54]. Selective inactivation of *Wt1* in Sertoli cells at E14.5 resulted in disruptions in testis cord formation due to the breakdown of the basal membrane and aberrant development of the seminiferous tubules. Consequently, progressive loss of Sertoli cells and germ cells led to severely atrophic testes. Inactivation of *Wt1* in Sertoli cells at 8 weeks of age causes massive germ cell loss and Sertoli-cell-only seminiferous epithelial histology, suggesting that WT1 plays a critical role in normal Sertoli cell function in adult testes [54]. Further analyses suggested that *Wt1* inactivation in adult testes causes disruptions in BTB and Sertoli cell polarity, indicating an essential role of WT1 in Sertoli cell functions in adulthood. These KO studies demonstrate that WT1 is a master transcription factor in Sertoli cells and plays an essential role in all stages during testicular development. Further efforts to define WT1 target genes in Sertoli cells at different developmental stages are needed to fully understand the WT1-mediated transcriptional program.

Doublesex and *mab-3* related transcription factor 1 (DMRT1) is a transcription factor that is ubiquitously expressed in both germ and Sertoli cells in the testis [17]. *Dmrt1*-null mutant mice have severely dysgenic testes in which Sertoli cells and germ cells both fail to differentiate properly after birth [17]. Selective depletion of *Dmrt1* in Sertoli cells causes maturation defects characterized by Sertoli cell overproliferation and incomplete differentiation, which, in turn, compromises germ cell development [42]. These KO studies demonstrate that *Dmrt1* is required for postnatal differentiation of Sertoli cells and for germ line maintenance and meiotic progression.

KLF4 is a pleiotropic zinc finger transcription factor that regulates genes involved in differentiation and cell cycle control [44]. Sertoli-cell-specific *Klf4* cKO mice display degenerated seminiferous tubules characterized by delayed lumen formation, germ cell depletion, highly vacuolized Sertoli cell cytoplasm, and impaired apical secretion [44], suggesting that KLF4 is required for proper Sertoli cell differentiation during postnatal testicular development.

ETS translocation variant 5 (ETV5) belongs to the E26 transformation-specific (ETS) family of transcription factors, which all share a unique ETS DNA-binding domain and participate in a variety of developmental processes [98]. Interestingly, global KO of *Etv5*, also called "*Emr*," leads to partial prenatal lethality [20]. Among the *Etv5* KO mice that survive, *Etv5* inactivation causes a phenotype that is mainly characterized by a loss of maintenance of spermatogonial stem cell self-renewal without a blockage of normal spermatogenic differentiation, thus leading to progressive germ cell depletion and Sertoli-cell-only seminiferous epithelium. Consequently, *Etv5*-null male mice display atrophic testes, azoospermia, loss of spermatogonia, and complete infertility. Because Sertoli cells provide niches for spermatogonial stem cells in adult testes, it has been suggested that ETV5 may be involved in the transcriptional control of the spermatogonial stem cell niche. Alternatively, ETV5 may control the expression of the paracrine signals required for spermatogonial stem cell self-renewal. These signals have been implicated in several follow-up studies [20,99–102]. AF4/FMR2 family member 4 (AFF4) belongs to the AF4 family of transcription factors that is involved in leukemia [103] and has been identified as a component of the positive transcription elongation factor b complex. A small fraction (approximately 13%) of *Aff4*-deficient male mice are sterile due to azoospermia [6], suggesting that AFF4 may function as a transcriptional regulator in Sertoli cells and is essential for male germ cell differentiation and survival.

CCR4-NOT transcription complex, subunit 7 (CNOT7) is a coregulator of RXRB in Sertoli cells and, thus, is involved in spermatogenesis [104]. Sertoli-cell-specific inactivation causes defective spermiogenesis, leading to oligoteratozoospermia, impaired sperm motility, unsynchronized spermatid maturation, and male sterility [16,104].

SOX8 is a member of the high-mobility group of transcription factors. *Sox8* global inactivation leads to male infertility due to increased BTB permeability and age-dependent dysregulation of spermatogenesis [25,105,106]. Because *Sox8* is exclusively expressed in Sertoli cells, abnormalities in both Sertoli cells and germ cells suggest that SOX8 directly regulates Sertoli cell function and indirectly affects male germ cell development [25].

In summary, a wide variety of transcription factors are required for either Sertoli cell development or normal function. Interestingly, some of these factors are not required for Sertoli cell development but are essential for normal spermatogenesis, suggesting that these Sertoli cell transcription factors control the expression of genes that control male germ cell development indirectly. Despite the increasingly long list of transcription factors identified as essential for Sertoli cell development and/or function, their target genes remain largely unknown. It is essential to identify these target genes so that the sophisticated networks of Sertoli cell transcriptional control can be defined in the future.

E. Phagocytosis and endocytosis

Phagocytosis is an important function of Sertoli cells, which are responsible for clearing apoptotic male germ cells during spermatogenesis and digesting cytoplasmic contents (i.e., residual body) shed by spermatozoa upon spermiation [1]. Increasing lines of evidence also suggest that Sertoli cell phagocytosis may target more than the two types of cells or cellular structures because developing male germ cells produce numerous vesicles containing unwanted or recyclable materials that can be phagocytized by Sertoli cells for processing and recycling [107]. The morphological observation of Sertoli cell phagocytosis through light or electron microscopy has been well documented [107–110]. However, the underlying molecular mechanism is only starting to be unveiled.

ELMO1 is cytoplasmic engulfment protein known to be involved in engulfment/phagocytosis in somatic cell lineages [111]. ELMO1 functions as a downstream effector of the phosphatidylserine receptor BAI1, and along with DOCK1 and the GTPase RAC1, it promotes internalization of the dying cells. Interestingly, global inactivation of *Elmo1* in mice leads to male infertility due to the failure of Sertoli cells to clear apoptotic spermatogenic cells [19]. *Elmo1*-deficient testes display disrupted seminiferous epithelium, multinucleated giant cells, accumulation of apoptotic germ cells, and decreased sperm output. The fact that Sertoli-cell-specific inactivation of *Elmo1* generates a phenotype that is less severe than that of *Elmo1* global KO suggests that either *Elmo1* has a role in Sertoli cell development prior to E14.5 (the onset of Cre expression from the *Amh-Cre* allele) or the expression of *Elmo1* in other testicular cell types also participates in phagocytosis or engulfment [19].

Amphiphysin is a protein associated with the cytoplasmic surface of synaptic vesicles [7]. In *Amph1*-null Sertoli cells, phagocytosis and phosphatidylinositol-4, 5-bisphosphate-induced actin polymerization are both inhibited [7–11]. Although the underlying mechanism remains elusive, this KO study indicates that amphiphysin 1 is important in the regulation of the actin dynamics required for phagocytosis. Moreover, reports suggest that autophagy may be an integral part of the Sertoli cell phagocytosis pathway [112,113]. However, direct evidence is still missing.

F. Signaling molecules

CTNNB1 is an essential component of the WNT/CTNNB1 signaling pathway. Sertoli-cell-specific inactivation of *Ctnnb1* in mice causes sterility because of testicular atrophy associated with degeneration of seminiferous tubules and the progressive loss of germ cells [36,37]. Incomplete differentiation of Sertoli cells coincides with germ cell loss, suggesting that germ cell depletion is secondary to deficiency in Sertoli cells due to poor differentiation. Interestingly, 70–90% of Sertoli-cell-specific *Ctnnb1* cKO males develop Sertoli cell tumors after 8 months of age [36]. These data suggest that the WNT/CTNNB1 pathway is essential for normal Sertoli cell proliferation and differentiation, which, in turn, are critical to these cells' capacity to support spermatogenesis in the postnatal testis.

G. Protein quality control

Chaperones are a group of proteins that mediate folding of *de novo* synthesized proteins, refolding of misfolded proteins, translocation of proteins into organelles, assembly of proteins, and protein degradation [114]. DNAJA1 (also called DjA1) is a member of HSP70 chaperone protein family. Inactivation of *Dnaja1* causes spermiogenic arrest, leading to a complete lack of elongating or elongated spermatids [18]. Interestingly, DNAJA1 appears to be regulated by androgen signaling because in the testes with AR-null Sertoli cells, DNAJA1 expression is lacking, suggesting that chaperone-mediated protein quality control is critical to successful spermiogenesis.

H. Lipid homeostasis

Peroxins (PEXs) are proteins that are essential for the assembly of functional peroxisomes [115]. Selective elimination of peroxisomes in Sertoli cells by inactivation of *Pex5* results in male infertility due to lipid inclusions and progressive degeneration of the seminiferous tubules, suggesting that peroxisome is essential for lipid homeostasis in the testis and for male fertility [47,48]. Peroxisomal multifunctional protein 2, also called "D-bifunctional enzyme," is involved in the peroxisomal β-oxidation of 2-methyl branched chain fatty acids (pristanic acid); C27-bile acid intermediates; very-long-chain fatty acids; and the synthesis of certain PUFAs, such as docosahexaenoic acid (C22: $6n-3$) and docosapentaenoic acid (C22: $5n-6$) [116]. *Mpf2*-null testes display extensive accumulations of neutral lipids in Sertoli cells starting during prepubertal testicular development and eventually leading to severe depletion of spermatocytes and spermatids and male infertility [24]. Clearly, peroxisomal β oxidation is essential for lipid homeostasis in the testis and for male fertility.

These global and conditional KO studies unveiled numerous genes that are essential for normal Sertoli cell development and/or functions, which, in turn, indirectly control normal male germ cell development. Sertoli cells express tens of thousands of genes, and, thus, the efforts to define Sertoli cell gene functions have been limited. Because most of the Sertoli cell genes are ubiquitously expressed, global KOs are of limited value because of embryonic/perinatal lethality and compound phenotypes, which hinder the analyses of Sertoli-cell-specific effects. Therefore, conditional KO represents an ideal way to define the Sertoli-cell-specific roles. However, the development of such models will depend on the availability of loxp lines and the efforts of Sertoli cell research in the future.

III. Lessons learned from the gene KO studies

A. Cell fate control (i.e., proliferation vs. differentiation) is critical to normal Sertoli cell function

During fetal testicular development, pro-Sertoli cells proliferate to increase their number within the testis cord in mice [117–119]. In newborn murine testes, most pro-Sertoli cells already migrate to and reside on the basal membrane, although some remain in the center of the precursor seminiferous tubules (Figure 15.1). During pubertal development, Sertoli cells continue to proliferate, and the number of basally located Sertoli cells continue to increase until approximately postnatal days 10–15, when Sertoli cells cease proliferation and undergo terminal differentiation (Figure 15.1). The signaling required for this transition remains unknown. However, this fate decision appears to be critical because in several Sertoli cell cKO mouse lines, Sertoli cells fail to differentiate and/or overproliferate during prepubertal testicular development, leading to catastrophic depletion of both Sertoli cells and developing germ cells. These genes include transcriptional factors (e.g., *Nr5a1/Sf1*, *Rb*, *Wt2*, *Klf4*, *Sox8*, *Wt1*, *Lkb*, *Dmrt1*, and *Aff4*), hormonal receptor (e.g., *Fshr*), signaling transducers (e.g., *Ctnnb1*), and small RNA biogenetic factors (e.g., *Dicer*) (Tables 15.1 and 15.2). Some of these cKO lines display defects in Sertoli cell polarity (e.g., *Wt1*, *Lkb*, and *Dmrt1*; Tables 15.1 and 15.2). These KO studies strongly suggest that temporal activation of a cascade of transcriptional programs (via transcriptional factors) in response to hormonal stimulation (e.g., via FSH) and proper posttranscriptional regulation of gene expression (e.g., via small noncoding RNAs) are critical for proper fate control of developing Sertoli cells (proliferation vs. differentiation). Further dissection of the temporal relationship among those transcription factors and the identification of target genes of miRNAs that are critical for Sertoli cell fate control will help elucidate the molecular mechanism underlying Sertoli cell fate control.

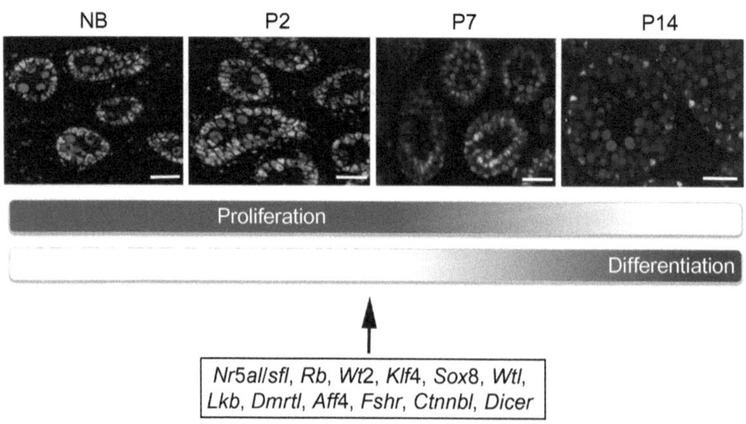

Figure 15.1 Sertoli cells actively proliferate until approximately postnatal day 10, and after that, Sertoli cells become terminally differentiated. In the upper panels, cross sections of the testes at birth (NB), postnatal day 2 (P2), P7, and P14 were stained with anti-WT1 antibodies (in green for Sertoli cells) and anti-germ cell nuclear antigen (GCNA) antibodies (in red for germ cells) (scale bars = 20 μm). Knockout (KO) studies have identified numerous genes that are essential for proper fate control of Sertoli cells (proliferation vs. differentiation) during postnatal testicular development (framed). (For interpretation of the references to color in this figure legend, the reader is referred to the web version of this book.)

B. Determinants of normal homeostasis of adult Sertoli cells

In several Sertoli cell cKO lines, although Sertoli cell proliferation and differentiation occur normally, the cKO Sertoli cells cannot perform some of their fundamental functions, including proper BTB formation (e.g., in *Sox8, Cx43*, and *Basigin* KO lines), lipid metabolism (e.g., in *Pex5* and *Mpf2* KO lines), and phagocytosis/engulfment of apoptotic spermatogenic cells or residual bodies (e.g., in *Amphl* and *Elmo1* KO lines), suggesting that a complex gene regulatory network exists to maintain the general health and support the basic functions of Sertoli cells (Figure 15.2). BTB formation

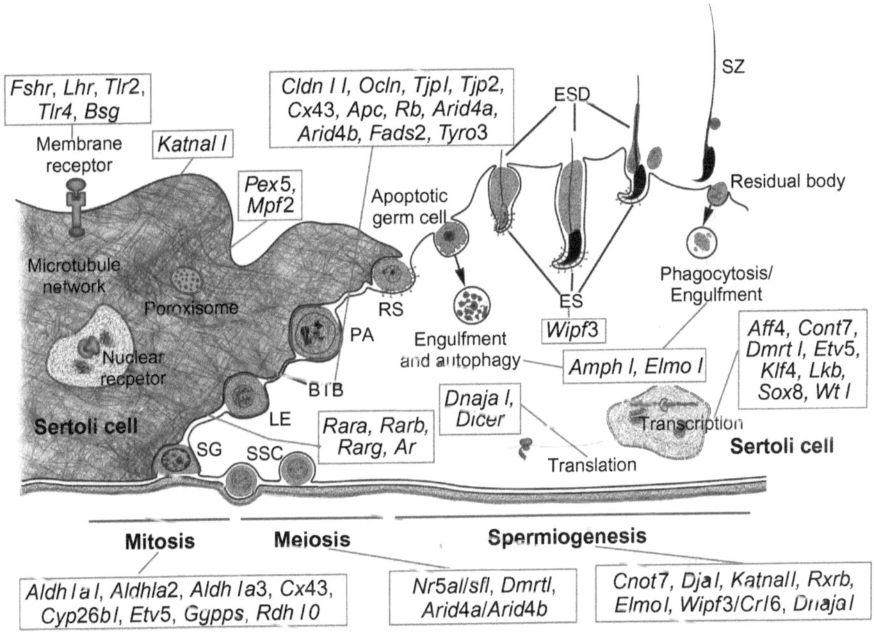

Figure 15.2 A schematic summary of genes that are essential for normal Sertoli cell function during spermatogenesis. Developing male germ cells, including spermatogonial stem cells (SSCs); differentiated spermatogonia (SG); spermatocytes at the preleptotene, leptotene (LE), zygotene, pachytene (PA), and diplotene stages of meiosis; round spermatids (RSs); and elongated spermatids (ESDs) are all enveloped by the cytoplasmic membrane of Sertoli cells. Spermatogenic cells attach to the Sertoli cell membrane through cell junctions, which exist in between Sertoli cells and spermatogonia as the blood–testis barrier (BTB) and in between Sertoli cells and developing spermatids as the ectoplasmic specialization (ES). During spermatogenesis, Sertoli cells direct all three phases of male germ cell development (i.e., the mitotic, meiotic, and haploid phases, also called spermiogenesis) via endocrine and paracrine mechanisms. Upon spermiation, spermatozoa (SZ) are released from the seminiferous epithelium, and residual bodies containing the shed cytoplasmic contents are phagocytosized by Sertoli cells. In addition, Sertoli cells clear apoptotic germ cells through phagocytosis/engulfment in conjunction with the autophagy pathway. Gene expression in Sertoli cells is regulated at both transcriptional and posttranscriptional levels, a process involving numerous transcriptional factors/coregulators and posttranscriptional regulators. Genes that are essential for these processes are framed, and details on their physiological roles are described in the text.

involves the activation of certain transcriptional programs (e.g., in *Sox8* and *Basigin* KO lines), so that structural components, such as connexin 43, can be synthesized (Tables 15.1 and 15.2). Healthy lipid metabolism within Sertoli cells is also critical to the overall health of Sertoli cells. Interestingly, the engulfment/phagocytosis functions of Sertoli cells appear to be controlled by the signaling pathway that is highly conserved among somatic cell lineages (e.g., the ELMO–Dock–Rac pathway).

C. Sertoli cells control all three phases of spermatogenesis

As discussed above, disrupted Sertoli cell fate control and/or homeostasis affect the overall health of Sertoli cells, which, in turn, disrupt spermatogenesis because dysfunctional Sertoli cells cannot fully execute their supporting roles. Interestingly, in many cases in which Sertoli cells display normal proliferation and differentiation as well as general health, germ cell development is disrupted. The apparent indirect effects reflect specific supporting roles that Sertoli cell exert on germ cells in each of the three phases of spermatogenesis.

Inactivation of transcriptional factor ETV5 in Sertoli cells leads to a complete loss of spermatogonial stem cells, probably due to the lack of availability of the spermatogonial stem cell niche or of paracrine factors required for spermatogonial stem cell self-renewal [20,99–102]. This finding suggests that a transcriptional program controlled by ETV5 is essential for spermatogonia stem cell fate control, and further dissection of the target genes controlled by this transcriptional factor should shed more light on this issue. RA signaling in Sertoli cells is critical for prepubertal spermatogonial development based upon the phenotype of Sertoli-cell-specific inactivation of *Aldh1a1-3*, *Rdh10*, and *Cyp26b1*, which display an arrest of spermatogonial development at the type Aal spermatogonial stage (*Aldh1a1-3*, *Rdh10* cKO) [30,51] and a loss of prospermatogonia (*Cyp26b1* cKO) [39]. In addition, failure to build structurally and functionally sound BTB during testicular development is incompatible with full spermatogenesis, as evidenced by the spermatogonial block phenotype in Sertoli-cell-specific *Ggpps* and *Cx43* cKO mice [38,43].

Inactivation of *Nr5a1/Sf1*, *Dmrt1*, or *Arid4a/Arid4b* specifically in Sertoli cells causes a meiotic arrest phenotype, suggesting that these transcriptional programs are required for the progress of spermatogenesis through meiosis (Figure 15.2). Further identification of target genes of these transcriptional factors will reveal the paracrine factors that control the meiotic phase of spermatogenesis. Similarly, Sertoli-cell-specific inactivation of a number of genes causes disrupted spermiogenesis, characterized by spermiogenic arrest (*Cnot7*, *Dja1*, and *Katnal1* cKOs), spermiation failure (*Rxrb* and *Elmo1* cKOs), and defective Sertoli cell/spermatid gap junction (*Wipf3/Cr16* and *Dnaja1*) (Figure 15.2).

These Sertoli-cell-specific cKOs demonstrate that male germ cell development is largely dependent on Sertoli cells, and Sertoli cells control all three phases of spermatogenesis through direct interactions (e.g., via cell–cell junctions) or in a paracrine manner. Of course, there are germ-cell-autonomous programs, but none of the three phases can occur normally without the support of Sertoli cells under physiological conditions. Based on these KO studies, two fundamentally important questions arise: (1) How is the fate of Sertoli cells determined during testicular development? (2) What are the paracrine factors that control each of the three phases of spermatogenesis? More detailed molecular characterization of existing or novel Sertoli cell cKO mouse models will shed more light on these outstanding questions. Next-generation technologies, including high-throughput deep sequencing and single-cell analyses, have made it possible to analyze genome-wide changes at both transcriptomic and epigenomic levels. Because of the increasing application of these powerful technologies, it is anticipated that more mechanistic insights will become available in the near future.

IV. Approaches to gene ablation in Sertoli cells

Global KO mice deficient in genes expressed in multiple organs often display an embryonic lethality phenotype that precludes further analyses of gene functions in Sertoli cells during postnatal development and adulthood. cKOs overcome such obstacles and make it possible to define the Sertoli-cell-specific functions. The Cre–Loxp system represents the most commonly used approach. In this approach, a Cre deletor line and a loxp line need to be generated and crossed to achieve cell-specific inactivation of a gene [120,121]. In addition to these classic methods, lentivirus-mediated gene delivery and EMU mutagenesis have been used to inactivate genes, although the former suffers from an incomplete penetrance and the latter needs verification of the mutations identified using gene KOs *in vivo* [23,122,123]. Although out of the scope of this chapter, overexpression of the dominant-negative or native form of the genes of interest specifically in Sertoli cells have also been used to study the roles of these genes [122,124–129].

In the Cre–loxp system, the Cre deletor line determines the timing and cell specificity of gene inactivation. Three Sertoli-cell-specific Cre lines have been utilized to inactivate genes in Sertoli cells so far, including *Amh-Cre*, *Amhr2-Cre*, and *Dhh-Cre* [130–133]. *Amh-Cre* mice are viable and fertile, and they carry the transgene-expressing Cre recombinase directed by a Sertoli-cell-specific promoter element from the anti-Müllerian hormone (*Amh*) gene [130,131]. Cre-recombinase activity is detected in Sertoli cells during fetal testicular development as early as E14.5, and non-Sertoli-cell *Cre* expression has never been reported. Although *Amhr2* is known to be expressed in Leydig cells, the particular promoter element used in *Amhr2-Cre* mice appears to direct Cre expression predominantly in Sertoli cells and weakly in Leydig cells starting from E14.5–16.5 [31,37,46,134,135]. Some studies used efficient recombination in both Leydig and Sertoli cells, whereas the results of others suggested that the Cre-mediated recombination occurred only in Sertoli cells. For example, both Leydig and Sertoli cells were affected in *Amhr2-Cre; Ar*$^{lox/lox}$ testes [136], whereas *Amhr2-Cre*-mediated *Gata4* cKO mice displayed

Sertoli-cell-specific ablation of the gene in postnatal testes [135]. Overall, the extent of deletion in Sertoli versus Leydig cells appears to be dependent upon the genes studied [37,128,132,136–138]. Given the variable efficiency of Cre recombination in Leydig cells, this deletor line may not be ideal for dissecting gene functions specific to Sertoli cells.

Dhh-Cre mice express *Cre* selectively in Sertoli cells, and the onset of Cre expression was detected in E18.5 pro-Sertoli cells [139]. However, the *Dhh-Cre* transgene is also expressed in other cell types (e.g., endothelial cells and Schwann cells of the peripheral nervous system) [17,42,133]. Therefore, one has to take other cell types into consideration when evaluating the Sertoli cell phenotype. In all three Cre deletor lines, the onset of Cre expression occurs in pro-Sertoli cells during fetal testicular development. If the genes to be inactivated have a role in pro-Sertoli cell development, then Sertoli cell proliferation and/or differentiation will be compromised, leading to germ cell abnormalities secondary to primary Sertoli cell defects, as discussed above. To define the gene function in fully developed Sertoli cells during spermatogenesis in adult testes, novel conventional or inducible Cre deletor lines that allow for Cre expression at any time during postnatal testicular development or in adult testes need to be generated.

Gene-targeting technologies involve construction of targeting vectors, embryonic stem cell targeting and screening, generation of chimeras through blastocyst injection, and germ line transmission via breeding [120,121]. Any mishaps during this lengthy and complex process add significant amounts of time to the successful generation of a KO line. Next-generation genome-editing technologies are now available [140,141], bringing hope of manipulating the mouse genome with much higher efficiency. These technologies are all based on the concept that eukaryotic cells fix a DNA double-strand break (DSB) generated by a nuclease using two mechanisms: (1) nonhomologous end joining (NHEJ), through which the break ends are directly ligated without the need for a homologous template; and (2) homologous recombination-mediated DSB repair, which requires a donor sequence containing two arms that are homologous to the regions flanking the breaking point. NHEJ is error prone and thus can be used to generate deletion mutations. Through donor-sequence-mediated homologous recombination, desired sequences (e.g., loxp sequence, cDNAs, or reporter genes) can be inserted into the genome in a targeted manner. The key to these methods is the nuclease, and the currently used ones include zinc finger nuclease, transcription activator-like effectors-based nuclease, and clustered regularly interspaced short palindromic repeats-guided Cas9 nuclease [142]. By co-injecting mRNAs encoding these nucleases together with donor sequences into the pronuclei of fertilized eggs, one can directly generate mice bearing deletions or insertions at desired locations in the genome. Due to the omission of steps, including embryonic stem cell targeting, chimera generation, and germ line transmission, which are usually unpredictable and low in efficiency, next-generation genome-editing technologies will allow for the generation of loxp and Cre lines at a much faster rate and, thus, speed up progress in the study of Sertoli cell biology or biomedicine in general.

V. Conclusions and perspectives

All KO studies conducted so far support the notion that Sertoli cells participate in almost all aspects of spermatogenesis by providing structural support (e.g., BTBs, ES, microtubule networks, and spermatogonial stem cell niches); paracrine signals for progression through mitotic, meiotic, and haploid phases; and "waste management and recycling" of apoptotic germ cells and residual bodies. The multitasking nature of Sertoli cells requires that these cells have an active transcriptional program that involves numerous transcriptional factors/regulators so that accurate spatiotemporal expression of multiple genes can be achieved to fulfill diverse functions of Sertoli cells. Indeed, numerous transcriptional regulators expressed in Sertoli cells have been found to play an essential role in supporting normal spermatogenesis (Figures 15.1 and 15.2; Tables 15.1 and 15.2). Sertoli cells in different stages of the seminiferous tubules support the development of multiple male germ cell types that are in different phases of spermatogenesis (e.g., mitotically proliferating or differentiating spermatogonia, spermatocytes undergoing meiosis, and spermatids undergoing spermiogenesis), and these developing male germ cells form specific cellular associations, also called "stages." It is conceivable that Sertoli cells at different stages of the seminiferous epithelial cycle have distinct transcriptomes, depending on the specific cell developmental stages that they are supporting. Therefore, it would be interesting to see how the stage-specific Sertoli cell transcriptomes are programmed and how such an elegant and sophisticated gene network operates so precisely during the progression of spermatogenic waves from one stage to the next. Dissection of the stage-specific roles of genes in Sertoli cells would require a strategy that allows for stage-specific inactivation of genes during spermatogenesis, which remains unavailable currently.

Early disruption of Sertoli cell development usually leads to depletion of both Sertoli cells and germ cells during testicular development. Therefore, malfunctions of the genes essential for early Sertoli cell development may cause male infertility with severely atrophic testes. Many genes have been found not to affect Sertoli cell development, but their inactivation in Sertoli cells leads to spermatogenic disruptions confined specifically to the meiotic and/or haploid phases. These genes or their products represent excellent drug targets for the development of male contraceptives. In fact, several male contraceptive drugs under development (e.g., adjudin [143] and gamendazole [144]) target Sertoli cell proteins and, thus, indirectly affect the development of spermatocytes and/or spermatids, leading to induced azoospermia in a reversible manner. Therefore, further identification of Sertoli cell genes that play an essential role exclusively in either late meiotic or haploid phases of spermatogenesis will help identify more contraceptive drug targets.

In vitro differentiation of sperm or sperm-like cells has been accomplished in several species [145–148]. Successful production of sperm or sperm-like cells in the absence of Sertoli cells appears to suggest that Sertoli cells are not essential for sperm production, and male germ cells can develop autonomously *in vitro*. Indeed, germ-cell-autonomous programs exist, but support from Sertoli cells is also necessary, which may explain why all *in vitro* differentiation methods have extremely low efficiency. Some coculture systems do not seem to be superior to the regular germ cell-only culture systems, which may be due to the associations between Sertoli cells and germ cells *in vitro* that are not physiological. It is also plausible that Sertoli cells *in vivo* mainly control the synchronized development of male germ cells (i.e., stages), which is one of the characteristics of mammalian spermatogenesis. Once male germ cells develop *in vitro*, synchronization may not be as important as it is *in vivo*. Nevertheless, whether or not Sertoli cells are needed in efficient spermatogenesis *in vitro* remains debatable.

In summary, next-generation genome-editing technologies will allow for the production of more Sertoli-cell-specific cKO lines. When these models become available, more functional insights into Sertoli cell function will be gained.

Acknowledgments

The author would like to apologize to colleagues whose work is relevant but not cited in this chapter due to page limits enforced by the publisher. Ms. Tingting Zhu is acknowledged for the artwork in Figure 15.2. Research in the Yan laboratory is supported by the NIH grants (HD060858, HD071736, and HD074573) and funds from the University of Nevada School of Medicine to W.Y.

References

[1] Hermo L, Pelletier RM, Cyr DG, Smith CE. Surfing the wave, cycle, life history, and genes/proteins expressed by testicular germ cells. Part 5: intercellular junctions and contacts between germs cells and Sertoli cells and their regulatory interactions, testicular cholesterol, and genes/proteins associated with more than one germ cell generation. Microsc Res Tech 2010;73(4):409–94.

[2] Mruk DD, Cheng CY. Sertoli–Sertoli and Sertoli–germ cell interactions and their significance in germ cell movement in the seminiferous epithelium during spermatogenesis. Endocrine Rev 2004;25(5):747–806.

[3] Guerrero-Bosagna C, Savenkova M, Haque MM, Nilsson E, Skinner MK. Environmentally induced epigenetic transgenerational inheritance of altered Sertoli cell transcriptome and epigenome: molecular etiology of male infertility. PLoS One 2013;8(3):e59922.

[4] McLean DJ, Friel PJ, Pouchnik D, Griswold MD. Oligonucleotide microarray analysis of gene expression in follicle-stimulating hormone-treated rat Sertoli cells. Mol Endocrinol 2002;16(12):2780–92.

[5] Huhtaniemi I. Mutations along the pituitary-gonadal axis affecting sexual maturation: novel information from transgenic and knockout mice. Mol Cell Endocrinol 2006;254–255:84–90.

[6] Urano A, Endoh M, Wada T, Morikawa Y, Itoh M, Kataoka Y, et al. Infertility with defective spermiogenesis in mice lacking AF5q31, the target of chromosomal translocation in human infant leukemia. Mol Cell Biol 2005;25(15):6834–45.

[7] Watanabe M, Tsutsui K, Hosoya O, Tsutsui K, Kumon H, Tokunaga A. Expression of amphiphysin I in Sertoli cells and its implication in spermatogenesis. Biochem Biophys Res Commun 2001;287(3):739–45.

[8] Yamada H, Padilla-Parra S, Park SJ, Itoh T, Chaineau M, Monaldi I, et al. Dynamic interaction of amphiphysin with N-WASP regulates actin assembly. J Biol Chem 2009;284(49):34244–56.

[9] Nakanishi A, Abe T, Watanabe M, Takei K, Yamada H. Dynamin 2 cooperates with amphiphysin 1 in phagocytosis in Sertoli cells. Acta Med Okayama 2008;62(6): 385–91.

[10] Yamada H, Ohashi E, Abe T, Kusumi N, Li SA, Yoshida Y, et al. Amphiphysin 1 is important for actin polymerization during phagocytosis. Mol Biol Cell 2007;18(11): 4669–80.

[11] Kusumi N, Watanabe M, Yamada H, Li SA, Kashiwakura Y, Matsukawa T, et al. Implication of amphiphysin 1 and dynamin 2 in tubulobulbar complex formation and spermatid release. Cell Struct Funct 2007;32(2):101–13.

[12] Wu RC, Jiang M, Beaudet AL, Wu MY. ARID4A and ARID4B regulate male fertility, a functional link to the AR and RB pathways. Proc Natl Acad Sci USA 2013;110(12): 4616–21.

[13] Bi J, Li Y, Sun F, Saalbach A, Klein C, Miller DJ, et al. Basigin null mutant male mice are sterile and exhibit impaired interactions between germ cells and Sertoli cells. Dev Biol 2013;380(2):145–56.

[14] Toyama Y, Maekawa M, Kadomatsu K, Miyauchi T, Muramatsu T, Yuasa S. Histological characterization of defective spermatogenesis in mice lacking the basigin gene. Anat Histol Embryol 1999;28(3):205–13.

[15] Mazaud-Guittot S, Meugnier E, Pesenti S, Wu X, Vidal H, Gow A, et al. Claudin 11 deficiency in mice results in loss of the Sertoli cell epithelial phenotype in the testis. Biol Reprod 2010;82(1):202–13.

[16] Ogawa T, Ito C, Nakamura T, Tamura Y, Yamamoto T, Noda T, et al. Abnormal sperm morphology caused by defects in Sertoli cells of Cnot7 knockout mice. Arch Histol Cytol 2004;67(4):307–14.

[17] Raymond CS, Murphy MW, O'Sullivan MG, Bardwell VJ, Zarkower D. Dmrt1, a gene related to worm and fly sexual regulators, is required for mammalian testis differentiation. Genes Dev 2000;14(20):2587–95.

[18] Terada K, Yomogida K, Imai T, Kiyonari H, Takeda N, Kadomatsu T, et al. A type I DnaJ homolog, DjA1, regulates androgen receptor signaling and spermatogenesis. EMBO J 2005;24(3):611–22.

[19] Elliott MR, Zheng S, Park D, Woodson RI, Reardon MA, Juncadella IJ, et al. Unexpected requirement for ELMO1 in clearance of apoptotic germ cells *in vivo*. Nature 2010;467(7313):333–7.

[20] Chen C, Ouyang W, Grigura V, Zhou Q, Carnes K, Lim H, et al. ERM is required for transcriptional control of the spermatogonial stem cell niche. Nature 2005;436(7053): 1030–4.

[21] Stoffel W, Holz B, Jenke B, Binczek E, Gunter RH, Kiss C, et al. Delta6-desaturase (FADS2) deficiency unveils the role of omega3- and omega6-polyunsaturated fatty acids. EMBO J 2008;27(17):2281–92.

[22] Dierich A, Sairam MR, Monaco L, Fimia GM, Gansmuller A, LeMeur M, et al. Impairing follicle-stimulating hormone (FSH) signaling in vivo: targeted disruption of the FSH receptor leads to aberrant gametogenesis and hormonal imbalance. Proc Natl Acad Sci USA 1998;95(23):13612–7.

[23] Smith LB, Milne L, Nelson N, Eddie S, Brown P, Atanassova N, et al. KATNAL1 regulation of Sertoli cell microtubule dynamics is essential for spermiogenesis and male fertility. PLoS Genet 2012;8(5):e1002697.
[24] Huyghe S, Schmalbruch H, De Gendt K, Verhoeven G, Guillou F, Van Veldhoven PP, et al. Peroxisomal multifunctional protein 2 is essential for lipid homeostasis in Sertoli cells and male fertility in mice. Endocrinology 2006;147(5):2228–36.
[25] Kennedy CL, Koopman P, Mishina Y, O'Bryan MK. Sox8 and Sertoli-cell function. Ann N Y Acad Sci 2007;1120:104–13.
[26] Zhang Y, Li N, Chen Q, Yan K, Liu Z, Zhang X, et al. Breakdown of immune homeostasis in the testis of mice lacking Tyro3, Axl and Mer receptor tyrosine kinases. Immunol Cell Biol 2013;91(6):416–26.
[27] Xu J, Anuar F, Ali SM, Ng MY, Phua DC, Hunziker W. Zona occludens-2 is critical for blood-testis barrier integrity and male fertility. Mol Biol Cell 2009;20(20):4268–77.
[28] Zhang X, Wang T, Deng T, Xiong W, Lui P, Li N, et al. Damaged spermatogenic cells induce inflammatory gene expression in mouse Sertoli cells through the activation of Toll-like receptors 2 and 4. Mol Cell Endocrinol 2013;365(2):162–73.
[29] Suetsugu S, Banzai Y, Kato M, Fukami K, Kataoka Y, Takai Y, et al. Male-specific sterility caused by the loss of CR16. Genes Cells 2007;12(6):721–33.
[30] Raverdeau M, Gely-Pernot A, Feret B, Dennefeld C, Benoit G, Davidson I, et al. Retinoic acid induces Sertoli cell paracrine signals for spermatogonia differentiation but cell autonomously drives spermatocyte meiosis. Proc Natl Acad Sci USA 2012;109(41): 16582–7.
[31] Tanwar PS, Zhang L, Teixeira JM. Adenomatous polyposis coli (APC) is essential for maintaining the integrity of the seminiferous epithelium. Mol Endocrinol 2011;25(10): 1725–39.
[32] Wang RS, Yeh S, Chen LM, Lin HY, Zhang C, Ni J, et al. Androgen receptor in Sertoli cell is essential for germ cell nursery and junctional complex formation in mouse testes. Endocrinology 2006;147(12):5624–33.
[33] Yu Z, Dadgar N, Albertelli M, Scheller A, Albin RL, Robins DM, et al. Abnormalities of germ cell maturation and Sertoli cell cytoskeleton in androgen receptor 113 CAG knock-in mice reveal toxic effects of the mutant protein. Am J Pathol 2006;168(1): 195–204.
[34] Willems A, Batlouni SR, Esnal A, Swinnen JV, Saunders PT, Sharpe RM, et al. Selective ablation of the androgen receptor in mouse Sertoli cells affects Sertoli cell maturation, barrier formation and cytoskeletal development. PLoS One 2010;5(11): e14168.
[35] Wang G, Weng CC, Shao SH, Zhou W, de Gendt K, Braun RE, et al. Androgen receptor in Sertoli cells is not required for testosterone-induced suppression of spermatogenesis, but contributes to Sertoli cell organization in Utp14b jsd mice. J Androl 2009;30(3): 338–48.
[36] Boyer A, Paquet M, Lague MN, Hermo L, Boerboom D. Dysregulation of WNT/CTNNB1 and PI3K/AKT signaling in testicular stromal cells causes granulosa cell tumor of the testis. Carcinogenesis 2009;30(5):869–78.
[37] Boyer A, Hermo L, Paquet M, Robaire B, Boerboom D. Seminiferous tubule degeneration and infertility in mice with sustained activation of WNT/CTNNB1 signaling in Sertoli cells. Biol Reprod 2008;79(3):475–85.
[38] Brehm R, Zeiler M, Ruttinger C, Herde K, Kibschull M, Winterhager E, et al. A Sertoli cell-specific knockout of connexin43 prevents initiation of spermatogenesis. Am J Pathol 2007;171(1):19–31.

[39] Li H, MacLean G, Cameron D, Clagett-Dame M, Petkovich M. Cyp26b1 expression in murine Sertoli cells is required to maintain male germ cells in an undifferentiated state during embryogenesis. PLoS One 2009;4(10):e7501.

[40] Papaioannou MD, Lagarrigue M, Vejnar CE, Rolland AD, Kuhne F, Aubry F, et al. Loss of Dicer in Sertoli cells has a major impact on the testicular proteome of mice. Mol Cell Proteomics: MCP 2011;10(4): M900587MCP200

[41] Kim GJ, Georg I, Scherthan H, Merkenschlager M, Guillou F, Scherer G, et al. Dicer is required for Sertoli cell function and survival. Int J Dev Biol 2010;54(5):867–75.

[42] Kim S, Bardwell VJ, Zarkower D. Cell type-autonomous and non-autonomous requirements for Dmrt1 in postnatal testis differentiation. Dev Biol 2007;307(2):314–27.

[43] Wang XX, Ying P, Diao F, Wang Q, Ye D, Jiang C, et al. Altered protein prenylation in Sertoli cells is associated with adult infertility resulting from childhood mumps infection. J Exp Med 2013;210(8):1559–74.

[44] Godmann M, Katz JP, Guillou F, Simoni M, Kaestner KH, Behr R. Kruppel-like factor 4 is involved in functional differentiation of testicular Sertoli cells. Dev Biol 2008;315(2): 552–66.

[45] Tanwar PS, Kaneko-Tarui T, Zhang L, Teixeira JM. Altered LKB1/AMPK/TSC1/TSC2/mTOR signaling causes disruption of Sertoli cell polarity and spermatogenesis. Hum Mol Genet 2012;21(20):4394–405.

[46] Kato T, Esaki M, Matsuzawa A, Ikeda Y. NR5A1 is required for functional maturation of Sertoli cells during postnatal development. Reproduction 2012;143(5):663–72.

[47] Baes M, Van Veldhoven PP. Generalised and conditional inactivation of *Pex* genes in mice. Biochim Biophys Acta 2006;1763(12):1785–93.

[48] Xiao Y, Karnati S, Qian G, Nenicu A, Fan W, Tchatalbachev S, et al. Cre-mediated stress affects sirtuin expression levels, peroxisome biogenesis and metabolism, antioxidant and proinflammatory signaling pathways. PLoS One 2012;7(7):e41097.

[49] Vernet N, Dennefeld C, Guillou F, Chambon P, Ghyselinck NB, Mark M. Prepubertal testis development relies on retinoic acid but not rexinoid receptors in Sertoli cells. EMBO J 2006;25(24):5816–25.

[50] Nalam RL, Andreu-Vieyra C, Braun RE, Akiyama H, Matzuk MM. Retinoblastoma protein plays multiple essential roles in the terminal differentiation of Sertoli cells. Mol Endocrinol 2009;23(11):1900–13.

[51] Tong MH, Yang QE, Davis JC, Griswold MD. Retinol dehydrogenase 10 is indispensible for spermatogenesis in juvenile males. Proc Natl Acad Sci USA 2013;110(2): 543–8.

[52] Mark M, Ghyselinck NB, Chambon P. Function of retinoid nuclear receptors: lessons from genetic and pharmacological dissections of the retinoic acid signaling pathway during mouse embryogenesis. Ann Rev Pharmacol Toxicol 2006;46:451–80.

[53] Gao F, Maiti S, Alam N, Zhang Z, Deng JM, Behringer RR, et al. The Wilms tumor gene, *Wt1*, is required for Sox9 expression and maintenance of tubular architecture in the developing testis. Proc Natl Acad Sci USA 2006;103(32):11987–92.

[54] Wang XN, Li ZS, Ren Y, Jiang T, Wang YQ, Chen M, et al. The Wilms tumor gene, *Wt1*, is critical for mouse spermatogenesis via regulation of Sertoli cell polarity and is associated with non-obstructive azoospermia in humans. PLoS Genet 2013;9(8): e1003645.

[55] Ihle JN. The challenges of translating knockout phenotypes into gene function. Cell 2000;102(2):131–4.

[56] Chambon P. The nuclear receptor superfamily: a personal retrospect on the first two decades. Mol Endocrinol 2005;19(6):1418–28.

[57] Cammas L, Romand R, Fraulob V, Mura C, Dolle P. Expression of the murine retinol dehydrogenase 10 (*Rdh10*) gene correlates with many sites of retinoid signalling during embryogenesis and organ differentiation. Dev Dyn 2007;236(10): 2899–908.

[58] Niederreither K, Dolle P. Retinoic acid in development: towards an integrated view. Nat Rev Genet 2008;9(7):541–53.

[59] Shackelford DB, Shaw RJ. The LKB1-AMPK pathway: metabolism and growth control in tumour suppression. Nat Rev Cancer 2009;9(8):563–75.

[60] Lizcano JM, Goransson O, Toth R, Deak M, Morrice NA, Boudeau J, et al. LKB1 is a master kinase that activates 13 kinases of the AMPK subfamily, including MARK/PAR-1. EMBO J 2004;23(4):833–43.

[61] Dugan JM, Allen CM. Changes in protein prenylation and prenyltransferase activity in the rat seminiferous epithelium during early stages of spermatogenesis. Biol Reprod 1995;53(4):958–73.

[62] Ashby MN. CaaX converting enzymes. Curr Opin Lipidol 1998;9(2):99–102.

[63] Shen N, Shao Y, Lai SS, Qiao L, Yang RL, Xue B, et al. GGPPS, a new EGR-1 target gene, reactivates ERK 1/2 signaling through increasing Ras prenylation. Am J Pathol 2011;179(6):2740–50.

[64] Ballew JW, Masters WH. Mumps; a cause of infertility. I. Present considerations. Fertil Steril 1954;5(6):536–43.

[65] Merritt WM, Bar-Eli M, Sood AK. The dicey role of Dicer: implications for RNAi therapy. Cancer Res 2010;70(7):2571–4.

[66] Jaskiewicz L, Filipowicz W. Role of Dicer in posttranscriptional RNA silencing. Curr Top Microbiol Immunol 2008;320:77–97.

[67] Song R, Hennig GW, Wu Q, Jose C, Zheng H, Yan W. Male germ cells express abundant endogenous siRNAs. Proc Natl Acad Sci USA 2011;108(32):13159–64.

[68] O'Shaughnessy PJ, Monteiro A, Abel M. Testicular development in mice lacking receptors for follicle stimulating hormone and androgen. PLoS One 2012;7(4):e35136.

[69] Muramatsu T. Basigin: a multifunctional membrane protein with an emerging role in infections by malaria parasites. Exp Opin Ther Targets 2012;16(10):999–1011.

[70] Kanekura T, Chen X. CD147/basigin promotes progression of malignant melanoma and other cancers. J Dermatol Sci 2010;57(3):149–54.

[71] Muramatsu T, Miyauchi T. Basigin (CD147): a multifunctional transmembrane protein involved in reproduction, neural function, inflammation and tumor invasion. Histol Histopathol 2003;18(3):981–7.

[72] Hedger MP. Toll-like receptors and signalling in spermatogenesis and testicular responses to inflammation—a perspective. J Reprod Immunol 2011;88(2):130–41.

[73] Riccioli A, Starace D, Galli R, Fuso A, Scarpa S, Palombi F, et al. Sertoli cells initiate testicular innate immune responses through TLR activation. J Immunol 2006;177(10): 7122–30.

[74] Greschik H, Schule R. Germ cell nuclear factor: an orphan receptor with unexpected properties. J Mol Med 1998;76(12):800–10.

[75] Wang RS, Yeh S, Tzeng CR, Chang C. Androgen receptor roles in spermatogenesis and fertility: lessons from testicular cell-specific androgen receptor knockout mice. Endocr Rev 2009;30(2):119–32.

[76] Walters KA, Simanainen U, Handelsman DJ. Molecular insights into androgen actions in male and female reproductive function from androgen receptor knockout models. Hum Reprod Update 2010;16(5):543–58.

[77] Hermo L, Pelletier RM, Cyr DG, Smith CE. Surfing the wave, cycle, life history, and genes/proteins expressed by testicular germ cells. Part 3: developmental changes in spermatid flagellum and cytoplasmic droplet and interaction of sperm with the zona pellucida and egg plasma membrane. Microsc Res Tech 2010;73(4):320–63.
[78] Cheng CY, Mruk DD. Cell junction dynamics in the testis: Sertoli–germ cell interactions and male contraceptive development. Physiol Rev 2002;82(4):825–74.
[79] Mitic LL, Van Itallie CM, Anderson JM. Molecular physiology and pathophysiology of tight junctions I. Tight junction structure and function: lessons from mutant animals and proteins. Am J Physiol Gastrointest Liver Physiol 2000;279(2):G250–4.
[80] Gow A, Southwood CM, Li JS, Pariali M, Riordan GP, Brodie SE, et al. CNS myelin and Sertoli cell tight junction strands are absent in Osp/claudin-11 null mice. Cell 1999;99(6):649–59.
[81] Saitou M, Furuse M, Sasaki H, Schulzke JD, Fromm M, Takano H, et al. Complex phenotype of mice lacking occludin, a component of tight junction strands. Mol Biol Cell 2000;11(12):4131–42.
[82] Gonzalez-Mariscal L, Betanzos A, Avila-Flores A. MAGUK proteins: structure and role in the tight junction. Semin Cell Dev Biol 2000;11(4):315–24.
[83] Katsuno T, Umeda K, Matsui T, Hata M, Tamura A, Itoh M, et al. Deficiency of zonula occludens-1 causes embryonic lethal phenotype associated with defected yolk sac angiogenesis and apoptosis of embryonic cells. Mol Biol Cell 2008;19(6): 2465–75.
[84] Xu J, Kausalya PJ, Phua DC, Ali SM, Hossain Z, Hunziker W. Early embryonic lethality of mice lacking ZO-2, but Not ZO-3, reveals critical and nonredundant roles for individual zonula occludens proteins in mammalian development. Mol Cell Biol 2008;28(5):1669–78.
[85] Dasgupta C, Escobar-Poni B, Shah M, Duncan J, Fletcher WH. Misregulation of connexin43 gap junction channels and congenital heart defects. Novartis Found Symp 1999;219:212–21 [discussion 21-5].
[86] Li MW, Mruk DD, Lee WM, Cheng CY. Connexin 43 and plakophilin-2 as a protein complex that regulates blood-testis barrier dynamics. Proc Natl Acad Sci USA 2009;106(25):10213–8.
[87] Aoki K, Taketo MM. Adenomatous polyposis coli (APC): a multi-functional tumor suppressor gene. J Cell Sci 2007;120(Pt 19):3327–35.
[88] Yan W, West A, Toppari J, Lahdetie J. Stage specific expression and phosphorylation of retinoblastoma protein (pRb) in the rat seminiferous epithelium. Mol Cell Endocrinol 1997;132(1-2):137–48.
[89] Yan W, Kero J, Suominen J, Toppari J. Differential expression and regulation of the retinoblastoma family of proteins during testicular development and spermatogenesis: roles in the control of germ cell proliferation, differentiation and apoptosis. Oncogene 2001;20(11):1343–56.
[90] Gong W, Zhou T, Mo J, Perrett S, Wang J, Feng Y. Structural insight into recognition of methylated histone tails by retinoblastoma-binding protein 1. J Biol Chem 2012;287(11): 8531–40.
[91] Lemercier C, Duncliffe K, Boibessot I, Zhang H, Verdel A, Angelov D, et al. Involvement of retinoblastoma protein and HBP1 in histone H1(0) gene expression. Mol Cell Biol 2000;20(18):6627–37.
[92] Cho HP, Nakamura M, Clarke SD. Cloning, expression, and fatty acid regulation of the human delta-5 desaturase. J Biol Chem 1999;274(52):37335–9.

[93] Lu Q, Gore M, Zhang Q, Camenisch T, Boast S, Casagranda F, et al. Tyro-3 family receptors are essential regulators of mammalian spermatogenesis. Nature 1999;398(6729): 723–8.
[94] Kato M, Miki H, Kurita S, Endo T, Nakagawa H, Miyamoto S, et al. WICH, a novel verprolin homology domain-containing protein that functions cooperatively with N-WASP in actin-microspike formation. Biochem Biophys Res Commun 2002;291(1): 41–7.
[95] Krzewski K, Chen X, Orange JS, Strominger JL. Formation of a WIP-, WASp-, actin-, and myosin IIA-containing multiprotein complex in activated NK cells and its alteration by KIR inhibitory signaling. J Cell Biol 2006;173(1):121–32.
[96] Vogl AW, Vaid KS, Guttman JA. The Sertoli cell cytoskeleton. Adv Exp Med Biol 2008;636:186–211.
[97] Quarmby L. Cellular Samurai: katanin and the severing of microtubules. J Cell Sci 2000;113(Pt 16):2821–7.
[98] Ramesh V. Merlin and the ERM proteins in Schwann cells, neurons and growth cones. Nat Rev Neurosci 2004;5(6):462–70.
[99] Simon L, Ekman GC, Garcia T, Carnes K, Zhang Z, Murphy T, et al. ETV5 regulates Sertoli cell chemokines involved in mouse stem/progenitor spermatogonia maintenance. Stem Cells 2010;28(10):1882–92.
[100] Tyagi G, Carnes K, Morrow C, Kostereva NV, Ekman GC, Meling DD, et al. Loss of Etv5 decreases proliferation and RET levels in neonatal mouse testicular germ cells and causes an abnormal first wave of spermatogenesis. Biol Reprod 2009;81(2): 258–66.
[101] Schlesser HN, Simon L, Hofmann MC, Murphy KM, Murphy T, Hess RA, et al. Effects of ETV5 (ets variant gene 5) on testis and body growth, time course of spermatogonial stem cell loss, and fertility in mice. Biol Reprod 2008;78(3):483–9.
[102] Morrow CM, Hostetler CE, Griswold MD, Hofmann MC, Murphy KM, Cooke PS, et al. ETV5 is required for continuous spermatogenesis in adult mice and may mediate blood testes barrier function and testicular immune privilege. Ann N Y Acad Sci 2007;1120:144–51.
[103] Lin C, Smith ER, Takahashi H, Lai KC, Martin-Brown S, Florens L, et al. AFF4, a component of the ELL/P-TEFb elongation complex and a shared subunit of MLL chimeras, can link transcription elongation to leukemia. Mol Cell 2010;37(3):429–37.
[104] Nakamura T, Yao R, Ogawa T, Suzuki T, Ito C, Tsunekawa N, et al. Oligo-astheno-teratozoospermia in mice lacking Cnot7, a regulator of retinoid X receptor beta. Nat Genet 2004;36(5):528–33.
[105] Barrionuevo F, Georg I, Scherthan H, Lecureuil C, Guillou F, Wegner M, et al. Testis cord differentiation after the sex determination stage is independent of Sox9 but fails in the combined absence of Sox9 and Sox8. Dev Biol 2009;327(2):301–12.
[106] O'Bryan MK, Takada S, Kennedy CL, Scott G, Harada S, Ray MK, et al. Sox8 is a critical regulator of adult Sertoli cell function and male fertility. Dev Biol 2008;316(2): 359–70.
[107] Murakami M, Yokoyama R, Nishida T, Shiromoto M, Sato H. Scanning and transmission electron microscope observations of the terminal segment of the cat seminiferous tubule: epithelial phagocytosis of spermatozoa and latex beads. Arch Histol Cytol 1988;51(2):185–92.
[108] Otsuka A, Abe T, Watanabe M, Yagisawa H, Takei K, Yamada H. Dynamin 2 is required for actin assembly in phagocytosis in Sertoli cells. Biochem Biophys Res Commun 2009;378(3):478–82.

[109] Tokuda N, Mano T, Levy RB. Phagocytosis by the murine testicular TM4 Sertoli cell line in culture. J Urol 1992;147(1):278–82.
[110] Kerr JB, de Kretser DM. Proceedings: the role of the Sertoli cell in phagocytosis of the residual bodies of spermatids. J Reprod Fertil 1974;36(2):439–40.
[111] Ravichandran KS, Lorenz U. Engulfment of apoptotic cells: signals for a good meal. Nat Rev Immunol 2007;7(12):964–74.
[112] Bustamante-Marin X, Quiroga C, Lavandero S, Reyes JG, Moreno RD. Apoptosis, necrosis and autophagy are influenced by metabolic energy sources in cultured rat spermatocytes. Apoptosis 2012;17(6):539–50.
[113] Zhang M, Jiang M, Bi Y, Zhu H, Zhou Z, Sha J. Autophagy and apoptosis act as partners to induce germ cell death after heat stress in mice. PLoS One 2012;7(7):e41412.
[114] Dun MD, Aitken RJ, Nixon B. The role of molecular chaperones in spermatogenesis and the post-testicular maturation of mammalian spermatozoa. Hum Reprod Update 2012;18(4):420–35.
[115] Van Veldhoven PP, Baes M. Peroxisome deficient invertebrate and vertebrate animal models. Front Physiol 2013;4:335.
[116] Huyghe S, Mannaerts GP, Baes M, Van Veldhoven PP. Peroxisomal multifunctional protein-2: the enzyme, the patients and the knockout mouse model. Biochim Biophys Acta 2006;1761(9):973–94.
[117] Tarulli GA, Stanton PG, Meachem SJ. Is the adult Sertoli cell terminally differentiated? Biol Reprod 2012;87(1):13 1-1
[118] Johnson L, Thompson Jr DL, Varner DD. Role of Sertoli cell number and function on regulation of spermatogenesis. Anim Reprod Sci 2008;105(1-2):23–51.
[119] Jegou B. The Sertoli- germ cell communication network in mammals. Int Rev Cytol 1993;147:25–96.
[120] Kos CH. Cre/loxP system for generating tissue-specific knockout mouse models. Nutr Rev 2004;62(6 Pt 1):243–6.
[121] Kuhn R, Torres RM. Cre/loxP recombination system and gene targeting. Methods Mol Biol 2002;180:175–204.
[122] Hasegawa K, Saga Y. Retinoic acid signaling in Sertoli cells regulates organization of the blood-testis barrier through cyclical changes in gene expression. Development 2012;139(23):4347–55.
[123] Jamsai D, O'Bryan MK. Genome-wide ENU mutagenesis for the discovery of novel male fertility regulators. Syst Biol Reprod Med 2010;56(3):246–59.
[124] Hazra R, Corcoran L, Robson M, McTavish KJ, Upton D, Handelsman DJ, et al. Temporal role of Sertoli cell androgen receptor expression in spermatogenic development. Mol Endocrinol 2013;27(1):12–24.
[125] Wang F, Zhang Q, Cao J, Huang Q, Zhu X. The microtubule plus end-binding protein EB1 is involved in Sertoli cell plasticity in testicular seminiferous tubules. Exp Cell Res 2008;314(1):213–26.
[126] Fleming SL, Shank PR, Boekelheide K. Gamma-Tubulin overexpression in Sertoli cells *in vivo*: I. Localization to sites of spermatid head attachment and alterations in Sertoli cell microtubule distribution. Biol Reprod 2003;69(1):310–21.
[127] Fleming SL, Shank PR, Boekelheide K. Gamma-Tubulin overexpression in Sertoli cells *in vivo*. II: Retention of spermatids, residual bodies, and germ cell apoptosis. Biol Reprod 2003;69(1):322–30.
[128] Tanwar PS, Kaneko-Tarui T, Zhang L, Rani P, Taketo MM, Teixeira J. Constitutive WNT/beta-catenin signaling in murine Sertoli cells disrupts their differentiation and ability to support spermatogenesis. Biol Reprod 2010;82(2):422–32.

[129] Fumel B, Guerquin MJ, Livera G, Staub C, Magistrini M, Gauthier C, et al. Thyroid hormone limits postnatal Sertoli cell proliferation *in vivo* by activation of its alpha1 isoform receptor (TRalpha1) present in these cells and by regulation of Cdk4/JunD/c-myc mRNA levels in mice. Biol Reprod 2012;87(1):16 1–9.

[130] Lecureuil C, Fontaine I, Crepieux P, Guillou F. Sertoli and granulosa cell-specific Cre recombinase activity in transgenic mice. Genesis 2002;33(3):114–8.

[131] Holdcraft RW, Braun RE. Androgen receptor function is required in Sertoli cells for the terminal differentiation of haploid spermatids. Development 2004;131(2):459–67.

[132] Jamin SP, Arango NA, Mishina Y, Hanks MC, Behringer RR. Requirement of Bmpr1a for Mullerian duct regression during male sexual development. Nat Genet 2002;32(3):408–10.

[133] Jaegle M, Ghazvini M, Mandemakers W, Piirsoo M, Driegen S, Levavasseur F, et al. The POU proteins Brn-2 and Oct-6 share important functions in Schwann cell development. Genes Dev 2003;17(11):1380–91.

[134] Arango NA, Kobayashi A, Wang Y, Jamin SP, Lee HH, Orvis GD, et al. A mesenchymal perspective of Mullerian duct differentiation and regression in Amhr2-lacZ mice. Mol Reprod Dev 2008;75(7):1154–62.

[135] Kyronlahti A, Euler R, Bielinska M, Schoeller EL, Moley KH, Toppari J, et al. GATA4 regulates Sertoli cell function and fertility in adult male mice. Mol Cell Endocrinol 2011;333(1):85–95.

[136] Xu Q, Lin HY, Yeh SD, Yu IC, Wang RS, Chen YT, et al. Infertility with defective spermatogenesis and steroidogenesis in male mice lacking androgen receptor in Leydig cells. Endocrine 2007;32(1):96–106.

[137] Archambeault DR, Yao HH. Activin A, a product of fetal Leydig cells, is a unique paracrine regulator of Sertoli cell proliferation and fetal testis cord expansion. Proc Natl Acad Sci USA 2010;107(23):10526–31.

[138] Jeyasuria P, Ikeda Y, Jamin SP, Zhao L, De Rooij DG, Themmen AP, et al. Cell-specific knockout of steroidogenic factor 1 reveals its essential roles in gonadal function. Mol Endocrinol 2004;18(7):1610–9.

[139] Lindeboom F, Gillemans N, Karis A, Jaegle M, Meijer D, Grosveld F, et al. A tissue-specific knockout reveals that Gata1 is not essential for Sertoli cell function in the mouse. Nucleic Acids Res 2003;31(18):5405–12.

[140] Zhang M, Wang F, Li S, Wang Y, Bai Y, Xu X. TALE: a tale of genome editing. Prog Biophys Mol Biol 2013.

[141] Perez-Pinera P, Ousterout DG, Gersbach CA. Advances in targeted genome editing. Curr Opin Chem Biol 2012;16(3-4):268–77.

[142] Gaj T, Gersbach CA, Barbas 3rd CF. ZFN, TALEN, and CRISPR/Cas-based methods for genome engineering. Trends Biotechnol 2013;31(7):397–405.

[143] Qian X, Cheng YH, Jenardhanan P, Mruk DD, Mathur PP, Xia W, et al. Adjudin disrupts spermatogenesis by targeting drug transporters: lesson from the breast cancer resistance protein (BCRP). Spermatogenesis 2013;3(2):e24993.

[144] Tash JS, Chakrasali R, Jakkaraj SR, Hughes J, Smith SK, Hornbaker K, et al. Gamendazole, an orally active indazole carboxylic acid male contraceptive agent, targets HSP90AB1 (HSP90BETA) and EEF1A1 (eEF1A), and stimulates Il1a transcription in rat Sertoli cells. Biol Reprod 2008;78(6):1139–52.

[145] Kerkis A, Fonseca SA, Serafim RC, Lavagnolli TM, Abdelmassih S, Abdelmassih R, et al. *In vitro* differentiation of male mouse embryonic stem cells into both presumptive sperm cells and oocytes. Cloning Stem Cells 2007;9(4):535–48.

[146] Li P, Hu H, Yang S, Tian R, Zhang Z, Zhang W, et al. Differentiation of induced pluripotent stem cells into male germ cells *in vitro* through embryoid body formation and retinoic acid or testosterone induction. Biomed Res Int 2013;2013:608728.

[147] Lee DR, Kim KS, Yang YH, Oh HS, Lee SH, Chung TG, et al. Isolation of male germ stem cell-like cells from testicular tissue of non-obstructive azoospermic patients and differentiation into haploid male germ cells *in vitro*. Hum Reprod 2006;21(2): 471–6.

[148] Lacham-Kaplan O. *In vivo* and *in vitro* differentiation of male germ cells in the mouse. Reproduction 2004;128(2):147–52 [145–148].

Kolibri 360 GmbH
Ferdinand-Jühlke-Straße 7
99095 Erfurt, Deutschland
produktsicherheit@kolibri360.de